The Semiclassical Way

to Dynamics and Spectroscopy

The Semiclassical Way

to Dynamics and Spectroscopy

ERIC J. HELLER

PRINCETON UNIVERSITY PRESS
Princeton and Oxford

Published by Princeton University Press,
41 William Street, Princeton, New Jersey 08540
In the United Kingdom: Princeton University Press,
6 Oxford Street, Woodstock, Oxfordshire OX20 1TR

press.princeton.edu

Library of Congress Control Number: 2018933993

ISBN 978-0-691-16373-4

British Library Cataloging-in-Publication Data is available

This book has been composed in Minion Pro and Universe Lt STd

Printed on acid-free paper. ∞

Typeset by Nova Techset Pvt Ltd, Bangalore, India
Printed in the United States of America

10 9 8 7 6 5 4 3 2 1

For Hong Wei

Contents

Preface

The biggest show on earth is forever locked away, and seemingly in a room just next door. An unsolved mystery of the universe looms there: How does reality emerge from the infinity of simultaneous, mutually exclusive quantum probability amplitudes? How does this happen, apparently everywhere, continuously? (Think Schrödinger cat paradox.) The totality of "decoherence theory" does not even touch the basic question, one as deep as that of consciousness or any in theology. Indeed, the answer may be entwined with both and with our very existence. Some "answers" are impossible to disprove, like the "many worlds" hypothesis, wherein the world splits into infinitely many realities every moment. Also impossible to disprove is the notion that we were all created just a second ago complete with memories. The author finds the many worlds hypothesis abhorrent. Earth never existed in all but a set of measure zero of these worlds. The author has clearly just disqualified himself as a philosopher. Peering over the edge of this sort of abyss is thrilling, but not the main point of this book. We look away from the abyss. As Feynman is reputed to have said: "If you think you understand quantum mechanics, you don't understand quantum mechanics." I don't claim to understand quantum mechanics.

Just short of the abyss is a challenging and fascinating world, where people try to piece together how it is the classical world we experience daily connects with the quantum world. This endeavor has three major components: (1) to understand quantum mechanics, including measurement theory, as well as possible without treading on Feynman's maxim; (2) developing a deeper understanding for quantum eigenstates and quantum dynamics; and (3) developing powerful computational tools that substitute classical mechanics for wave mechanics.

Quantum mechanics governs the real world. Every seemingly classical system, including a marathon runner, is completely quantum mechanical. The runner is made of waves, and stepping on waves. It is rewarding to understand how it is that big quantum systems can appear to be classical. But big systems should not be mistaken as the only arena for a semiclassical description. Even the smallest coherent quantum systems yield their best secrets aided by an advanced knowledge of classical mechanics and semiclassical connections.

Everyone thinks classically, embedded as we all are in the macroscopic world with millions of years of evolution to shape our sense of reality. Indeed, it is difficult for people to think in terms of waves in more than two or three dimensions. Given something so elementary as atom A collides with diatomic molecule BC in free space, who among us thinks of that process as a wave in nine-dimensional coordinate space (or even six dimensions if the center of mass is removed)? Yet we can imagine or watch a computer animation of the collision happening classically, with real comprehension.

Thus, it is not a stretch to say that it is powerful and enabling if you (1) understand classical mechanics and (2) know how it carries over and indeed forms the backbone of quantum mechanics.

Many concepts in quantum mechanics depend on a classical baseline. Tunneling and diffraction, for example, are meaningless without a reference to classical behavior.

Exact quantum mechanics is given by the Feynman path integral, which is entirely specified by the classical action functional $S = \int L\, dt$ (although taken not just on classical paths but rather on "all" paths). Boiling "all" Feynman paths down to classical paths and their immediate vicinity is an extremely powerful approximation, both numerically and intuitively. It falls out that way once the stationary phase approximation to integration is applied. The crucial aspect of interference is preserved as a sum of amplitudes, one for each stationary action (that is, classical) path, which allows classical paths to interfere.

Great progress has been made in the last twenty or thirty years in the development of semiclassical methods, proving classical trajectories sufficient to accurately calculate many things that were thought to be purely quantum in nature.

How was it that so much of nonrelativistic quantum mechanics was worked out by a few people in three or four years? Granted that Born, Bohr, Schrödinger, and so on were giants, but they had help: they knew classical mechanics cold! A look at M. Born, *Mechanics of the Atom* (1927) makes this clear.

I end here with a note about an important related book, *Introduction to Quantum Mechanics: A Time-dependent Perspective* by my friend and colleague, David Tannor [1]. I view *The Semiclassical Way*, written 10 years later, as an extension of that important work. The two books share an enthusiasm for the insights and tools that both time-dependent quantum mechanical and semiclassical approaches to problems bring. In fact, Tannor's book makes this book possible, because so many topics we both hold dear, including those I learned from him, are found only in his book.

Acknowledgments

The author is greatly indebted for the hospitality and financial support of the Max Planck Institute for Complex Systems (MPIPKS) in Dresden, Germany, where much of this book was written. Special thanks goes to Dr. Prof. J. M. Rost, Director of MPIPKS, for this support and for scientific collaborations. Indeed, the author owes a large debt to the Federal Republic of Germany, which through supported sabbaticals, postdoctoral fellows, predoctoral fellows, and many visits, has provided a large fraction of the resources for researching and writing this book. The author also owes a great debt to Harvard University, especially the Faculty of Arts and Sciences and the Department of Chemistry and Chemical Biology, for generous financial support.

I have benefited from many wonderful interactions and learning experiences involving colleagues and students. Notable among them are Prof. Sir Michael Berry, Prof. R. Coalson, Dr. M. Davis, B. Drury, Dr. D. Eigler, Prof. W. Gelbart, Dr. F. Grossman, Dr. Martin Gutzwiller, Prof. W. Harter, Dr. D. Imry, Prof. L. Kaplan, Dr. A. Klales, Dr. L. Kocia, Dr. Tobias Kramer, Dr. K. C. Kulander, Dr. B. Landry, Prof. S.-Y. Lee, Dr. S. Ling, Prof. R. Littlejohn, Dr. P. Luukko, Prof. N. Maitra, Dr. D. Mason, Prof. W. Miller, Dr. A. Mody, Dr. S. Pittman, Prof. E. Räsänen, Prof. J. Reimers, Prof. Bill Reinhardt, Dr. S. Sanders, Dr. M. Schram, Dr. M. Sepúlveda, Dr. S. Shaw, Dr. E. Stechel, Dr. R. Sundberg, Prof. D. Tannor, Prof. S. Tomsovic, Dr. J. Vaishnav, Prof. J. Vanicek, Dr. T. Van Voorhis, S. Vardan, Prof. A. Wasserman, and Dr. Y. Yang. These scientists and friends, and many more, made this work possible. It has been a great joy and privilege to work with them.

Introduction

Classical mechanics is the backbone of quantum mechanics. The highest level of insight into quantum systems can be had only through a good grasp of classical mechanics. On that principle, this book first presents the key foundations of classical mechanics necessary to semiclassical insight. In this, we loosely follow the best book on the subject, *Mechanics*, by Landau and Lifshitz [2]. While we do not give a full treatment of classical mechanics here, neither do we offer just a subset of the usual topics. Included, for example, is a discussion of the time evolution of classical manifolds and continuous phase space distributions, classical processes formulated as probabilities, and classical resonance theory not found in most texts.

Empowered by classical mechanics, the middle part of this book develops the Feynman path integral, passing to the stationary phase approximations linking quantum mechanics with classical mechanics, giving birth to semiclassical theory. Here we learn to *construct* much of quantum mechanics, qualitatively and sometimes quantitatively, out of classical mechanics. Much deeper insight into quantum tunneling, interference, entanglement, decoherence, spectroscopy, and quantum dynamics in general becomes possible. Even when classical and quantum mechanics are differing qualitatively and perhaps by orders of magnitude, the underlying answers are often still classical in origin, using classically determined amplitude and phase interference.

It is very satisfying to gain a deeper appreciation and intuition for quantum mechanics through a semiclassical understanding. The highest reward is the computational and intuitive insight into important experimental realms. The last part of this book is about applications, especially to spectroscopy of several kinds, including molecular spectroscopy, for which we need the Born-Oppenheimer approximation. We also take up subjects important to nano-physics, such as electron imaging, and the modern understanding of the connections between classical chaos and the quantum mechanics of the same systems.

Tunneling and diffraction are by definition both nonclassical phenomena, but we can include them as semiclassical by analytic continuation of classical mechanics, and the trick of "uniformization" (replacing a key stationary phase integral by an exact or numerical one). This often yields great insight still underpinned by classical mechanics.

In several places we focus on tunneling phenomena, including diffraction, called "dynamical tunneling," to distinguish it from the ubiquitous potential barrier tunneling. Diffraction entails quantum allowed, classically forbidden processes without a potential barrier to burrow through.

Semiclassical approximations are often amazingly accurate, as seen in many places in this book, and as they are always guided by classical trajectories, a new level of understanding of quantum mechanics is born.

This book is written for a wide audience, but not in the sense of the general public. It is wide in the sense of the kind of mathematics, physics, and chemistry background one might bring to making use of this book. Admittedly, this should not be the reader's first book on classical or quantum mechanics. There are quite a few "special topics" that are treated at the back end of the book, but they are often not given an exhaustive literature context and preview. The back-end topics are not necessarily to be read through linearly or exhaustively, but rather the topics are meant as a menu to choose from that the reader might already know a bit about. Nonetheless, if some topics are new to the reader, they are written (I hope) so as to interest, engage, and form a launching pad for further study for those who find them interesting.

This book is not a textbook in the sense of chapter summaries, exercises, and lesson plans given to instructors. Nor is it a monograph exactly. I hope it will find itself on the shelf of graduate students and their professors alike as a reference and for enrichment of courses taught or research undertaken. Much of the book is meant to be suggestive of areas of research interest and of pregnant topics where progress is imminent.

PART I

Classical Mechanics with an Eye to Quantum Mechanics

Chapter 1

The Lagrangian and the Action

1.1 Extremal Action and Equations of Motion

Classical mechanics can be introduced in several ways; there is no "proof" of the equations of motion, only a statement of experience. One road to the equations of motion for classical systems is a variational principle, finding extremes (maxima or minima; usually the latter) of a quantity called the "action" along a path subject to given end points. Looking ahead, the classical action plays a central role in quantum mechanics via the Feynman path integral, which involves sums over complex exponentials of the action for all possible paths (not just classically allowed paths). The action makes its appearance in semiclassical approximations by stationary phase evaluation of Feynman path integrals. Stationary phase evaluation means exploiting extremal action (stationary or unchanging action against small path changes), which causes a buildup of the result with a phase of that action and an amplitude depending on just how stable the action is to changes. Classical trajectories are paths with stationary action, and they appear naturally out of quantum mechanics in this way. This is the best way to understand the correspondence principle,[1] the best way to build intuition for quantum systems having a classical analog. In what follows we focus on the action principles of classical mechanics, making the future transition to quantum and semiclassical mechanics almost seamless.

Landau and Lifshitz, in their famously short but spot-on book, *Mechanics* [2], introduce the Lagrangian $L = L(\mathbf{q}, \dot{\mathbf{q}}, t)$, saying it should be a function of the coordinates $\mathbf{q}$, their velocities $\dot{\mathbf{q}}$, and the time only, banking on the observation that specifying the coordinates and velocities completely determines the subsequent motion of any classical system. The action S is defined as the time integral ("action integral") of the Lagrangian:

$$\boxed{S = \int_0^t L(\mathbf{q}, \dot{\mathbf{q}}, \tau)\, d\tau} \tag{1.1}$$

[1] Bohr's correspondence principle asserts that quantum systems will behave classically in the limit that quantum numbers become "large." Thus, taking the "classical limit" of a quantum system means going to high energy, or if you are a theorist, small $\hbar$. Either way gets you to large quantum numbers, but of course, the dynamics usually changes with energy. A theorist can stay near fixed energy, as $\hbar$ is taken smaller. There are subtleties though, and these will be discussed in context

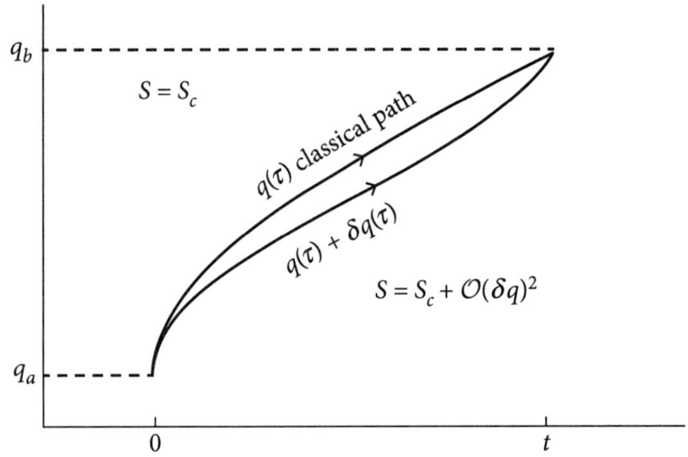

Figure 1.1. The classical path $q(\tau)$ going from $q(0) = q_a$ to $q(t) = q_b$ in time t, with action $S = S_c$, has no first-order difference in action compared to a slightly nonclassical path, $q(\tau) + \delta q(\tau)$—that is, $S = S_c + \mathcal{O}(\delta q)^2$ for small deviations $\delta q(\tau)$. The second-order differences in action provide a minimum by which we can find the genuine classical path.

Without even knowing the explicit form of the Lagrangian, the form of the equations of motion for the coordinates and velocities can be derived by introducing the *principle of least action*, also called "Hamilton's principle": the action defined earlier must be stationary against small changes in the coordinates and their velocities along the path starting at time 0 with positions $\boldsymbol{q}(0)$, and ending at time t with position $\boldsymbol{q}(t)$. (see figure 1.1). The initial and final positions are fixed, and for the moment not to be varied. The resulting equations can be used to determine the path (or paths—there may be more than one stationary solution differing in their initial velocities) between $\boldsymbol{q}(0)$ and $\boldsymbol{q}(t)$. We derive the dynamics, called Lagrange's equations, using one-dimensional notation next.

LAGRANGE'S EQUATION

We consider the path $q(\tau) + \delta q(\tau)$, where $q(\tau)$ is a classical path having action stationary against small changes, and $\delta q(\tau)$ is a small, arbitrary deviation from this path. The goal is to get from $q(0)$ to $q(t)$ in time t, so the end points are held fixed—that is, $\delta q(0) = \delta q(t) = 0$. Imposing the extremum property, we have to first order in $\delta q(\tau)$

$$\delta S = S[q(\tau) + \delta q(\tau)] - S[q(\tau)] = 0. \tag{1.2}$$

Then, from equation 1.1, we have

$$\begin{aligned} S[q(\tau) + \delta q(\tau)] &= \int_0^\tau L(q + \delta q, \dot{q} + \delta \dot{q}, \tau)\, d\tau, \\ &= \int_0^\tau \left[L(q, \dot{q}, \tau) + \delta \dot{q} \frac{\partial L}{\partial \dot{q}} + \delta q \frac{\partial L}{\partial q} \right] d\tau, \\ &= S[q(\tau)] + \int_0^t \left(\delta \dot{q} \frac{\partial L}{\partial \dot{q}} + \delta q \frac{\partial L}{\partial q} \right) d\tau. \end{aligned} \tag{1.3}$$

Integrating the $\dot{q}$ term by parts,

$$\delta S = \delta q \frac{\partial L}{\partial \dot{q}} \Big|_0^t - \int_0^t \delta q(\tau) \left[\frac{d}{dt} \left(\frac{\partial L}{\partial \dot{q}} \right) - \frac{\partial L}{\partial q} \right] d\tau = 0. \tag{1.4}$$

The starting and ending positions are not allowed to vary (yet), $\delta q(0) = \delta q(t) = 0$, so the first term on the right vanishes. Since $\delta q(\tau)$ is an arbitrary function of τ, the only way to guarantee $\delta S = 0$ is for

$$\frac{d}{dt} \left(\frac{\partial L}{\partial \dot{q}} \right) - \frac{\partial L}{\partial q} = 0. \tag{1.5}$$

This is the Lagrange equation of motion, which must be obeyed by the trajectory if the action is to be extremal. We still haven't needed to say what L is, apart from some very general properties. For N coordinates $(q_1, q_2, \ldots, q_N)$, N equations like equation 1.5 hold—for example,

$$\boxed{\frac{d}{dt} \left(\frac{\partial L}{\partial \dot{q}_i} \right) - \frac{\partial L}{\partial q_i} = 0, \; i = 1, 2, \ldots, N,} \tag{1.6}$$

where $L = L(q_1, q_2, \ldots, \dot{q}_1, \dot{q}_2, \ldots, t)$ and the N Lagrange equations must be solved simultaneously.

Since no restrictions on the coordinates used in deriving equation 1.5 or 1.6 were imposed, Lagrange's equations evidently look the same in any coordinate system $(q_1, q_2, \ldots)$—that is, they are invariant in form. We do not yet know how to construct the Lagrangian for a particular system, apart from the fact that it depends on its coordinates, their velocities, and time. We consider this next.

1.2 Form of the Lagrangian

Landau and Lifshitz [2], using the principles of inertial frames and transformations between them, show that the nonrelativistic Lagrangian for a free particle must be proportional to the square of the velocities:

$$L = \frac{1}{2} \sum_i m_i v_i^2. \tag{1.7}$$

The proportionality factors m_i are the particle masses. Ordinary Cartesian velocities are meant here. The form of the Lagrangian in some other coordinate system follows by transforming its form to the new coordinates. For example, for a free particle in two dimensions, we can write

$$L = \frac{1}{2} m \left(\dot{x}^2 + \dot{y}^2 \right) \equiv T. \tag{1.8}$$

Lagrange's equations are

$$\frac{d}{dt} \left(\frac{\partial T}{\partial \dot{x}} \right) - \frac{\partial T}{\partial x} = 0, \tag{1.9}$$

and similarly for y, yielding

$$m\ddot{x} = 0; \quad m\ddot{y} = 0. \tag{1.10}$$

In polar coordinates we have

$$L = T = \frac{1}{2}m\left(r^2\dot{\theta}^2 + \dot{r}^2\right), \tag{1.11}$$

and

$$\frac{d}{dt}\left(\frac{\partial T}{\partial \dot{\theta}}\right) - \frac{\partial T}{\partial \theta} = 0 = mr^2\ddot{\theta} + 2mr\dot{r}\dot{\theta}, \tag{1.12}$$

$$\frac{d}{dt}\left(\frac{\partial T}{\partial \dot{r}}\right) - \frac{\partial T}{\partial r} = m\ddot{r} - mr\dot{\theta}^2 = 0. \tag{1.13}$$

These equations of motion are recognized as $ma = F$, where a is the acceleration and the force $F = 0$.

Forces F_i can be introduced into Lagrange's equations as

$$\frac{d}{dt}\left(\frac{\partial T}{\partial \dot{q}_i}\right) - \frac{\partial T}{\partial q_i} = F_i, \tag{1.14}$$

where F_i is the ith component of the force. Presumably, the F_i can be functions of the q's, the $\dot{q}$'s, and time, like the Lagrangian. We encounter velocity-dependent forces in section 1.7. If the force components $F_i(\boldsymbol{q})$ depend only on coordinates $\boldsymbol{q}$, they can be derived as gradients of a potential V—that is, $F_i = \partial V/\partial q_i$:

$$L(\boldsymbol{q}, \dot{\boldsymbol{q}}, t) = \sum_i \frac{1}{2}m_i\dot{\boldsymbol{q}}_i{}^2 - V(\boldsymbol{q}_1, \boldsymbol{q}_2, \ldots), \text{ or}$$

$$\boxed{L = T - V,} \tag{1.15}$$

where $\boldsymbol{q}_i$ is the coordinate of the ith particle, and $\dot{\boldsymbol{q}}_i$ is its velocity. We take $\boldsymbol{q}$ to represent all the particles and their coordinates—that is, $\boldsymbol{q} = (\boldsymbol{q}_1, \boldsymbol{q}_2, \ldots)$. (The lumping of coordinates into "particles" is not necessary, but is very useful when 3D rotations are considered.)

In one dimension, for the familiar form of kinetic energy $T = \frac{1}{2}m\dot{q}^2$ and potential energy $V(q, t)$,

$$L(q, \dot{q}, t) = \frac{1}{2}m\dot{q}^2 - V(q, t), \tag{1.16}$$

$$\frac{\partial L}{\partial \dot{q}} = m\dot{q}, \quad \frac{\partial L}{\partial q} = -\frac{\partial V}{\partial q}, \tag{1.17}$$

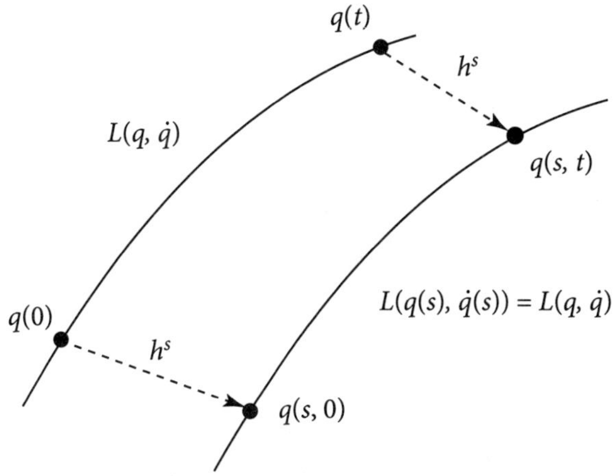

Figure 1.2. A trajectory $q(t)$ can be transformed at any time by h^S to $q(s, t)$. If everywhere $L(q, \dot{q}) = L(q(s), \dot{q}(s))$, then h^S corresponds to a conserved quantity.

and

$$m\ddot{q} = -\frac{\partial V}{\partial q}, \tag{1.18}$$

that is, $ma = F$.

1.3 Noether's Theorem

One of the most important theorems in mechanics is a relatively recent contribution by Emmy Noether, a leading member of the Göttingen Mathematics Department until 1933. We suppose a set of coordinates $\boldsymbol{q}$, and a transformation of coordinates h^s

$$\boldsymbol{q}(s, t) = h^s(\boldsymbol{q}(t)) \tag{1.19}$$

that leaves the Lagrangian invariant (see figure 1.2):

$$L(\boldsymbol{q}, \dot{\boldsymbol{q}}) = L(\boldsymbol{q}(s), \dot{\boldsymbol{q}}(s)). \tag{1.20}$$

Example:

$$h^s : \; x \to x\cos s - y\sin s,$$

$$y \to x\sin s + y\cos s.$$

Then,

$$L = \frac{1}{2}m(\dot{x}^2 + \dot{y}^2) - \frac{1}{2}m\omega^2(x^2 + y^2)$$

is clearly invariant to the transformation h^s—that is,

$$\frac{dL}{ds} = 0.$$

Since $\boldsymbol{q}(s)$ is a perfectly good coordinate system, the Euler-Lagrange equations of motion hold in them for fixed s,

$$\frac{d}{dt}\left(\frac{\partial L(\boldsymbol{q}(s),\dot{\boldsymbol{q}}(s))}{\partial \dot{q}_i(s)}\right)=\frac{\partial L(\boldsymbol{q}(s),\dot{\boldsymbol{q}}(s))}{\partial q_i(s)}. \tag{1.21}$$

If L is invariant to s, i.e. $dL/ds=0$, then

$$\frac{dL}{ds}=0=\sum_i\left[\frac{\partial L}{\partial q_i(s)}\frac{dq_i(s)}{ds}+\frac{\partial L}{\partial \dot{q}_i(s)}\frac{d\dot{q}_i(s)}{ds}\right]. \tag{1.22}$$

This equals, after substituting from equation 1.21,

$$\begin{aligned}\frac{dL}{ds}&=\sum_i\left[\frac{d}{dt}\left(\frac{\partial L}{\partial \dot{q}_i(s)}\right)\frac{dq_i(s)}{ds}+\frac{\partial L}{\partial \dot{q}_i(s)}\frac{d}{dt}\left(\frac{dq_i(s)}{ds}\right)\right],\\ &=\frac{d}{dt}\left[\sum_i\frac{\partial L}{\partial \dot{q}_i(s)}\frac{dq_i(s)}{ds}\right].\end{aligned} \tag{1.23}$$

Therefore,

$$I=\sum_i\left.\frac{\partial L}{\partial \dot{q}_i}\frac{dq_i}{ds}\right|_{s=0} \tag{1.24}$$

is a constant of the motion. One may as well evaluate the expression at $s=0$ as at any other s.

Applying this to our example, we have

$$\begin{aligned}I&=\left.\frac{\partial L}{\partial \dot{x}}\frac{dx}{ds}\right|_{s=0}+\left.\frac{\partial L}{\partial \dot{y}}\frac{dy}{ds}\right|_{s=0},\\ &=m(\dot{y}x-y\dot{x})=l_z=\text{const.},\end{aligned}$$

that is, the z-component of the angular momentum is a constant of the motion.

BEAD ON A WIRE

As a physical example of the Lagrangian approach to mechanics, we consider a bead confined frictionlessly to a wire (figure 1.3). We suppose for concreteness that the equation defining the shape of the wire is

$$r(\theta)=1+\frac{1}{2}\cos\theta, \tag{1.25}$$

that is,

$$x(\theta)=r(\theta)\cos\theta=\left(1+\frac{1}{2}\cos\theta\right)\cos\theta;\quad y(\theta)=r(\theta)\sin\theta=\left(1+\frac{1}{2}\cos\theta\right)\sin\theta. \tag{1.26}$$

Writing the Cartesian velocities in terms of θ results in

$$\dot{x}=-\left(\sin\theta+\frac{1}{2}\sin 2\theta\right)\dot{\theta};\qquad \dot{y}=\left(\cos\theta+\frac{1}{2}\cos 2\theta\right)\dot{\theta}. \tag{1.27}$$

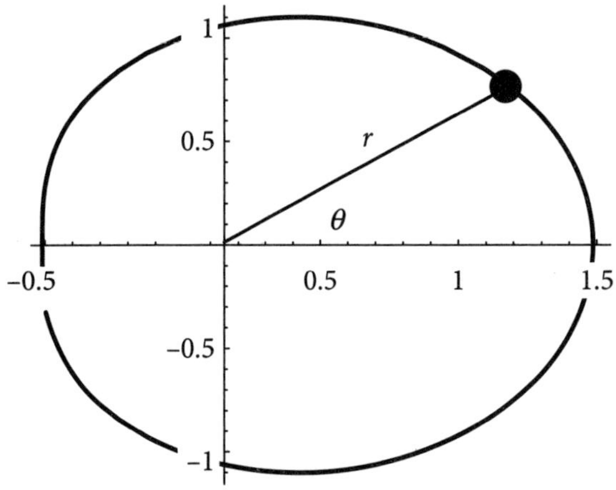

Figure 1.3. A bead slides frictionlessly on a wire of this shape.

Then, the kinetic energy (there is no potential energy), the Lagrangian for the bead on the wire, can be written as

$$\frac{1}{2}m(\dot{x}^2+\dot{y}^2)=\frac{1}{2}m\left(\frac{5}{4}+\cos\theta\right)\dot{\theta}^2=L(\theta,\dot{\theta}). \tag{1.28}$$

The Lagrange equation of motion is

$$\frac{d}{dt}\left(\frac{\partial L}{\partial\dot{\theta}}\right)-\left(\frac{\partial L}{\partial\theta}\right)=0, \tag{1.29}$$

that is,

$$\left(2\cos\theta+\frac{5}{2}\right)\ddot{\theta}-\sin\theta\,\dot{\theta}^2=0. \tag{1.30}$$

This is not readily solved directly. Instead, we use the fact that the kinetic energy T of the bead remains constant, as is easily checked by calculating

$$\frac{d}{dt}L=\frac{d}{dt}T=0, \tag{1.31}$$

so

$$\frac{m}{2}\left(\frac{5}{4}+\cos\theta\right)\dot{\theta}^2=L_0=\text{const}. \tag{1.32}$$

This is called a "first integral," a relation of the form

$$f(q_1,q_2,\ldots,q_N,\dot{q}_1,\dot{q}_2,\ldots,\dot{q}_N;t)=0.$$

As can often be done, we will use the first integral to help solve the full problem. Here, we solve for $\dot{\theta}$, which is better than dealing directly with equation 1.30,

$$\dot{\theta}=\sqrt{\frac{2}{m}\frac{L_0}{\left(\frac{5}{4}+\cos\theta\right)}}, \tag{1.33}$$

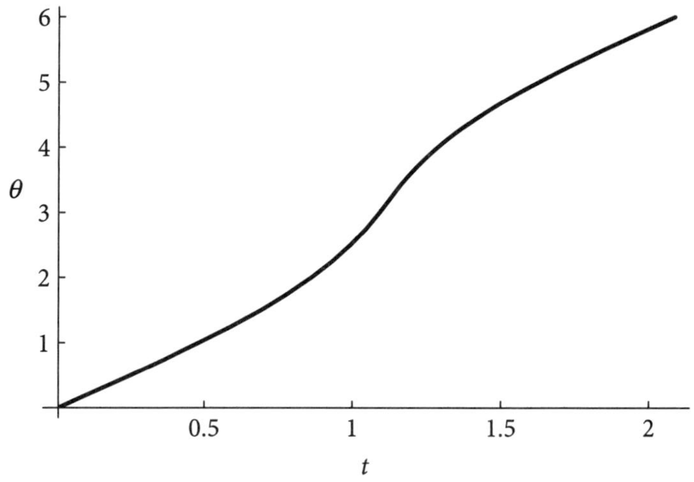

Figure 1.4. Progress of the bead on the wire defined earlier, in angle θ.

where L_0 is the initial and constant value of the Lagrangian. Equation 1.33 implies (assume $\theta = 0$ at $t = 0$)

$$t = \int_0^\theta \sqrt{\frac{\left(\frac{5}{4} + \cos\theta\right)}{\frac{2}{m}L_0}}\, d\theta. \tag{1.34}$$

This integral is of the elliptic form, and we have reduced the problem to a quadrature. We have

$$t = 3\sqrt{\frac{m}{2L_0}}\; E\left(\frac{\theta}{2}, \frac{8}{9}\right), \tag{1.35}$$

where E is an elliptic integral of the second kind. The equation for $t(\theta)$ can be inverted to give $\theta(t)$ (figure 1.4).

1.4 Hamiltonian Formulation

We define new *generalized momenta* p_i as

$$p_i = \frac{\partial L(\boldsymbol{q}, \dot{\boldsymbol{q}}, t)}{\partial \dot{q}_i} \tag{1.36}$$

in any coordinates—for example, $P_i = \partial L(\boldsymbol{Q}, \dot{\boldsymbol{Q}}, t)/\partial \dot{Q}_i$. There is a path to a powerful alternative formulation of mechanics by the seemingly benign substitution of the momenta p_i in place of the coordinate velocities $\dot{q}_i$ as independent variables. We define the Hamiltonian $H(\boldsymbol{p}, \boldsymbol{q}, t)$ as

$$H(\boldsymbol{p}, \boldsymbol{q}, t) = \sum_i p_i \dot{q}_i - L(\boldsymbol{q}, \dot{\boldsymbol{q}}, t), \tag{1.37}$$

where the $\dot{q}_i$ are to be regarded as functions of the p_i and q_i. The definition of H is in the standard form of a Legendre transformation (see p. 12), here eliminating the $\dot{\boldsymbol{q}}$ in favor of the momenta $\boldsymbol{p}$. The term $\sum_i p_i\dot{q}_i = \sum_i \dot{q}_i \partial L/\partial \dot{q}_i$ is twice the kinetic energy T, as may be seen in Cartesian coordinates, where $T = \sum_i m_i \dot{q}_i^2/2$. Thus, we may write

$$H = 2T - L = T + V; \tag{1.38}$$

this is a constant for a time-independent Lagrangian. We call $T + V$ the Hamiltonian, H, and

$$\boxed{H = T + V.} \tag{1.39}$$

Note that since $H(\boldsymbol{p}, \boldsymbol{q}, t)$ doesn't depend on the $\dot{\boldsymbol{q}}$ explicitly, we may write

$$\frac{\partial H(\boldsymbol{p}, \boldsymbol{q}, t)}{\partial \dot{q}_i} = 0 = p_i - \frac{\partial L(\boldsymbol{q}, \dot{\boldsymbol{q}}, t)}{\partial \dot{q}_i}; \tag{1.40}$$

this is just equation 1.36.

It follows by differentiation on equation 1.37 that

$$\boxed{\dot{q}_i = \frac{\partial H}{\partial p_i}; \ \dot{p}_i = -\frac{\partial H}{\partial q_i}, \quad i = 1, 2, \ldots N.} \tag{1.41}$$

These are *Hamilton's equations of motion*, derived from the Hamiltonian $H(\boldsymbol{p}, \boldsymbol{q}, t)$. We have not used any special properties of the coordinates $\boldsymbol{q}$, which can therefore be quite general. It follows that the form of Hamilton's equations of motion is the same in any coordinate system, providing the momenta are defined by equation 1.36 from the Lagrangian expressed in the same coordinates.

Suppose we follow this procedure and obtain a Hamiltonian $H(\boldsymbol{p}, \boldsymbol{q}, t)$ with equations of motion in the "canonical" form—namely, equations 1.41. The skew-symmetrical ("skew" because of the minus sign) canonical form of the equations of motion feature $\dot{\boldsymbol{q}}$ as the derivative of H with respect to $\boldsymbol{p}$, and $\dot{\boldsymbol{p}}$ as the negative of the derivative of H with respect to $\boldsymbol{q}$. If now an arbitrary and mathematically well defined transformation of coordinates and momenta is given—that is, $(\boldsymbol{p}, \boldsymbol{q}) \to (\boldsymbol{P}, \boldsymbol{Q})$—the canonical form of the equations 1.41 will *not* hold in general for $(\boldsymbol{P}, \boldsymbol{Q})$ using $H(\boldsymbol{P}, \boldsymbol{Q}, t) = H(\boldsymbol{p}(\boldsymbol{P}, \boldsymbol{Q}), \boldsymbol{q}(\boldsymbol{P}, \boldsymbol{Q}), t)$. There is however a systematic way to define "canonical transformations" to new variables that do preserve the canonical form of Hamilton's equations, as discussed in chapter 2. This will not patch up some arbitrary definition of new coordinates, but does provide a route to a wide set of new canonical coordinates preserving Hamilton's equations of motion.

If the Lagrangian (and thus the Hamiltonian) does not depend explicitly on time, the value of the Hamiltonian, called the "energy," remains constant at its initial value ("conservation of energy") under time evolution of a trajectory:

$$\frac{dH\boldsymbol{p}, \boldsymbol{q})}{dt} = \sum_i \left(\frac{\partial H}{\partial p_i}\dot{p}_i + \frac{\partial H}{\partial q_i}\dot{q}_i \right) = \sum_i \dot{q}_i\dot{p}_i - \dot{q}_i\dot{p}_i = 0 \tag{1.42}$$

LEGENDRE TRANSFORMATIONS

Equation 1.37 is an example of a *Legendre transformation*, accomplishing the change from $\dot{q}$ to p as an independent variable. To see what has happened, consider the function $f(x, z)$ and the replacement of x with y as an independent variable. Crucially, y is the derivative of f with respect to x (z is coming along for the ride so to speak). The Legendre transformation is

$$g(y, z) = yx - f(x, z), \tag{1.43}$$

and since g does not depend explicitly on x,

$$\frac{\partial g}{\partial x} = 0 = y - \frac{\partial f}{\partial x}. \tag{1.44}$$

Under equation 1.43, the new variable y is indeed the derivative of f with respect to x. The Legendre transform supplies its own inverse in the sense that

$$f(x, z) = xy - g(y, z), \tag{1.45}$$

with the condition $\partial f/\partial y = 0 = x - \partial g/\partial y$.

Legendre transformations are familiar from thermodynamics: start with the internal energy $U(S, V, N)$, and suppose temperature T is preferred as an independent variable, rather than entropy S. The following Legendre transformation yields the Helmholtz free energy function A:

$$A(T, V, N) = TS - U(S, V, N), \tag{1.46}$$

with $T = \partial U/\partial S$. We will encounter Legendre transformations repeatedly in this book.

There is one caveat, however. In the simple example for $f(x, z)$ earlier, we must insist that $\partial^2 f/\partial x^2 \neq 0$. The reason is simple: if $\partial^2 f/\partial x^2 = 0$, a range of x values map onto the same value of y, to first order in small changes δx. This is bad, and the transformation of coordinates fails. In the case of our Lagrangian, a function of many variables, the condition guaranteeing that the $\dot{q}_i$ can be written in terms of the new momenta $\boldsymbol{p}$ becomes

$$\det \left| \frac{\partial^2 L}{\partial \dot{q}_i \partial \dot{q}_j} \right| \neq 0. \tag{1.47}$$

LAGRANGE MULTIPLIER METHOD

The Lagrange multiplier method is introduced in the next section in the context of Hamilton's principle and the action. This is an advanced application, in the sense that an infinity of Lagrange multipliers appear, as we shall see. To see the essence of the Lagrange multiplier method, we first consider maximizing a function of only two variables with one constraint; introducing a single Lagrange multiplier. The example reveals that the trick is to augment the function to be maximized by adding the constraint condition multiplied by an unknown Lagrange multiplier to the function to be maximized. The key is that this augmented function may now be extremized without regard to constraints, and the extremum is in fact the original constrained solution we sought.

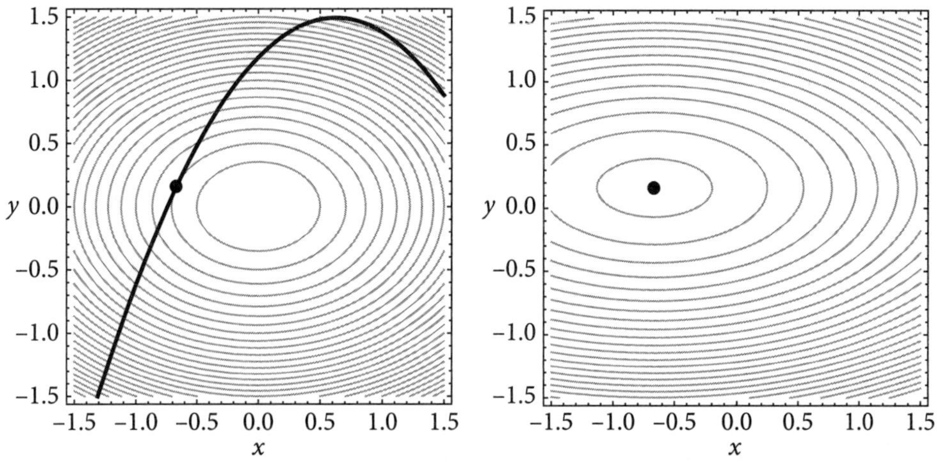

Figure 1.5. (Left) The unconstrained extremum is clearly at $(x, y) = (0, 0)$, but the new constraint (thick black line) forces a constrained maximum to lie somewhere along the line. The constrained maximum is shown by the dot. (Right) The contours of the function $G(x, y, \lambda)$ with Lagrange multiplier $\lambda = -0.6444$ have restored the problem to an extremum problem in two (formally three) dimensions. The new, global, unconstrained extremum of $G(x, y, \lambda)$ falls at the extremum of the original constrained problem.

The function $g(x, y) = -x^2 - 2y^2$ has the contours shown in figure 1.5, left. g maximizes at (0,0), but this is no longer relevant if the constraint

$$h(x, y) = y - (x + 1.18 - 0.8\,x^2) = 0 \tag{1.48}$$

is imposed (see the line cutting across the contours). Reading the contours of $g(x, y)$ along the line $y = (x + 1.18 - 0.8\,x^2)$, it appears that the constrained function maximizes at around $(x^*, y^*) \approx (-0.66, 0.17)$. (We notice another, locally minimizing solution around (0.79, 1.5) that we won't examine.)

The strategy is to find a new function $G(x, y, \lambda)$, augmented with a new variable λ, with an *unconstrained* extremum at (x^*, y^*, λ^*), where (x^*, y^*) is the *previously constrained* coordinate maximum we seek. How do we construct such a function?

Suppose we are on the constraint curve $h(x, y) = 0$. The gradient of g is

$$\vec{\nabla} g = \frac{\partial g}{\partial s}\hat{s} + \frac{\partial g}{\partial u}\hat{u}, \tag{1.49}$$

where $\hat{s}$ points along the constraint curve and $\hat{u}$ is perpendicular to it. The desired constrained maximum is specified simply by $\partial g/\partial s = 0$, and thus $\vec{\nabla} g$ lies perpendicular to the constraint curve at the constrained solution point. Also, $\vec{\nabla} h$ is everywhere perpendicular to the constraint curve $h(x, y) = 0$. Consider the function

$$G(x, y, \lambda) = g(x, y) + \lambda h(x, y), \tag{1.50}$$

obtained by adding the constraint function h to g with a proportionality λ, the "Lagrange multiplier." Then we require that

$$\vec{\nabla}G = \vec{\nabla}g + \lambda\vec{\nabla}h = 0. \tag{1.51}$$

The only way the two gradients on the right can sum to 0 for some nonvanishing value of λ is if they point in the same or opposite directions, but as we said, $\vec{\nabla}h$ always is perpendicular to the constraint curve. Therefore, when equation 1.51 holds, $\vec{\nabla}g$ must also be perpendicular to the constraint curve, meaning we have found the desired extremum of f subject to the constraint. Of course, λ needs to be set to

$$\lambda = \pm\frac{|\vec{\nabla}g|}{|\vec{\nabla}h|}, \tag{1.52}$$

where the sign depends on whether the two vectors point in the same or opposite directions.

Thus, the complete solution is specified by

$$\begin{aligned}\frac{\partial G(x, y, \lambda)}{\partial x} &= \frac{\partial g(x, y)}{\partial x} + \lambda\frac{\partial h(x, y)}{\partial x} = 0,\\ \frac{\partial G(x, y, \lambda)}{\partial y} &= \frac{\partial g(x, y)}{\partial y} + \lambda\frac{\partial h(x, y)}{\partial y} = 0,\\ \frac{\partial G(x, y, \lambda)}{\partial \lambda} &= h(x, y) = 0.\end{aligned} \tag{1.53}$$

We have three equations in the three unknowns x, y, and λ.

In the specific example chosen, the parameters obtained numerically are

$$x \to -0.665, \qquad y \to 0.161, \qquad \lambda \to -0.6444, \tag{1.54}$$

and the function $G(x, y, \lambda)$ when plotted as a function of (x, y) for the solution value of λ appears as in figure 1.5, right.

1.5 Constrained Systems and Forces of Constraint

We often need to understand dynamical systems under constraints, such as a bead confined to a wire, or a pendulum of fixed length.[2] If you are building a roller coaster for example, you need to know the constraint forces in order to determine the load on the superstructure. Forces of constraint are revealed by the method of Lagrange multipliers. So-called holonomic constraints take the form of equations obeyed by the coordinates, such as $r - r_0 = 0$ for a pendulum constrained to fixed length r_0. More generally, we write $h_i(q_1, q_2, \ldots, q_n) = 0$; $i = 1, \ldots, c < n$ for n coordinates $(q_1, q_2, \ldots, q_n)$ and c constraint conditions. It is often possible to find a new set of only $n - c$ coordinates, in terms of which the Lagrangian is uniquely expressed and the constraints automatically incorporated. This is just what was done for the bead on

[2] It is fascinating that constrained systems are *intrinsically ambiguous* to quantize directly [3]. One can define the constraint as the limit of stiff coordinates becoming infinitely rigid, but the analytic form of the stiffness matters!

the wire earlier, where the constraint equation was $r - \frac{1}{2}\cos\theta - 1 = 0$. It was simple to express the Lagrangian in terms of θ alone with the constraint built in. Unfortunately, using a system of coordinates adapted to the constraint obscures the forces that are at work accomplishing the constraint.

In analogy to the simple example given earlier, we include the constraints in the Lagrangian with c time-dependent Lagrange multipliers, an independent one for each constraint for every time t:

$$L_\lambda(\boldsymbol{q}, \dot{\boldsymbol{q}}, t) = L(\boldsymbol{q}, \dot{\boldsymbol{q}}, t) + \sum_{i=1}^{c} \lambda_i(t) h_i(\boldsymbol{q}, t). \tag{1.55}$$

The action S is the integral of L_λ over all times, and we have added the Lagrange multiplier terms $\lambda_i(t)h_i(\boldsymbol{q}, t)$ to the function to be extremized—that is, the action S. There are an infinity of them, because every time t has its own set of Lagrange multipliers. Now we are free to treat Hamilton's principle as an unconstrained problem, obtaining

$$\sum_k \int_{t_1}^{t_2} \left[\frac{\partial L}{\partial q_k} - \frac{d}{dt}\left(\frac{\partial L}{\partial \dot{q}_k}\right) + \sum_{i=1}^{M} \lambda_i(t) \frac{\partial h_i}{\partial q_k} \right] \delta q_k \, dt = 0. \tag{1.56}$$

The δq_k are free to vary unconstrained, thanks to the addition of the constraint terms and Lagrange multipliers. The new Lagrange equations, with δq_k arbitrary, are

$$\boxed{\frac{d}{dt}\left(\frac{\partial L}{\partial \dot{q}_k}\right) - \frac{\partial L}{\partial q_k} = \sum_{i=1}^{c} \lambda_i(t) \frac{\partial h_i}{\partial q_k},} \tag{1.57}$$

that clearly casts $\sum_i \lambda_i(t)\partial h_i/\partial q_k$ into the role of a generalized force $f_k(t)$ on the kth coordinate. There are in addition an infinity of new equations to "solve," but we know these are just the constraint equations

$$h_i(\boldsymbol{q}, t) = 0; \text{ all } i,\ t. \tag{1.58}$$

CONSTRAINED PENDULUM IN A GRAVITATIONAL FIELD

As an example, we consider a pendulum constrained to have length R_0 in a gravitational field. As with the bead on a wire, the problem could also be solved by writing the Lagrangian in terms of the pendulum angle θ, but here we use Lagrange multipliers. The reward will be finding the forces holding the pendulum bob at fixed length R_0 from the pivot.

The potential energy is $V = -mgR\cos\theta$. We have

$$\begin{aligned} L_\lambda &= T - V + \lambda(t)(R - R_0), \\ &= \frac{1}{2}mR^2\dot{\theta}^2 + \frac{1}{2}m\dot{R}^2 + mgR\cos\theta + \lambda(t)(R - R_0). \end{aligned} \tag{1.59}$$

The equations of motion are

$$\frac{d}{dt}\left(\frac{\partial L}{\partial \dot{\theta}}\right) - \frac{\partial L}{\partial \theta} = 0\text{—that is, } mR^2\ddot{\theta} + mgR\sin\theta = 0; \tag{1.60}$$

$$\frac{d}{dt}\left(\frac{\partial L}{\partial \dot{R}}\right) - \frac{\partial L}{\partial R} = 0\text{—that is,}$$

$$m\ddot{R} - mR\,\dot{\theta}^2 - mg\cos\theta - \lambda(t) = 0. \tag{1.61}$$

Since we know R will remain fixed at R_0, we can solve equation 1.60 in exactly the way it would have appeared in the "direct" approach without the Lagrange multiplier. Then equation 1.61 can be used to determine the constraint forces, once $\theta(t)$ is known. With the constraint, since $\ddot{R} = 0$, and $R - R_0 = 0$, we have

$$mR_0\,\dot{\theta}(t)^2 + mg\cos\theta(t) = -\lambda(t), \tag{1.62}$$

and $-\lambda(t)$ is the force directed along its length toward the pivot point that the pendulum must exert to keep length constant. The force comes in two parts, a centripetal term $mR\,\dot{\theta}(t)^2$ due to the rotation speed, mass, and radius, and a gravitational potential term, $mg\cos\theta(t)$ due to the radial component of the vertical gravitation force mg. This is the force required to keep the pendulum at fixed length R_0.

1.6 Derivatives and Legendre Transformations of the Action

COORDINATE DERIVATIVES OF $S(q_t, q_0, t)$ GIVE THE MOMENTA

The first equation in this book defined the action S as $S = \int_0^t L(\boldsymbol{q}, \dot{\boldsymbol{q}}, \tau)\, d\tau$. It might seem most natural that S depend on initial position $\boldsymbol{q}_0$ and initial momentum $\boldsymbol{p}_0$—these determine the trajectory. But integration by parts in equation 1.4 made $\boldsymbol{q}_t$ and $\boldsymbol{q}_0$ most natural, leaving us with a "shooting" problem, searching for the right momentum at the initial position $\boldsymbol{q}_0$ to reach the final position $\boldsymbol{q}_t$ in time t. The shooting problem causes inconvenience later in semiclassical applications. Giving any initial coordinate or momentum (but not both) and any final coordinate or momentum leads to the shooting problem.

Now we return to equation 1.4 and allow the initial and final positions to vary, by δq_0 and δq_t, respectively. We reproduce that equation here for convenience:

$$\delta S = \delta q \frac{\partial L}{\partial \dot{q}}\Big|_0^t - \int_0^t \delta q(\tau)\left[\frac{d}{dt}\left(\frac{\partial L}{\partial \dot{q}}\right) - \frac{\partial L}{\partial q}\right] d\tau = 0. \tag{1.63}$$

Since we still have

$$\frac{d}{dt}\left(\frac{\partial L}{\partial \dot{q}}\right) - \frac{\partial L}{\partial q} = 0,$$

we now use the first term on the right-hand side to get, remembering our definition of the momenta,

$$\frac{\partial S(\boldsymbol{q}_t, \boldsymbol{q}_0, t)}{\partial q_{i,0}} = -\frac{\partial L}{\partial \dot{q}_{i,0}} = -p_{i,0},$$
$$\frac{\partial S(\boldsymbol{q}_t, \boldsymbol{q}_0, t)}{\partial q_{i,t}} = \frac{\partial L}{\partial \dot{q}_{i,t}} = p_{i,t}. \tag{1.64}$$

In words, the initial and final momenta are given by the partial derivative of the action $S(\boldsymbol{q}_t, \boldsymbol{q}_0, t)$ with respect to initial and final position, noting the signs:

$$\boxed{\frac{\partial S(\boldsymbol{q}_t, \boldsymbol{q}_0, t)}{\partial \boldsymbol{q}_0} = -\boldsymbol{p}_0, \quad \frac{\partial S(\boldsymbol{q}_t, \boldsymbol{q}_0, t)}{\partial \boldsymbol{q}_t} = \boldsymbol{p}_t.} \tag{1.65}$$

PARTIAL TIME DERIVATIVE OF THE ACTION LEADS TO THE ENERGY

Next we consider the partial derivative of $S(\boldsymbol{q}_t, \boldsymbol{q}_0, t)$ with respect to time. By the fundamental theorem of calculus and the definition of the action, the *total* time derivative is given by

$$\frac{d}{dt} S(\boldsymbol{q}_t, \boldsymbol{q}_0, t) = L(\boldsymbol{q}_t, \dot{\boldsymbol{q}}_t, t) = T - V, \tag{1.66}$$

and of course we can also write

$$\frac{d}{dt} S(\boldsymbol{q}_t, \boldsymbol{q}_0, t) = \sum_i \frac{\partial S}{\partial q_{i,t}} \dot{q}_{i,t} + \frac{\partial S}{\partial t} = 2T + \frac{\partial S}{\partial t}. \tag{1.67}$$

Thus, comparing equations 1.66 and 1.67, we have

$$\boxed{\frac{\partial S(\boldsymbol{q}_t, \boldsymbol{q}_0, t)}{\partial t} + H = 0.} \tag{1.68}$$

There may be more than one trajectory going from $\boldsymbol{q}_0$ to $\boldsymbol{q}_t$ in time t. If so, there is a distinct action function $S_k(\boldsymbol{q}_t, \boldsymbol{q}_0, t)$ and generally distinct energy E_k (for a time-independent Hamiltonian) for the kth such trajectory.

REPLACING TIME WITH ENERGY: THE ABBREVIATED ACTION

Suppose we are more interested in trajectories that go from q_0 to q_t with a given energy E, rather than in a given time t—a perfectly reasonable preference, often motivated by physical circumstance. We'd like to swap t for E as an independent variable, which sounds familiar: a job for another Legendre transformation, especially since we just learned that the Hamiltonian is minus the derivative of the action with respect to time.

We define the new Legendre transformation (replacing t with E), yielding what is called the "abbreviated action" $S_0(\boldsymbol{q}_t, \boldsymbol{q}_0, E)$:

$$S_0(\boldsymbol{q}_t, \boldsymbol{q}_0, E) = S(\boldsymbol{q}_t, \boldsymbol{q}_0, t) + Et. \tag{1.69}$$

This is consistent with what we know about the time derivative of $S(\boldsymbol{q}_t, \boldsymbol{q}_0, t)$: take $\partial/\partial t$ on both sides of equation 1.69; we obtain $0 = \partial S(\boldsymbol{q}_t, \boldsymbol{q}_0, t)/\partial t + E$. By taking $\partial/\partial E$ on both sides of equation 1.69, we get

$$\boxed{\frac{\partial S_0(\boldsymbol{q}_t, \boldsymbol{q}_0, E)}{\partial E} = t.} \tag{1.70}$$

This is the relation used to fix the transformation from t as independent to E as independent. With E fixed, we can write, using multidimensional notation and relabeling subscripts for the coordinates (to emphasize independence on time in favor of the energy)

$$S_0(\boldsymbol{q}_b, \boldsymbol{q}_a, E) = \int_0^t (\boldsymbol{p}\dot{\boldsymbol{q}} - H)\, dt + Et = \int_0^t \boldsymbol{p}\dot{\boldsymbol{q}}\, dt - Et + Et,$$

$$\boxed{S_0(\boldsymbol{q}_a, \boldsymbol{q}_b, E) = \int_{\boldsymbol{q}_a}^{\boldsymbol{q}_b} \boldsymbol{p}(\boldsymbol{q}, E) \cdot d\boldsymbol{q}.} \tag{1.71}$$

In fact, the abbreviated action $S_0(\boldsymbol{q}_b, \boldsymbol{q}_a, E)$ is the solution to a variational problem for the *shape* of classical orbits at fixed E, called *Maupertuis's principle.* This principle states that if changes in S_0 due to all possible small changes to the shape of a trajectory are second order in those changes (the actual progress of the trajectory in time is not specified, but the trial shape changes need to keep energy fixed), then the trajectory in question is the physical, classical orbit in space. That is, $S_0(\boldsymbol{q}_b, \boldsymbol{q}_a, E)$ is stationary against small changes in shape around the correct classical orbit, for fixed energy.

MOMENTUM AS INDEPENDENT VARIABLES IN THE ACTION

Since momentum is the slope of the action with respect to $\boldsymbol{q}_0$ or $\boldsymbol{q}_t$, we can perform Legendre transformations replacing position as independent variables with momenta $\boldsymbol{p}_0$ or $\boldsymbol{p}_t$, respectively. For example, set

$$S(\boldsymbol{q}_t, \boldsymbol{p}_0, t) = S(\boldsymbol{q}_t, \boldsymbol{q}_0, t) + \boldsymbol{q}_0 \boldsymbol{p}_0, \tag{1.72}$$

with

$$\frac{\partial S(\boldsymbol{q}_t, \boldsymbol{p}_0, t)}{\partial \boldsymbol{q}_0} = \frac{\partial S(\boldsymbol{q}_t, \boldsymbol{q}_0, t)}{\partial \boldsymbol{q}_0} + \boldsymbol{p}_0 = 0. \tag{1.73}$$

It is understood that $S(\boldsymbol{q}_t, \boldsymbol{q}_0(\boldsymbol{q}_t, \boldsymbol{p}_0), t) \equiv S(\boldsymbol{q}_t, \boldsymbol{p}_0, t)$ has of course a different functional dependence on its arguments than $S(\boldsymbol{q}_t, \boldsymbol{q}_0, t)$. The new functional dependence is found by eliminating $\boldsymbol{q}_0$ in favor of $\boldsymbol{p}_0$ using $\partial S(\boldsymbol{q}_t, \boldsymbol{q}_0, t)/\partial \boldsymbol{q}_0 + \boldsymbol{p}_0 = 0$.

We note that nothing in this Legendre transformation disturbs the partial derivative of S with respect to time, so that also

$$\boxed{\frac{\partial S(\boldsymbol{q}_t, \boldsymbol{p}_0, t)}{\partial t} + H = 0.} \tag{1.74}$$

The same arguments may be made as those following equation 1.68—that is, the action boils down to individual trajectories with definite $\boldsymbol{p}_0$ and $\boldsymbol{q}_0$, although now with $\boldsymbol{p}_0$ specified as fixed, it is $\boldsymbol{q}_0$ that must be adjusted slightly if time is incremented. The same substitution of the $\boldsymbol{p}_0$ for the $\boldsymbol{q}_0$ as independent variables may be made with the abbreviated action as well.

1.7 Velocity-Dependent Potentials

Another possibility preserving the form of Lagrange's equations is a special type of *v*elocity-dependent potential $U(\boldsymbol{q}, \dot{\boldsymbol{q}}, t)$ from which the force F_i is derived as

$$F_i = -\frac{\partial U}{\partial q_i} + \frac{d}{dt}\frac{\partial U}{\partial \dot{q}_i}. \tag{1.75}$$

Again, it is straightforward to show that if we take $L = T - U$, we have

$$\frac{d}{dt}\left(\frac{\partial L}{\partial \dot{q}_i}\right) - \frac{\partial L}{\partial q_i} = 0. \tag{1.76}$$

A KEY EXAMPLE OF A VELOCITY-DEPENDENT POTENTIAL: ELECTROMAGNETIC FORCES

Miraculously, such a potential can be constructed for particles in electromagnetic fields. We begin with Maxwell's equations in Gaussian units:

$$\nabla \cdot \boldsymbol{E} = 4\pi\rho, \tag{1.77}$$

$$\nabla \times \boldsymbol{B} - \frac{1}{c}\frac{\partial \boldsymbol{E}}{\partial t} = \frac{4\pi \boldsymbol{J}}{c}, \tag{1.78}$$

$$\nabla \times \boldsymbol{E} + \frac{1}{c}\frac{\partial \boldsymbol{B}}{\partial t} = 0, \tag{1.79}$$

$$\nabla \cdot \boldsymbol{B} = 0. \tag{1.80}$$

The force (Lorentz force) on charge e is

$$\boldsymbol{F} = e\left\{\boldsymbol{E} + \frac{1}{c}\boldsymbol{v} \times \boldsymbol{B}\right\}. \tag{1.81}$$

Since $\nabla \cdot \boldsymbol{B} = 0$, $\boldsymbol{B}$ can be written as the curl of the *vector potential* $\boldsymbol{A}$:

$$\boldsymbol{B} = \nabla \times \boldsymbol{A}. \tag{1.82}$$

Then it follows

$$\nabla \times \boldsymbol{E} + \frac{1}{c}\frac{\partial}{\partial t}(\nabla \times \boldsymbol{A}) = \nabla \times \left(\boldsymbol{E} + \frac{1}{c}\frac{\partial \boldsymbol{A}}{\partial t}\right) = 0. \tag{1.83}$$

We can write what is inside the parentheses as the gradient of a potential—that is,

$$E + \frac{1}{c}\frac{\partial A}{\partial t} = -\nabla\phi. \tag{1.84}$$

From this it can be shown that

$$F_x = e\left[-\frac{\partial}{\partial x}\left(\phi - \frac{1}{c}v \cdot A\right) - \frac{1}{c}\frac{d}{dt}\left(\frac{\partial}{\partial v_x}(A \cdot v)\right)\right], \tag{1.85}$$

or

$$F_x = -\frac{\partial U}{\partial x} + \frac{d}{dt}\frac{\partial U}{\partial v_x}, \tag{1.86}$$

where

$$U = e\phi - \frac{e}{c}A \cdot v. \tag{1.87}$$

This means the Lagrangian of a particle in an electromagnetic field is

$$\boxed{L = T - e\phi + \frac{e}{c}A \cdot v.} \tag{1.88}$$

Lagrange's equations hold, and the trajectory of charged particles in electromagnetic fields can be determined from them.

THE NONUNIQUENESS OF THE LAGRANGIAN

The Lagrangian may be altered by adding terms that are the total time derivative of any function f of coordinates (but not their velocities) and time, without altering the motion. For example, if

$$L \to L' = L + \frac{d}{dt}f(q, t) = L + \frac{\partial f(q, t)}{\partial q}\dot{q} + \frac{\partial f(q, t)}{\partial t}, \tag{1.89}$$

then

$$S \to S' = S + f(q_t, t) - f(q_0, 0). \tag{1.90}$$

The new terms contribute nothing to variation of the path, since the starting and end points are fixed. The result of the variation—that is, Lagrange's equation—is thus the same as before the addition of df/dt. The momenta are altered, since f affects the end point derivatives. Since the actual motion is not affected, the momenta have simply been redefined. It will presently become clear that such alterations to the Lagrangian are intimately related to so-called gauge transformations.

The redefinition of the momenta hints at the possibility of adding total time derivatives in order to accomplish a full-throated change of coordinates, involving both the positions and the momenta. A step in this direction is taken by supposing that the function f depends on other, possibly time-dependent variables $\mathbf{Q}$ (there is no harm in having a new f for each new $\mathbf{Q}$)—as in,

$$L \to L' = L + \frac{d}{dt}f(q, Q, t), \tag{1.91}$$

then

$$S \to S' = S + f(\boldsymbol{q}_t, \boldsymbol{Q}_t, t) - f(\boldsymbol{q}_0, \boldsymbol{Q}_0, 0). \tag{1.92}$$

In spite of this new complexity, df/dt leaves the physical motion untouched. We develop this idea further in the next chapter, on canonical transformations.

GAUGE TRANSFORMATIONS

We have just seen that the derivatives of vector and scalar potentials $\boldsymbol{A}$ and ϕ are used to construct the physical, measurable quantities $\boldsymbol{E}$ and $\boldsymbol{B}$, the electric and magnetic fields. $\boldsymbol{A}$ and ϕ are not measurable, physical quantities; only their derivatives are. This means that whole families of choices for $\boldsymbol{A}$ and ϕ can lead to the same, physical $\boldsymbol{E}$ and $\boldsymbol{B}$. Moving around in these families is called *gauge transformation*. The electromagnetic gauge transformation is defined through

$$\begin{aligned} \boldsymbol{A} &\to \boldsymbol{A} + \nabla\psi(\boldsymbol{r}, t), \\ \phi &\to \phi - \frac{1}{c}\frac{\partial\psi(\boldsymbol{r}, t)}{\partial t}, \end{aligned} \tag{1.93}$$

where $\psi(\boldsymbol{r}, t)$ is a "gauge potential," a scalar function of space and time chosen for convenience. If we add the "extra" gauge terms to the Lagrangian L,

$$L = T - e\phi + \frac{e}{c}\boldsymbol{A}\cdot v \to L' = L + \frac{e}{c}\frac{\partial\psi}{\partial t} + \frac{e}{c}\nabla\psi\cdot\boldsymbol{v}, \tag{1.94}$$

we notice that we can write this as the total time derivative of a function of coordinates and time:

$$\boxed{L' = L + \frac{e}{c}\frac{d}{dt}\psi(\boldsymbol{r}, t).} \tag{1.95}$$

Thus, the motion is the same after the gauge transformation, as promised in electromagnetic theory.

PROBLEM: PARTICLE IN A MAGNETIC AND ELECTRIC FIELD—TWO GAUGES

Suppose we have the vector potential $\boldsymbol{A} = A_0(-y, x, 0)$ and scalar potential $\phi = -\phi_0 x$. Then the magnetic field $\vec{B}$ points in the z direction, $\boldsymbol{B} = \nabla\times\boldsymbol{A} = 2A_0\hat{z}$, and the electric field is $\boldsymbol{E} = -\nabla\phi = \phi_0\hat{x}$. The Lagrangian is

$$\begin{aligned} L &= T - e\phi + \frac{e}{c}\boldsymbol{A}\cdot\boldsymbol{v}, \\ &= \frac{1}{2}m\left(\dot{x}^2 + \dot{y}^2 + \dot{z}^2\right) - e\phi_0 x - \frac{eA_0}{c}(x\dot{y} - y\dot{x}); \end{aligned} \tag{1.96}$$

we find

$$\begin{aligned} m\ddot{x} + 2\frac{eA_0}{c}\dot{y} + e\phi_0 &= 0, \\ m\ddot{y} - 2\frac{eA_0}{c}\dot{x} &= 0, \\ m\ddot{z} &= 0. \end{aligned} \tag{1.97}$$

We see the x and y motions are coupled. Given that there are three second-order differential equations, we must have six constants, coming from three initial positions and three initial velocities. The general solution to this problem is

$$x_t = \alpha + \beta \cos(\omega t) + \gamma \sin(\omega t),$$
$$y(t) = \delta - \gamma \cos(\omega t) + \beta \sin(\omega t) - \frac{e\phi_0}{m\omega} t, \tag{1.98}$$

and

$$z(t) = z_0 + \dot{z}_0 t, \tag{1.99}$$

where the so-called cyclotron frequency is

$$\omega = \frac{2e A_0}{mc} = \frac{|\boldsymbol{B}|e}{mc}.$$

Chapter 2

Canonical Transformations

In chapter 1, we noted that the spatial coordinates $\boldsymbol{q}$ may be chosen quite freely. Lagrange's equations and Hamilton's equations of motion always take on the same, "canonical" form, whether we start with one set ($\boldsymbol{q}$) as coordinates, or another set ($\boldsymbol{Q}$). The momenta are derived as $p_i = \partial L(\boldsymbol{q}, \dot{\boldsymbol{q}})/\partial q_i$ or $P_i = \partial L(\boldsymbol{Q}, \dot{\boldsymbol{Q}})/\partial Q_i$ accordingly. The form of Hamilton's equations is the same in both systems of coordinates—that is, in the $(\boldsymbol{q}, \boldsymbol{p})$ system,

$$\dot{q}_i = \frac{\partial H(\boldsymbol{p}, \boldsymbol{q}, t)}{\partial p_i}, \quad \dot{p}_i = -\frac{\partial H(\boldsymbol{p}, \boldsymbol{q}, t)}{\partial q_i}, \quad i = 1, 2, \ldots N, \tag{2.1}$$

and in the $(\boldsymbol{Q}, \boldsymbol{P})$ coordinates,

$$\dot{Q}_i = \frac{\partial K(\boldsymbol{P}, \boldsymbol{Q}, t)}{\partial P_i}, \quad \dot{P}_i = -\frac{\partial K(\boldsymbol{P}, \boldsymbol{Q}, t)}{\partial Q_i}, \quad i = 1, 2, \ldots N. \tag{2.2}$$

This is not the most general type of "canonical transformation"—that is, a transformation from $(\boldsymbol{q}, \boldsymbol{p})$ to a new set of variables $(\boldsymbol{Q}, \boldsymbol{P})$ that preserves the form of Hamilton's equations of motion. We changed spatial coordinates $\boldsymbol{q}$ to the set $\boldsymbol{Q}$; the associated momenta followed obsequiously so to speak. This is called a *point transformation.* More general transformations are possible that leave Hamilton's equations in canonical form, treating the coordinates and momenta of the $2N$ variables $(\boldsymbol{q}, \boldsymbol{p})$ on an equal footing. They become $2N$ variables $(u_1, \ldots u_{2N})$ to be transformed to a new set $(U_1, \ldots U_{2N})$.

We seek the most general transformation of coordinates $\{\boldsymbol{q}, \boldsymbol{p}\} \to \{\boldsymbol{Q}, \boldsymbol{P}\}$ that preserves the canonical form of Hamilton's equations—that is, equations 2.1 become equations 2.2 in the new coordinates $\{\boldsymbol{Q}, \boldsymbol{P}\}$. First, the Lagrangian and Hamilton's principle is expressed in terms of the Hamiltonian, viz.,

$$\delta \int_0^t \left[\sum_i p_i \dot{q}_i - H(\boldsymbol{p}, \boldsymbol{q}, t) \right] dt = 0. \tag{2.3}$$

In the new system $\{\boldsymbol{Q}, \boldsymbol{P}\}$, we also need

$$\delta \int_0^t \left[\sum_i P_i \dot{Q}_i - K(\boldsymbol{P}, \boldsymbol{Q}, t) \right] dt = 0. \tag{2.4}$$

Because of the nonuniqueness of either Lagrangian, $L = \sum_i p_i \dot{q}_i - H(\boldsymbol{p}, \boldsymbol{q}, t)$ or $L' = \sum_i P_i \dot{Q}_i - K(\boldsymbol{P}, \boldsymbol{Q}, t)$, the two are not required to be the same at corresponding values of their arguments. Instead, they can differ by a total time derivative of a function f, since each is nonunique in that sense. But what should the arguments of f be? Apart from time t, it is useful for f to have one argument from the "old" variables $(\boldsymbol{q}, \boldsymbol{p})$ and one from the new set $(\boldsymbol{Q}, \boldsymbol{P})$.

We already had some practice with f using the old coordinate only, except we added an "extra" variable $\boldsymbol{Q}$, as in $f(\boldsymbol{q}, \boldsymbol{Q}, t)$ (at that time $\boldsymbol{Q}$ had no special significance and just labeled a possibly different function $f(\boldsymbol{q}, \boldsymbol{Q}, t)$ for each $\boldsymbol{Q}$; see equation 1.91). Now $\boldsymbol{Q}$ will become the new coordinate, but without the restriction that it be a transformation of the old coordinates (and not the old momenta) only. If f depends on the old $\boldsymbol{q}$'s and new $\boldsymbol{Q}$'s, we call it an f_1 type generator $f_1(\boldsymbol{q}, \boldsymbol{Q}, t)$ of the canonical transformation. We now express the difference between the two Lagrangians in the form (that is, they have to differ in this way or else the equations of motion in $\boldsymbol{P}, \boldsymbol{Q}$ will not turn out to be canonical)

$$\boldsymbol{p} \cdot \dot{\boldsymbol{q}} - H(\boldsymbol{p}, \boldsymbol{q}, t) = \boldsymbol{P} \cdot \dot{\boldsymbol{Q}} - K(\boldsymbol{P}, \boldsymbol{Q}, t) + \frac{d}{dt} f_1(\boldsymbol{q}, \boldsymbol{Q}, t). \tag{2.5}$$

This ensures that equation 2.4 will hold if equation 2.3 does. Expanding this as

$$\boldsymbol{p} \cdot \dot{\boldsymbol{q}} - H(\boldsymbol{p}, \boldsymbol{q}, t) = \boldsymbol{P} \cdot \dot{\boldsymbol{Q}} - K(\boldsymbol{P}, \boldsymbol{Q}, t) + \frac{\partial f_1}{\partial \boldsymbol{q}} \cdot \dot{\boldsymbol{q}} + \frac{\partial f_1}{\partial \boldsymbol{Q}} \cdot \dot{\boldsymbol{Q}} + \frac{\partial f_1}{\partial t}, \tag{2.6}$$

we need to set

$$\boxed{\boldsymbol{p} = \frac{\partial f_1(\boldsymbol{q}, \boldsymbol{Q}, t)}{\partial \boldsymbol{q}}; \qquad \boldsymbol{P} = -\frac{\partial f_1(\boldsymbol{q}, \boldsymbol{Q}, t)}{\partial \boldsymbol{Q}},} \tag{2.7}$$

or else $\boldsymbol{Q}$ would depend on $\boldsymbol{q}$, but these must be independent:

$$\boxed{K(\boldsymbol{P}, \boldsymbol{Q}, t) = H(\boldsymbol{p}, \boldsymbol{q}, t) + \frac{\partial f_1(\boldsymbol{q}, \boldsymbol{Q}, t)}{\partial t}.} \tag{2.8}$$

The transformation to the new variables will be canonical. Canonical transformations that do not depend on time are called *restricted*.

For example, the trivial generator $f_1(\boldsymbol{q}, \boldsymbol{Q}, t) = \boldsymbol{q}\boldsymbol{Q}$ generates $\boldsymbol{p} = \boldsymbol{Q}$; $\boldsymbol{P} = -\boldsymbol{q}$, and the role of the position and momentum variables has been reversed. This possibility reflects the formally *symplectic* structure, a symmetry seen in Hamilton's equations of motion, about which we will elaborate later.

The caveat is that equation 2.7 can be solved for $(\boldsymbol{Q}, \boldsymbol{P})$ in terms of $(\boldsymbol{p}, \boldsymbol{q})$ only if

$$\det\left(\frac{\partial^2 f_1(\boldsymbol{q}, \boldsymbol{Q})}{\partial \boldsymbol{q} \partial \boldsymbol{Q}}\right) \neq 0 \tag{2.9}$$

(the nondegeneracy condition). Exactly analogous conditions apply below for f_2, f_3, and f_4.

We can see why the nondegeneracy condition is needed by supposing (in one degree of freedom), for example, that $\partial^2 f_1(q, Q)/\partial q \partial Q = -(\partial P/\partial q) = 0$. This implies that there is a range of q and Q where P does not change even though q does. The change of variables fails there.

It is natural to start with $(\boldsymbol{q}, \boldsymbol{Q})$ as the independent variables in f, since as we just saw, terms arise in df/dt that correspond to others already present in the Lagrangians. The $f_2(\boldsymbol{q}, \boldsymbol{P}, t)$ generator can then be reached with a Legendre transformation (see section 1.4):

$$f_1(\boldsymbol{q}, \boldsymbol{Q}, t) = -\boldsymbol{Q} \cdot \boldsymbol{P} + f_2(\boldsymbol{q}, \boldsymbol{P}, t). \tag{2.10}$$

Now we can write

$$\boldsymbol{p} \cdot \dot{\boldsymbol{q}} - H(\boldsymbol{p}, \boldsymbol{q}, t) = \boldsymbol{P} \cdot \dot{\boldsymbol{Q}} - K(\boldsymbol{P}, \boldsymbol{Q}, t) + \frac{d}{dt}\left[-\boldsymbol{Q} \cdot \boldsymbol{P} + f_2(\boldsymbol{q}, \boldsymbol{P}, t)\right], \tag{2.11}$$

or expanding,

$$\begin{aligned} \boldsymbol{p} \cdot \dot{\boldsymbol{q}} - H(\boldsymbol{p}, \boldsymbol{q}, t) &= -\boldsymbol{Q} \cdot \dot{\boldsymbol{P}} - K(\boldsymbol{P}, \boldsymbol{Q}, t) + \frac{\partial f_2(\boldsymbol{q}, \boldsymbol{P}, t)}{\partial \boldsymbol{q}} \cdot \dot{\boldsymbol{q}} \\ &\quad + \frac{\partial f_2(\boldsymbol{q}, \boldsymbol{P}, t)}{\partial \boldsymbol{P}} \cdot \dot{\boldsymbol{P}} + \frac{\partial f_2(\boldsymbol{q}, \boldsymbol{P}, t)}{\partial t}. \end{aligned} \tag{2.12}$$

From this we learn, reassuringly,

$$\boxed{\boldsymbol{p} = \frac{\partial f_2(\boldsymbol{q}, \boldsymbol{P}, t)}{\partial \boldsymbol{q}}; \qquad \boldsymbol{Q} = \frac{\partial f_2(\boldsymbol{q}, \boldsymbol{P}, t)}{\partial \boldsymbol{P}},} \tag{2.13}$$

and

$$\boxed{K(\boldsymbol{P}, \boldsymbol{Q}, t) = H(\boldsymbol{p}, \boldsymbol{q}, t) + \frac{\partial f_2(\boldsymbol{q}, \boldsymbol{P}, t)}{\partial t}.} \tag{2.14}$$

In the same way, we discover, for $f_3(\boldsymbol{p}, \boldsymbol{Q}, t)$, and $f_4(\boldsymbol{p}, \boldsymbol{P}, t)$,

$$\boxed{\boldsymbol{q} = -\frac{\partial f_3(\boldsymbol{p}, \boldsymbol{Q}, t)}{\partial \boldsymbol{p}}; \qquad \boldsymbol{P} = -\frac{\partial f_3(\boldsymbol{p}, \boldsymbol{Q}, t)}{\partial \boldsymbol{Q}};} \tag{2.15}$$

$$\boxed{K(\boldsymbol{P}, \boldsymbol{Q}, t) = H(\boldsymbol{p}, \boldsymbol{q}, t) + \frac{\partial f_3(\boldsymbol{p}, \boldsymbol{Q}, t)}{\partial t};} \tag{2.16}$$

and

$$\boxed{\boldsymbol{q} = -\frac{\partial f_4(\boldsymbol{p}, \boldsymbol{P}, t)}{\partial \boldsymbol{p}}; \qquad \boldsymbol{Q} = \frac{\partial f_4(\boldsymbol{p}, \boldsymbol{P}, t)}{\partial \boldsymbol{P}};} \tag{2.17}$$

$$\boxed{K(\boldsymbol{P}, \boldsymbol{Q}, t) = H(\boldsymbol{p}, \boldsymbol{q}, t) + \frac{\partial f_4(\boldsymbol{p}, \boldsymbol{P}, t)}{\partial t}.} \tag{2.18}$$

We have said that defining a new coordinate set in the Lagrangian formalism results in a point transformation (a transformation among the coordinates only, with the momenta following suit). The f_2 generator $f_2(\boldsymbol{q}, \boldsymbol{P}) = u(\boldsymbol{q}, t) \cdot \boldsymbol{P}$ is seen to accomplish the point transformation $\partial f_2(\boldsymbol{q}, \boldsymbol{P})/\partial \boldsymbol{P} = \boldsymbol{Q} = u(\boldsymbol{q}, t)$ with $\boldsymbol{P} = [\partial u(\boldsymbol{q}, t)/\partial \boldsymbol{q}]^{-1} \cdot \boldsymbol{p}$.

2.1 Hamilton-Jacobi Idea

The idea behind the Hamilton-Jacobi equation is to stop the motion altogether in a new coordinate system. That is, we want

$$K(P, Q, t) = H(p(P, Q), q(P, Q), t) + \frac{\partial F_1(q, Q)}{\partial t} = 0, \tag{2.19}$$

where an F_1 generator is used. Then clearly

$$\begin{aligned} \dot{P} &= -\frac{\partial K}{\partial Q} = 0, \\ \dot{Q} &= \frac{\partial K}{\partial P} = 0. \end{aligned} \tag{2.20}$$

The Hamilton-Jacobi idea is not a panacea for solving problems but rather a useful formal development. We already know one solution to the Hamilton-Jacobi idea—namely, if we take

$$F_1(q, Q) \equiv S(q_2, q_1, t),$$

where q_2 plays the role of q and q_1 plays the role of Q, then equation 2.19 is satisfied. To turn this idea into an equation, we can write

$$H\left(\frac{\partial S}{\partial q}, q, t\right) = -\frac{\partial S}{\partial t}; \tag{2.21}$$

this is the *Hamilton-Jacobi equation.* It renders the new Hamiltonian $K = 0$, as seen in equation 2.19. For example, in the case of the simplest form $H = p^2/2m + V(q)$, we have

$$\frac{1}{2m}\left(\frac{\partial S}{\partial q}\right)^2 + V(q) = -\frac{\partial S}{\partial t}. \tag{2.22}$$

The Hamilton-Jacobi (H-J) equation for the "abbreviated" action S_0 is then

$$H\left(\frac{\partial S_0}{\partial q}, q\right) = E, \tag{2.23}$$

where we have taken the Hamiltonian to be time independent.

The action $S(q, q_1, t)$ is a solution to the Hamilton-Jacobi equation. Note that we said "a" solution, not "the" solution. The Hamilton-Jacobi equation is a partial differential equation involving only the q and t variables explicitly. One should ask about boundary conditions. Indeed, these are specified if we use $S(q, q_1, t)$—that is, if q_1 is initially specified. But one may want to specify other things as given at $t = 0$—for example, the initial momentum p_1 instead of the position q_1.

As a simple example, consider a free particle. The action is $S(q_1, q_0, t) = m(q_1 - q_0)^2/2t$, and the energy is $m(q_1 - q_0)^2/2t^2$. But $\partial S(q_1, q_0, t)/\partial t = -m(q_1 - q_0)^2/2t^2$, so $K = 0$ indeed.

Since $\boldsymbol{p}_t = \partial S(\boldsymbol{q}_t, \boldsymbol{q}_0, t)/\partial \boldsymbol{q}_t$, we can write

$$\frac{\partial S}{\partial t} + H\left(\frac{\partial S(\boldsymbol{q}_t, \boldsymbol{q}_0, t)}{\partial \boldsymbol{q}_t}, \boldsymbol{q}_t, t\right) = 0. \tag{2.24}$$

This raises some interesting questions leading to important new directions. Fixing the N coordinates $\boldsymbol{q}_0$, we could try to solve this equation for S. For example, in one dimension,

$$\frac{\partial S(q_t, q_0, t)}{\partial t} + H\left(\frac{\partial S(q_t, q_0, t)}{\partial q_t}, q_t, t\right) = 0. \tag{2.25}$$

This is a perfectly viable equation to try to solve for $S(q_t, q_0, t)$. For the 1D harmonic oscillator, we have

$$\frac{1}{2m}\left(\frac{\partial S}{\partial q_t}\right)^2 + \frac{1}{2}m\omega^2 q_t^2 + \frac{\partial S}{\partial t} = 0. \tag{2.26}$$

It is a simple exercise to show that $S(q_t, q_0, t)$ of equation 2.30, later, solves this partial differential equation for the simple harmonic oscillator.

2.2 Action as the Generating Function for Dynamics

The action $S(\boldsymbol{q}_t, \boldsymbol{q}_0, t)$ is perfectly qualified as an f_1 generator. After all, classical generators do not belong to a very exclusive club; the requirement is only that they have to depend on coordinates in a nondegenerate way (equation 2.9). We take $f_1(\boldsymbol{q}, \boldsymbol{Q}, t) = S(\boldsymbol{q}_t, \boldsymbol{q}_0, t)$ with $\boldsymbol{q}_t \to \boldsymbol{q}$ and $\boldsymbol{q}_0 \to \boldsymbol{Q}$, taking the old variable to be the final position and the new variable as the initial position. We have already seen (equation 1.65)

$$\frac{\partial S(\boldsymbol{q}_t, \boldsymbol{q}_0, t)}{\partial \boldsymbol{q}_0} = -\boldsymbol{p}_0, \quad \frac{\partial S(\boldsymbol{q}_t, \boldsymbol{q}_0, t)}{\partial \boldsymbol{q}_t} = \boldsymbol{p}_t. \tag{2.27}$$

These are precisely the requirements determining the trajectory starting with the information of the initial and final position, using an f_1 generating function.

Since $S(\boldsymbol{q}_t, \boldsymbol{q}_0, t)$ encodes the transformation from (q_0, p_0) into (q_t, p_t), the action $S(\boldsymbol{q}_t, \boldsymbol{q}_0, t)$ is the generator of the canonical transformation corresponding to dynamical evolution for a time t (this can be any time, with a different $S(\boldsymbol{q}_t, \boldsymbol{q}_0, t)$ for each time). That is, if you are handed $S(\boldsymbol{q}_t, \boldsymbol{q}_0, t)$, you know the trajectories: Specify a $(\boldsymbol{q}_0, \boldsymbol{q}_t)$ and take the end point derivatives of S to get $(\boldsymbol{p}_0, \boldsymbol{p}_t)$, and thus you know $(\boldsymbol{q}_0, \boldsymbol{p}_0) \to (\boldsymbol{q}_t, \boldsymbol{p}_t)$. The transformation is area preserving, since it is manifestly canonical, being given by a generating function. The area preservation is Liouville's theorem (see section 2.3).

COMPOSING ACTION GENERATORS FOR LONGER TIME PROPAGATION

We may apply what we just learned about composing two canonical transformations to create a generator for dynamics at twice the time, $2t$—that is,

$$S(\boldsymbol{q}_{2t}, \boldsymbol{q}_0, 2t) = S(\boldsymbol{q}_{2t}, \boldsymbol{q}_t, t) + S(\boldsymbol{q}_t, \boldsymbol{q}_0, t), \tag{2.28}$$

by imposing the condition

$$\frac{\partial S(\boldsymbol{q}_{2t}, \boldsymbol{q}_0, 2t)}{\partial \boldsymbol{q}_t} = \frac{\partial S(\boldsymbol{q}_{2t}, \boldsymbol{q}_t, t)}{\partial \boldsymbol{q}_t} + \frac{\partial S(\boldsymbol{q}_t, \boldsymbol{q}_0, t)}{\partial \boldsymbol{q}_t} = \boldsymbol{p}_t(\boldsymbol{q}_{2t}, \boldsymbol{q}_t) - \boldsymbol{p}_t(\boldsymbol{q}_t, \boldsymbol{q}_0) = 0. \tag{2.29}$$

This requires the position $\boldsymbol{q}_t$ and momentum $\boldsymbol{p}_t$ to coincide at the "handoff" from one generator to another, taking place at time t. See section 2.4 for a more general discussion of composition of canonical transformations.

EXAMPLE: THE HARMONIC OSCILLATOR

We illustrate these points using the generator of harmonic oscillator motion, which for frequency ω, mass m, and time t is

$$S(q_t, q_0, t) = f_1(q_t, q_0, t) = \frac{m\omega}{2\sin\omega t}\left[(q_t^2 + q_0^2)\cos\omega t - 2\,q_0 q_t\right]. \tag{2.30}$$

This is a quadratic form in q_0 and q_t corresponding to a linear transformation, here a pure rigid clockwise rotation of the whole phase space about the origin, progressing steadily with angle ωt. The reader should verify that the usual motion $p_t = p_0 \cos\omega t - m\omega q_0 \sin\omega t$ and $q_t = q_0 \cos\omega t + p_0/m\omega \sin\omega t$ follows from this generator, and that $S(q_{2t}, q_0, 2t) = S(q_{2t}, q_t, t) + S(q_t, q_0, t)$ with $\partial S(q_{2t}, q_0, 2t)/\partial q_t = 0$ (this follows from the composition rule for two successive canonical transformations).

The harmonic oscillator action also contains the free particle, by taking the limit $\omega \to 0$.

2.3 Phase Space and Canonical Transformations

Phase space, the $2N$ dimensional collection of canonically conjugate coordinates and momenta $(q_1, q_2, \ldots, q_N, p_1, p_2, \ldots p_N)$ is an extremely important construct in both classical and quantum mechanics.

QUESTION OF DISTANCE IN PHASE SPACE

Position q and momentum p have different units, making the notion of distance in phase space ill defined. Areas in phase space have no problem: they have units of action—that is, momentum times position. Canonical transformations distort phase space in such a way as to preserve areas. Defining a generating function $f_2(q, P) = \gamma q P$, where γ has units of $\sqrt{\text{kg/s}}$, gives $P = \gamma^{-1} p$, $Q = \gamma q$, and gives both new variables dimensions of $\sqrt{\text{action}}$, so lines would have that dimension too. This does not resolve the problem, because the magnitude of γ is arbitrary, scaling Q while also scaling P by the inverse amount. Distances ds measured along a line, $ds = \sqrt{dQ^2 + dP^2}$ will differ according to the choice of γ. Once an (arbitrary) choice is made, however, distances can be measured with reference to that choice.

AREA PRESERVATION UNDER CANONICAL TRANSFORMATION

We first recall a few things about transformations, canonical or not. Under the coordinate change $(\boldsymbol{p}, \boldsymbol{q}) \to (\boldsymbol{P}, \boldsymbol{Q})$, integrals transform as

$$\alpha = \iint h(\boldsymbol{p}, \boldsymbol{q}) d\boldsymbol{p} d\boldsymbol{q} = \iint g(\boldsymbol{P}, \boldsymbol{Q})\, J(\boldsymbol{P}, \boldsymbol{Q})\, d\boldsymbol{P} d\boldsymbol{Q}, \tag{2.31}$$

where $g(\boldsymbol{P}, \boldsymbol{Q}) = h(\boldsymbol{p}(\boldsymbol{P}, \boldsymbol{Q}), \boldsymbol{q}(\boldsymbol{P}, \boldsymbol{Q}))$ and $J(\boldsymbol{P}, \boldsymbol{Q})$ is the Jacobian of the transformation, needed to give proper weighting of volume elements in the new coordinates

$$J(\boldsymbol{P}, \boldsymbol{Q}) = \left| \frac{\partial(\boldsymbol{p}, \boldsymbol{q})}{\partial(\boldsymbol{P}, \boldsymbol{Q})} \right| = \begin{vmatrix} \left(\dfrac{\partial \boldsymbol{p}}{\partial \boldsymbol{P}}\right)_{\boldsymbol{Q}} & \left(\dfrac{\partial \boldsymbol{p}}{\partial \boldsymbol{Q}}\right)_{\boldsymbol{P}} \\ \left(\dfrac{\partial \boldsymbol{q}}{\partial \boldsymbol{P}}\right)_{\boldsymbol{Q}} & \left(\dfrac{\partial \boldsymbol{q}}{\partial \boldsymbol{Q}}\right)_{\boldsymbol{P}} \end{vmatrix}, \tag{2.32}$$

or

$$J(\boldsymbol{P}, \boldsymbol{Q}) = \left| \left(\frac{\partial \boldsymbol{p}}{\partial \boldsymbol{P}}\right)_{\boldsymbol{Q}} \left(\frac{\partial \boldsymbol{q}}{\partial \boldsymbol{Q}}\right)_{\boldsymbol{P}} - \left(\frac{\partial \boldsymbol{q}}{\partial \boldsymbol{P}}\right)_{\boldsymbol{Q}} \left(\frac{\partial \boldsymbol{p}}{\partial \boldsymbol{Q}}\right)_{\boldsymbol{P}} \right|. \tag{2.33}$$

For canonical transformations, the Jacobian is always 1, as we now show. Using one spatial coordinate for simplicity, consider the following term in equation 2.33:

$$\left(\frac{\partial p}{\partial P}\right)_Q = \left(\frac{\partial P}{\partial p}\right)_Q^{-1} = \left(\frac{\partial q}{\partial Q}\right)_P^{-1}, \tag{2.34}$$

where the last equality follows using an $f_3(p, Q)$ generating function to write

$$\left(\frac{\partial P}{\partial p}\right)_Q = \frac{\partial f_3(p, Q)}{\partial p \partial Q} = \left(\frac{\partial q}{\partial Q}\right)_p. \tag{2.35}$$

In the same way, using an $f_4(p, P)$ generating function chosen to accomplish the same canonical transformation, it may be seen that

$$\left(\frac{\partial Q}{\partial p}\right)_P = \frac{\partial f_4(p, P)}{\partial p \partial P} = -\left(\frac{\partial q}{\partial P}\right)_p. \tag{2.36}$$

Thus, we have

$$J = \left(\frac{\partial q}{\partial Q}\right)_P \left(\frac{\partial Q}{\partial q}\right)_p + \left(\frac{\partial q}{\partial P}\right)_Q \left(\frac{\partial P}{\partial q}\right)_p. \tag{2.37}$$

This is easily seen to be 1, since it is the expression for $(\partial q/\partial q)_P$—that is,

$$dq = \left(\frac{\partial q}{\partial Q}\right)_P dQ + \left(\frac{\partial q}{\partial P}\right)_Q dP, \tag{2.38}$$

and when divided by dq with p held fixed, this becomes $(\partial q/\partial q)_p \equiv 1$.

Consider integration over some closed area in (p, q) coordinates. Since the Jacobian of a canonical transformation is 1, we have

$$\alpha = \iint_A dp dq = \iint_{A'} dP dQ, \tag{2.39}$$

where A' is the mapping of the region A into (P, Q). This area is preserved under the transformation.

Working backward, we can relate the area preservation directly to the existence of a total differential of a generating function, as follows. Using the Stokes theorem, we have

$$\iint_A dpdq - \iint_{A'} dPdQ = \oint_A pdq - \oint_{A'} PdQ = 0. \tag{2.40}$$

If we take q and Q as independent variables, the area in the (q, p) plane is specified by a closed curve in the (q, Q) plane—that is, $(q, p(q, Q))$, and likewise for $(Q, P(q, Q))$, allowing us to write the result as a line integral in the (q, Q) plane:

$$\oint_A pdq - \oint_{A'} PdQ = \oint_C [pdq - PdQ] = 0, \tag{2.41}$$

$$= \oint_C df_1(q, Q), \tag{2.42}$$

where the total differential ensures the null result of the integral, and

$$df_1(q, Q) = \frac{\partial f_1}{\partial q}dq + \frac{\partial f_1}{\partial Q}dQ = pdq - PdQ. \tag{2.43}$$

Earlier, area preservation was derived from the properties of canonical transformations and generating functions. Just now, *assuming* area preservation of a transformation, birth was given to a generating function $f_1(q, Q)$ and the derivatives that define p and Q, as before.

2.4 Transformation of Manifolds, Areas, and Densities

LINEAR TRANSFORMATIONS

A subset of canonical transformations are linear, which are important because they describe some key model systems (like a harmonic oscillator), and because they are extremely useful as local approximations to more general and ubiquitous globally nonlinear transformations. The generating function for a linear transformation is always a quadratic form in its variables. The most general quadratic form for an f_1 generator is

$$f_1(q, Q, t) = a\,Q^2 + b\,qQ + c\,q^2 + d\,Q + e\,q + g, \tag{2.44}$$

where the six parameters (a, b, c, d, e, g) are possibly functions of time. Linear relations result when first derivatives of $f_1(q, Q, t)$ are taken with respect to q and Q to obtain the relations between the old and new variables—that is,

$$P = \left(\frac{2a}{b}\right)p + \left(b - \frac{4ac}{b}\right)q + \left(d - \frac{2ae}{b}\right), \tag{2.45}$$

$$Q = \left(\frac{1}{b}\right)p - \left(\frac{2c}{b}\right)q - \frac{e}{b}. \tag{2.46}$$

A linear transformation takes linear manifolds into new linear manifolds, which may be stretched or compressed, displaced, and rotated relative to pretransformation.

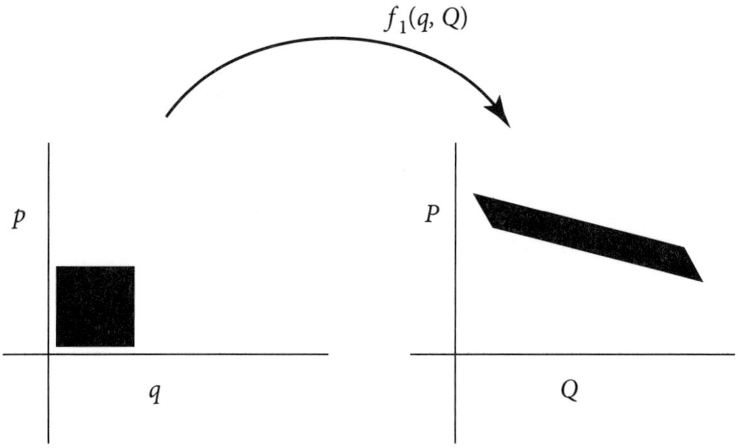

Figure 2.1. Phase space diagram of a linear canonical transformation: the square at the left is taken into the rotated and translated parallelogram of the same area on the right.

In this example of a two-dimensional phase space (generalization to many dimensions is straightforward), there are five parameters generating a quadric form, not counting a sixth parameter (g) not affecting the transformation. It takes two parameters to define the translation in phase space, one to define a rotation, and two to define a stretching along some direction. (The stretch factor is one parameter, its orientation in angle is another. Since the stretching must be area preserving, a stretch in some direction is compensated by a mandatory compression in a perpendicular direction, in order to keep area constant, so only two parameters are involved.) All five parameters are needed. Under a typical linear canonical transformation, the square area in figure 2.1, left, can take on the appearance seen at the right, with the same area.

MANIFOLD WEIGHTING AND NONLINEAR TRANSFORMATIONS

Under more general, nonlinear transformations, a manifold such as a straight line in (p, q) (figure 2.2, top left) space is transformed into a curve such as the one in figure 2.2, top right. But there is something misleading about this way of plotting the result. The manifold has zero width, so it seems natural that we draw a line as narrow as possible. After the transformation, it still has zero width perpendicular to its tangents, and we draw the same thickness line. This is deceiving, because unless the transformation is linear, it is certain to weight the manifold nonuniformly along its length. What this means is clear from figure 2.2, bottom left, where the initial manifold has been represented by a narrow rectangle of finite uniform width on the left. This region is propagated (not schematically, but quantitatively point by point in the whole rectangle) for a time. It is shown on the right and is clearly no longer uniform in width, having a zone near the "kink" that collects more of the phase space points comprising the initial data (the thin vertical rectangle) than anywhere else. The probability density along the line is related to the initial manifold and the stability matrix $\boldsymbol{M}$ (see sections 3.8, and equations 3.73 and 7.49, later). The logarithmic measure of the stretching and compression along the distorting manifold is called the *rarefaction exponent*

$$r(t) = \log[\boldsymbol{M}_t \cdot \hat{d}], \tag{2.47}$$

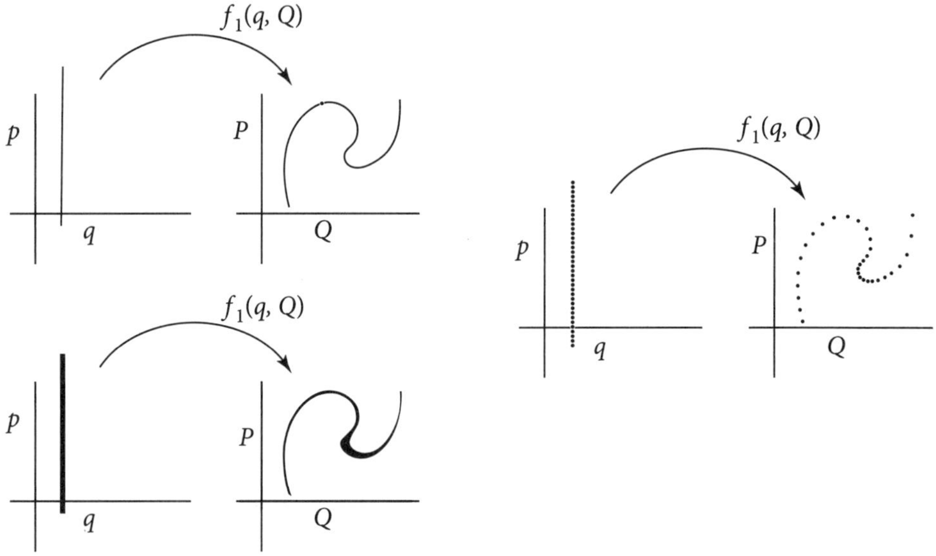

Figure 2.2. Phase space diagram of a nonlinear canonical transformation. (Top left) Under a nonlinear canonical transformation, the vertical (position-like) linear Lagrangian manifold on the left is distorted into the nonlinear Lagrangian manifold on the right. Both left and right lines are representing manifolds of zero thickness. (Bottom left) Under the same area preserving, canonical transformation, the thin rectangular region is mapped according to the canonical transformation, with the result shown to the right. The "line," now a properly transformed area, has acquired a variable thickness. The total area is the same as the starting rectangle. Another useful representation is to use initially equally spaced dots to comprise the manifold (right), taking each one to its new location under the transformation. The compression and expansion of dot distance reveals the local weighting of the transformed manifold.

where $\hat{d}$ is a unit vector along the initial manifold and $\boldsymbol{M}_t$ is the time-dependent stability matrix [4, 5].

A more extreme example hints at practical applications. In figure 2.3, we see a closed manifold and its transformation pictured in the same two ways seen in figure 2.2, except now the thickness variation has become rather extreme. The phase space points in the black island on the lower left have largely been "cooled" having found their way to the region of the origin, where energy is low. In spite of this, of course all area preservation rules have been followed, particularly the area of the interior black region and the area of the enclosed white region. It may seem that area preservation has been obeyed in the letter of the law but not the spirit! This example was inspired by a transformation introduced by Prof. Chris Jarzynski. So much of his work is indeed inspiring—for example, see reference [6].

COMPOSING TWO (OR MORE) CANONICAL TRANSFORMATIONS

Suppose we want to string together two canonical transformations to make a third (figure 2.4). This is a perfectly reasonable idea, since two successive area preserving transformations in phase space remain area preserving, and therefore canonical. This idea is important in several contexts, including dynamics in phase space (already

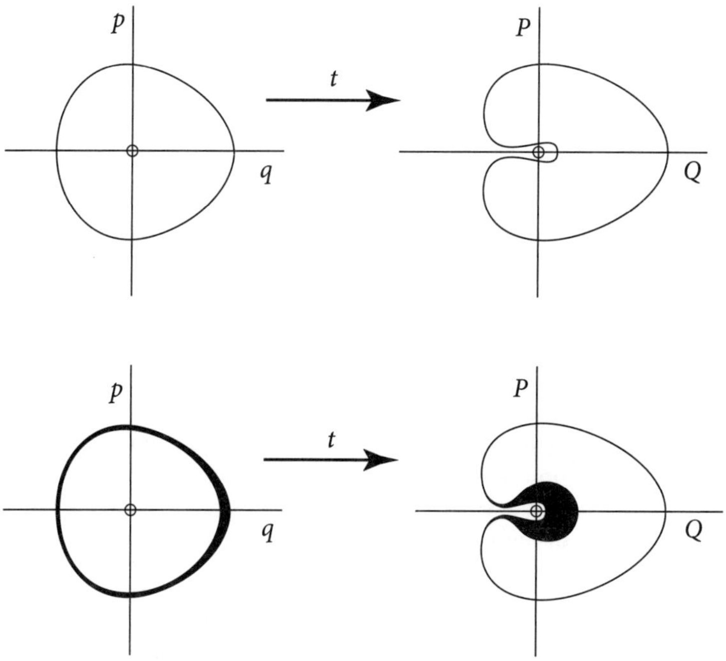

Figure 2.3. Phase space diagram of a nonlinear canonical transformation. (Top row) The oval manifold, left, typical of a constant energy manifold of an anharmonic oscillator, is distorted into the shape on the right under the transformation. Both left and right upper diagrams represent classical manifolds of zero thickness. (Bottom row) Under the same area-preserving canonical transformation, the black oval band of finite thickness, bottom left, is transformed into the result shown at the bottom right. The band is seen to have developed extreme thickness variations. The area of the original black band is the same as the black area in the transformed version, and the initial enclosed white area remains constant under the transformation (an extremely thin line on the lower right frame encloses the white interior area). The transformation has managed to "cool," or lower the energy, of the vast majority of the phase space points within the black oval, assuming the energy increases with distance from the origin at the intersection of the p and q axes. The thickening and its implication (implying lowering of the energy of the vast majority of the initial phase space points) is completely hidden in the figures at the top. Example inspired by Prof. Chris Jarzynski; see [6] and references therein.

considered; see equation 2.29), which is merely a succession of infinitesimal area preserving transformations. We take the case of a transformation $f_2(q, P')$ and $f_4(P', P)$: What the new transformation $f_2'(q, P)$ accomplishes in one step is the same result as successive application of $f_2(q, P')$ and $f_4(P', P)$. It ought to be

$$f_2'(q, P) = f_2(q, P') + f_4(P', P), \tag{2.48}$$

since then, there being no P' dependence in $f_2'(q, P)$,

$$\frac{\partial f_2'(q, P)}{\partial P'} = 0 = \left(\frac{\partial f_2(q, P')}{\partial P'} + \frac{\partial f_4(P', P)}{\partial P'}\right) = Q'(q, P') - Q'(P', P). \tag{2.49}$$

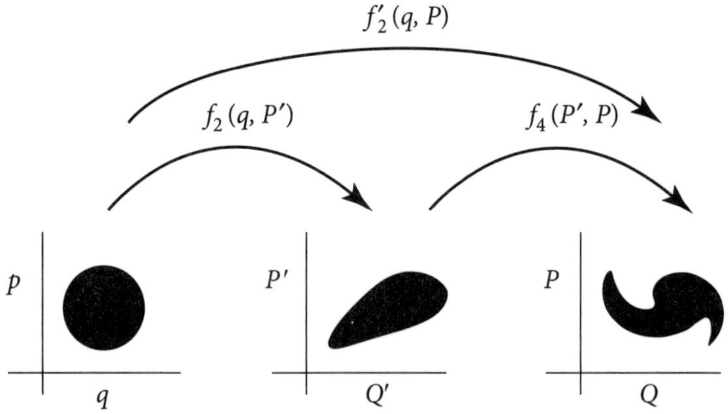

Figure 2.4. Phase space diagram of the composition of two canonical transformations. A succession of two canonical transformations, controlled by $f_2(q, P')$ and $f_4(P', P)$ generating functions, can be effected in one step using a generator $f_2'(q, P)$. It is constructed as the sum of the two generators, with a condition linking them. All three black patches have the same area.

The condition $0 = Q'(q, P') - Q'(P', P)$ is certainly reasonable; it says that at the "handoff" from the first transformation to the second we are using the same value of Q' (the same P' is guaranteed by construction). These conditions set the dependence of P' on (q, P):

$$f_2'(q, P) = f_2(q, P'(q, P)) + f_4(P'(q, P), P). \tag{2.50}$$

Since

$$\frac{\partial f_2'(q, P)}{\partial q} = \frac{\partial f_2(q, P')}{\partial q} + \left(\frac{\partial f_2(q, P')}{\partial P'} + \frac{\partial f_4(P', P)}{\partial P'}\right)\frac{\partial P'}{\partial q},$$

$$\frac{\partial f_2'(q, P)}{\partial P} = \frac{\partial f_4(P', P)}{\partial P} + \left(\frac{\partial f_2(q, P')}{\partial P'} + \frac{\partial f_4(P', P)}{\partial P'}\right)\frac{\partial P'}{\partial P},$$

with equation 2.49 we have

$$\begin{aligned}\frac{\partial f_2'(q, P)}{\partial q} &= p,\\ \frac{\partial f_2'(q, P)}{\partial P} &= Q.\end{aligned} \tag{2.51}$$

Chapter 3

Time Evolution in Phase Space

3.1 Time Evolution of Point Trajectories

The distribution $\delta_0(q, p; q_0, p_0) \equiv \delta(q - q_0)\delta(p - p_0)$ represents a point in phase space. It can be thought of as a probability distribution, or density in phase space, albeit an infinitely localized one. The total probability is unity, since

$$\iint \delta_0(q, p; q_0, p_0)\, dq\, dp = \iint \delta(q - q_0)\delta(p - p_0)\, dq\, dp = 1. \tag{3.1}$$

Under time evolution $\delta_0(q, p; q_0, p_0)$ becomes $\delta_t(q, p; q_0, p_0) \equiv \delta(q - q_t(q_0, p_0))\delta(p - p_t(q_0, p_0))$, where $q_t(q_0, p_0)$ and $p_t(q_0, p_0)$ are the classical trajectories that evolve from initial conditions q_0 and p_0. The *total* derivative $d\delta_t(q, p; q_0, p_0)/dt$ measures the local density *streaming with the flow*:

$$\begin{aligned}\frac{d}{dt}\delta_t(q, p; q_0, p_0) = {} & \frac{\partial}{\partial t}\delta_t(q, p; q_0, p_0) + \dot{q}\ \delta'(q - q_t(q_0, p_0))\delta(p - p_t(q_0, p_0)) \\ & + \dot{p}\ \delta(q - q_t(q_0, p_0))\delta'(p - p_t(q_0, p_0)),\end{aligned} \tag{3.2}$$

but since $\dot{q} \to \dot{q}_t$ and $\dot{p} \to \dot{p}_t$ (the delta functions impose this) and

$$\begin{aligned}\frac{\partial}{\partial t}\delta_t(q, p; q_0, p_0) = {} & -\dot{q}_t\ \delta'(q - q_t(q_0, p_0))\delta(p - p_t(q_0, p_0)) \\ & -\dot{p}_t\ \delta(q - q_t(q_0, p_0))\delta'(p - p_t(q_0, p_0)),\end{aligned} \tag{3.3}$$

we have

$$\frac{d}{dt}\delta_t(q, p; q_0, p_0) = 0. \tag{3.4}$$

Thus, the density of a point streaming with the flow is time independent. If we didn't know about Liouville's theorem before, this observation about the constant streaming density (and the one following about extended distributions) reveals it, since if area (or volume) of some patch of phase space were not preserved under the flow, while total probability stays the same, the density streaming with the flow would have to change.

TIME EVOLUTION OF MANIFOLDS

Under time evolution, the manifold $\mu_0(p,q;q_0) = \delta(q-q_0) = \int dp_0\ \delta(q-q_0)\,\delta(p-p_0)$ becomes

$$\mu_t(p,q;q_0) = \int_{-\infty}^{\infty} dp_0\ \delta(q - q_t(q_0,p_0))\delta(p - p_t(q_0,p_0)). \tag{3.5}$$

The more general parametric manifold γ, defined by $\int_0^1 du\ \delta(q-q_0(u))\delta(p-p_0(u))$, evolves according to

$$\mu_t(p,q;\gamma) = \int_0^1 du\ \delta(q - q_t(q_0(u),p_0(u)))\delta(p - p_t(q_0(u),p_0(u))). \tag{3.6}$$

We have already considered the fate of manifolds under general canonical transformations in connection with figure 2.2, earlier. Since dynamics also generate a canonical transformation, that figure still applies, including the evolution of bends and regions along the manifold that are *relatively* "thin" and "thick."

TIME EVOLUTION OF SMOOTH DISTRIBUTIONS: LIOUVILLE'S EQUATION

We can construct smooth phase space densities by "smearing" point trajectory solutions. Such smooth phase space distributions are useful in a number of contexts. One can think of smooth densities as made up of a continuous distribution of *weighted* δ point trajectories uniformly distributed in phase space, or a nonuniform distribution of equally weighted delta points. The following integral for the density $\rho_0(q,p)$ has either interpretation, with the weighting given by $\rho_0(q_0,p_0)$ over a uniform distribution suggested by the first form, or alternatively a nonuniform density of points suggested by the second:

$$\begin{aligned}\rho_0(q,p) &= \iint dq_0 dp_0\ \rho_0(q_0,p_0)\ \delta(q-q_0)\delta(p-p_0),\\ &= \iint \rho_0(q_0,p_0)\ dq_0 dp_0\ \delta(q-q_0)\delta(p-p_0).\end{aligned} \tag{3.7}$$

We normalize ρ_0—that is, $\iint \rho_0(q_0,p_0)\,dq_0\,dp_0 = 1$. The initial smooth density $\rho_0(q,p)$ becomes, under time evolution, $\rho_t(q,p)$:

$$\begin{aligned}\rho_t(q,p) &= \iint dq_0 dp_0\ \rho_0(q_0,p_0)\ \delta(q - q_t(q_0,p_0))\delta(p - p_t(q_0,p_0)),\\ &= \iint dq_0 dp_0\ \rho_0(q_0,p_0)\ \delta_t(q,p;q_0,p_0).\end{aligned} \tag{3.8}$$

The time evolution is entirely due to the motion of the delta trajectories with time. Now we can write, since $d\delta_t(q,p;q_0,p_0)/dt = 0$,

$$\frac{d}{dt}\rho_t(q,p) = \iint dq_0 dp_0\ \rho_0(q_0,p_0)\frac{d}{dt}\delta_t(q,p;q_0,p_0) = 0. \tag{3.9}$$

This declares the streaming local density to be time invariant—that is, another manifestation of Liouville's theorem, stressing *incompressibility of the flow in phase*

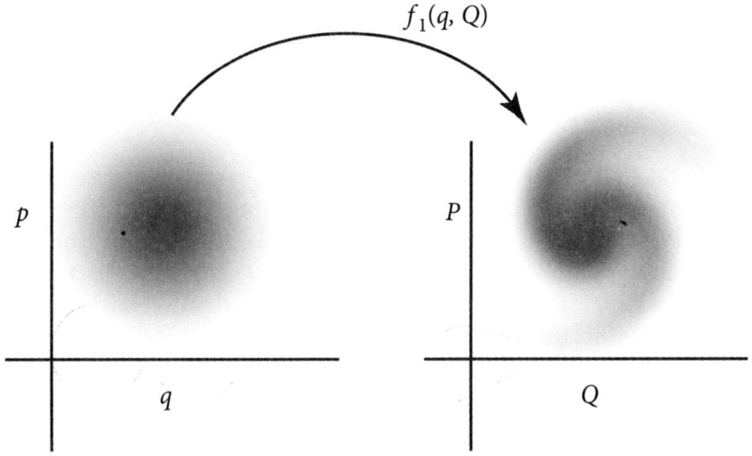

Figure 3.1. This figure shows the evolution of a smooth density in phase space (left) under a nonlinear canonical transformation, leading to the right-hand phase space density. A black dot helps track the environs of one region, seen to have maintained the same gray density after the transformation.

space. In other words, the density in the environs of any point trajectory moving with the flow does not change. Figure 3.1 depicts the evolution of a smooth density in phase space under a nonlinear canonical transformation. A black dot is also included; its environs are seen to have maintained the same gray density after the transformation.

For any $\rho_t(q, p)$ we have

$$
\begin{aligned}
\frac{d}{dt}\rho_t(q, p) = 0 &= \frac{\partial \rho_t(q, p)}{\partial q}\dot{q} + \frac{\partial \rho_t(q, p)}{\partial p}\dot{p} + \frac{\partial \rho_t(q, p)}{\partial t}, \\
&= \frac{\partial \rho_t(q, p)}{\partial t} - \left(\frac{\partial H(q, p)}{\partial q}\frac{\partial \rho_t(q, p)}{\partial p} - \frac{\partial H(q, p)}{\partial p}\frac{\partial \rho_t(q, p)}{\partial q}\right), \text{ or} \\
0 &\equiv \frac{\partial \rho_t(q, p)}{\partial t} - \{H, \rho_t\},
\end{aligned}
\tag{3.10}
$$

where we have defined the *Poisson bracket*

$$
\{A, B\} = \frac{\partial A}{\partial q}\frac{\partial B}{\partial p} - \frac{\partial A}{\partial p}\frac{\partial B}{\partial q}, \tag{3.11}
$$

where here $A(q, p) = H(q, p)$, and $B(q, p) = \rho_t(q, p)$. We have derived Liouville's equation for the explicit time evolution of a phase space density $\rho_t(q, p)$:

$$
\boxed{\frac{\partial \rho_t(q, p)}{\partial t} = \{H, \rho_t\}.} \tag{3.12}
$$

The unit Jacobian of a canonical transformation ensures preservation of the local density in phase space. We write

$$1 = \iint h(p, q)\, dp\, dq = \iint g(P, Q)\, dP dQ, \tag{3.13}$$

where we have taken the "gray" density $h(p, q)$ to be a normalized distribution having the interpretation of a classical probability distribution. The new density $g(P, Q) = h(p(P, Q), q(P, Q))$ local to the point (P, Q) remains unchanged from what it was at the corresponding point (p, q) that mapped to (P, Q), since the Jacobian of the transformation is 1. This is true of any canonical transformation, including dynamical evolution of a phase space density.

The overlap of two smooth densities under time propagation is mentioned now and again in chapter 21. Consider the overlap $\int d\Gamma \rho_1 \rho_2 \equiv \int d\boldsymbol{q} d\boldsymbol{p}\ \rho_1(\boldsymbol{q}, \boldsymbol{p})\rho_2\boldsymbol{q}, \boldsymbol{p})$. We can easily prove

$$\int d\Gamma \rho_1 \rho_2 = \int d\Gamma \rho_{1,t} \rho_{2,t}. \tag{3.14}$$

This is often a nonintuitive result, in the sense that ρ_1 can be centered on a different part of phase space than ρ_2 in a chaotic system where for individual trajectories such differences in initial condition diverge exponentially, yet the overlap of the two distributions is time independent.

3.2 Generating Functions and Joint Probabilities

Why resort to probabilities in classical mechanics? Why average over or remain ignorant of position, or momentum, (or some other classical coordinate) when, at least in principle, you can have precise, point-like trajectories at your service, corresponding to certainties, not probabilities?

It often happens that we choose to describe classical outcomes as probabilities, and just as often we are forced to do so by experimental reality. There is a lot of unnecessary data generated if every trajectory is followed and reported, when only statistical results are required. For example, classical collision problems are often formulated in terms of momentum transfer, averaging over *impact parameter* (distance of closest approach of projectile to target, if there were no interaction between them). This average corresponds to a Lagrangian manifold description of the initial data, in the sense of $\delta(p - p_0) = \int dq_0 \delta(q - q_0)\delta(p - p_0)$. $\delta(q - q_0)\delta(p - p_0)$ is a trajectory point, $\delta(p - p_0)$ is a manifold, and the integral is the average.

For practical reasons (for example, the target particles are too small and the aiming is random or imprecise), we often must forego trajectory descriptions, even though theoretical calculations can of course be done with trajectories, using the very same action functions (for example, dynamical canonical transformations) we will use shortly to get the probability distributions. Another example: consider an impact on a body rotating with some angular momentum. In practice we very often can't be sure of when the impact will occur with respect to orientation of the body. Predictions about the collision must then necessarily be given as probability densities.

There is another motivation for developing classical probability density descriptions: analogy and correspondence with quantum mechanics, where answers are always given as probabilities.

We now get down to brass tacks concerning classical joint probability distributions. Suppose we ask the question, what is the probability of finding a particle at position q_f, regardless of its momentum there, given that a distribution of particles began a time t earlier at position q_0 with all momenta equally probable? The initial manifold

$$\mu_0(q, p; q_0) = \delta(q - q_0) = \int_{-\infty}^{\infty} dp_0\, \delta(q - q_0)\delta(p - p_0) \tag{3.15}$$

is constructed as a sum of points in phase space with fixed q_0 and all possible momenta uniformly. The manifold evolves as

$$\mu_t(q, p; q_0) = \int_{-\infty}^{\infty} dp_0\, \delta(q - q_t(q_0, p_0))\delta(p - p_t(q_0, p_0)). \tag{3.16}$$

The joint probability density we seek, $P(q_f, q_0, t)$, is simply given by the phase space overlap of the propagated manifold $\mu_t(q, p; q_0)$ with the final "probe" manifold $\delta(q - q_f)$:

$$\begin{aligned} P(q_f, q_0, t) &= \int dq\, dp \int_{-\infty}^{\infty} dp_0\, \delta(q - q_f)\, \delta(q - q_t(q_0, p_0))\delta(p - p_t(q_0, p_0)), \\ &= \int dp_0\, \delta(q_f - q_t(q_0, p_0)), \\ &= \sum_k \frac{1}{\left|\frac{\partial q_t}{\partial p_0}\right|_{q_0}} = \sum_k \left|\frac{\partial p_0}{\partial q_t}\right|_{q_0}. \end{aligned}$$

This can be written (note: $q_t = q_f$ now) as

$$\boxed{P(q_f, q_0, t) = \sum_k \left|\frac{\partial^2 S_k(q_f, q_0, t)}{\partial q_f \partial q_0}\right|.} \tag{3.17}$$

We see that the mixed second derivative of the generating function $S(q_f, q_0, t)$ determines the joint probability density for starting at q_0 and being detected later at q_f. The sum is over all initial q_0 that lead to arrival at q_f at time t.

We emphasize that the same plan applies to generating functions of all sorts. For example, the mixed second derivative of the generating function $S(q_f, p_0, t)$ determines the joint probability density for starting with momentum p_0 and being detected later at q_f. Or, the mixed second derivative of the generating function $f_2(q, P)$ determines the joint probability density for having momentum P initially, with Q uniformly distributed, and being detected at time t with position q and any momentum p.

The classical probability density as just introduced is closely related to quantum mechanical probability densities. Specifying an initial state such as $|q_0\rangle$ with definite position renders the conjugate variable momentum p_0 completely uncertain. Likewise, specifying final position as in $|q_f\rangle$ makes final momentum p_f uniformly uncertain. The question of getting from q_0 to q_f under the influence of a Hamiltonian H in time

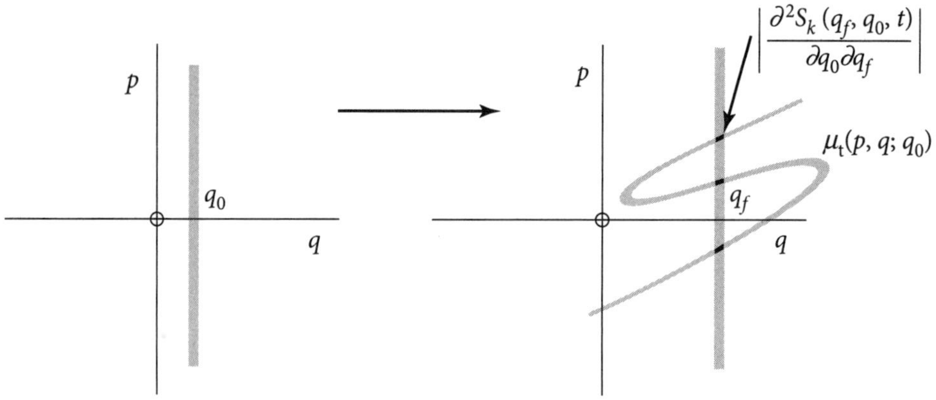

Figure 3.2. On the left is a representation of a classical manifold $\delta(q - q_0)$; on the right, the manifold $\mu_t(q, p; q_0)$ that has evolved from the initial manifold overlapping with a "probe" manifold $\delta(q - q_f)$. The size and number of overlaps (black, three in this zone) of the evolved and probe manifolds determine the probability density for starting at position q_0 and ending at position q_f after time t.

t is formulated quantum mechanically as

$$P_{quantum}(q_f, q_0, t) = |\langle q_f | q_0(t)\rangle|^2 = |\langle q_f | e^{-iHt/\hbar} | q_0\rangle|^2, \tag{3.18}$$

where $e^{-iHt/\hbar}$ is the quantum propagator. The quantum amplitude for propagation from $|q_0\rangle$ to $|q_f\rangle$ a time t later is called the *Green function*, $G(q_f, q_0, t)$:

$$G(q_f, q_0, t) = \langle q_f | e^{-iHt/\hbar} | q_0\rangle. \tag{3.19}$$

The Green function is intimately related to the classical probability $P(q_f, q_0, t)$, as we shall see in subsequent chapters. For example, it contains the factors

$$G(q_f, q_0, t) \sim \left|\frac{\partial^2 S_k(q_f, q_0, t)}{\partial q_f \partial q_0}\right|^{1/2}.$$

The classical action $S(q_f, q_0, t)$, or better its mixed second derivatives, is playing a role analogous to a quantum Green function. See also chapter 7.

DIAGRAMS FOR JOINT PROBABILITIES

We have seen in section 2.4 and following that depicting classical manifolds using finite width reveals how the classical density evolves on the manifold. If two finite-width manifolds are depicted in one phase space plot, they overlap in patches that represent the joint probability of belonging to both manifolds, as seems intuitively obvious. For example, on the left in figure 3.2 is a representation of a classical manifold $\mu_0(q, p; q_0) = \delta(q - q_0)$ depicting a condition of definite initial position q_0; on the right another manifold $\mu_t(q, p; q_0)$ is taken to be the time evolution of the manifold on the left. The overlap with a "probe" manifold corresponding to being at position q_f (with any momentum), $\delta(q - q_f)$, is

$$P(q_f, q_0, t) = \iint dq\, dp\, \mu_0(q, p; q_f)\, \mu_t(q, p; q_0). \tag{3.20}$$

In figure 3.2, the manifold $\mu_t(q, p; q_0)$ that evolved from the initial manifold $\mu_0(q, p; q_0) = \delta(q - q_0)$ and the final "probe" manifold $\delta(q - q_t)$ are shown. The total area of overlapping regions of the two manifolds (shown as three black zones in this example) are proportional to the probability density for starting at position q_0 and ending at position q_t after time t.

As another example we consider a situation motivated by the classical analog of a quantum eigenstate of energy E'. The appropriate classical manifold is $\delta(H(q, p) - E')$, where $H(q, p)$ is the classical Hamiltonian for one degree of freedom. The (unnormalized) joint probability density for possessing energy E' and also being located at position q' is the phase space overlap

$$P(q, E) = \iint dpdq\ \delta(q - q')\delta(E - H(p, q)), \tag{3.21}$$

$$= \int dp\ \delta(E - H(p, q')) = \sum_k \left|\frac{\partial H}{\partial p}\right|_{q'}^{-1}, \tag{3.22}$$

$$= \sum_k \frac{m}{|p_k|}.$$

The sum is over momenta $p = p_k$ that satisfy $H(q', p_k) = E'$. If $H(q, p) = p^2/2m + V(q)$, there are two such momenta, satisfying $p_\pm = \pm\sqrt{2m(E' - V(q'))}$ for q' in the realm visited by the trajectory at energy E' (the "classically allowed" region). The result that the probability of finding the particle is inversely proportional to its speed (and momentum) is well known of course (see figure 3.3).

Caveat: When the overlap of two manifolds represented approximately by finite thickness zones get sloppy, by which we mean the overlap region shows some curvature of one or both of the participating densities, it is necessary to reduce the thickness to render the overlaps locally linear, yielding a good correspondence between the overlap region and the integral of the δ manifolds over phase space.

The joint probability density becomes singular when two parallel manifolds touch. For example, equation 3.21 blows up at the "classical turning points" where $p_E(q_f) = 0$. The classical manifold $\delta(E - H(q_f, p))$ runs vertically at such places, parallel to $\delta(q - q_f)$.

We can put a useful spin on this, connecting the classical probabilities with canonical transformations, and further solidifying the semiclassical analogies mentioned earlier: Suppose we have a "new" momentum variable P, and an "old" position variable q. Then the conditional probability that we are at q given that P is definite (or vice versa) is

$$P_P(q) \equiv P_q(P) = \int dp'dq'\ \delta(q - q')\delta(P - P') = \sum |\partial P(p', q)/\partial p'|_{P=P'}^{-1},$$

$$= \sum \left|\frac{\partial p'}{\partial P}\right|,$$

$$= \sum \left|\frac{\partial^2 F_2(q, P)}{\partial P \partial q}\right|, \tag{3.23}$$

where $P_Q(q)$ is the classical probability density for finding the value q given that Q is definite, and so on. Thus, the mixed second derivative of the F_1 generating function is a classical probability as described earlier.

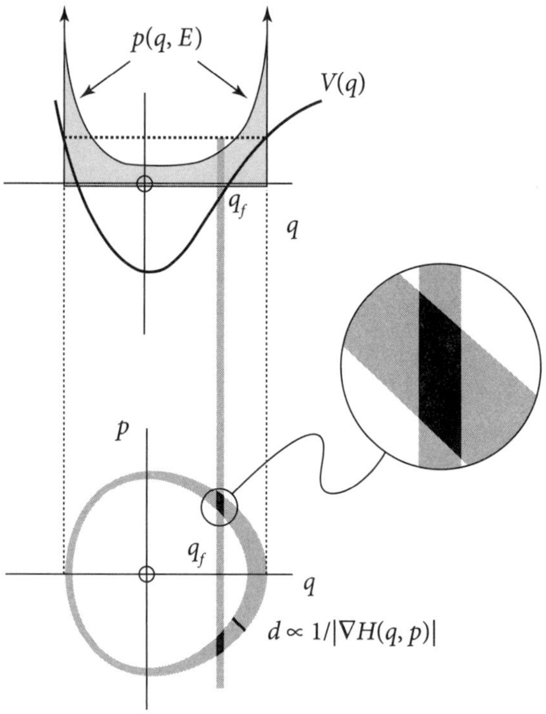

Figure 3.3. The potential energy $V(q)$ of an anharmonic oscillator is shown at the top. The coordinate range of a given trajectory of energy E' is depicted as a dotted line. A thin line shows the classical momentum as a function of position. At the bottom, the thickness (being mindful not to make absolute statements about distance in phase space) of the band corresponding to energy E' varies according to the gradient of the Hamiltonian—that is, $d \propto 1/|\nabla H(q, p)|$. The area of the black parallelograms (one magnified in the inset) highlighting the intersection of the position manifold (thin vertical band) and the energy manifold (oval) is proportional to the joint classical probability density of both possessing energy E' and being at position q.

Now we derive the meaning of the intersection area directly by constructing two finite thickness zones using pairs of manifolds differing by a small value of the quantity defining the manifold. For example, the manifold $\delta(q' - q)$ is shown in figure 3.4 as a vertical line at q and $q + dq$. The P manifold, $p(q) = \partial F_2(q, P)/\partial q$ is shown at P and $P + dP$. The parallelogram of intersection (figure 3.4) has an area da given by the magnitude of the cross product of $dp_1\, \hat{p}$ and $d\vec{v} = \hat{p}\ dp_2 + \hat{q}\ dq$—that is, $dp_1\, \hat{p} \times (dp_2\, \hat{p} + dq\, \hat{q}) \rightarrow da = \partial^2 F_2(q, P)/\partial q \partial P\ dq\, dP$. This checks with figure 3.4 and the cross product giving the area of the parallelogram,

$$\begin{vmatrix} \hat{z} & \hat{p} & \hat{q} \\ 0 & \dfrac{\partial^2 F_2(q, P)}{\partial q \partial P} dP & 0 \\ 0 & \dfrac{\partial^2 F_2(q, P)}{\partial q^2} dq & dq \end{vmatrix} \rightarrow \frac{\partial^2 F_2(q, P)}{\partial q \partial P}\, dq\, dP = \left(\frac{\partial p}{\partial P}\right)_q dP dq.$$

Equation 3.23, derived with delta function methods, is confirmed.

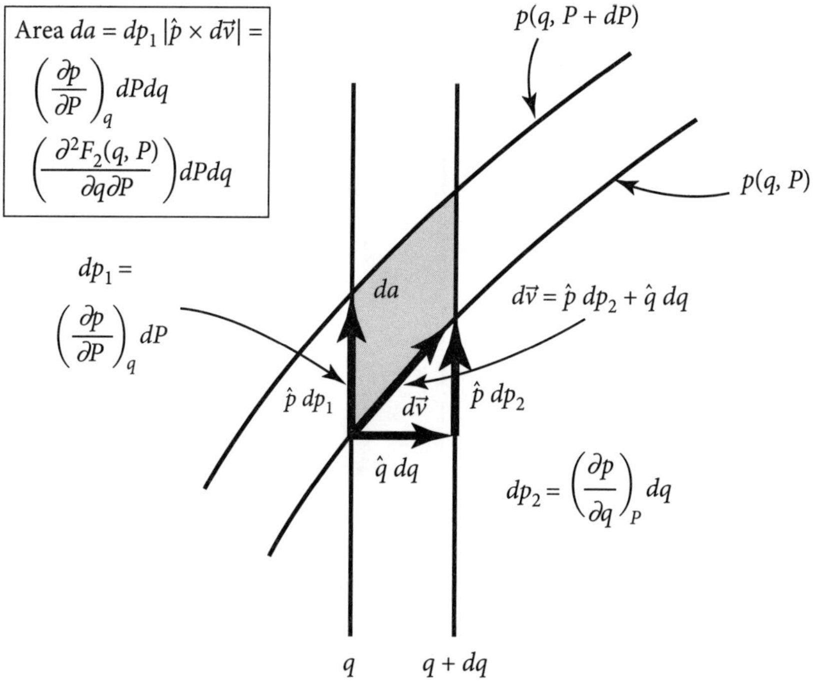

Figure 3.4. Phase space picture for overlap of two areas bounded by the manifolds $\delta(q' - q)$, $\delta(q' - q - dq)$, $\delta(P' - P)$, and $\delta(P' - P - dP)$.

ENCLOSED AREAS

Suppose we have a classical manifold defined by an F_2 generating function with a fixed "momentum" P—that is,

$$p(q) = \frac{\partial F_2(q, P)}{\partial q}. \tag{3.24}$$

It is then clear that the area under the curve shown in figure 3.5 is

$$\text{Area } \boldsymbol{A} = F_2(q_2, P) - F_2(q_1, P). \tag{3.25}$$

LOOKING AHEAD TO SEMICLASSICS

We shall see (in chapter 7) that the "Golden Rule" of semiclassical approximations is "Quantum amplitudes are to be approximated as the sum of square roots of classical probabilities or probability densities, with phases given by classical actions." We have been dealing with several examples of classical probability densities, including ones corresponding to dynamics in figure 3.2. Recall that the action S is an F_1 generator of canonical transformation corresponding to dynamics:

$$F_1(q, q', t) = S(q, q', t). \tag{3.26}$$

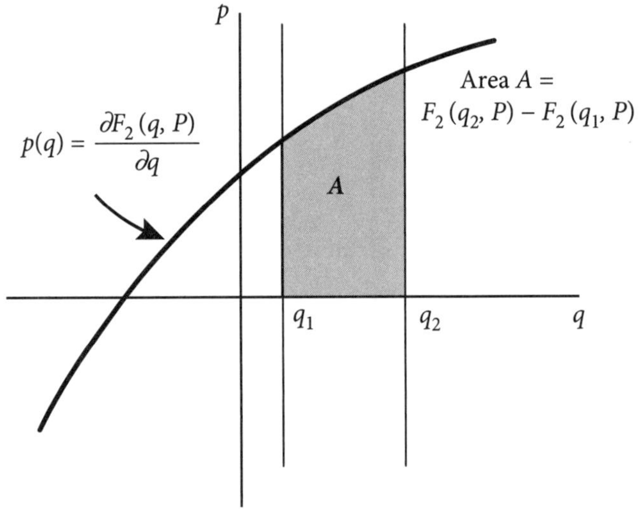

Figure 3.5. The area of the shaded region is given in terms of the generating function that defines the curve $p(q) \equiv p_P(q) = (\partial F_2(q, P)/\partial q)$.

The semiclassical propagator (to be derived by stationary phase on the Feynman path integral in chapter 7) is

$$G(q, q', t) = \frac{1}{\sqrt{2\pi i \hbar}} \sum_k \left| \frac{\partial^2 S_k(q, q', t)}{\partial q \partial q'} \right|^{\frac{1}{2}} e^{\frac{i}{\hbar}(S_k(q,q',t) + i\nu_k)}. \tag{3.27}$$

It is seen that amplitude prefactors in this expression, the *Van Vleck determinants* as they are called, are indeed square roots of dynamical classical probability densities. The phase space pictures appropriate to this example are shown in figure 3.2.

3.3 Poincaré-Cartan Invariants

The Poincaré-Cartan theorem involves *an extended phase space* formed by adding time to the N momenta and N positions, making the space $2N + 1$ dimensional. Forming a one-dimensional closed loop γ_0 embedded in the extended phase space, each point on the loop generally lies on a different trajectory, with time, position, and momentum specified. The trajectories on the loop may also have different energies. As we move all the points further along in time, the collection of all points on the loop traces out a tube. We consider closed line integrals of the form $\oint_\gamma \boldsymbol{p} \cdot d\boldsymbol{q} - H dt$ lying on the surface of this tube. The Poincaré-Cartan theorem states that such integrals that can be deformed into one another while remaining on the surface are invariant to their path (see figure 3.6):

$$\oint_{\gamma_1} \boldsymbol{p} \cdot d\boldsymbol{q} - H dt = \oint_{\gamma_2} \boldsymbol{p} \cdot d\boldsymbol{q} - H dt. \tag{3.28}$$

Note that in general the path γ spans a range of times, and since the energy may vary from point to point on the surface of the tube (although not along the streamlines—that is, the trajectories) the line integral $\oint_{\gamma_1} H dt$ is different for different paths γ_i.

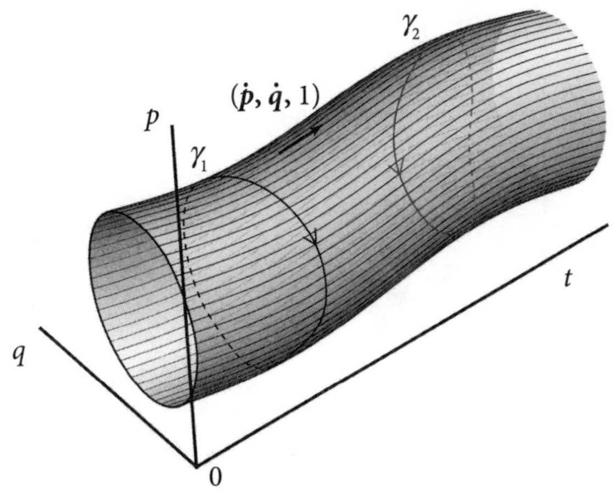

Figure 3.6. Illustration of Poincaré-Cartan integral invariants. A tube is formed by time evolution of an initial closed one-dimensional path (dark circle) constructed at time 0. Two additional arbitrary closed paths, labeled γ_1 and γ_2, are shown. The vectors defining the streamlines (trajectories) are $(\dot{\boldsymbol{p}}, \dot{\boldsymbol{q}}, 1)$.

The proof of the invariance of the line integral hinges on Stokes' theorem; see Arnold [7]. We prove it now for a time-dependent Hamiltonian of one degree of freedom, having the advantage of faithful depiction of the idea and pointing the way to the general N dimensional derivation.

Consider evolution under $H(p, q, t)$ in the extended phase space (p, q, t), shown in figure 3.7. We construct a vector $2\vec{V} = (-q, p, -H)$, then

$$\vec{\nabla} \times \vec{V} = \left(\hat{p}\frac{\partial}{\partial p}, \hat{q}\frac{\partial}{\partial q}, \hat{t}\frac{\partial}{\partial t}\right) \times \vec{V} = \left(-\frac{\partial H}{\partial q}, \frac{\partial H}{\partial p}, 1\right) = (\dot{p}, \dot{q}, \dot{t}). \tag{3.29}$$

This vector evidently lies tangent to the streamlines of the trajectories in the extended phase space. The vector $d\vec{S} = \hat{n} dS$, where $\hat{n}$ is a unit vector normal to the surface of the tube and dS is an elemental area on the surface, is perpendicular to $\vec{\nabla} \times \vec{V}$.

The surface integral of $\vec{\nabla} \times \vec{V} \cdot d\vec{S}$ vanishes and by Stokes' theorem,

$$\begin{aligned}\iint \vec{\nabla} \times \vec{V} \cdot d\vec{S} &= 0,\\ &= \oint_{\gamma_1} \vec{V} \cdot d\boldsymbol{r} + \oint_{\gamma_2} \vec{V} \cdot d\boldsymbol{r},\end{aligned} \tag{3.30}$$

where $d\boldsymbol{r}$ lies along the surface, and thus $\vec{V} \cdot d\boldsymbol{r} = p dq - H dt$. Equation 3.30 reveals the Poincaré invariant integral

$$\oint_{\gamma_1} p dq - H dt = -\oint_{\gamma_3} p dq - H dt, \tag{3.31}$$

where γ_3 is merely γ_2 taken in the opposite sense (see figure 3.7).

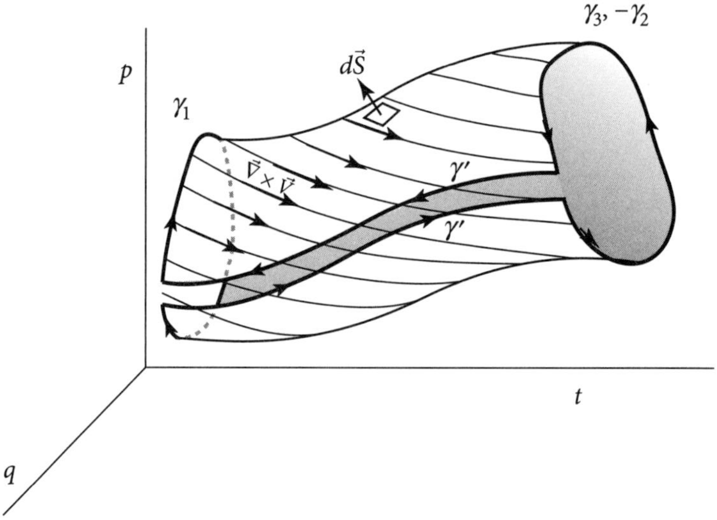

Figure 3.7. $\vec{\nabla} \times \vec{V}$ lies on this surface, and $d\vec{S}$ is perpendicular to it. Therefore, $\vec{\nabla} \times \vec{V} \cdot d\vec{S} = 0 = \iint \vec{\nabla} \times \vec{V} \cdot d\vec{S}$.

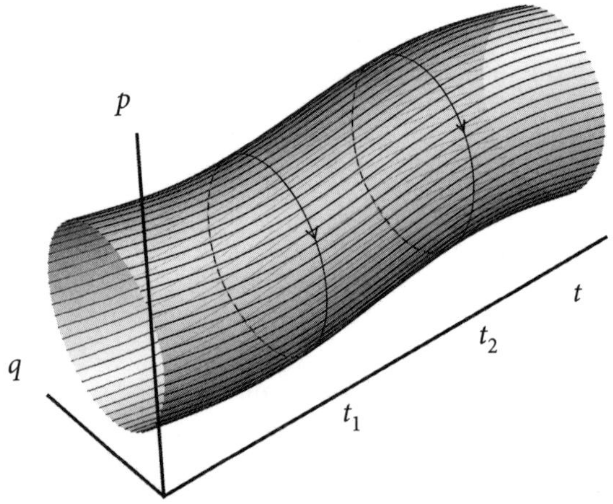

Figure 3.8. Illustration of the Poincaré integral invariant, showing the tube formed by time evolution of an initial closed path constructed at fixed time. Two closed paths on the surface of the tube at subsequent fixed times t_1 and t_2 are shown.

A special case of the Poincaré-Cartan integral invariants is obtained by taking paths γ_1 and γ_2 at fixed times t_1 and t_2 (figure 3.8). Since $dt = 0$ along such paths, we have

$$\oint_{\gamma_1} \boldsymbol{p} \cdot d\boldsymbol{q} = \oint_{\gamma_2} \boldsymbol{p} \cdot d\boldsymbol{q}.$$

Evidently the sum of the projected areas of the loop on N phase planes p_k, q_k:

$$\oint_{\gamma_1} \boldsymbol{p} \cdot d\boldsymbol{q} = \sum_k \oint_{\gamma_{1k}} p_k dq_k.$$

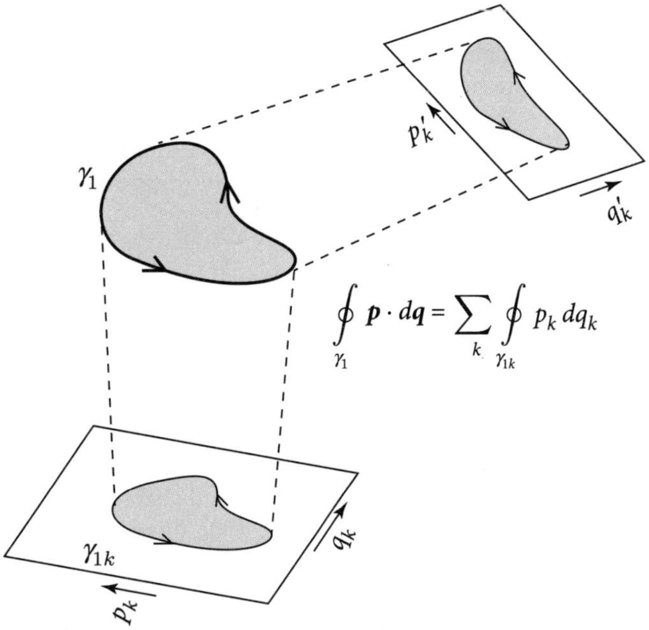

Figure 3.9. Construction of the Poincaré integral invariant. A closed loop in the full phase space is shown schematically, together with the decomposition of its line integral into the sum of the projections of the loop onto the various orthogonal phase planes.

Figure 3.9 depicts a closed loop in the full phase space and the decomposition of its line integral into the sum of the projections of the loop onto the various orthogonal phase planes.

The constancy of $A = \sum_k \oint_{\gamma_{1k}} p_k dq_k$ in time is also revealed by computing its time derivative,

$$\begin{aligned}\frac{dA}{dt} &= \sum_k \oint_{\gamma_1} (\dot{p}_k dq_k + p_k d\dot{q}_k), \\ &\overset{parts}{=} \sum_k \oint_{\gamma_1} (\dot{p}_k dq_k - \dot{q}_k dp_k) + p_k \dot{q}_k - p_k \dot{q}_k, \end{aligned} \tag{3.32}$$

where the boundary terms cancel because the path γ_1 begins and ends at the same place. Then we write, using Hamilton's equations of motion,

$$\begin{aligned}\frac{dA}{dt} &= \sum_k \oint_{\gamma_1} \left(-\frac{\partial H}{\partial q_k} dq_k - \frac{\partial H}{\partial p_k} dp_k \right) \\ &= -\oint_{\gamma_1} dH = 0 \end{aligned} \tag{3.33}$$

The Poincaré-Cartan theorem applies to dynamical structures called invariant tori. The tori are traced out by single trajectories over time in integrable systems (see section 3.5, and see figure 3.15, later), possessing N constant actions for N degrees of freedom. The surface segment just discussed was generated more generally by an arbitrary closed curve together with its subsequent time evolution, and applies even to chaotic systems. But the invariant tori if they exist qualify as Poincaré-Cartan surfaces because the initial closed curve could be chosen to lie entirely on any torus; the Poincaré-Cartan theorem then guarantees the conservation of the action integrals against deformations of the integration path on the tori.

3.4 Poincaré Surface of Section

Visualizing the motion of trajectories or manifolds in phase space is simple for one degree of freedom—that is, a two-dimensional phase space. Things become more interesting if the Hamiltonian is time dependent—for example, the "kicked" rotor, equation 3.64 (section 3.6), since resonance islands and chaos can ensue.

For two degrees of freedom, phase space is four dimensional, and resistant to depiction, but plotting slices of phase space is especially informative. Suppose we have a time-independent Hamiltonian $H(p_x, p_y, x, y)$. Energy $H(p_x, p_y, x, y) = E$ is conserved, restricting the motion of a single trajectory to at most a three-dimensional surface embedded in the four-dimensional space. Suppose we select a value of y often visited by the trajectory, say $y = 0$, and construct a *Poincaré surface of section* there, as follows. There are now two constraints, $H(p_x, p_y, x, y) = E$ and $y = 0$, leaving a region in the two-dimensional (x, p_x) plane consistent with fixed y and E. Every allowed point x, p_x in this plane corresponds to a unique point in the four-dimensional phase space, apart from a sign: x, p_x had been picked, $y = 0$, and p_y is specified through $p_y = \pm\sqrt{2E - p_x^2 - 2V(x, 0)}$. Choosing one of the signs, that point lies on a unique trajectory, leading to a unique next intersection point on the surface of section, and so on. In this way the dynamics provides the surface of section with a point-to-point mapping of the surface onto itself. We now show that this mapping is area preserving, using the Poincaré-Cartan theorem.

In figure 3.10, points within the dark area enclosed by the path γ_0 on the surface of section (s.o.s.) map into the area enclosed by the path γ_1 on the surface, obtained by following every enclosed point in γ_0 and noting its next intersection with the s.o.s.. The points in the γ_1 region have not all returned to the surface of section at the same time. We show the time evolution of γ_0 schematically as the shape $\gamma_t \equiv \gamma_0(t)$, a "snapshot" taken at time t. Some trajectories have already passed through the s.o.s., a few (0%) are on it, and some have yet to pass through. Thus, the area inside γ_1 is not an exact time evolution of γ_0 but rather a record of where γ_t penetrated the surface of section.

Every point in the perimeter γ_1 has a unique value of all four phase space coordinates. Each of these points penetrated the s.o.s. at a specific time. Suppose we take the line integral

$$\oint_{\gamma_0} p_x dx - H dt = A_0 \tag{3.34}$$

at time $t = 0$. Then $dt = 0$ and the integral corresponds to the area A_0 of the initial dark region. Equation 3.34 is a Poincaré-Cartan line integral, which we take around

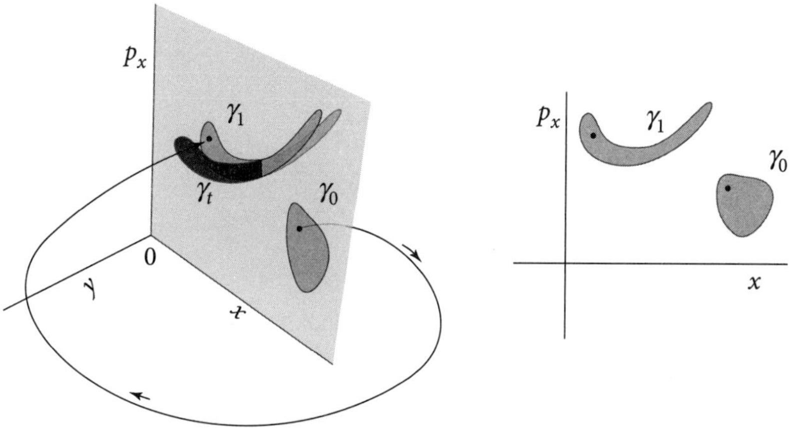

Figure 3.10. Construction of a Poincaré surface of section, showing the mapping of an area γ_0 on the (x, p_x) surface to a region γ_1 on the surface. The time evolution of the region γ_0 becomes γ_t (shown darker before it has intersected the surface of section), penetrating the (x, p_x) surface over a range of times. The region γ_1 lies on the (x, p_x) surface and is the locus of all intersections of γ_t with that surface as t progresses through this pass. Every point (x', p_x') on the surface of section corresponds to a unique point in the full phase space, $(x', y = 0, p_x', p_y = \sqrt{2E - p_x^2 - 2V(x, y)})$.

the perimeter of the area γ_1—that is,

$$\oint_{\gamma_1} p_x dx - H dt = A_1, \tag{3.35}$$

where now time varies along the perimeter, and $dt \neq 0$. The paths around γ_0 and γ_1 are simply two different paths on a trajectory tube in the Poincaré-Cartan sense, so

$$\oint_{\gamma_1} p_x dx - H dt = \oint_{\gamma_0} p_x dx - H dt, \tag{3.36}$$

but the time integrals vanish because $H = \text{const.}$ and $\oint_{\gamma_1} dt = \oint_{\gamma_0} dt = 0$; thus, $A_0 = A_1$ and the map defined by the dynamics passing through the s.o.s. is area preserving.

Hard-walled billiards are used often to study properties of nontrivial classical systems. The dynamics can be chaotic, yet at least things are simple away from any walls, and the collisions with the walls are also simple (that is, specular reflection, as in a beam of light on a mirror). Making things even more comfortable are the Birkhoff coordinates—that is, the distance along the perimeter of the billiard—and the momentum component along the perimeter. Clearly, these coordinates define a point-to-point mapping, since specifying where a trajectory intersects the wall and with what angle also specifies the next point and angle. We have just shown earlier that this point-to-point mapping is area preserving, so it is a surface of section (figure 3.11).

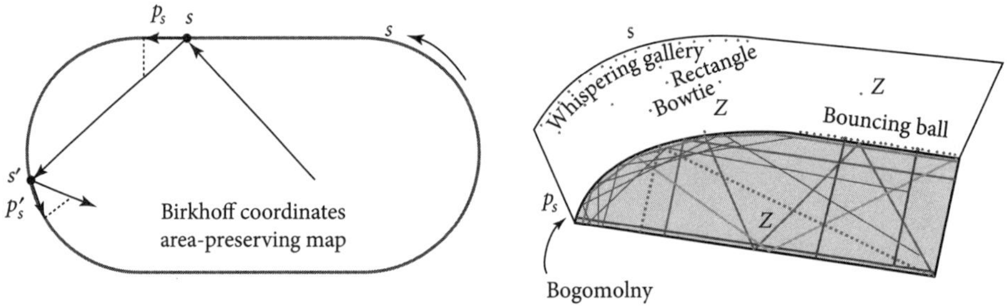

Figure 3.11. Definition (left) and use (right) of Birkhoff coordinates and some periodic orbits and where they lie on the surface of section. The whispering gallery mode bounces parallel to the wall with high p_s (as unstable periodic orbits); there are rectangular and bowtie-shaped unstable periodic orbits that can be seen among the ones drawn in space (the bowtie in this quarter stadium portion of it is the dotted line), and the bouncing ball orbits hit normal to the straight section of the wall, showing up as $p_s = 0$. The Bogomolny periodic orbit hits normal to the center of the end caps, also at $p_s = 0$. Another unstable periodic orbit, called "z" for its shape, is shown. All types have their names written approximately where they appear on the surface of section.

3.5 Action-Angle Coordinates

The appropriate choice of independent variables depends on physical boundary conditions: what is the initial state of the system, and what property will be measured? For example, suppose we know the momentum p_0 of the system and either (1) we know nothing of the initial position or (2) we want to average over it. If we measure final position q_t, then $S(q_t, p_0, t)$ is the appropriate action generating function. We can get it from $S(q_t, q_0, t)$, by Legendre transform (equation 1.72), making initial momentum an independent variable. If instead we specify the final momentum, p_t, then $S(p_t, p_0, t)$ is the appropriate action function, and so on.

Consider figure 3.12. Typical manifolds (top) will move and distort under time evolution, but not the one at the bottom, based on a conserved quantity (the energy, bottom).

INTRODUCING ACTION-ANGLE VARIABLES—ONE DIMENSION

We have been exploring integrals of the form $\oint p \cdot dq$ on surfaces in phase space.[1] This suggests using something like

$$\boxed{J = \frac{1}{2\pi} \oint p \cdot dq} \tag{3.37}$$

as a new canonical variable; we call this the action. The momentum $p = \sqrt{2m(E - V(q))}$; thus, the action at energy E is $1/2\pi$ times the area of the inner region in figure 3.13. The contour lines in figure 3.13 are of constant energy—that is, each contour is the path a trajectory follows in phase space. By increasing the energy by δE, the action changes by δJ. Since all points on the curves have the same energy, $H = H(J)$ does not depend on the variable conjugate to J, the angle we call θ.

[1] This section follows Percival and Richards [8].

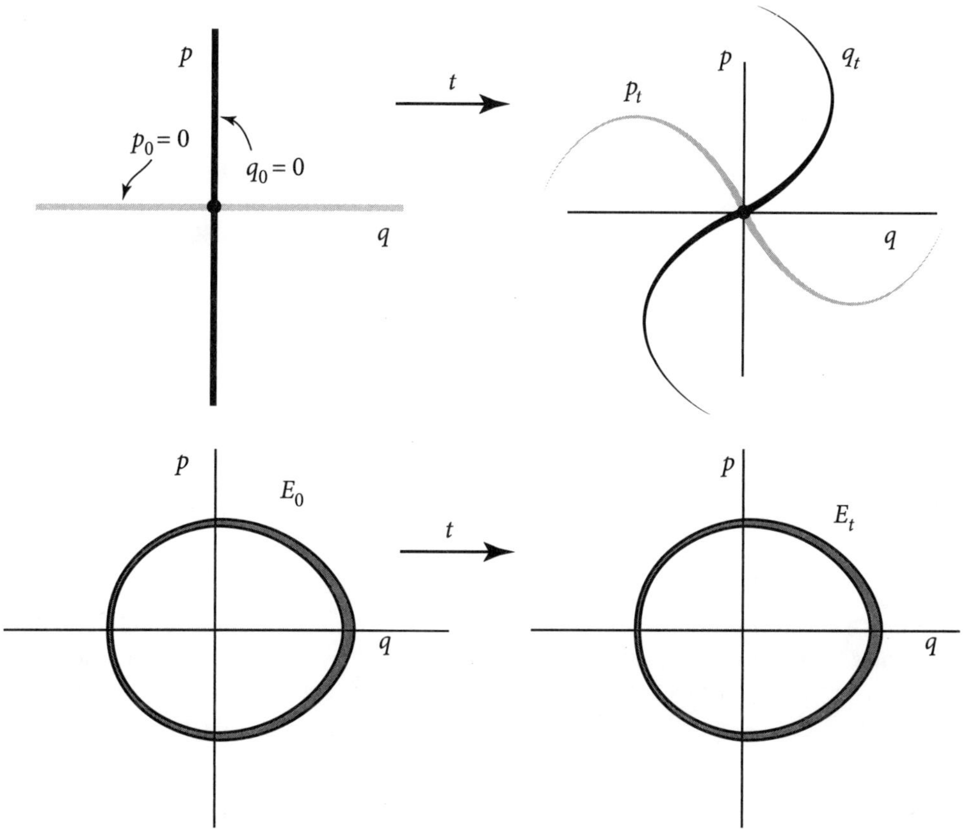

Figure 3.12. (Top) A position (black) and momentum (gray) manifold, centered here at $q_0 = 0$ and $p_0 = 0$, are brought forward in time under a Hamiltonian governed by a potential well (not shown). This dynamics is controlled by $S(q_t, q_0, t)$ and $S(q_t, p_0, t)$, respectively. (Bottom) A new momentum coordinate E generates a manifold that does not change with time. It is represented as a band, uniformly dense, with a width proportional to $1/\nabla H(q, p)$. In time δt, a trajectory point in the band will move "distance" $\delta t\sqrt{\dot{q}^2 + \dot{p}^2} = \delta t\sqrt{(\partial H/\partial p)^2 + (-\partial H/\partial q)^2} = |\nabla H|\,\delta t$. Perpendicular to that path, the thickness of the band is proportional to $1/\,|\nabla H|$. Thus, the area (length $\times$ width) of a little segment of the band is constant in time as it moves along, consistent with an incompressible fluid in the channel or band. This ensures that pieces of the band take each other's place under the dynamics, resulting in no change of this manifold with time.

Given a generating function $S(\mathrm{J}, q)$ that takes $(p, q) \to (\mathrm{J}, \theta)$, Hamilton's equations read

$$
\begin{aligned}
\dot{\theta} &= \omega = \frac{\partial H(\mathrm{J})}{\partial \mathrm{J}}, \\
\dot{\mathrm{J}} &= \frac{\partial H(\mathrm{J})}{\partial \theta} = 0.
\end{aligned}
\tag{3.38}
$$

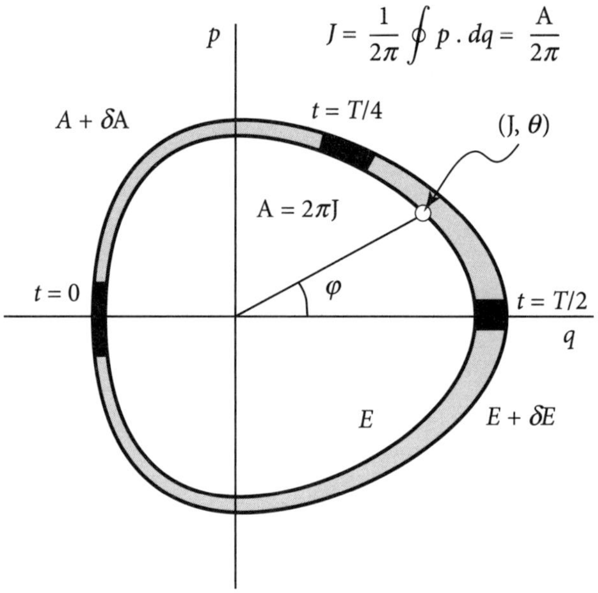

Figure 3.13. The construction of the action J in one dimension. The black region on the left flows to become the region of the same area labeled by time *T*/4 (in one-quarter of a period), and likewise flows to the region of the same area labeled by time *T*/2 (in one-half of a period).

The change in the angle θ upon one complete orbit is

$$\Delta\theta = \omega T = \oint \frac{\partial\theta}{\partial q}\, dq = \oint \frac{\partial^2 S(\mathrm{J}, q)}{\partial \mathrm{J}\partial q}\, dq,$$
$$= \frac{\partial}{\partial \mathrm{J}} \oint \frac{\partial S(\mathrm{J}, q)}{\partial q}\, dq = \frac{\partial}{\partial J} \oint p \cdot dq = 2\pi. \tag{3.39}$$

The generating function $S(J, q)$ is the solution of the Hamilton-Jacobi equation,

$$\boxed{\frac{1}{2m}\left(\frac{\partial S(\mathrm{J}, q)}{\partial q}\right)^2 + V(q) = H(\mathrm{J}),} \quad \text{or} \tag{3.40}$$

$$S(\mathrm{J}, q) = \int \sqrt{2m(H(\mathrm{J}) - V(q))}\, dq = \int_0^q p(\mathrm{J}, q)\, dq, \tag{3.41}$$

with $\partial S(\mathrm{J}, q)/\partial q = p(\mathrm{J}, q)$. As noted in figure 3.14, right, the hatched area is $\delta a = \delta\mathrm{J}\, \theta(\mathrm{J}, q)$; on the left it is seen to be

$$\delta a = \int_0^q dq\, [p(\mathrm{J} + \delta\mathrm{J}, q) - p(\mathrm{J}, q)] = \delta\mathrm{J} \int_0^q dq\, \frac{\partial p(\mathrm{J}, q)}{\partial \mathrm{J}}. \tag{3.42}$$

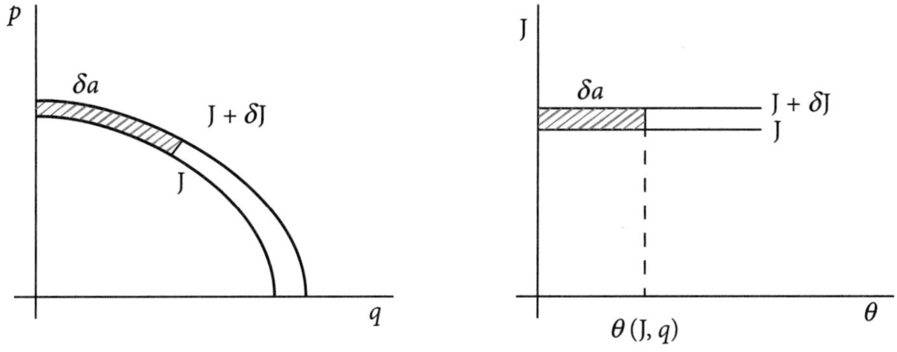

Figure 3.14. The action and phase space paths in (p, q) space and in action-angle (J, θ) space. Because the transformation to (J, θ) is canonical, the hatched area must be the same on both sides. Calling the hatched area δa, we see on the right that $\delta a = \delta J\theta$.

Comparing the two expressions for δa, we have

$$\theta(J, q) = \int_0^q dq\, \frac{\partial p(J, q)}{\partial J} = \frac{\partial}{\partial J}\int_0^q dq\; p(J, q),$$
$$= \frac{\partial S(J, q)}{\partial J} \qquad (3.43)$$

as should be the case, for a transformation using a generating function, $S(J, q)$.

The black region on the left in figure 3.13, earlier, becomes the black region of the same area shown at a time $t = T/4$ later, where T is the period of the (anharmonic) oscillator, and again at time $t = T/2$ at one-half the period.

The black area must speed up in its progress clockwise around the region enclosed by the two constant action curves whenever those curves lie close together, and slow down (such as on the right-hand side) where they are farther apart. It behaves as a slug of incompressible fluid. True, we are here carelessly using the concepts of distance (and speed—that is, change of distance with time) in phase space. There is a mitigating factor to help us complete this discussion. Once a metric has been chosen to make a picture such as figure 3.13, we can measure physical distance in the sense of centimeters on the paper used to make the plot. Such distances are not invariant with respect to canonical transformations, but have meaning for any given plot.

EXAMPLE: HARMONIC OSCILLATOR AND ACTION-ANGLE VARIABLES

The harmonic oscillator generating function $S(q_t, q_0, t)$ was discussed earlier without reference to the action, in section 2.2. Now, with action as a coordinate, we follow the transparent approach of Percival and Richards [8]. We want to find the generating function $S(J, q)$ for the linear oscillator with Hamiltonian

$$H(p, q) = \frac{1}{2m}p^2 + \frac{1}{2}m\omega^2 q^2. \qquad (3.44)$$

Then

$$J = \frac{1}{2\pi}\oint p \cdot dq = \frac{2}{2\pi}\int_{-q'}^{q'} dq \left[2m\left(E - \frac{1}{2}m\omega^2 q^2\right)\right]^{\frac{1}{2}} = E/\omega. \tag{3.45}$$

The Hamiltonian in action-angle variables must therefore be

$$H(J) = \omega J. \tag{3.46}$$

Comparing equations 3.44 and 3.46, we see

$$p(J, q) = [2m\omega J - m^2\omega^2 q^2]^{\frac{1}{2}}, \tag{3.47}$$

and, choosing $q = 0$ as our origin,

$$\begin{aligned} S(J, q) &= \int_0^q [2m\omega J - m^2\omega^2 q^2]^{\frac{1}{2}}\, dq, \\ &= J\sin^{-1}\left[q\left(\frac{m\omega}{2J}\right)^{1/2}\right] + \frac{1}{2}q(2Jm\omega - m^2\omega^2 q^2)^{1/2}. \end{aligned} \tag{3.48}$$

We can compute $\theta(J, q)$ as $\partial S(J, q)/\partial J$ or directly do the integral

$$\begin{aligned} \theta(J, q) &= \int_0^q \frac{\partial p(J, q)}{\partial J}\, dq = \int_0^q dq \left[\frac{m\omega}{2J - m\omega q^2}\right]^{\frac{1}{2}}, \\ &= \sin^{-1}\left[q\left(\frac{m\omega}{2J}\right)^{\frac{1}{2}}\right], \text{ that is,} \end{aligned} \tag{3.49}$$

$$q = \left(\frac{2J}{m\omega}\right)^{1/2} \sin\theta. \tag{3.50}$$

From this and the Hamiltonian, one deduces

$$p = (2Jm\omega)^{1/2}\cos\theta. \tag{3.51}$$

MORE DIMENSIONS—INTEGRABLE SYSTEMS

How does the attempt to define time invariant coordinates like the action J extend to several or many dimensions? If the Hamiltonian is completely separable, and reads, for example,

$$H(\boldsymbol{q}, \boldsymbol{p}) = H_1(p_1, q_1) + H_2(p_2, q_2) + \ldots, \tag{3.52}$$

that is, noninteracting degrees of freedom, then there are N separate actions conserved for N degrees of freedom. Actions J_1 and J_2, and so on, will all be conserved. An abbreviated action $S(q_1, q_2, \ldots, E_1, E_2, \ldots, q_{01}, q_{02}, \ldots)$ can be constructed.

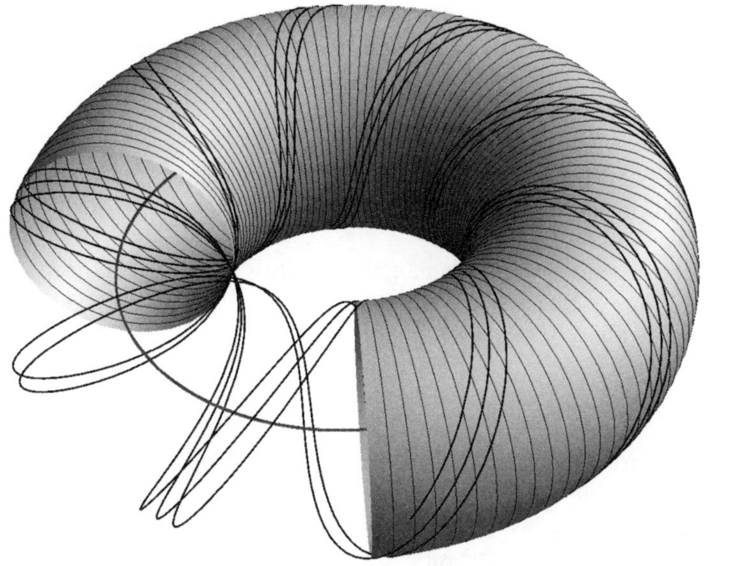

Figure 3.15. A two-dimensional toroidal manifold formed by the collection of smaller circles centered on the larger one. This represents (unfaithfully, necessarily) a 2-torus embedded in a four-dimensional phase space—that is, a Lagrangian submanifold. A trajectory path on the surface of the manifold is shown, obtained by allowing both one-dimensional bound oscillators to evolve freely.

This obvious fact is less trivial than it seems, if its topological implications are explored. Consider the constant energies or actions corresponding to a separable problem with two degrees of freedom. There are four phase space dimensions, but a Lagrangian manifold results from the two constraints: energies J_1 and J_2, reducing the motion to two-dimensional surfaces embedded in four dimensions.

Suppose that the motion is bounded and lies on a closed curve in each phase plane separately—for example, a circle in the (q_1, p_1) plane and another in the (q_2, p_2) plane. Each circle is navigated with period τ_1 and τ_2, respectively, by a trajectory tracing out a one-dimensional track in time through the four-dimensional phase space. This applies to a pair of harmonic oscillators, for example. Suppose too that the period of the motion around each of the two closed orbits is incommensurate—that is, the ratio of the periods is an irrational number, which is normally the case if the frequencies are randomly chosen. All pairs of angles θ_1, θ_2 are then approached arbitrarily closely over time. (If the ratio is rational—say, n/d where n and d are integers—then an initial pair of angles θ_1, θ_2 will exactly repeat after a definite time, and of all possible pairs θ_1, θ_2 only a set of measure zero is ever accessed, with finite gaps in between.) This is aptly described as "the product of two circles." To draw it schematically, we embed one large circle in three dimensions and use it as the roving center of a "perpendicular" set of smaller circles embedded in the sense of figure 3.15. Each circle is a one-dimensional manifold, but stacked together they form the surface of a two-dimensional toroidal manifold, a Lagrangian submanifold.

INTEGRABILITY

We can extend the restrictive idea of a trivially separable system to the notion of Liouville integrability, by supposing there exist N independent functions $I_m(\boldsymbol{p}, \boldsymbol{q})$, $m = 1, \ldots N$, on the phase space (each of which could involve all the phase space coordinates and momenta) not changing with time—that is, whose Poisson brackets (see equation 3.11) with the Hamiltonian vanish, for N degrees of freedom. The mutual Poisson brackets between the N independent functions $\{I_m(\boldsymbol{p}, \boldsymbol{q}), I_n(\boldsymbol{p}, \boldsymbol{q})\}$ must also vanish.

For a time-independent Hamiltonian, the energy is a constant of the motion. If there are more functions of phase space with vanishing Poisson brackets, but fewer than a total of N, we say the system is *partially integrable.*

A system that is Liouville integrable has the same type of foliation of phase space that a completely separable system does (that is also trivially Liouville integrable)—namely, a dense set of N-dimensional tori. (However, for unbounded motion in free space the toroidal "leaves" are not the right geometry, rather flat geometry applies. Another point is that "superintegrable" systems exist, the hydrogen atom being one of them, that live on one-dimensional tori. (All orbits are closed one-dimensional ellipses.) Specifying a precise point in phase space is possible by first specifying which torus the point is on; this requires N "actions" in toriodal space as defined later, and then by specifying the location on the torus with N "angles."

HIERARCHY OF ERGODICITY

The actual motion on the toroidal surface depends on the relative periods τ_1/τ_2. If they are incommensurate (irrational ratio), then over time the entire surface, a Lagrangian manifold, will be uniformly covered, passing as close as desired to any point on the manifold in a way that uniformly populates the surface of the torus, as we now discuss.

Let a trajectory run for a very long time. (How long is needed depends on the ratio τ_1/τ_2. The closer this ratio falls to some rational ratio, especially one involving small integers, the longer we will have to wait to get a quasi-uniform coverage such that a bit of "blurring" or coarse graining will render it uniform on the manifold.) We collect the trajectory as a one-dimensional track in phase space (see the short section of spiral path on the torus in figure 3.5, earlier):

$$\rho(\boldsymbol{q}, \boldsymbol{p}, T) = \int_0^T dt\, \delta(\boldsymbol{q} - \boldsymbol{q}_t(\boldsymbol{q}_0, \boldsymbol{p}_0))\delta(\boldsymbol{p} - \boldsymbol{p}_t(\boldsymbol{q}_0, \boldsymbol{p}_0)). \tag{3.53}$$

All points on the surface are approached by the trajectory to within some small error after a long time T. The distribution is constructed from a trajectory that, when averaged over its long history, becomes a dense winding path on the Lagrangian manifold, repeatedly approaching each point on the surface. We coarse grain or blur the trajectory so the individual trajectory paths lying densely on the surface become smoothed. The one-dimensional paths don't change if the whole path is propagated forward in time, since the points comprising it just move in the same way the trajectory creating it did, taking each other's place. Therefore, the smoothed density created as described from trajectories on the Lagrangian manifold cannot change under time evolution, and cannot be nonuniform on the manifold. We say that the averaged trajectory is *ergodic on the toriodal manifold,* in the general case involving

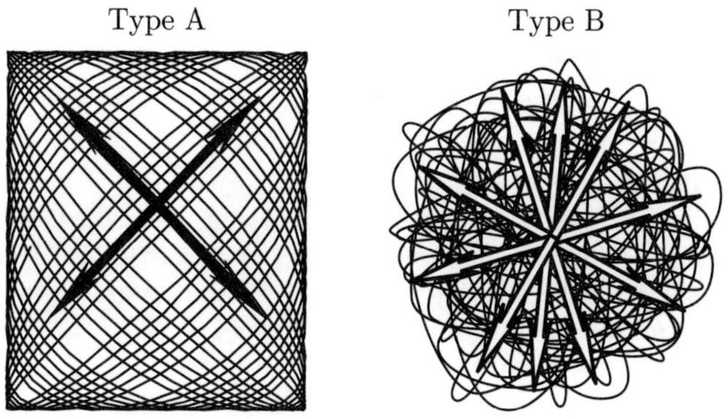

Figure 3.16. Einstein's Type A and Type B motion illustrated in two spatial dimensions, using graphics capability not available to him.

incommensurate frequency ratios. If there are two conserved quantities (two "good" actions I_1 and I_2), each is a constraint, together reducing freedom of movement by two dimensions.

Suppose there are no good actions, and the energy is the only conserved quantity, constituting a single restriction of the motion—that is, a three-dimensional constant energy manifold embedded in four dimensions, or generally $2N-1$ dimensions in a $2N$-dimensional phase space. By an argument analogous to the one used for the coverage of the torus, the constant energy surface must be uniformly covered in the sense that the time average of a very long trajectory, suitably smoothed in a fine-grained way, is a smooth distribution that is time invariant and equivalent to the phase space distribution

$$\rho_E(\boldsymbol{p}, \boldsymbol{q}) = \delta(E - H(\boldsymbol{p}, \boldsymbol{q})). \tag{3.54}$$

EINSTEIN AND ERGODICITY

In 1917 Einstein pointed out [9] that what we now call classical chaos (and Einstein called "type B" classical motion) raises serious issues for Bohr-Sommerfeld (or Wantzel-Kramers-Brillouin [WKB]) quantization, to be discussed in section 12.2. Type B motion (see figure 3.16) would not be quantizable in the "Sommerfeld and Epstein" sense, but even the motion of three or more bodies is quantizable if it is of type A (integrable) motion. The 1917 paper is often cited as the first ever in the field of *quantum chaology*, or the study of the implications of classical chaos for quantum mechanics.

If there are no good actions and energy is the only constant of motion, the loose statement that the "system is ergodic" applies. Statements about ergodicity are made in relation to a trajectory's full exploration of phase space, where the available phase space is subject only to the "obvious" constraints like energy conservation, or total angular momentum, and so on. Finding that there are "extra" constants of the motion beyond what we knew about to begin with causes us to label the system *nonergodic*. If we account for newly discovered constants of motion, we will find that the system is generally ergodic on appropriate submanifolds where they are constant! For example, we argued earlier that motion on the 2-torus is ergodic on its surface in the case of incommensurate frequencies. If the ratio of periods associated with the actions, τ_1/τ_2, is in fact rational, trajectories will eventually retrace themselves and fail ergodicity on the 2-torus—that is, fail to visit all but a one-dimensional track on the two-dimensional toroidal Lagrangian manifold. You guessed it: the average trajectory is then ergodic *on, or with respect to, this one-dimensional manifold.*

CONSTANTS OF THE MOTION

All the usual one-dimensional examples are trivial because energy is conserved, giving one constant of the motion. The central force problem in 3D has three constants of the motion (that can be taken to be energy E, the square of the total angular momentum $|J|^2$, and the z component of angular momentum J_z). Why not add in J_x and J_y? Aren't they constants of the motion? Yes, but they are not *in involution* with the ones already defined, meaning their Poisson brackets must vanish, but

$$\{J_x, J_z\} = J_y \neq 0. \tag{3.55}$$

Why does it matter if constants are "in involution"? Take the example of a Kepler orbit, three degrees of freedom and three constants of the motion, just mentioned above. The Hamiltonian H, the square J^2 of the total angular momentum, and one component of angular momentum—say, J_z, all commute with each other—for example, $\{J^2, J_z\} = 0$; they are in involution. Moreover, specifying all three constants specifies a unique Kepler orbit. J_z,J_x, J_y are also three constants of the motion, but they are not in involution. They do not specify a unique orbit, either: Say $J_x = J_y = 0$, and J_z is specified. There are many different orbits with all three properties, since for example we can have the same J_z for a low-energy orbit or a high-energy one.

The deeper meaning of two constants A and B being in involution is that they are symmetries of each other. The transformation induced by B in $\{A, B\} = 0$ leaves B invariant, and vice versa. For example, since $\partial\rho/\partial t = \{\rho, H\}$, H induces time translation. It is easy to see that $\partial\rho/\partial\theta = \{\rho, J_z\}$, since $\{\rho, J_z\} = (\partial\rho/\partial\theta)(\partial J_z/\partial J_z) - (\partial\rho/\partial J_z)(\partial J_z/\partial\theta)$, and thus J_z induces rotations about the z axis. Then $\{H, J_z\} = 0$ if H has rotational symmetry about the z axis, or equivalently J_z is invariant under time evolution. N constants of the motion in involution "play nice" with each other; they can all remain fixed under the action of any of the others. Suppose there were N good actions $I_1, I_2, \ldots, I_N$, with each action inducing translation along its own conjugate angle—that is, $\partial\rho/\partial\theta_i = \{\rho, I_i\}$. Then, for example, $\partial I_j/\partial\theta_i = \{I_j, I_i\} = 0$. The actions are constant as their angles are changed. Knowing all the actions determines the unique torus the motion lies on; there is no ambiguity.

The spherical pendulum has two coordinates (two angles) and two constants in involution; these can be taken to be the energy E and the z component of angular momentum $J_z = m$. Systems with N coordinates and N constants in involution are called *completely integrable*.

Most "real life" systems are *not* completely integrable. For example, the stadium billiard (a particle in a football stadium–shaped hard-walled box, desymmetrized by using only the upper quarter) is known to have energy as the only constant of motion even though there are four phase space freedoms. The stadium is therefore an ergodic system. This requires a proof (Bunimovich). The wall has a discontinuous second derivative where the circular end caps meet the straight wall. To the author's knowledge, no billiard has been shown to be completely chaotic without some sort of nonanalytic behavior at the walls. Eigenstates can be labeled by a single quantum number in the desymmetrized quarter-stadium. A part of a typical classical trajectory in a quarter-stadium is shown in figure 3.17.

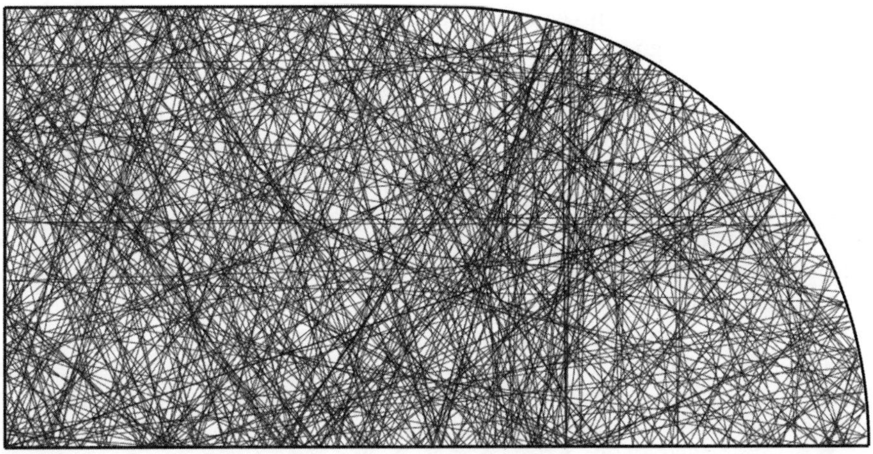

Figure 3.17. A single classical trajectory is followed for many bounces in the quarter-stadium billiard. This is chaotic motion, working its way toward sampling all the available phase space uniformly. This trajectory is typical of almost all orbits in the stadium. The exceptions are periodic orbits, of zero measure (0% of all orbits, just as the rationals are 0% of all the real numbers), and are all unstable to small perturbations. Also homoclinic and heteroclinic orbits (leaving one fixed point vicinity and approaching another) are zero measure exceptions, falling away from a periodic orbit after a tiny push away from it, eventually returning to the vicinity of the same (homoclinic) or another (heteroclinic) periodic orbit for the rest of time. If it begins the infinite reapproach on the stable axis on the first return to the vicinity, it is a "primary" homoclinic or heteroclinic orbit; the second try is secondary; and so on.

In quantum mechanics, there is one "good quantum number" for each classical constant in involution. For example, in one degree of freedom we have the integer n that specifies the energy level E_n. For the hydrogen atom, in the center of mass coordinates, there are three degrees of freedom and three constants in involution classically. Quantum mechanically we have three good quantum numbers (n, ℓ, m) corresponding to energy, angular momentum, and one component of angular momentum (usually taken to be the z component).

The H atom is superintegrable, as mentioned earlier, so it has also the Laplace-Runge-Lenz vector (giving the direction of the major semiaxis) as an independent constant of motion in involution. Indeed, on a deeper level, this explains why the quantum spectrum of H does not depend on quantum number L, but only on n. Interestingly, the classical superintegrability was used by Pauli to derive the H spectrum algebraically, without any differentiation, at the same time as (or even slightly earlier than) Schrödinger, who derived it with the tedious differential equation approach that is taught in 99.9% of textbooks today.[2]

In quantum mechanics textbooks, one often passes from hydrogen to helium, noting that helium has no exact quantum solution. One resorts to numerical methods to get estimates of energies and eigenfunctions. With the center of mass removed, helium has six degrees of freedom. Should we then expect six constants of the

[2] Thanks go to Prof. Jiri Vanicek for the remarks in this paragraph, and for catching many typos and some errors in a thorough reading of the draft manuscript.

motion? We know of only three good quantum numbers: $|L|^2$, giving the total angular momentum, $L_z = m$, giving its z component; and n, labeling the energies E_n. In fact, these good quantum numbers, having direct classical counterparts, arise from isotropy of space (for an isolated helium atom) and homogeneity of time. These are derivable from Noether's theorem. We have "lost" three other quantum numbers, just as in the classical version we have three, not six, conserved functions of the phase space coordinates in involution. Classically, the motion of helium is known from trajectory studies to be partly chaotic—implying an absence of any other constants of motion. In fact, helium ionizes spontaneously and rapidly for most classical initial conditions, even if both electrons are initiated in reasonable orbits corresponding to the quantum ground state of helium. From the point of view of quantum mechanics, the electron remaining behind has "illegally" dropped below the quantum ground state energy of the He^+ ion, thus providing the energy to eject the other electron. In lithium we still have only three good quantum numbers, and thus six "missing" quantum numbers, and so on.

Almost certainly there are no other exact quantum numbers in helium waiting to be found. There is nonetheless a large effort and equally large reward for finding *approximate* symmetries in the sense of Noether, or by finding approximately conserved actions, and corresponding approximate quantum numbers. The rewards are large because four or more degree of freedom systems can be difficult to describe, requiring gigantic numerical effort. If approximate constants of the motion exist, one can hope to reduce to the subspaces defined by that "constant," thus reducing the dimensionality and numerical effort (that typically goes up by more than a factor of 10 for each degree of freedom that must be mixed in), and increasing intuition and understanding.

3.6 Area-Preserving Maps

We know that area-preserving transformations correspond to canonical transformations, but suppose we build these based on Hamiltonian dynamics. For small but finite time steps δt, we might have numerical integration in mind, and a canonical form would ensure the desirable trait of area preservation under each time step. For larger time steps, things get interesting with resonances and chaos entering the picture, even in one dimension; we explore this in the following.

Hamilton's equations of motion for one degree of freedom can be written as

$$\begin{aligned}
\lim_{\delta t \to 0} q(t+\delta t) &= q(t) + p(t)/m \cdot \delta t, \\
\lim_{\delta t \to 0} p(t+\delta t) &= p(t) - \left.\frac{\partial V(q)}{\partial q}\right|_{q=q(t+\delta t)} \cdot \delta t, \\
&= p(t) - \frac{\partial V(q(t) + p(t)/m \cdot \delta t)}{\partial q} \cdot \delta t, \qquad (3.56)
\end{aligned}$$

for a Hamiltonian $H = p^2/2m + V(q)$, to order δt^2 as $\delta t \to 0$. There is an important subtlety in this form of the equations in that the potential is evaluated at the "updated" position $q(t+\delta t)$ in the second line. Now let's calculate something called the *stability matrix* (see section 3.8) for moving forward the short time step δt:

$$M = \begin{pmatrix} m_{11} & m_{12} \\ m_{21} & m_{22} \end{pmatrix} = \begin{pmatrix} \dfrac{\partial p(t+\delta t)}{\partial p(t)} & \dfrac{\partial p(t+\delta t)}{\partial q(t)} \\ \dfrac{\partial q(t+\delta t)}{\partial p(t)} & \dfrac{\partial q(t+\delta t)}{\partial q(t)} \end{pmatrix} = \begin{pmatrix} 1+\dfrac{\partial^2 V}{\partial q^2}\dfrac{\delta t}{m} & \dfrac{\partial^2 V}{\partial q^2} \\ \left(\dfrac{\delta t}{m}\right) & 1 \end{pmatrix}. \tag{3.57}$$

It is easily seen that the determinant of M is 1, and just as easily seen that it wouldn't be if we hadn't used the "updated" position in the second line. This is called a symplectic integrator if the $\lim_{\delta t \to 0}$ is not taken and a finite δt is used in equation 3.56. Equation 3.56 can be regarded as mapping points $[q(t), p(t)]$ to new points $[q(t+\delta t), p(t+\delta t)]$. The unit determinant guarantees that the map is area preserving.

The area preservation implies that this finite time transformation is given by a generating function, changing notation slightly to reflect possible large time steps Δt,

$$F_3(p_n, q_{n+1}) = -q_{n+1} p_n + \frac{1}{2} p_n^2 \, \Delta t + V(q_{n+1}) \, \Delta t, \tag{3.58}$$

giving, for any Δt,

$$\begin{aligned} q_{n+1} &= q_n + p_n/m \cdot \Delta t; \\ p_{n+1} &= p_n - \frac{\partial V(q_{n+1})}{\partial q_{n+1}} \cdot \Delta t. \end{aligned} \tag{3.59}$$

By a similar transformation one can also update the momentum first and use it in updating the position—that is,

$$\begin{aligned} p_{n+1} &= p_n - \frac{\partial V(q_n)}{\partial q} \cdot \Delta t; \\ q_{n+1} &= q_n + p_{n+1}/m \cdot \Delta t. \end{aligned} \tag{3.60}$$

These equations can be generalized to

$$\begin{aligned} p_{n+1} &= p_n - \kappa f(q_n); \\ q_{n+1} &= q_n + p_{n+1}. \end{aligned} \tag{3.61}$$

This is the form of the famous "standard map,"

$$p_{n+1} = p_n + k \sin(\theta_n), \tag{3.62}$$

$$\theta_{n+1} = \theta_n + p_{n+1}. \tag{3.63}$$

θ_{n+1} is taken to be modulo 2π, requiring the standard map to lie on a cylinder. By taking momentum also to be modulo 2π, the map is converted to lie on a torus. Without going into detail, we examine the standard map for various "kick strengths" k and various initial (q_0, p_0). A very useful code, StdMap, has been created and maintained by Prof. J. D. Meiss of the University of Colorado, easily downloadable from the Web, but it works on Macintosh systems only. There are many more maps and analytical and numerical tools in this code than we are showing here.

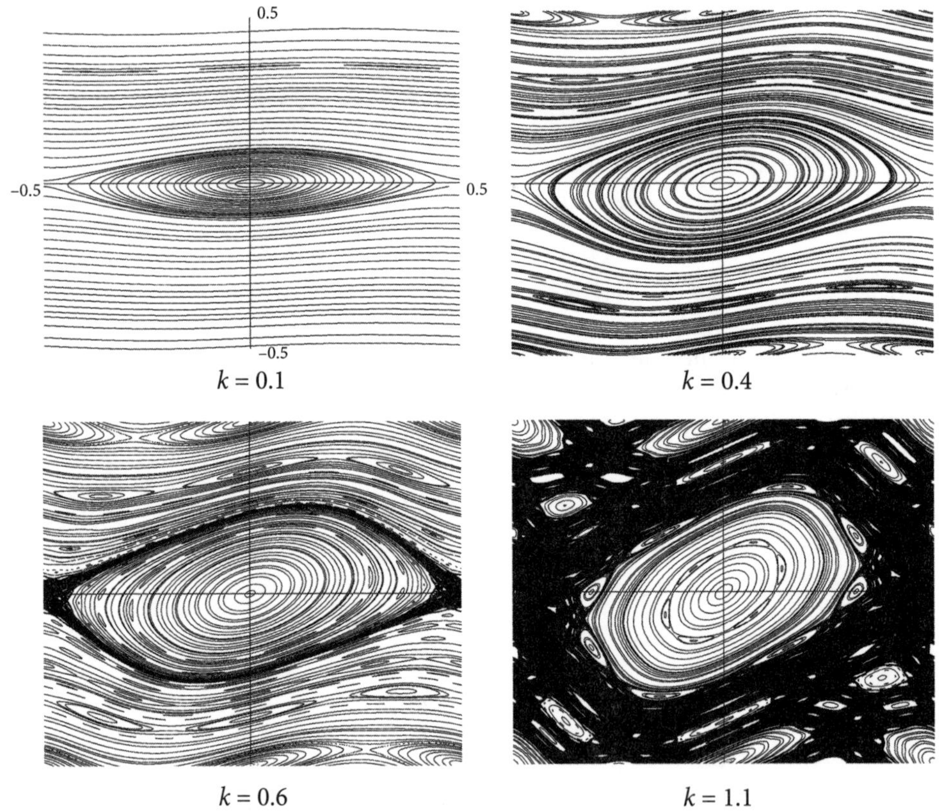

Figure 3.18. Standard map, equations 3.62 and 3.63, for four different kick parameters k. Increasing k leads to an increasing presence of island resonance zones and chaotic regions. On the order of 50 to 100 initial points were iterated 70,000 times, filling in the dark regions densely in the case of chaotic motion. This is a "mixed" system, by far the most common type of dynamics. Fully integrable systems are rare (although you wouldn't know it from reading most textbooks), as are fully chaotic systems, at least in low dimension.

A Poincaré surface of section for a two degree of freedom system defines an area preserving point-to-point mapping on the plane—that is, on the surface of section. Suppose we dispense with the machinations between intersections with the surface of section and directly define an area-preserving, nonlinear map of the plane into itself. One quite general way of doing this culminated in equation 3.59. Such area-preserving maps are a much more convenient way of generating classical phase space evolution; they exhibit the same phenomenology as the more laboriously created s.o.s. maps (see an example in figure 3.18). *Also, such maps will be just as useful for quantum studies when we quantize them, later.*

The standard map is equivalent to a time-dependent Hamiltonian, also known as the *δ-kicked rotor*:

$$H(p, \theta, t) = \frac{1}{2}p^2 + k\cos(\theta) \sum_{n=-\infty}^{\infty} \delta(t-n). \tag{3.64}$$

Setting up Hamilton's equations of motion, the "kick" is seen to jump the momentum by $k\sin(\theta)$ as required, at integer times.

3.7 Poincaré-Birkhoff Fixed-Point Theorem

An important step in understanding the structure of phase space is to see how resonance islands, and the stable and unstable regions around them, are born. A construction due to Poincaré and Birkhoff accomplishes this nicely. We illustrate the idea using the standard map equations 3.62 and 3.63 (section 3.6).

Note that if $k=0$, the momentum $p_{n+1}=p_n$ doesn't change from one iteration to the next, while $\theta_{n+1}=\theta_n+p_n$. Some momenta are special: if $p_0=2\pi r/s$, where r and s are relatively prime integers (that is, the only positive integer that divides them both is 1), then the whole line $p=2\pi r/s$ consists of fixed points, with any initial θ returning periodically, after s iterations. We say the line $p_0=2\pi r/s$ consists of only fixed points with period s, a so-called invariant curve. Such resonant lines (in general they certainly don't need to be straight) are ripe for the development of a resonance island chain when a perturbation (that is, $k\neq 0$) is introduced. This is illustrated for the standard map in figure 3.19.

To show this, the Poincaré-Birkhoff idea is to find, for small values of k, loci of points (p_n,θ_n) such that $\theta_n=\theta_{n+s}$. For sufficiently small k, the flow near the invariant lines is a shear flow, just as it is for $k=0$. The more p moves above the invariant line, the more the iterated θ tends to move to the right under the mapping; lowering p moves θ more to the left; for some value of p therefore, θ doesn't move, even with $k\neq 0$. Thus, for each initial θ, there is a p such that $\theta_n=\theta_{n+s}$. This defines a "constant θ line" (figure 3.20). Now we map the whole, undulating, constant θ line; the area under it must be preserved in the mapping. After s iterations, the undulations that lay above the unperturbed invariant line must now lie below it, because that is the only way the opposite shear above and below can be employed to keep θ constant. The minimum that can happen is one zone moving up, balanced by another moving down. Since the map is periodic with period 2π, this implies there must be at least two θ for which momentum does not move at all, in the transition from up to down; these are fixed points; both p and θ don't change. Thus, the existence of at least two fixed points is proven.

The fixed points have compensated shear in θ, but they are special and rare in that p also returns to itself, so above the fixed point there is right-trending shear, and below, left-trending; this is true for all fixed points. If the part of the θ fixed curve to the left of the fixed point lies below the unperturbed invariant line, it must move above it upon returning θ to itself, and to the right of the fixed point the θ fixed curve must lie above the invariant line and move down below it upon returning. As figure 3.20 shows, this implies clockwise circulation about that fixed point.

If on the other hand the part of the θ fixed curve to the left of the fixed point lies above the unperturbed invariant line, it must move below, and to the right the movement must be from below to above. This is not a circulation, but a *hyperbolic* flow, as seen on the right side of figure 3.20. We illustrate this for the standard map for period 1 and 2 resonances, for $k=0.1$. We reproduce the mapping equation here, as

$$p_{n+1}=p_n+k\sin(\theta_n), \tag{3.65}$$

$$\theta_{n+1}=\theta_n+p_{n+1}=\theta_n+p_n+k\sin(\theta_n). \tag{3.66}$$

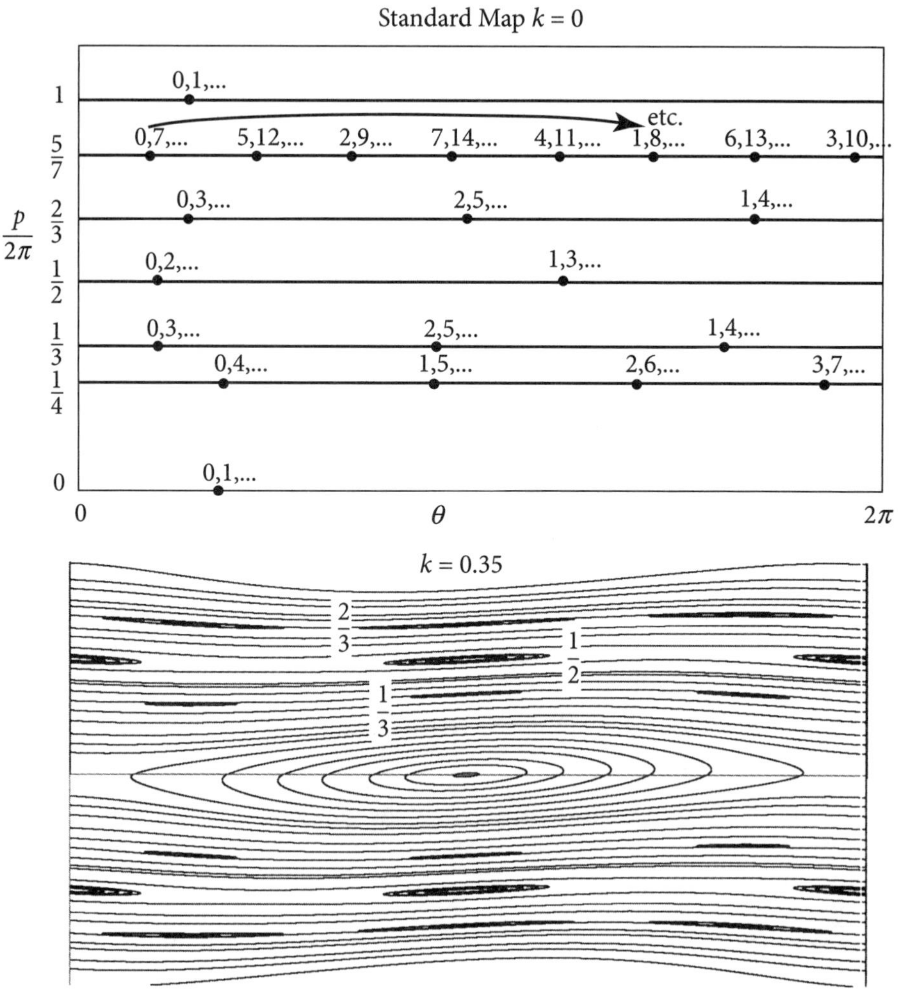

Figure 3.19. (Top) Sample lines of fixed points for the standard map, strength $k = 0$. Black dots show the history of an arbitrary point on each line, but the whole line consists of such fixed points. All momenta with $p/2\pi = r/s$ are invariant lines with period s. Irrational multiples of 2π remain on a line but never visit the same point twice. (Bottom) 50,000 iterates of various hand-picked initial momenta for $k = 0.35$ reveal some of the resonances predicted by the Poincaré-Birkhoff theorem. The resonances quickly become very narrow as s increases, and most were missed by the random choices.

Figure 3.20 shows the vicinity of the $k = 0$ period 1 invariant line, for $k = 0.1$. It has become a resonance island chain of alternating stable and unstable fixed points lying near the former invariant line. The condition $\theta_{n+1} = \theta_n$ imposes $p_n = -k\sin(\theta_n)$; this is the equation of the heavy dotted line in figure 3.20 showing the locus of points that map vertically in one iteration to the same angle.

The middle and upper insets show the regions surrounding the stable and unstable fixed points. The right insets show the stable and unstable manifolds meeting at the fixed point (see also figure 3.21). The stable manifolds have black arrows pointing in toward the fixed point; the unstable manifolds have black arrows pointing away

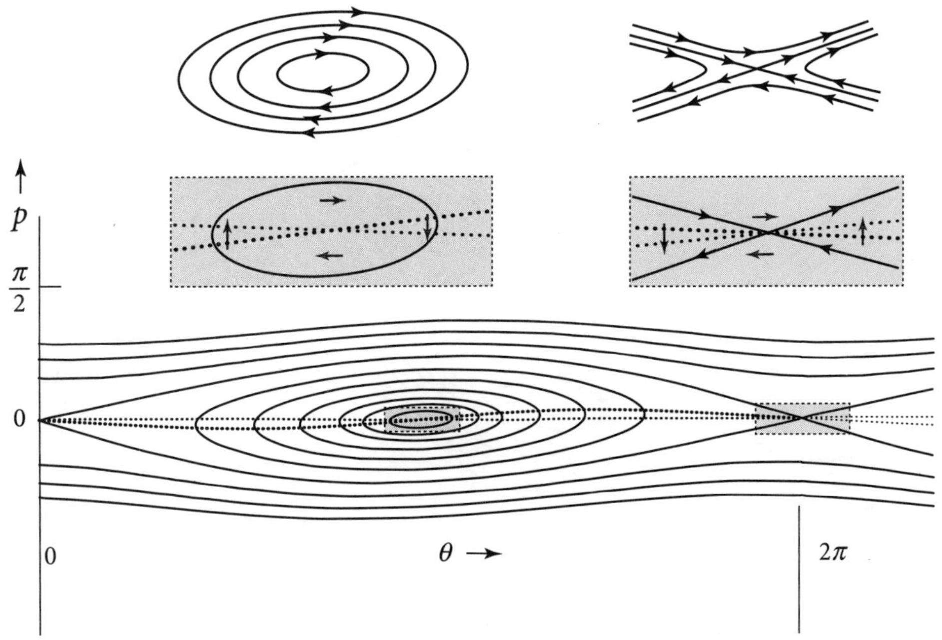

Figure 3.20. The existence of an even number of fixed points (here, two) arise from an invariant line of period 1 fixed points by addition of a perturbation. The alternating stability that results by raising the kicking parameter above 0 is seen here for the standard map with kicking parameter $k = 0.1$. The solid curved lines at the bottom of the diagram are the result of iterating 14 different initial points many thousands of times each. The heavy dotted line represents a set of initial points that, after one iteration of the map, become the lighter dots, each dot moving vertically up or down (no change in θ). The location of the heavy dotted line and the one-step mapping of it, the lighter dotted line, is unique. The area below the heavy dotted line is preserved in the mapping, forcing the mapped line to intersect the original line an even number of times, since the heavy dotted line is periodic with period 2π. The local phase space flow (insets, middle) near the intersection points ("fixed points"), that changed neither p nor θ, shows the nearby displacements in momentum p (leaving θ fixed) for the dots on the initial line. This, together with the shear flow at other nearby places, determines the topology of the local flow, seen to be alternating stable "elliptic" and unstable "hyperbolic" motion, mirroring the motion in phase space for a local potential minimum and maximum, respectively.

from the fixed point, showing the direction of flow there. Starting on the unstable manifold minutely off the fixed point, repeated iterates map along the manifold with exponentially increasing increments away from the fixed point, along a line that is the unstable manifold. Reversing the map would change this motion into an approach to the fixed point with exponentially decreasing steps, never reaching that point in a finite number of iterations. This is just what happens along the stable manifolds in the forward mapping.

Figure 3.21 shows a similar construction for a magnified (in momentum—the angle still goes from 0 to 2π) region near $p = 0.5(2\pi)$ and $k = 0.1$.

The Poincaré-Birkhoff construction *proves* the existence of the island chain (the elliptical regions are the "islands"). There are $2s$ fixed points generated when the

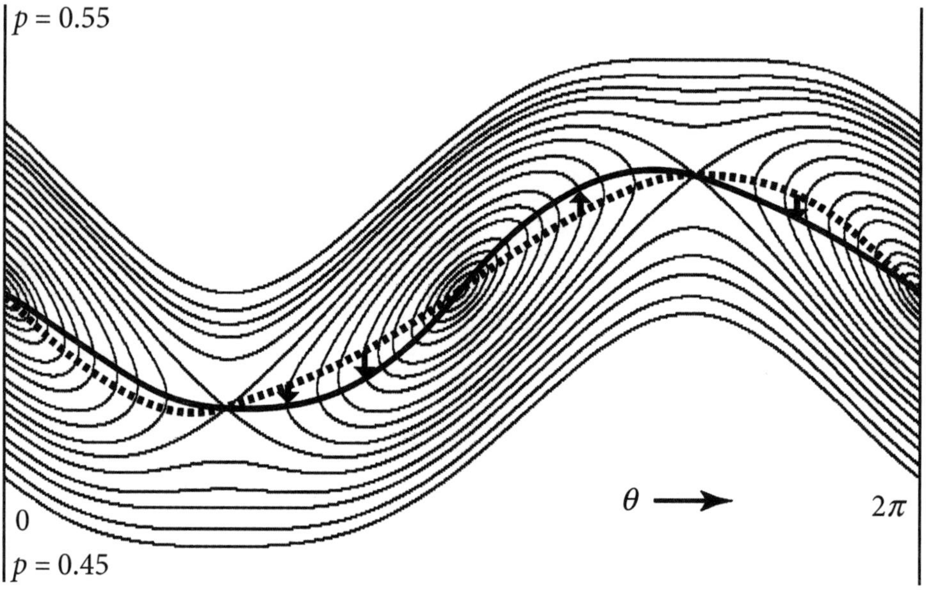

Figure 3.21. The vicinity of the period 2 line (for $k = 0$) is shown for the standard map at $k = 0.1$, plotted from $p = 0.45$ to $p = 0.55$ and θ on $(0, 2\pi)$. The Poincaré-Birkhoff line having unchanged θ is shown dashed; it maps in two iterations to the line shown solid. A few arrows show the constant θ movement after two iterations.

invariant curve of period s is perturbed, half of them stable, half unstable. In our present example, the width of the resonance zone, measured across the stable zone from one stable manifold to the other, is only abut $1/5$th as great for the $s = 2$ period 2 resonance as it is for the $s = 1$ period 1. This is a general trend; higher order resonances grow thinner, usually exponentially fast in their period. The resonance island chain corresponding to $p = 2\pi/4$ is hard to find because it is so thin for $k = 0.1$; it is about $1/500$th as wide as the period $s = 1$ resonance island.

3.8 Stability Analysis

SYMPLECTIC MATRICES

We define a matrix $\boldsymbol{J}$ with the special structure [5, 10]

$$\boldsymbol{J} = \begin{pmatrix} \mathbf{0} & -\boldsymbol{I} \\ \boldsymbol{I} & \mathbf{0} \end{pmatrix}, \tag{3.67}$$

where $\boldsymbol{I}$ is the $N \times N$ identity matrix, and $\mathbf{0}$ is an $N \times N$ matrix filled with zeros. The matrix $\boldsymbol{J}$ is used to define *symplectic matrices* $\boldsymbol{A}$ through

$$\begin{aligned} \boldsymbol{A}^T \boldsymbol{J} \boldsymbol{A} &= \boldsymbol{J}, \\ \boldsymbol{A} \boldsymbol{J} \boldsymbol{A}^T &= \boldsymbol{J}. \end{aligned} \tag{3.68}$$

It is easy to show the matrix $\boldsymbol{J}$ is itself symplectic.

Hamilton's equations of motion have a symplectic structure, since

$$\frac{d\boldsymbol{X}_t}{dt} = \boldsymbol{J} \cdot \vec{\nabla} H(\boldsymbol{X}_t), \tag{3.69}$$

where

$$\boldsymbol{X}_t = (q_1(t), q_2(t), \ldots, q_N(t); p_1(t), \ldots p_N(t)), \tag{3.70}$$

and

$$\vec{\nabla} \equiv \left(\frac{\partial}{\partial q_1}, \ldots \frac{\partial}{\partial q_N}; \frac{\partial}{\partial p_1}, \ldots \frac{\partial}{\partial p_N}\right). \tag{3.71}$$

STABILITY MATRIX

One of the most important topics in dynamics is the sensitivity of trajectories to their initial conditions. This subject is called *stability analysis.* Stability has direct bearing on quantum mechanics, controlling for example the spreading of wavepackets, as we will see.

Starting with a given reference trajectory $[\boldsymbol{q}(t), \boldsymbol{p}(t)]$, we ask what happens if we start nearby that trajectory. In a $2N$-dimensional phase space, there are $2N$ possible small deviations to take near the initial location in phase space, $[\boldsymbol{q}(0), \boldsymbol{p}(0)]$. These deviations live in a *tangent space*, a $2N$-dimensional vector space attached to each point $\boldsymbol{q}$.

Let Ω_t be the (generally nonlinear) operator that takes initial phase space points $\boldsymbol{X}_0$ into current points $\boldsymbol{X}_t$:

$$\boldsymbol{X}_t = \Omega_t[\boldsymbol{X}_0]. \tag{3.72}$$

A trajectory evolving from the initial deviation $\boldsymbol{X}_0 + \delta\boldsymbol{X}_0$ is $\boldsymbol{X}_t + \delta\boldsymbol{X}_t$. This leads to the definition of the $2N \times 2N$ *stability matrix* $\boldsymbol{M}_t(\boldsymbol{X}_0)$ as

$$\boxed{\delta\boldsymbol{X}_t = \boldsymbol{M}_t(\boldsymbol{X}_0) \cdot \delta\boldsymbol{X}_0,} \tag{3.73}$$

with

$$\boldsymbol{M}_t(\boldsymbol{X}_0) = \left(\frac{\partial \boldsymbol{X}_t}{\partial \boldsymbol{X}_0}\right) = \begin{pmatrix} \frac{\partial q_{1t}}{\partial q_{10}} & \cdots & \frac{\partial q_{1t}}{\partial p_{N0}} \\ \vdots & \vdots & \vdots \\ \frac{\partial p_{Nt}}{\partial q_{10}} & \cdots & \frac{\partial p_{Nt}}{\partial p_{N0}} \end{pmatrix} = \begin{pmatrix} m_{1,1} & \cdots & m_{1,2N} \\ \vdots & \vdots & \vdots \\ m_{2N,1} & \cdots & m_{2N,2N} \end{pmatrix}. \tag{3.74}$$

Equation 3.73 explains the key role of the stability matrix: to take any initial infinitesimal deviation $\delta\boldsymbol{X}_0$ from a reference trajectory $\boldsymbol{X}_0$ and turn it into what that deviation becomes, $\delta\boldsymbol{X}_t$ a time t later. Unless the system is linear, $\boldsymbol{M}_t(\boldsymbol{X}_0)$ depends on the starting point $\boldsymbol{X}_0$. $\boldsymbol{M}_t(\boldsymbol{X}_0)$ has its own equation of motion:

$$\boxed{\frac{d}{dt}\boldsymbol{M}_t(\boldsymbol{X}_0) = \boldsymbol{K}(\boldsymbol{X}_t)\, \boldsymbol{M}_t(\boldsymbol{X}_0).} \tag{3.75}$$

where

$$\boldsymbol{K}(\boldsymbol{X}_t) = \begin{pmatrix} \dfrac{\partial^2 H}{\partial \boldsymbol{q} \partial \boldsymbol{p}} & \dfrac{\partial^2 H}{\partial \boldsymbol{q}^2} \\ -\dfrac{\partial^2 H}{\partial \boldsymbol{p}^2} & -\dfrac{\partial^2 H}{\partial \boldsymbol{q} \partial \boldsymbol{p}} \end{pmatrix}. \tag{3.76}$$

Equation 3.75 can be derived by direct differentiation of matrix elements of $\boldsymbol{M}_t(\boldsymbol{X}_0)$ with respect to time, as in

$$\frac{\partial \dot{q}_t}{\partial q_0} = \frac{\partial^2 H}{\partial p_t \partial q_0} = \frac{\partial}{\partial p_t}\left(\frac{\partial H}{\partial q_t}\frac{\partial q_t}{\partial q_0} + \frac{\partial H}{\partial p_t}\frac{\partial p_t}{\partial q_0}\right). \tag{3.77}$$

For a Hamiltonian of the form $H = \boldsymbol{p}^2/2m + V(\boldsymbol{q})$, $\boldsymbol{K}$ reads

$$\boldsymbol{K}(\boldsymbol{X}_t) = \begin{pmatrix} 0 & -\dfrac{\partial^2 V}{\partial \boldsymbol{q}^2} \\ \dfrac{1}{m} & 0 \end{pmatrix}. \tag{3.78}$$

By a property of linear time-dependent operators, as in quantum mechanics, we have

$$\boldsymbol{M}_t(\boldsymbol{X}_0) = T\, e^{-\int_0^t K(t')\, dt'}\, \boldsymbol{M}_0(\boldsymbol{X}_0), \tag{3.79}$$

where T is the Dyson time-ordering operator.

EIGENVALUES AND EIGENVECTORS OF M_t

What are the restrictions on the eigenvalues of the stability matrix? An eigenvector $\vec{Y}_k$ obeys

$$\boldsymbol{M}_t(\boldsymbol{X}_0)\vec{Y}_k = \mu_k \vec{Y}_k, \tag{3.80}$$

meaning that the vector $\vec{Y}_k$ is stretched or compressed but doesn't change direction, provided μ_k is real. Eigenvectors corresponding to complex μ_k correspond to spiral or circulating motion around the reference trajectory (for real μ).

First, we note that the so-called characteristic equation

$$P(\mu) = \det(\boldsymbol{M}_t(\boldsymbol{X}_0) - \mu 1) = \prod_k (\mu - \mu_k) \tag{3.81}$$

has to be real, since the elements of $\boldsymbol{M}_t$ are real. Thus, every complex eigenvalue μ_k has a partner complex conjugate μ_k^* that is also an eigenvalue. The stability matrix $\boldsymbol{M}_t(\boldsymbol{X}_0)$ is symplectic, as seen by the linearized dynamics $\boldsymbol{X}_t = \boldsymbol{M}_t(\boldsymbol{X}_0) \cdot \boldsymbol{X}_0$ and the Poisson

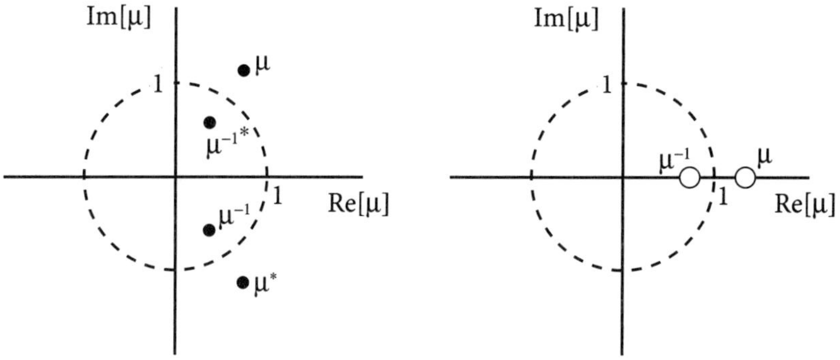

Figure 3.22. (Left) The generic case of a complex eigenvalue configuration for a given complex μ. (Right) Another possible case, where the open circle represents a doubly degenerate eigenvalue.

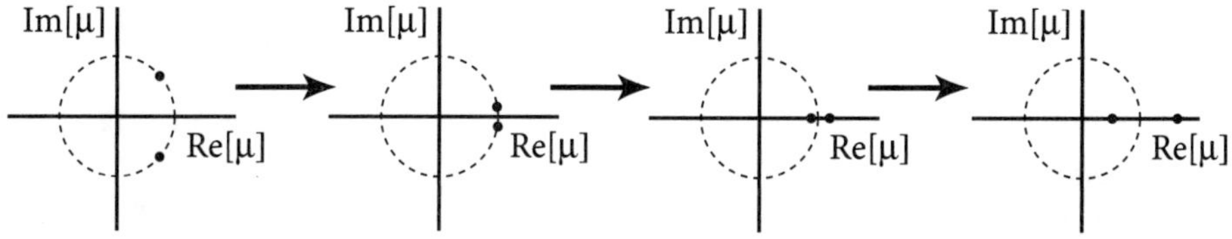

Figure 3.23. Complex eigenvalues of unit modulus collide and become real, as a function of driving frequency or amplitude, corresponding to the transition from stable to unstable dynamics.

bracket relations that apply to the elements of $\boldsymbol{X}_t$:

$$\{q_{it}, q_{jt}\} = 0; \;\; \{p_{it}, p_{jt}\} = 0; \;\; \{q_{it}, p_{jt}\} = \delta_{ij}, \tag{3.82}$$

which implies that

$$\begin{aligned} \boldsymbol{M}_t^T \boldsymbol{J} \boldsymbol{M}_t &= \boldsymbol{J}, \\ \boldsymbol{M}_t \boldsymbol{J} \boldsymbol{M}_t^T &= \boldsymbol{J}. \end{aligned} \tag{3.83}$$

The transpose of $\boldsymbol{M}_t$, $\boldsymbol{M}_t^T$, has the same eigenvalues as $\boldsymbol{M}_t$. Let

$$\boldsymbol{M}_t^T \vec{W}_k = \mu_k \vec{W}_k. \tag{3.84}$$

Then from equations (3.83) and (3.84), we can write

$$\boldsymbol{M}_t \boldsymbol{J} \boldsymbol{M}_t^T \vec{W}_k = \mu_k \boldsymbol{M}_t \left[\boldsymbol{J} \vec{W}_k\right] = \boldsymbol{J} \vec{W}_k, \tag{3.85}$$

and therefore $\boldsymbol{J} \vec{W}_k$ is an eigenvector of $\boldsymbol{M}_t$ with eigenvalue μ_k^{-1}. Now we know that given an eigenvalue μ_k, there must be eigenvalues μ_k^* and μ_k^{-1}. But since we have an eigenvalue μ_k^{-1}, we necessarily have its complex conjugate eigenvalue μ_k^{-1*}. On the real axis eigenvalues come in pairs μ_k and μ_k^{-1}; on the unit circle they also come in pairs μ_k, μ_k^* (see figure 3.22). The various possible cases transform into one another at *bifurcations* and *collisions* of eigenvalues. For the case of two phase space dimensions (corresponding to a one-dimensional driven system such as the parametric oscillator) a collision is shown in figure 3.23. The fact that μ and $1/\mu$ are both eigenvalues means the characteristic polynomial is *reflexive*—that is,

$$P(\mu) = \mu^{2n} P(1/\mu). \tag{3.86}$$

LYAPUNOV EXPONENTS

In general it may happen that the trajectories near a reference (or "guiding") trajectory may separate exponentially fast from it—that is, the distance $d_t = \|\delta \boldsymbol{X}_t\|$ increases exponentially with time. To follow and document this deviation, the initial deviation needs to be so small that the current deviation is still small with respect to nonlinearities in the Hamiltonian, in spite of the exponential separation. (This means that the matrix $\boldsymbol{M}_t$ propagates the initial small deviation $\delta \boldsymbol{X}_0$ with small error. There are fixes for this that involve gathering the separating trajectory and restarting the computation.) The maximal *Lyapunov exponent* λ is defined as

$$\lambda = \lim_{t\to\infty} \lim_{\delta \boldsymbol{X}_0 \to 0} \frac{1}{t} \ln \left[\frac{|\delta \boldsymbol{X}_t|}{|\delta \boldsymbol{X}_0|} \right], \tag{3.87}$$

where $\delta \boldsymbol{X}_0$ is an initial deviation vector. This definition works for "almost all" initial $\delta \boldsymbol{X}_0$ in the technical sense. Both limits are required for this definition to make sense: the initial deviation $\boldsymbol{X}_0$ must go to zero so that the pair of trajectories never separate very far, invalidating the linearization assumption. Time must go to infinity to ensure that all the phase space that is going to be explored *is* explored. There are some very special $\delta \boldsymbol{X}_0$ for which one will get 0 for the limit in equation 3.87, as we will see shortly. There are $2N$ Lyapunov exponents for N degrees of freedom, but the one with fastest growth will dominate at large times.

NEUTRAL DIRECTIONS

The vector along the flow is $\vec{\xi} = \boldsymbol{J} \cdot \vec{\nabla} H(\boldsymbol{X}_t)$. Specifically, in two degrees of freedom, it is

$$\vec{\xi} = \begin{pmatrix} 0 & 0 & 0 & 1 \\ 0 & 0 & 1 & 0 \\ 0 & -1 & 0 & 0 \\ -1 & 0 & 0 & 0 \end{pmatrix} \begin{pmatrix} \dfrac{\partial H}{\partial x} \\ \dfrac{\partial H}{\partial y} \\ \dfrac{\partial H}{\partial p_x} \\ \dfrac{\partial H}{\partial p_y} \end{pmatrix} = \begin{pmatrix} \dfrac{\partial H}{\partial p_x} = \dot{x} \\ \dfrac{\partial H}{\partial p_y} = \dot{y} \\ -\dfrac{\partial H}{\partial x} = \dot{p}_x \\ -\dfrac{\partial H}{\partial y} = \dot{p}_y \end{pmatrix}. \tag{3.88}$$

This is a special direction, the path of the trajectory through phase space. We consider a small deviation $\delta \boldsymbol{X}_0$ given by an initial condition $\boldsymbol{X}_0$ and its evolution for a small time τ:

$$\delta \boldsymbol{X}_0 \equiv \boldsymbol{X}_\tau - \boldsymbol{X}_0, \tag{3.89}$$

$$d_\tau(t) \equiv \|\boldsymbol{X}_{t+\tau} - \boldsymbol{X}_t\|. \tag{3.90}$$

Then

$$d_\tau(t) = \|\boldsymbol{X}_t + \boldsymbol{J} \cdot \vec{\nabla} H[\boldsymbol{X}_t]\tau - \boldsymbol{X}_t\| = \|\boldsymbol{J} \cdot \vec{\nabla} H[\boldsymbol{X}_t]\|\tau. \tag{3.91}$$

$\boldsymbol{J} \cdot \vec{\nabla} H[\boldsymbol{X}_t]$ is the flow vector at position $\boldsymbol{X}_t$; its length is just a local property of the space.

Is there a canonically conjugate coordinate to $\boldsymbol{J} \cdot \vec{\nabla} H[\boldsymbol{X}_t]$? There is, and it has a compelling interpretation. Indeed $\boldsymbol{J} \cdot \vec{\nabla} H[\boldsymbol{X}_t]$ corresponds to time variation, and its conjugate coordinate is expected to correspond to energy variation. We use $\boldsymbol{J}$ for writing the Poisson bracket $\{A, B\}$ as

$$\{A, B\} = (\vec{\nabla} A, \boldsymbol{J} \cdot \vec{\nabla} B), \tag{3.92}$$

where (,) is the inner product. Thus, the Poisson bracket of A and B becomes the skew (because of the intervention of $\boldsymbol{J}$) inner product of their gradient vectors, $\vec{\nabla} A$ and $\vec{\nabla} B$. We normalize $\boldsymbol{J} \cdot \vec{\nabla} H[\boldsymbol{X}_t]$ as

$$\hat{\xi} \equiv \frac{\boldsymbol{J} \cdot \vec{\nabla} H[\boldsymbol{X}_t]}{\|\nabla H[\boldsymbol{X}_t]\|}, \tag{3.93}$$

and define the normalized vector $\hat{\eta} = \vec{\nabla} H[\boldsymbol{X}_t]/\|\nabla H[\boldsymbol{X}_t]\|$. $\hat{\eta}$ is the gradient of the Hamiltonian (that is, direction of maximum change in energy). It is a conjugate variable to $\hat{\xi}$. The Poisson bracket of these two vectors is

$$\{\vec{\xi}, \vec{\eta}\} = \left(\frac{\boldsymbol{J} \cdot \vec{\nabla} H[\boldsymbol{X}_t]}{\|\nabla H[\boldsymbol{X}_t]\|}, \frac{\boldsymbol{J} \cdot \vec{\nabla} H[\boldsymbol{X}_t]}{\|\nabla H[\boldsymbol{X}_t]\|} \right) = \frac{\|\nabla H[\boldsymbol{X}_t]\|^2}{\|\nabla H[\boldsymbol{X}_t]\|^2} = 1, \tag{3.94}$$

as befits a conjugate pair, just like like x and p_x—$\{x, p_x\} = 1$. That is, the motion along the orbit and the energy variation are canonically conjugate, if both are properly normalized.

3.9 Motion near a Periodic Orbit

We have just seen (section 3.7) that the addition of an arbitrary perturbation generally breaks up an invariant curve on a surface of section into an island chain with an even number $2s$ of fixed points ("periodic orbits") that return to themselves after s penetrations of the surface (or after s iterations if we are considering an area preserving point-to-point mapping of the plane into itself). Following the flow lines revealed that half of these fixed points are hyperbolic in their vicinity, and "unstable." Half are elliptic, and therefore stable against small initial deviations away from the fixed point (the future intersections do not move very far away).

Consider points given by small deviations δp and δq from a fixed point at p_0 and q_0, that without loss of generality we take to be at $(p_0, q_0) = (0, 0)$ (figure 3.24). For a smooth system, the point will return to the same vicinity after s iterations.

We characterize the new point by the stability matrix,

$$\delta \boldsymbol{X}_t = \begin{pmatrix} \delta p_t \\ \delta q_t \end{pmatrix} = \boldsymbol{M} \cdot \delta \boldsymbol{X}_0 = \begin{pmatrix} m_{11} & m_{12} \\ m_{21} & m_{22} \end{pmatrix} \begin{pmatrix} \delta p_0 \\ \delta q_0 \end{pmatrix}, \tag{3.95}$$

where $m_{11} = \partial p_s/\partial p|_{p_0,q_0}$, $m_{12} = \partial p_s/\partial q|_{p_0,q_0}$, $m_{21} = \partial q_s/\partial p|_{p_0,q_0}$, and $m_{22} = \partial q_s/\partial q|_{p_0,q_0}$. Figure 3.24 shows the small rectangle defined by its corners at $(0, 0)$, $(\delta p, 0)$, and $(0, \delta q)$, together with the parallelogram it maps into after s iterations.

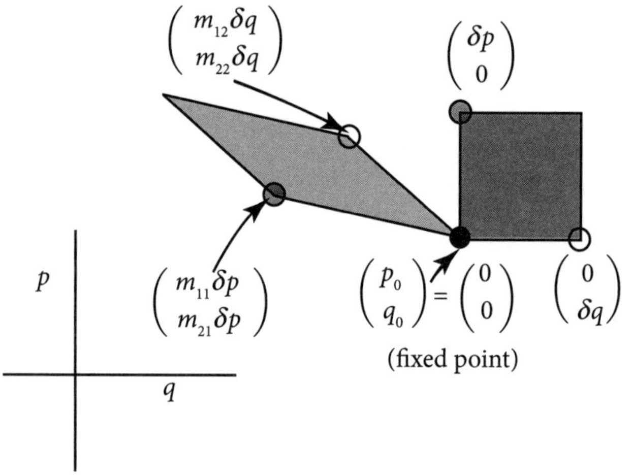

Figure 3.24. Evolution in phase space near the fixed point (p_0, q_0), taken to be at (0, 0).

The mapping of the rectangle into a parallelogram is a result of the linearity of the final conditions as a function of initial conditions near the periodic orbit, valid for sufficiently small δp and δq. The area of the parallelogram is given by the cross product of the two vectors in the plane defined by the periodic orbit and the mapping of the points $(\delta p, 0)$ and $(0, \delta q)$. This is $(m_{11}m_{22} - m_{12}m_{21})\delta p\delta q = \delta p\delta q$, since the map is area preserving, as already established. Therefore,

$$\det(\boldsymbol{M}) = \begin{vmatrix} m_{11} & m_{12} \\ m_{21} & m_{22} \end{vmatrix} = 1. \tag{3.96}$$

That is, the stability matrix $\boldsymbol{M}$ has unit determinant, required by area preservation of the mapping. This result generalizes to volume preservation under Hamiltonian dynamics requiring unit determinant of the $2N$-dimensional stability matrix for N degrees of freedom, as discussed presently.

We come now to the important topic of classical periodic orbits. Since a periodic orbit closes on itself, the stability matrix acquires some new and useful properties.

STABILITY MATRIX OF PERIODIC ORBITS: EIGENVALUES AND EIGENVECTORS

A periodic orbit with period τ traces out a closed path in phase space. The stability matrix elements for one period will depend on where we put a surface of section to intersect it, but the eigenvalues of the stability matrix do not. For two degrees of freedom, since $\det \boldsymbol{M}_t(\tau) = 1$, the eigenvalues of $\boldsymbol{M}_t(\tau) \equiv \mathcal{M}_\tau$ can be written

$$\mu_\pm = \frac{1}{2}\left[\mathrm{Tr}[\mathcal{M}_\tau] \pm \sqrt{\mathrm{Tr}[\mathcal{M}_\tau]^2 - 4}\right]. \tag{3.97}$$

If $|\,\mathrm{Tr}[\mathcal{M}_\tau]| > 2$, then the eigenvalues are real and can be written (since $\mu_+\mu_- = 1$) either as

$$\mu_\pm = \exp[\pm\lambda\tau] \tag{3.98}$$

or

$$\mu_\pm = -\exp[\pm\lambda\tau], \tag{3.99}$$

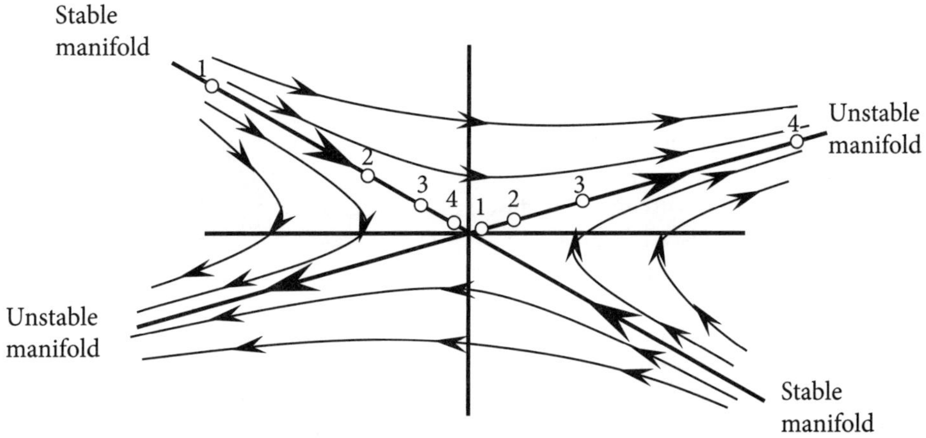

Figure 3.25. Structure of phase space near an unstable periodic orbit analyzed on a surface of section. If a given trajectory starts at the point labeled 1 on the unstable manifold, it penetrates the surface of section once each period, appearing next at 2 on the unstable manifold, and so on. If a given trajectory starts at the point labeled 1 on the stable manifold, it appears next at 2 on the stable manifold, and so on.

($\lambda > 0$). If $|\,\mathrm{Tr}[\mathcal{M}_\tau]| < 2$, then the eigenvalues are complex and can be written

$$\mu_\pm = \exp[\pm i\omega\tau]. \tag{3.100}$$

In the chaotic regime, all periodic orbits are of the unstable type, corresponding to cases 3.98 (hyperbolic) or 3.99 (hyperbolic with reflection). The eigenvectors $\boldsymbol{v}_\pm$ of $\mathcal{M}_\tau$ obey (in the hyperbolic case)

$$\mathcal{M}_\tau \boldsymbol{v}_\pm = \exp[\pm\lambda\tau]\boldsymbol{v}_\pm. \tag{3.101}$$

This means any point along the line $(x, p_x) = \eta\boldsymbol{v}_+$, where η is a proportionality constant, remains on the same line when iterated by $\mathcal{M}_\tau$, becoming $\eta\exp[\lambda\tau]\boldsymbol{v}_+$ after one period—that is, departing exponentially from the periodic orbit at (0, 0). That is, $\boldsymbol{v}_+$ lies along the initial direction of the unstable invariant manifold leading from the periodic orbit. Likewise, any point along the line $(x, p_x) = \eta\boldsymbol{v}_-$, where η is a proportionality constant, lies along the initial direction of the stable invariant manifold leading from the periodic orbit, becoming $\eta\exp[-\lambda\tau]\boldsymbol{v}_+$—that is, approaching exponentially to the periodic orbit at (0, 0). The manifolds are invariant because they map into themselves after a period. Individual points on the manifolds hop each iteration, landing back somewhere on the line, but the line remains. The manifolds bend and curl without intersecting themselves (but the stable manifold freely intersects the unstable one). A point on the unstable manifold that happens to land on the stable manifold at a point of intersection of the two is a homoclinic orbit, and will forever approach the unstable periodic orbit in the future.

An arbitrary point (x, p_x) near (0, 0) will lie on neither invariant manifold, and have a component of both eigenvectors. The unstable component of the coordinate of the point will cause it to soon recede exponentially away from the fixed point. It cannot recede forever if the phase space is finite, and will make a close approach to the periodic point at some time in the future, and then later again, and so on. The motion in the vicinity of the unstable periodic orbit is depicted in figure 3.25.

Only the stroboscopic progress along the manifolds is seen in the surface of section. The stable and unstable manifolds for a given $\mathcal{M}_\tau$ are a set of points in phase space mapping back onto the manifold after after any number of intersections with the surface of section.

In chaotic systems, periodic orbits are everywhere dense. Anywhere a trajectory is launched there are infinitely many periodic orbits nearby. Still, the measure of such orbits is zero, and no truly random choice of initial conditions will ever be a periodic orbit. The situation is analogous and very closely related to the measure and the construction of reals among all the rationals. Most of the periodic orbits are extremely long; a few are short. The number of periodic orbits grows exponentially with their length in a chaotic system. The exponential parameter governing the rate of increase is proportional to the topological entropy. And what is the *topological entropy*? It is a nonnegative real number that is a measure of the complexity of the system: the average amount of information (bytes if you wish) per iteration required to describe long iterations of the map. It would seem to be strongly related to the Lyapunov exponent.

3.10 Classical Baker's Map

The baker's map is simple and efficiently instructive about chaos, periodic orbits, homoclinic orbits (soon to be introduced), manifolds, and more. It is a bona fide fully chaotic system. It gets its name honestly, being very close to what happens in stretching and folding certain types of bread dough in a bakery. But a better name might have been the samurai map or katana map, because a very similar process is used to produce the famous katana, with a blade possessing over 30,000 fine layers of steel. While being forged, the sword ingot is pounded to twice its length with the width kept constant. Now thinner, it is cut into two equal pieces and folded on itself, becoming the original thickness and length, with two layers. The process of pounding out and folding repeats 15 times, giving 32,768 internal layers, accounting for the sword's combination of strength and flexibility. Different kinds of steel were used on one side of the ingot, which became the sharpened blade and were not mixed during the stretching and folding process. Incredibly high tech and ingenious. The samurai's foreign enemies, if in possession of a broken katana, might melt it down and pound out a new sword—without the folding and layering. The puzzling result was an inferior sword.

Consider a square ingot of metal, hammered and height reduced by a factor of 2 and lengthened by the same factor in the x direction, and cut symmetrically into two pieces at the middle. The right half is then placed (without inverting it) atop the left half (see figure 3.26).

A mathematical description of the evolution of a point on the original ingot is a chaotic map, as we now discuss. The map has a beautiful and simple representation in terms of binary numbers, a so-called symbolic dynamics. The iteration corresponds to the displacement of the decimal point to the right along an infinitely long binary sequence (although the katana map is a variation on this with 180-degree rotation of the cut piece before it is placed on top). To the right of the decimal point lies the address of the q coordinate of the point, and to the left of the decimal point lies the address of the p coordinate of the point, read backward.

The map is area preserving, taking the unit square into itself. The lower left-hand corner of the square has the address $(\ldots\,\overline{000.000}\,\ldots)$—that is, $(p, q) = (0, 0)$ in decimal notation; the upper left-hand corner of the square has the address $(\ldots\,\overline{111}.\overline{000}\,\ldots)$—that is, $(p, q) = (1, 0)$. The upper right-hand corner of the square

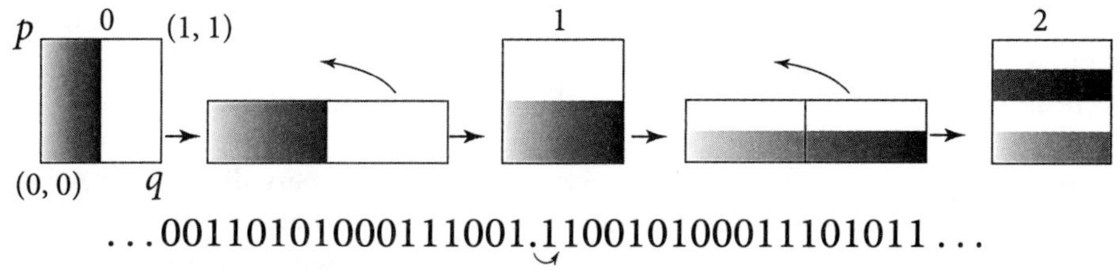

Figure 3.26. The unit square as it evolves for two iterations of the baker's map. (Bottom row) The p coordinate of a point in the unit square resides to the left of the decimal point, read backward, and the q coordinate is to the right, read normally. One step of the map changes the address of a point by shifting the decimal point to the right one position, as shown.

has the address $(\ldots\overline{111}.\overline{111}\ldots)$, $(p, q) = (1, 1)$. Finally, the lower right-hand corner of the square has the address $(\ldots\overline{000}.\overline{111}\ldots)$, $(p, q) = (0, 1)$.

The dynamics of one iteration of the map is exactly the shift of the decimal point one position to the right, giving the new p and q coordinates. Thus $(\ldots\overline{000}.\overline{000}\ldots)$ and $(\ldots\overline{111}.\overline{111}\ldots)$ are clearly fixed points or periodic orbits with period 1. The point $(\ldots\overline{111}.\overline{000}\ldots)$ is not a fixed point; the iteration of this point approaches the fixed point $(\ldots\overline{000}.\overline{000}\ldots)$ exponentially. Earlier, this point departed from the vicinity of the fixed point $(\ldots\overline{111}.\overline{111}\ldots)$.

The location $(\ldots\overline{001}.\overline{001}\ldots)$, is also a periodic orbit, with period 3. It lies at $p = 0.5714282\ldots$, $q = 0.14285707\ldots$, and hops to two new positions before returning exactly, on the third try.

Now consider

$$\ldots\overline{100}1110100011101011.111001110010\overline{100}\ldots.$$

One can see that in the distant past and in the distant future, these trajectory coordinates will be very close (but not exactly, because of errors in distant binary points) to $(\ldots\overline{100}.\overline{100}\ldots)$. That is, the trajectory left the periodic orbit of period 3, getting farther away exponentially, starting from a very small displacement. It later reapproaches the same periodic orbit, getting exponentially closer with each iteration, never quite reaching it. This is a homoclinic orbit.

It will not surprise the reader that

$$\ldots\overline{01}0001110101.1110011100101\overline{001}\ldots,$$

which left the vicinity of the period 2 orbit, has wandered and will wander around for a while, and will then approach the period 3 orbit for the rest of time, is called a heteroclinic orbit. The point $(\ldots\overline{111}.\overline{000}\ldots)$ just discussed also lies on a heteroclinic orbit.

Let a given finite sequence of consecutive 0's and 1's be called ν. Orbits in stable manifolds emanating from the periodic orbit $\ldots\overline{\nu}.\overline{\nu}\ldots$ can be written (anything $\nu\nu\nu\ldots$); the stable manifold itself is written as the set of points (everything $\nu\nu\nu\ldots$). Similarly, orbits in unstable manifolds emanating from the periodic orbit $\ldots\overline{\nu}.\overline{\nu}\ldots$ can be written ($\ldots\nu\nu\nu$ anything); the unstable manifold itself is written as the set of points ($\ldots\nu\nu\nu$ everything).

Homoclinic and heteroclinic orbits are rare denizens of the stable and unstable manifolds (but far more numerous than periodic orbits). Homoclinic orbits leave the near vicinity of a periodic orbit on the unstable manifold and lie on some of the intersections between the stable and unstable manifolds. By virtue of the orbit attaching to a stable manifold, it hops its way back toward the periodic orbit, eventually approaching it ever more closely for the rest of time, in the final, everlasting stage. It is more accurate to say that the homoclinic orbit on the unstable manifold was always attached to (perhaps the far reaches of) the stable manifold, since any given intersection between the stable and unstable manifolds can be iterated backward in time and will carry the intersection with it.

In the baker's map, every nearby pair of orbits, including the cases of one of them being a periodic orbit, shares the property that a small deviation in their q coordinates will grow exponentially, increasing by a factor of 2 each iteration, and small deviations in p coordinates will diminish by the same factor of 2. This is clear from the binary construction and from the physical nature of the map. The q and p axes are the directions of the unstable and stable axes near any fixed point. We have also just shown that stable and unstable axes flow in those same directions along with any arbitrary trajectory, viewed as a moving reference point.

The factor of 2 exponential increase in distance between almost all nearby orbits in the unit square means that the Lyapunov exponent λ for the baker's map is $\exp[\lambda] = 2$ or $\lambda = \log 2$.

The homoclinic and heteroclinic orbits can be classified as "primary," secondary, and so on. A primary homoclinic orbit approaches the periodic orbit on the primary branch of the stable manifold on its first attempt to revisit the vicinity of the periodic orbit. It is caught, so to speak, by the periodic orbit on the first pass. On the other hand, the orbit might pass relatively close to the periodic orbit without lying (yet) on the first part of its stable branch. If it attaches to that part on the second pass, it is a secondary homoclinic orbit, and so on. A point on a primary homoclinic orbit is $(\overline{01}1.\overline{01})$ (one "error" in the binary sequence) and a point on a secondary homoclinic orbit is $(\overline{01}11.\overline{01})$.

We have been considering some very special trajectories, with infinitely long repeating segments. Now the imagination can fearlessly run wild with other schemes for sequences of 0's and 1's. For example, think of all the possibilities of random binary numbers on both sides of the decimal point! This is the "typical" orbit, a type of orbit that characterizes almost all of the points in the unit square, like the irrationals among the rationals—that is, they have measure 1. Within any truly random sequence there must be infinitely many places with the property

$$\ldots \text{ anything } 01010101010101010.1 \text{ anything},$$

that is, an orbit that emerges from the chaotic sea and becomes "almost" a period 2 orbit for a few iterations, and then reenters the chaotic sea. A "random sequence" must embed infinitely many, even very long such sequences. It is easy to see from this that the amount of time that a random orbit happens to spend in N consecutive iterations near a given periodic orbit of period P diminishes as $2^{-N \cdot P}$.

It is clear from the binary representation that there are only two period 1 periodic orbits, only one period 2 orbit (but it visits two places of course), and two period

3 orbits—namely, $(\ldots\overline{001}.\overline{001}\ldots)$ and $(\ldots\overline{011}.\overline{011}\ldots)$. A little thought reveals that the number of periodic orbits of period N grows exponentially with N.

The baker's map has been studied quantum mechanically and semiclassically. We will not be able to pursue this here.

3.11 (Classical) Trace of a Periodic Orbit

The trace of a periodic orbit is an interesting quantity that plays a prominent role in the semiclassical theory of quantum eigenvalues (that is, the Gutzwiller "trace formula"). Here, we do a purely classical analysis. We define a delta function trajectory and its overlap with itself. This vanishes except when the initial point and mapped point (working on a surface of section or a map) are the same. The rub is that coincidence of the initial and final point is of course a matter of the initial conditions; you have to be dead on the periodic orbit to get coincidence. But what is interesting is how the final point sweeps past the periodic coincidence or periodic orbit as the initial conditions are varied near the periodic point. Sweeping fast—that is, the trajectory is very sensitive to initial conditions—would imply a smaller trace after integrating over initial conditions. This is clearly a measure of the instability of the orbit, with classical (and when we get to it, semiclassical) implications.

We integrate (trace) $\delta(p_0 - p_1(p_0, q_0))\ \delta(q_0 - q_1(p_0, q_0))$, and over initial points (p_0, q_0):

$$\begin{aligned}
\mathrm{Tr} &= \int dp_0 dq_0\, \delta(p_0 - p_1(p_0, q_0))\delta(q_0 - q_1(p_0, q_0)), \\
&= \sum_k \int \frac{dq_0}{\left|1 - \left(\frac{\partial p_1}{\partial p_0}\right)_{q_0}\right|}\delta(q_0 - q_1(p_0, q_0)), \\
&= \sum_k \frac{1}{\left|1 - \left(\frac{\partial p_1}{\partial p_0}\right)_{q_0}\right| \left|1 - \left(\frac{\partial q_1}{\partial q_0}\right)_{p_0} - \left(\frac{\partial q_1}{\partial p_0}\right)_{q_0}\frac{\partial p_0}{\partial q_0}\right|}.
\end{aligned} \tag{3.102}$$

Now, since $p_0 = p_1(p_0, q_0)$, we have

$$dp_0 = \left(\frac{\partial p_1}{\partial p_0}\right)_{q_0} dp_0 + \left(\frac{\partial p_1}{\partial q_0}\right)_{p_0} dq_0, \tag{3.103}$$

or

$$\frac{\partial p_0}{\partial q_0} = \frac{\left(\frac{\partial p_1}{\partial q_0}\right)_{p_0}}{\left(1 - \left(\frac{\partial p_1}{\partial p_0}\right)_{q_0}\right)}. \tag{3.104}$$

Thus,

$$\mathrm{Tr}_k = \sum_k \frac{1}{\left|1-\left(\frac{\partial p_1}{\partial p_0}\right)_{q_0}\right| \left|1-\left(\frac{\partial q_1}{\partial q_0}\right)_{p_0} - \left(\frac{\partial q_1}{\partial p_0}\right)_{q_0} \frac{\left(\frac{\partial p_1}{\partial q_0}\right)_{p_0}}{\left(1-\left(\frac{\partial p_1}{\partial p_0}\right)_{q_0}\right)}\right|},$$

$$= \sum_k \frac{1}{\left|\left(1-\left(\frac{\partial p_1}{\partial p_0}\right)_{q_0}\right)\left(1-\left(\frac{\partial q_1}{\partial q_0}\right)_{p_0}\right) - \left(\frac{\partial q_1}{\partial p_0}\right)_{q_0}\left(\frac{\partial p_1}{\partial q_0}\right)_{p_0}\right|},$$

$$= \sum_k \frac{1}{|1-\mathrm{Tr}(M_k)+\det(M_k)|} = \sum_k \frac{1}{|\det[M_k - I]|}, \tag{3.105}$$

where

$$M_k = \begin{pmatrix} \left(\frac{\partial p_1}{\partial p_0}\right)_{q_0} & \left(\frac{\partial p_1}{\partial q_0}\right)_{p_0} \\ \left(\frac{\partial q_1}{\partial p_0}\right)_{q_0} & \left(\frac{\partial q_1}{\partial q_0}\right)_{p_0} \end{pmatrix}, \tag{3.106}$$

and

$$\det(M_k) = 1, \tag{3.107}$$

since M_k gives an area-preserving transformation. The sum is over all periodic orbits of the map; we pick up only the period 1 orbits for one iteration of the map, but can obviously go to two iterations, and so on. The contribution for one periodic orbit is

$$\mathrm{Tr}_k = \frac{1}{|\det[M_k - I]|}, \tag{3.108}$$

where M_k is the stability matrix for that periodic orbit. This is a close analog of the famous semiclassical *Gutzwiller trace formula* of classically chaotic systems (see figure 7.1 and section 7.7, later).

3.12 Survival of (Most of) the Tori under Weak Interactions

The Poincaré-Birkhoff construction shows that rational invariant curves (section 3.7) can break up into resonance island chains under a perturbation. Rationals of the form r/s are measure 0 as compared to the irrationals that surround them. On the other hand, the rationals are dense on the number line, so why doesn't a small perturbation fill phase space with resonance islands? (Or worse, with chaos, since as we shall see chaos results from overlapping resonance islands.)

The answer lies in the exponential decrease in the resonance widths with the resonance order s. Essentially, the resonances shrink faster than they proliferate, saving phase space from an immediate chaotic catastrophe. Immediate chaos was once considered the probable fate of a system of several degrees of freedom as soon as nontrivial perturbations were switched on. In the famous pioneering work by Fermi,

Pasta, and Ulam at Los Alamos after World War II, on a many-particle nonlinear system defined by

$$\dot{p}_i = q_{i+1} + q_{i-1} - 2q_i + \beta[(q_{i+1} - q_i)^n - (q_i - q_{i-1})^n], \tag{3.109}$$

where n was taken to be 2 or 3, there was very little phase space mixing even after long runs. Ulam stated later: "The results of the calculation... were quite surprising to Fermi. He expressed to me the opinion that they really constituted a little discovery in that they provided intimations that the present beliefs in the universality of 'mixing and thermalization' in non-linear systems may not always be justified."

An estimate can be made of how much of the area between two unperturbed invariant lines, say a period 2 and period 3 line, survives intact (that is, does not end up within a resonance zone or, worse, a chaotic zone). The Chirikov resonance overlap criterion states that if two different resonance zones affect the same phase space region—that is, they have overlapping resonance widths. (see discussion surrounding figure 4.3, later—then chaos ensues there. The qualitative reason is that the trajectories get "confused" as to which resonance they should belong to. The Chirikov criterion goes back to 1959; more recent and readable accounts are found in references [11] and [12].

Returning to our estimate, we seek to show the situation as it appears to be in figure 3.19, earlier—that is, well separated, narrow resonance zones leaving much of phase space only weakly disturbed—is common in typical systems. The analysis serves also to assure us that chaos has not run amok on scales we cannot see easily in a system like that in figure 3.19. All rational lines m/n are subject to breaking up into n resonance islands with separatrices in between. As the order n goes up, the resonance width tends to decrease as $\sim \exp[-\alpha n]$. There are certainly exceptions, but smooth analytic functions serving as perturbations have Fourier components that fall off exponentially. Between the resonances 1/2 and 2/3 for example, there are higher order resonances 3/5, 4/7, 5/8, 5/9, 6/10*, 6/11, 7/11, 7/12, 7/13, 8/13, 8/14*, 9/14, 8/15, 9/15, ... The asterisks represent repeated rational ratios. The 8/14 resonance, for example, has 14 islands, and so is different than 4/7, but the islands are typically destroyed by the lower order 4/7 island chain.

We could apply a correction, using a theorem that only $6/\pi^2 = 61\%$ of random rational integer ratios involve co-prime integers. We will remain on the conservative side and keep all the rational integers in the interval. As the denominators n grow larger, eventually there will be $(2/3 - 1/2) \times \mathrm{n}$ rational integer ratio resonances of order n. For example, if $n = 150$, then $m = 76$ to 99, inclusive, gives a ratio between 1/2 and 2/3. This is a linear growth of resonances with order, but the widths are dying exponentially.

The area contained between 1/2 and 2/3 is determined by ratios of the rate of change of the frequencies with action, $\omega_i = \partial H(\boldsymbol{J})/\partial \mathrm{J}_i$. Two relevant factors leading to chaos appear in the prefactor $\sqrt{\frac{V}{G}}$ in the formula for the resonance width (see equation 4.74)

$$W_n \propto \sqrt{\frac{V}{G}}\, n e^{-\alpha n}, \tag{3.110}$$

where $Vne^{-\alpha n}$ is the Fourier strength for the smallest relevant n, and G controls how fast the frequency ratios change with action. The area in phase space to

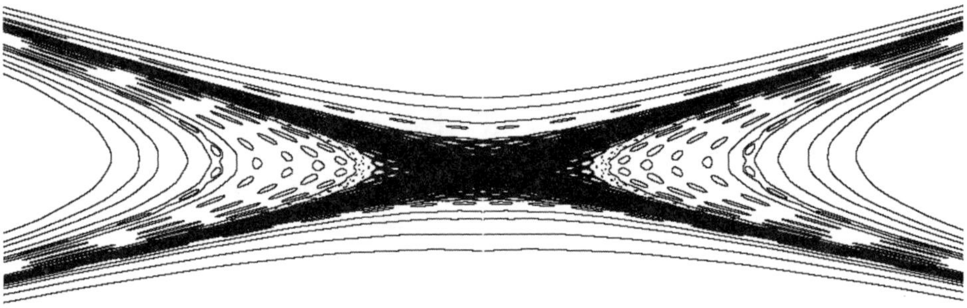

Figure 3.27. Highly magnified look at the zone near the unstable fixed point. The figure spans 0.004 in p, and 0.0125 in θ. The dense black region is the result of iteration of a single initial point.

accommodate the resonances goes as $1/G$, and the resonance widths go only as $1/\sqrt{G}$ (see section 4.5). Thus, both V and G pose a threat, causing resonance overlap as their ratio becomes large.

If the resonances are not overlapping at low order, then with resonance zone spacing dropping as $1/n$, but resonance width dropping as $ne^{-\alpha n}$, the ratio of clear area to resonance disturbed area increases in zones with only high n present. That is, if we take a microscope to a region with no naked eye resonance, we will not find the now visible high-order resonances overlapping.

The standard map (figures 3.20 and 3.21, earlier) is a surrogate for a surface of section analysis of a two degree of freedom Hamiltonian system. The smooth lines running across the map then mimic the intersection of the surface of section with the invariant tori foliating phase space. When an isolated resonance zone forms due to the addition of a generic perturbation, a new zone of foliation develops with new invariant tori. Inside the resonance zone, the topology of these tori is different; outside, the nearby tori have deformed but retain the same topology.

A very magnified look at the region surrounding the unstable fixed point (as in figure 3.27) reveals trouble ahead: points initiated near the unstable fixed point fill a region densely under unending iteration, rather than a Lagrangian manifold (figure 3.27). Instead of joining smoothly, the stable and unstable manifolds appear to break up, although they actually remain smooth and join smoothly across the fixed point. The structures near the fixed point are fascinating and include the famous *homoclinic tangle*, developing myriads of oscillations and long, thin tendrils. We take up this subject in section 3.13.

3.13 Structure of Phase Space

To see how a homoclinic tangle might arise, and with it a zone of chaos near a formerly integrable unstable fixed point, we start with a one-dimensional cubic potential. It has a well on the left and a barrier in the middle, and heads downhill ever more steeply to the right of the barrier. (The potential is shown in black in frame 4 of figure 3.28.) Using ordinary classical mechanics, there is an unstable fixed point and a stable one to its left. The stable and unstable manifolds emanating from the unstable fixed point meet smoothly; there is just one continuous manifold leaving and smoothly returning to the fixed point.

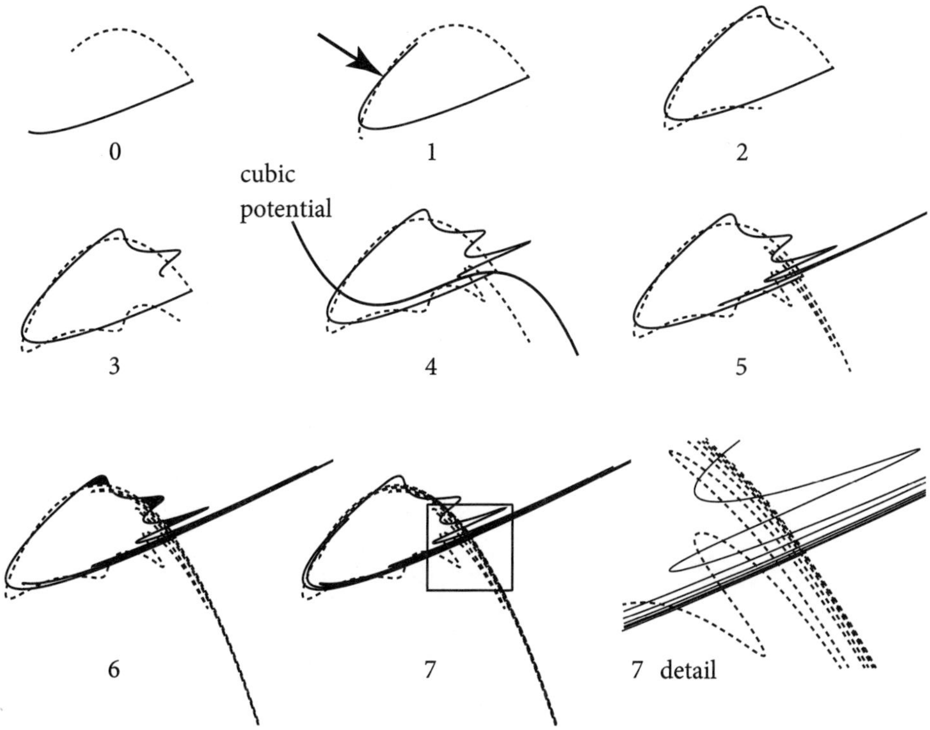

Figure 3.28. Starting at the upper left, a sequence of seven consecutive forward iterations of the unstable manifold (dashed lines) of the map of equation 3.111 for $\lambda = 1.451$ and time step of 1. Seven backward iterations of the section of stable manifold are shown as solid lines. The stable and unstable manifolds first cross in frame 1 as indicated by the arrow. Once a loop involving successive intersections of the stable and unstable manifolds develops, the area of that loop must be preserved under the map as it marches toward or away from the fixed point. This is highlighted in frame 6 for the unstable manifold. By frame 7, close inspection reveals a crossing of stable and unstable secondary "tendrils" or narrow, stretched homoclinic oscillations near the first intersection seen in frame 1. A detail of frame 7, giving a closer look at the developing homoclinic tangle, is shown at the lower right.

Not so if we turn the cubic potential problem into a map, by using area-preserving discrete time steps. The map starts with the Hamiltonian

$$H = \frac{p^2}{2m} + \lambda(q^2/2 - q^3/3), \tag{3.111}$$

and uses a backward time step of -1. If the potential is strong—that is, large λ—a unit time step gives a large "error" compared to continuous time dynamics. If λ is small enough, the dynamics is still nearly integrable in spite of the time step. Figure 3.28 shows a sequence of seven consecutive forward iterations of the unstable manifold, and backward iterations of the stable manifold (it becomes unstable for backward iterations). The iterations start at the top left (where the distribution is chosen to lie along the stable, solid, and unstable, dashed manifolds). An arrow shows the moment

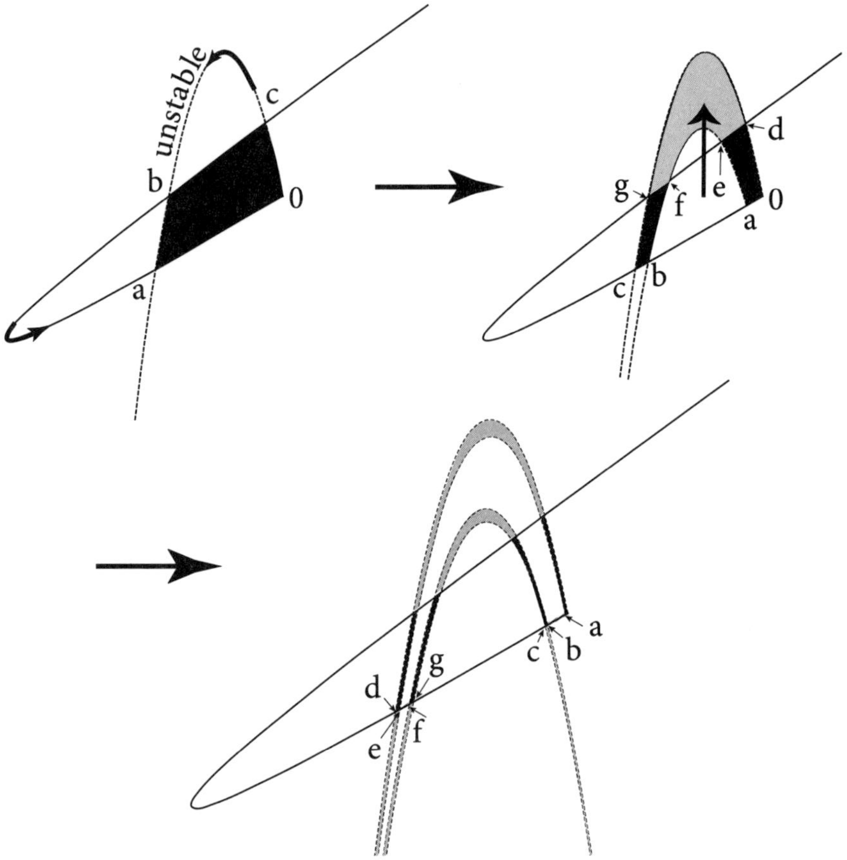

Figure 3.29. Successive iterations of the cubic map for $\lambda = 6.751$, starting at the upper left, where the stable and unstable manifolds are seen to intersect at a steep angle. In the next iteration (upper right), this map ejects much of the dark "inside" region by stretching and bending it (but in an area-preserving way) so that about 3/4 of it is placed irrevocably outside. Only two inside pillars remain, one at each end (still colored dark). This is a fully chaotic "Smale horseshoe" map, resulting in a zero-measure Cantor set (see Lichtenberg and Lieberman [12]) of zones remaining within the inside region after infinitely many iterations (see the "dust" beginning to form in the next iteration). The fixed point is labeled 0, and three intersections of the stable and unstable manifolds are labeled a, b, and c at the upper left. Since these points are on the stable manifold, they must approach the fixed point, as seen by their new positions in the upper right image after one more iteration. New intersections d, e, f, and g are labeled there, and their iterates are labeled on the somewhat magnified bottom figure. The areas are shaded as follows: area remaining "inside" remains dark; area having been expelled in the current iteration is lighter gray, and inside area emptied out by expulsion is white. The white emptied part of the inside region has the same area as the light gray expelled region (vertical arrow) outside, on the upper right.

the manifolds cross rather than meet smoothly, with irreversible consequences, including a homoclinic tangle and a chaotic zone near the fixed point (see figure 3.27). The lower right shows a detail of the developing homoclinic tangle ("homoclinic" because the tangle is gathering back upon the same place it started, rather than approaching another fixed point). The first intersection must be followed by infinitely many others, since the intersection point can be iterated forward or backward and therefore infinitely many repeats must lie along the stable and unstable axis as it approaches the fixed point (forever). The loop formed by the first intersection and the next one gives an area that has to be preserved in forward or backward iteration. As the intersection points accumulate ever closer together, the only way for the area to be preserved is for the shape to grow ever thinner and extended away from the intersection points; this is exactly what is happening in figure 3.28. A steeper intersection of the stable and unstable manifolds can lead to a full "Smale horseshoe", see figure 3.29.

Chapter 4

Classical Resonance Theory

Resonance tends to be introduced in restricted contexts, like the driven oscillator, but it applies much more broadly. The key to resonance is to ask what is happening to amplitudes, such as velocity amplitude (momentum), pressure amplitude (acoustics), or wavefunction amplitude (quantum mechanics). It is fair to say that what makes the world go round is the fact that energy and probability density go as the *square* of their amplitudes. Doubling the amplitude gives you four times the energy, or power, or probability.

Resonant buildup involves adding amplitude in phase with what is already there. The resonant addition does not have to be periodic. When pushing a child on a swing, resonant addition could be semi-sporadic, like pushing every so often but at the right time so as to increase velocity amplitude.

Resonance is a key concept in mechanics, be it classical or quantum mechanics. It will come up many times in this book. Here, we establish some important concepts, first mostly in a classical context. Later, resonance in the context of waves will become very important.

4.1 Paradigm System: Linearly Forced, Damped Oscillator

The canonical vehicle to introduce resonance in classical mechanics is the driven, damped harmonic oscillator. This is not as general as we might like (for example, a harmonic oscillator maintains the same frequency independent of its energy), but it does offer analytic results that are a starting point for more realistic systems.

Here, we will see that applying either very slow (adiabatic) or very fast periodic forces compared to the natural frequency of a system results in little change in its classical actions. In the adiabatic case the energy may change drastically during the time the force is applied, but since the action remains nearly constant, the energy returns very closely to its initial value if the slowly varying change finally reverses itself, returning the system to its original Hamiltonian.

We take the Hamiltonian to be

$$H_t = \frac{p^2}{2m} + \frac{1}{2}m\omega^2 x^2 - f(t)x \equiv H_0 - f(t)x, \tag{4.1}$$

where the varying force on the oscillator is $f(t)$. Suppose initially that the oscillator is at rest, and $f(-\infty) = 0$. Define

$$\alpha_t = \frac{1}{\sqrt{2m\omega}}(m\omega x_t + i p_t), \tag{4.2}$$

and also

$$\alpha_t^* = \frac{1}{\sqrt{2m\omega}}(m\omega x_t - i p_t). \tag{4.3}$$

The reader may want to define a generating function to accomplish the $(q, p) \to (\alpha, \alpha^*)$ canonical transformation, in the process deriving the equations of motion given here.

Now in terms of α, α^* we have

$$H_0 = \omega \alpha \alpha^*. \tag{4.4}$$

The parameters α_t and α_t^* obey

$$\dot{\alpha}_t = -i \frac{\partial H_0}{\partial \alpha^*} = -i\omega\alpha_t, \tag{4.5}$$

and

$$\dot{\alpha}_t^* = i \frac{\partial H_0}{\partial \alpha} = i\omega\alpha_t^*. \tag{4.6}$$

These equations are equivalent to the usual classical equations for x_t and p_t. The solutions are elementary:

$$\alpha_t = e^{-i\omega t}\alpha_0, \tag{4.7}$$

and

$$\alpha_t^* = e^{i\omega t}\alpha_0^*. \tag{4.8}$$

Including the driving term $-f(t) \cdot x$ gives:

$$H = \omega\alpha\alpha^* - \frac{1}{\sqrt{2m\omega}} f(t)(\alpha + \alpha^*). \tag{4.9}$$

The equation for α_t is now

$$\dot{\alpha}_t = -i \frac{\partial H}{\partial \alpha^*} = -i\omega\alpha_t + \frac{i}{\sqrt{2m\omega}} f(t). \tag{4.10}$$

This admits of a simple integration to give

$$\alpha_t = e^{-i\omega t}\alpha_0 + \frac{i}{\sqrt{2m\omega}} \int_{-\infty}^{t} e^{-i\omega(t-t')} f(t')\, dt', \tag{4.11}$$

and similarly for α^*. We see from equation (4.11) that the energy change of the oscillator rests in the Fourier transform of the time-dependent force $f(t)$ at the frequency ω of the oscillator. If the force or applied field is *resonant* with the oscillator, then the oscillator will be maximally disturbed for a given strength of the field. We should not neglect to emphasize the word *disturbed*—for example, if $|\alpha_0| \gg 0$, a strong resonant force with the right phasing could bring $|\alpha_t| \to 0$, indeed a big disturbance!

It is interesting to examine an oscillating-Gaussian pulse

$$f(t) = f_0 e^{-\Gamma^2 t^2} \exp[-i\omega_0 t]. \tag{4.12}$$

Then the second term in equation (4.11) at large times after the pulse has died down becomes

$$i f_0 \frac{\sqrt{\pi}}{\Gamma\sqrt{2m\omega}} e^{-\frac{(\omega-\omega_0)^2}{4\Gamma^2}}. \tag{4.13}$$

The energy absorbed if the system begins at rest is

$$\varepsilon = \omega|\alpha_f|^2 = |f_0|^2 \frac{\pi}{2m\Gamma^2} e^{-\frac{(\omega-\omega_0)^2}{2\Gamma^2}}. \tag{4.14}$$

As anticipated, the amount of energy the oscillator can absorb is limited only by the strength of the field, and the proximity of ω_0 to ω. $\omega_0 = \omega$ is the *resonance condition.*

ADDING FRICTION (DAMPING)

We now add friction, with a strength governed by a parameter λ, taking the equation of motion to be

$$\dot{\alpha}_t + (i\omega + \lambda)\alpha_t = \frac{i}{\sqrt{2m\omega}} f(t). \tag{4.15}$$

This has the solution

$$\alpha_t = e^{-i\omega t - \lambda t}\alpha_0 + \frac{i}{\sqrt{2m\omega}} \int_0^t e^{-i\omega(t-t') + \lambda(t'-t)} f(t')\, dt'. \tag{4.16}$$

The first term is a transient, dying out exponentially. Suppose we force sinusoidally—that is, $f(t) = f_0\, e^{-i\omega_0 t}$, then

$$\alpha_t = e^{-i\omega t - \lambda t}\alpha_0 + \frac{i f_0}{\sqrt{2m\omega}} e^{-i\omega t} \int_0^t e^{i(\omega-\omega_0)t' + \lambda(t'-t)}\, dt', \tag{4.17}$$

$$= e^{-i\omega t - \lambda t}\alpha_0 + \frac{i f_0}{2\sqrt{2m\omega}} e^{-i\omega t - \lambda t} \left[\frac{e^{i(\omega-\omega_0)t + \lambda t} - 1}{i(\omega - \omega_0) + \lambda} \right], \tag{4.18}$$

$$\to \frac{i f_0}{2\sqrt{2m\omega}} \left[\frac{e^{-i\omega_0 t}}{i(\omega - \omega_0) + \lambda} \right]. \tag{4.19}$$

Note that this oscillates strictly with the *drive* frequency ω_0! The energy absorbed per unit time will be just

$$\varepsilon = \omega|\alpha_t|^2 \to \frac{1}{8m}\frac{|f_0|^2}{(\omega-\omega_0)^2+\lambda^2}. \tag{4.20}$$

This corresponds to a resonance peak at $\omega - \omega_0$ whose height and width depends on the damping λ.

4.2 Slow External Time-Dependent Perturbations

We now consider both very slow and very fast time-dependent perturbations. If we make even a rather strong change to the Hamiltonian but do it very slowly, it might be that perturbation theory applies moment to moment, since there are only small changes to the Hamiltonian over many oscillations of the system, which is then a fast oscillation. There is a problem however if the system oscillation frequency (or a low-order integer combination of frequencies) becomes slow. We take the Hamiltonian to be, first thinking of η as constant,

$$H(p,q,\eta) = H_0 + \eta H_1. \tag{4.21}$$

Let (I,φ) be the action-angle variables appropriate to a fixed value η, through a generating function $F_2(q,I,\eta)$. Then, $H(p,q,\eta) = H(I,\eta)$. We now suppose time dependence enters through $\eta(t) = \epsilon t$, with the idea of ϵ small but t eventually being of order $1/\epsilon$. (Other slow time dependencies are of course also possible.) Let $F_2(q,I,\eta=\epsilon t)$ be the generating function that transforms from the initial (q,p) coordinates to the action-angle variables corresponding to the time t (and the parameter $\eta(t)$). Then,

$$K(I,\varphi,t) = H(I,\eta) + \epsilon\frac{\partial F_2}{\partial\eta} \tag{4.22}$$

has the Hamilton equations of motion

$$\begin{aligned} \dot I &= -\epsilon g(I,\varphi,\eta), \\ \dot\varphi &= \omega(I,\eta) + \epsilon f(I,\varphi,\eta), \end{aligned} \tag{4.23}$$

where

$$\begin{aligned} f(I,\varphi,\eta) &= \frac{\partial^2 F_2(q(I,\varphi),I,\eta)}{\partial I\partial\eta}, \\ g(I,\varphi,\eta) &= \frac{\partial^2 F_2(q(I,\varphi),I,\eta)}{\partial\eta\partial\varphi}. \end{aligned} \tag{4.24}$$

$f(I,\varphi,\eta)$ and $g(I,\varphi,\eta)$ are ripe for averaging, as in section (4.4), since φ cycles many times through 2π before η changes by much, if ϵ is small. Thus,

$$\begin{aligned} \dot I &= -\epsilon\bar g(I,\eta), \\ \dot\varphi &= \omega(I,\eta) + \epsilon\bar f(I,\eta). \end{aligned} \tag{4.25}$$

Now F_2 and, $\partial F_2/\partial\eta$ are manifestly periodic in φ, since φ is an angle variable:

$$F_2(q(I,\varphi), I, \eta) = \sum_n a_n(I,\varphi,\eta)e^{in\varphi}. \tag{4.26}$$

Then $g(I,\varphi,\eta)$ is not only periodic in φ, but averages to 0:

$$\bar{g} = \frac{1}{2\pi}\oint g(I,\varphi,\eta)\,d\varphi = 0, \tag{4.27}$$

and as a consequence, within the averaging we have done, the action

$$I = \text{const.} \tag{4.28}$$

This shows that the action I is an *adiabatic invariant*—namely, a quantity that doesn't change if we change the system very slowly (adiabatically).

Adiabatic motion in the phase plane means the system is undergoing many complete cycles (oscillations) before a change in the orbit is noticeable. Thus, the area enclosed by the orbit is well defined and approximately constant (exactly constant if we change the Hamiltonian infinitely slowly, unless we reach a point where $\omega = 0$).

Consider a harmonic oscillator with frequency ω_0, which we slowly increase to ω_1. Since the action I will be constant, the area enclosed by the orbit in phase space is constant. The energy of the oscillator will increase since $H = \omega I$ and since ω increases. If we slowly return the frequency back to ω_0, the action still remains constant and thus the energy will return to its old value (see figure 4.1).

4.3 Fast External Time-Dependent Perturbations

So far we have essentially used the system frequency $\omega(I)$ to provide a fast oscillation that averages out and nullifies static perturbations or slowly varying ones as far as $\dot{I}$ is concerned.

Now suppose we apply a sinusoidal perturbation $V_0(I,\varphi)\sin\omega' t$ with frequency $\omega' \gg \omega(I)$:

$$H = H_0(I) + V_0(I,\varphi)\sin\omega' t; \tag{4.29}$$

$\omega' \gg \omega = \partial H_0(I)/\partial I$. (More generally we can apply a nonperiodic potential $V(I,\varphi,t)$ such that all of its Fourier components are at high frequency compared to $\omega(I)$). We have

$$\begin{aligned}\dot{I} &= -\frac{\partial V_0}{\partial\varphi}\sin\omega' t,\\ \dot{\varphi} &= \omega(I) + \frac{\partial V_0}{\partial I}\sin\omega' t.\end{aligned} \tag{4.30}$$

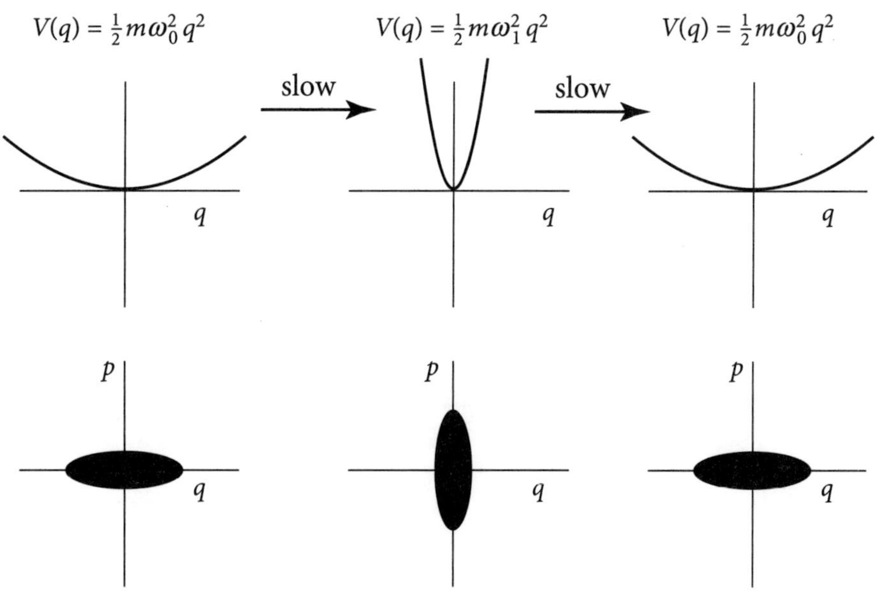

Figure 4.1. Adiabatic time evolution of a region of phase space shaded black up to a fixed energy in a harmonic oscillator (lower left). Under slow increase of the harmonic force constant, the black region at the middle is reached. It remains stationary if the force constant is not changed further. Every point in the initial region was propagated classically, oscillating as the potential changed. The new region necessarily has the same area by Liouville's theorem, under adiabatic or nonadiabatic change. However, the new area achieved adiabatically is bounded by a higher, fixed energy and fixed classical action; these properties are a consequence of the slow, adiabatic change. Reversing the slow change brings the distribution back to the starting region. *Adiabatic* here means the potential should not change by much in one period of the oscillator.

Averaging this over one period of the fast oscillation gives

$$\begin{aligned} \dot{I} &\approx -\int_0^{2\pi} \frac{\partial V_0}{\partial \varphi} \sin \omega' t \, dt = 0, \\ \dot{\varphi} &\approx \omega(I) + \int_0^{2\pi} \frac{\partial V_0}{\partial I} \sin \omega' t \, dt = \omega(I). \end{aligned} \tag{4.31}$$

In other words, the fast forcing does little to the system over one or more complete cycles.

These examples demonstrate the importance of *resonance* in classical mechanics. Large frequency disparities (far off resonance) mean action is approximately conserved, if the system frequencies do not vanish. Low-order resonances are a danger signal to perturbation theory, suggesting that the system will respond in a more drastic, nonperturbative way.

4.4 Averaging of Perturbations

Suppose we have found action-angle variables $(\vec{I}, \vec{\varphi})$ for a Hamiltonian system. Thus,

$$H_0 = H_0(\vec{I}) \tag{4.32}$$

is given and

$$\begin{aligned} \dot{I}_i &= -\frac{\partial H_0(\vec{I})}{\partial \varphi_i} = 0, \\ \dot{\varphi}_i &= \frac{\partial H_0(\vec{I})}{\partial I_i} = \omega_i(\vec{I}). \end{aligned} \tag{4.33}$$

Now we add a "small" term $\epsilon H_1(\vec{I}, \vec{\varphi})$ to H_0, $\epsilon \ll 1$, making

$$H = H_0(\vec{I}) + \epsilon H_1(\vec{I}, \vec{\varphi}). \tag{4.34}$$

The new Hamilton's equations of motion are

$$\begin{aligned} \dot{I}_i &= -\epsilon \frac{\partial H_1(\vec{I}, \vec{\varphi})}{\partial \varphi_i} \equiv \epsilon g_i(\vec{I}, \vec{\varphi}), \\ \dot{\varphi}_i &= \omega_i(\vec{I}) + \epsilon \frac{\partial H_1(\vec{I}, \vec{\varphi})}{\partial I_i}. \end{aligned} \tag{4.35}$$

In general this will be just as hard as any problem to solve exactly. Furthermore, there is no guarantee any set of actions exists. Importantly, the content of the celebrated Kolmolgorov-Arnold-Moser (KAM) theorem implies that the loss of good actions upon adding a small perturbation to an integrable system is rare; the conditions for the robustness of the actions are given in the KAM theorem (see sections 3.12 and 4.4). This is especially the case since we are supposing that "fast" perturbations means no low-order rational resonances with the system.

We now average over the angular dependence and work with the averaged system instead of the exact system. The idea behind the averaging is that $H_1(\vec{I}, \vec{\varphi})$ is necessarily a periodic function of the angles $\vec{\varphi}$, and $H_1(\vec{I}, \vec{\varphi})$ may quickly oscillate its influence on a given trajectory from positive to negative and back, and so on—that is, leave the actions nearly constant. The averaging leaves us with

$$\dot{I}_i \approx \epsilon \bar{g}_i(\vec{I}), \tag{4.36}$$

$$\dot{\varphi}_i \approx \omega_i(\vec{I}) + \epsilon \bar{w}_i(\vec{I}), \tag{4.37}$$

where

$$\bar{g}_i(\vec{I}) = -\left(\frac{1}{2\pi}\right)^N \int_0^{2\pi} \cdots \int_0^{2\pi} \frac{\partial H_1(\vec{I}, \vec{\varphi})}{\partial \varphi_i} \, d\varphi_1 \cdots d\varphi_N, \tag{4.38}$$

and

$$\bar{w}_i(\vec{I}) = \left(\frac{1}{2\pi}\right)^N \int_0^{2\pi} \cdots \int_0^{2\pi} \frac{\partial H_1(\vec{I}, \vec{\varphi})}{\partial I_i} \, d\varphi_1 \cdots d\varphi_N. \tag{4.39}$$

If we average the force,

$$\begin{aligned} \bar{g}_i(\vec{I}) &= -\left(\frac{1}{2\pi}\right)^N \int_0^{2\pi} \cdots \int_0^{2\pi} \frac{\partial H_1(\vec{I}, \vec{\varphi})}{\partial \varphi_i} \, d\varphi_1 \cdots d\varphi_N, \\ &\equiv -\left(\frac{1}{2\pi}\right) \int_0^{2\pi} \frac{\partial \tilde{H}_1(\vec{I}, \varphi_i)}{\partial \varphi_i} d\varphi_i, \\ &= 0, \end{aligned} \tag{4.40}$$

any constant $\vec{K} = \vec{0}$ term in the expansion

$$H_1(\vec{I}, \vec{\phi}) = \sum_{\vec{K}} h_{\vec{K}}(\vec{I}) e^{i\vec{K}\cdot\vec{\phi}} \tag{4.41}$$

vanishes because a derivative is taken with respect to angle, and all $\vec{K} \neq \vec{0}$ terms average out. Thus, in the averaged approximation,

$$\begin{aligned} \frac{d\vec{I}}{dt} &= 0, \\ \frac{d\vec{\varphi}}{dt} &= \vec{\omega}(\vec{I}) + \frac{\partial h_{\vec{0}}(\vec{I})}{\partial(\vec{I})}. \end{aligned} \tag{4.42}$$

The averaging makes sense physically, if the system is making rapid excursions over the "barriers" in H_1 so that they don't act too long in one direction or the other. In two degrees of freedom, the unperturbed angles are behaving like (setting initial angles to zero with no loss of generality)

$$\begin{aligned} \varphi_1 &= \left(\frac{\partial H_0}{\partial I_1}\right) t = \omega_1 t, \\ \varphi_2 &= \left(\frac{\partial H_0}{\partial I_2}\right) t = \omega_2 t, \end{aligned} \tag{4.43}$$

and so on. Suppose we consider a term in the Fourier expansion equation 4.41 of H_1—say,

$$h_{\vec{K}=2,-3} \exp[i(2\omega_1 - 3\omega_2)t]. \tag{4.44}$$

If it happened that

$$(2\omega_1 - 3\omega_2) \approx 0, \tag{4.45}$$

then we are in trouble! The reason is that the averaging is taking a very long time to happen. The condition $(2\omega_1 - 3\omega_2) \approx 0$ is a *low-order resonance condition*; low

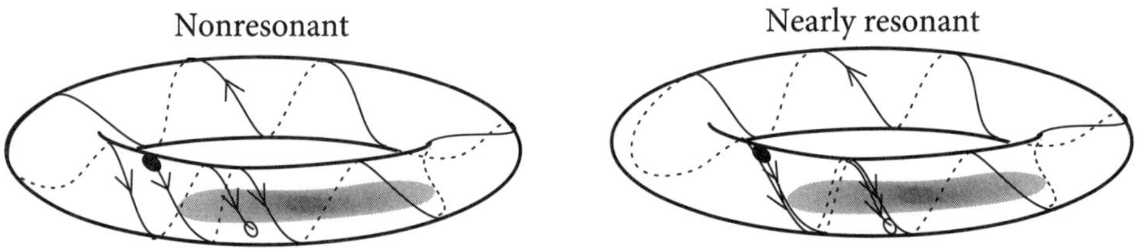

Figure 4.2. This figure reveals the problem with low-order resonances: the torus and perturbation are initially sparsely sampled. If the resonance instead was high order, involving large integers, then the torus would be fairly well sampled before the close coincidence of paths on the torus commenced.

order because small integers are involved. They are worse than high-order resonances because almost always the coefficients $h_{\vec{K}}$ are very small for high-order resonances. Evidently, low-order resonances are worrisome if we want to use perturbation theory.

We have seen that integrable motion takes place on a torus. If for example in two dimensions the frequencies ω_1 and ω_2 are irrationally related to each other—that is, $\omega_1/\omega_2 \neq n/m$, where n and m are integers, then every point on the torus will be accessed as closely as one wishes if one is patient. In fact, the torus will be uniformly covered by the dynamics in the long run, as measured by the time spent in a small spot in the (θ_1, θ_2) plane that we can move around. We say that the dynamics is ergodic *on the torus*, but not on the energy hypersurface $E =$ constant. If on the other hand $\omega_1/\omega_2 = n/m$, where n/m is in its lowest terms, then the motion will be periodic and the torus will be accessed on a manifold of lower dimension, but ergodic on that manifold in the absence of other rational frequency ratios.

Figure 4.2 makes it clear that the presence of a near resonance slows the sampling of a localized perturbation; the motion is not "fast" or averageable over a near cycle. More generally, if

$$\sum n_i \omega_i \approx 0, \tag{4.46}$$

there will be a problem with the averaging idea, since Fourier terms with this linear combination of frequencies will be slow rather than fast. In general, there may be $N-1$ such approximate (or exact) resonance conditions among N frequencies, without all of the frequencies individually vanishing.

4.5 Resonance Characteristics and Removal

The perturbation theory analysis seems to imply that all is lost due to resonances. However, a nonperturbative method presented here extracts a particular resonance (for example, $\omega_1 + \omega_2 - 3\omega_3 = 0$) and leaves an integrable problem with fast motion, under the assumption that only one resonance was doing any "damage." Consider a Hamiltonian in action-angle form

$$H(\boldsymbol{J}, \boldsymbol{\theta}) = H_0(\boldsymbol{J}) + V(\boldsymbol{J}, \boldsymbol{\theta}), \tag{4.47}$$

where $H_0(\boldsymbol{J})$ is integrable and a function of certain actions $\boldsymbol{J}$ only. Since the perturbation term $V(\boldsymbol{J}, \boldsymbol{\theta})$ is periodic in the angles, we may always express it as a Fourier series:

$$V(\boldsymbol{J},\boldsymbol{\theta}) = \eta \sum_{m,n} V_{m,n}(J_1, J_2) \exp[i(m\,\theta_1 - n\,\theta_2)]. \tag{4.48}$$

For example, we might have

$$\begin{aligned} H &= \alpha\, J_1 + \beta\, J_2 + \gamma\, J_1^2/2 \\ &\quad + \delta\, J_2^2/2 + \mu J_1\, J_2 + \eta\, J_1 \sqrt{J_2} \sum_{m,n} \exp[i(m\,\theta_1 - n\,\theta_2)]. \end{aligned} \tag{4.49}$$

Hamilton's equations of motion give

$$\dot{\theta}_1 = \frac{\partial H_0(\boldsymbol{J})}{\partial J_1} + \sum_{m,n} \frac{\partial V_{m,n}(\boldsymbol{J})}{\partial J_1} \exp[i(m\,\theta_1 - n\,\theta_2)], \tag{4.50}$$

$$\dot{\theta}_2 = \frac{\partial H_0(\boldsymbol{J})}{\partial J_2} + \sum_{m,n} \frac{\partial V_{m,n}(\boldsymbol{J})}{\partial J_2} \exp[i(m\,\theta_1 - n\,\theta_2)], \tag{4.51}$$

and

$$\dot{J}_1 = -\frac{\partial H}{\partial \theta_1} = -i \sum_{n,m} V_{m,n}(\boldsymbol{J})\, m\; \exp[i(m\,\theta_1 - n\,\theta_2)], \tag{4.52}$$

$$\dot{J}_2 = -\frac{\partial H}{\partial \theta_2} = i \sum_{n,m} V_{m,n}(\boldsymbol{J})\, n\; \exp[i(m\,\theta_1 - n\,\theta_2)]. \tag{4.53}$$

If

$$\frac{d}{dt}(m\,\theta_1 - n\,\theta_2) \approx 0, \quad \text{that is,} \quad m\,\omega_1 - n\,\omega_2 \approx 0, \tag{4.54}$$

in any term, we must pay special attention to it. The condition

$$s\,\omega_1 - r\,\omega_2 = 0; \;\; \frac{\omega_1}{\omega_2} = \frac{r}{s} \tag{4.55}$$

with integer r and s defines a *resonant* or *rational* torus with winding numbers $W = r/s$ or $W = s/r$, depending upon which angle you are tracking. The *resonance condition* for two degrees of freedom implies that the ratio of the two frequencies is rational. The harmonics of the resonance expressed in lowest integer terms are also slow—that is, $\ell\,(s\,\omega_1 - r\,\omega_2) \approx 0$.

We now suppose such a resonance condition holds for a given (s, r) at certain values of the actions J_0 (remember that the frequencies are functions of the actions and the resonance condition will hold only approximately in some vicinity of certain actions). Ignoring any other resonances except (s, r) and its harmonics $(\ell s, \ell r)$ for integer ℓ):

$$H = H_0(\boldsymbol{J}) + \sum_{\ell} V_{\ell s, \ell r} \exp\left[i\ell(s\,\theta_1 - r\,\theta_2)\right]. \tag{4.56}$$

Ignoring the other, nonresonant combinations can be justified (if V is not too large) by the fast oscillation approximation discussed earlier. We construct a canonical transformation

$$(\boldsymbol{J}, \boldsymbol{\theta}) \rightarrow (\boldsymbol{I}, \boldsymbol{\phi}), \tag{4.57}$$

with

$$F_2(\boldsymbol{\theta}, \boldsymbol{I}) = (\theta_1 - \frac{r}{s}\,\theta_2)\, I_1 + \theta_2 I_2. \tag{4.58}$$

Then

$$\phi_1 = \theta_1 - \frac{r}{s}\theta_2; \quad \phi_2 = \theta_2, \tag{4.59}$$

$$\mathrm{J}_1 = I_1; \quad \mathrm{J}_2 = I_2 - \frac{r}{s} I_1, \tag{4.60}$$

and now focusing on the $r : s$ resonance,

$$H(\boldsymbol{I}^{res}, \phi_1) = H_0\left(I_1^{res}, I_2^{res} - \frac{r}{s} I_1^{res}\right) + \sum_{\ell} V_{\ell s, \ell r}(\boldsymbol{I}^{res}) \exp[i(s\,\ell\,\phi_1)]. \tag{4.61}$$

Since $(\omega_1^{res} - \frac{r}{s}\omega_2^{res}) = 0$, we write

$$H_0 \approx H_0(\boldsymbol{I}^{res}) + \frac{\partial H_0}{\partial I_2}\,\delta I_2 + \frac{1}{2}\frac{\partial^2 H_0}{\partial I_1^2}\,\delta I_1^2 + \frac{\partial H_0}{\partial I_1 \partial I_2}\delta I_1 \delta I_2 + \frac{1}{2}\frac{\partial H_0}{\partial I_2^2}\delta I_2^2. \tag{4.62}$$

Recognizing the resonance condition and (for now) setting $\delta I_2 = 0$ (it is a constant of motion), we have

$$H_0 \equiv H_0(\boldsymbol{I}^{res}) + \frac{1}{2} G\,\delta I_1^2, \tag{4.63}$$

with

$$\begin{aligned} G = \frac{\partial^2 H_0}{\partial I_1^2} &= \left(\frac{\partial^2 H_0}{\partial \mathrm{J}_1^2} - 2\frac{r}{s}\frac{\partial^2 H_0}{\partial \mathrm{J}_1 \partial \mathrm{J}_2} + \frac{r^2}{s^2}\frac{\partial^2 H_0}{\partial \mathrm{J}_2^2}\right), \\ &= \left(\frac{\partial \omega_1}{\partial \mathrm{J}_1} - 2\frac{r}{s}\frac{\partial \omega_1}{\partial \mathrm{J}_2} + \frac{r^2}{s^2}\frac{\partial \omega_2}{\partial \mathrm{J}_2}\right). \end{aligned} \tag{4.64}$$

If the lowest ($\ell = 1$) Fourier component is the most important, we can write the Hamiltonian as

$$H^{eff} = H_0(\boldsymbol{I}^{res}) + \frac{1}{2} G\,\delta I_1^2 + V_0(\boldsymbol{I}^{res})\,\cos(s\,\phi_1). \tag{4.65}$$

We did not expand the perturbation in action because it is already supposed to be small. The analysis will not be valid unless $G \neq 0$, which we take as the nondegeneracy condition. It says that we want the frequencies to depend on the J's in such a way that they do not stay in resonance as the J's change, so that phases can drift. This allows the J's to wax and wane, restricting the domain of the resonance in phase space (see the following).

Equation 4.65 is recognized as a pendulum Hamiltonian (see figure 4.3). The reduction of a resonance to a pendulum Hamiltonian is an old trick. If we had retained the higher $\ell > 1$ harmonics, we would still have fundamentally the same pendulum-like problem, with a somewhat different potential:

$$V(\phi_1) = \sum_{\ell=1}^{\infty} V_{s,r}(\mathrm{I}^{res})\ \cos(\ell s\ \phi_1). \tag{4.66}$$

It is now a simple matter to find the fixed points (stable and unstable) and estimate the resonance width. One just finds the separatrices and stable islands for H^{eff}. For example, for the case $s = 3$, $r = 2$,

$$3\omega_1 \approx 2\omega_2, \tag{4.67}$$

we transform according to the generating function equation 4.58; the hyperbolic fixed points are found at

$$\phi_1 = \frac{2l\pi}{3} = 0, \quad \pm\frac{2\pi}{3}\ \ldots \tag{4.68}$$

and the elliptic fixed points are found at

$$\phi_1 = \frac{(2l+1)\pi}{3} = \pm\frac{\pi}{3}\ \pm\pi,\ \ldots. \tag{4.69}$$

Near $\phi_1 = 0$, we have

$$H \approx H_0 + \frac{1}{2}G\ \delta I_1^2 + V_{3,2}\ \left(1 - \frac{9}{2}\phi_1^2\right). \tag{4.70}$$

Near $\phi_1 = \pi/3$, we have

$$H \approx H_0 + \frac{1}{2}G\ \delta I_1^2 + V_{3,2}\ \left[-1 + \frac{9}{2}\left(\phi_1 - \frac{\pi}{3}\right)^2\right]. \tag{4.71}$$

We define the resonance width W as the action difference between the top and bottom excursion of the separatrices. It is approximately given by

$$\frac{1}{2}G\ \delta I_{1+}^2 + V_{s,r} = \frac{1}{2}G\ \delta I_{1-}^2 - V_{s,r}, \tag{4.72}$$

or

$$\Delta(\delta I_1)^2 = \delta I_{1-}^2 - \delta I_{1+}^2 = \frac{4V_{s,r}}{G}, \tag{4.73}$$

implying (see equation 3.110)

$$W = \Delta(\delta I_1) = 2\sqrt{\frac{V_{s,r}}{G}}. \tag{4.74}$$

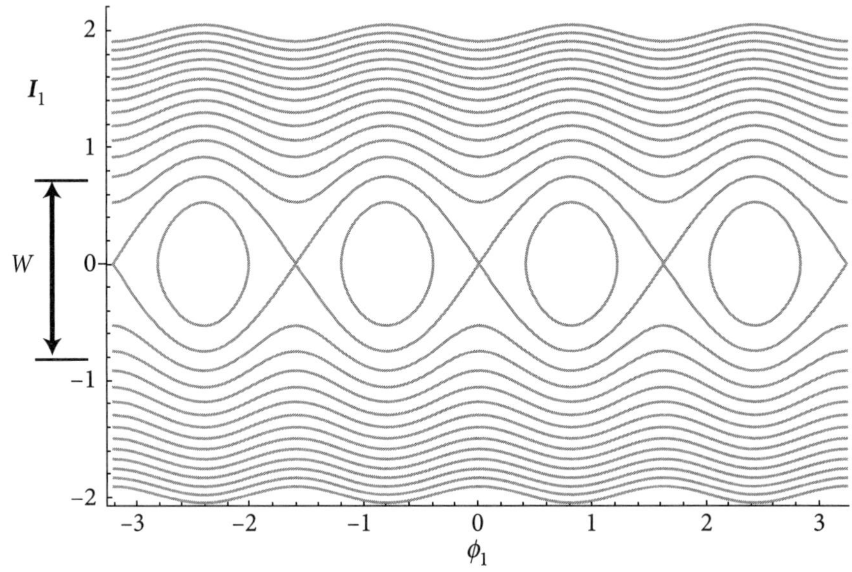

Figure 4.3. Contour plot of the resonance Hamiltonian H^{eff} of equation 4.65, from which the resonance width may be estimated as $W = \Delta(\delta I_1) = 2\sqrt{V_{s,r}/G}$.

4.6 Resonance and Classical Perturbation Theory

Suppose we have an integrable unperturbed Hamiltonian $H_0(\boldsymbol{I})$, having a good set of actions $\boldsymbol{I}$. Now introduce the full Hamiltonian, not too different, as

$$H(\boldsymbol{I}, \boldsymbol{\phi}) = H_0(\boldsymbol{I}) + \epsilon H_1(\boldsymbol{I}, \boldsymbol{\phi}). \tag{4.75}$$

Question: Can we find a canonical transformation $S(\boldsymbol{J}, \boldsymbol{\phi})$ that generates a new action angle set such that

$$H = H(\boldsymbol{J}) + O(\epsilon^2)\,? \tag{4.76}$$

Let

$$\begin{aligned} \boldsymbol{\nabla}_I &\equiv \left(\frac{\partial}{\partial I_1}, \ldots, \frac{\partial}{\partial I_N}\right), \\ \boldsymbol{\nabla}_\phi &\equiv \left(\frac{\partial}{\partial \phi_1}, \ldots, \frac{\partial}{\partial \phi_N}\right). \end{aligned} \tag{4.77}$$

Then

$$\begin{aligned} \boldsymbol{I} &= \boldsymbol{\nabla}_\phi S(\boldsymbol{J}, \boldsymbol{\phi}), \\ \boldsymbol{\theta} &= \boldsymbol{\nabla}_J S(\boldsymbol{J}, \boldsymbol{\phi}). \end{aligned} \tag{4.78}$$

Now we expand S as

$$S(\boldsymbol{J}, \boldsymbol{\phi}) = \boldsymbol{\phi} \cdot \boldsymbol{J} + \epsilon S_1(\boldsymbol{J}, \boldsymbol{\phi}). \tag{4.79}$$

Then

$$H_0(\boldsymbol{I}) + \epsilon H_1(\boldsymbol{I}, \boldsymbol{\phi}) = K_0(\boldsymbol{J}) + \epsilon K_1(\boldsymbol{J}) + O(\epsilon^2), \tag{4.80}$$

or

$$H_0(\boldsymbol{\nabla}_\phi S(\boldsymbol{J}, \boldsymbol{\phi})) + \epsilon H_1(\boldsymbol{\nabla}_\phi S(\boldsymbol{J}, \boldsymbol{\phi}), \boldsymbol{\phi}) = K_0(\boldsymbol{J}) + \epsilon K_1(\boldsymbol{J}) + O(\epsilon^2). \tag{4.81}$$

Using equation 4.79 we write

$$H_0(\boldsymbol{J} + \epsilon \boldsymbol{\nabla}_\phi S_1(\boldsymbol{\phi}, \boldsymbol{J})) + \epsilon H_1(\boldsymbol{J}, \boldsymbol{\phi}) = K_0(\boldsymbol{J}) + \epsilon K_1(\boldsymbol{J}), \tag{4.82}$$

and expand, keeping like terms in ϵ together, getting

$$\epsilon^0 : \quad H_0(\boldsymbol{J}) = K_0(\boldsymbol{J}), \tag{4.83}$$

$$\epsilon^1 : \quad \boldsymbol{\nabla}_\phi S_1 \cdot \boldsymbol{\nabla}_J H_0(\boldsymbol{J}) + H_1(\boldsymbol{\phi}, \boldsymbol{J}) = K_1(\boldsymbol{J}). \tag{4.84}$$

We next expand in a Fourier series,

$$S_1 = \sum_{\boldsymbol{K}} s_{\boldsymbol{K}}(\boldsymbol{J}) e^{i\boldsymbol{K}\cdot\boldsymbol{\phi}}, \tag{4.85}$$

$$H_1 = \sum_{\boldsymbol{K}} h_{\boldsymbol{K}}(\boldsymbol{J}) e^{i\boldsymbol{K}\cdot\boldsymbol{\phi}}. \tag{4.86}$$

Then the $\boldsymbol{K} = \boldsymbol{0}$ term gives simply

$$K_1(\boldsymbol{J}) = h_0(\boldsymbol{J}), \tag{4.87}$$

and the kth component reads, using $\boldsymbol{\nabla}_J H_0(\boldsymbol{J}) = \boldsymbol{\omega}_0(\boldsymbol{J})$,

$$(\boldsymbol{\omega}_0(\boldsymbol{J}) \cdot i\boldsymbol{K})\, s_{\boldsymbol{K}}(\boldsymbol{J}) = h_{\boldsymbol{K}}, \tag{4.88}$$

or

$$s_{\boldsymbol{K}}(\boldsymbol{J}) = \frac{i h_{\boldsymbol{K}}}{\boldsymbol{K} \cdot \boldsymbol{\omega}_0}, \tag{4.89}$$

and finally

$$S_1(\boldsymbol{\phi}, \boldsymbol{J}) = \sum_{\boldsymbol{K}} \frac{i h_{\boldsymbol{K}}}{\boldsymbol{K} \cdot \boldsymbol{\omega}_0} e^{i\boldsymbol{K}\cdot\boldsymbol{\phi}}. \tag{4.90}$$

Here, the famous problem of "resonant denominators" is revealed: for frequencies related by rational ratios, the denominator

$$\boldsymbol{K} \cdot \boldsymbol{\omega}_0 = 0 \tag{4.91}$$

is causing a serious problem. Even if the frequencies are not strictly resonant, they can be very close. For example, if $\omega_1 = \sqrt{2}\,\omega_2$, we might think we are safe because that ratio is irrational. Yet

$$\sqrt{2} = \frac{7}{5} + \mathcal{O}(10^{-2}),$$

so even irrational frequency ratios can be so closely approximated by rationals that the corresponding term in the perturbation series is possibly large, unless the Fourier component of h_1 is correspondingly small.

ARNOLD DIFFUSION

In 1964 Arnold published a key paper that established it is possible to wander far in classical action, in spite of the KAM theorem and most of phase space being filled with invariant tori, for nearly integrable systems of more than two degrees of freedom [13]. If a trajectory is on one of these tori, there is no Arnold diffusion. The diffusion occurs between the integrable tori required by the KAM theorem, in the analog of what we see in two dimensions as the small chaotic zone that appears near a separatrix. As stated, invariant tori block any attempt to leave in two dimensions, but not in three or more.

The scuttlebutt ever since has been that Arnold diffusion is slow, so slow that even quantum mechanical tunneling processes may outstrip it. That is to say, quantum systems may experience action diffusion by mechanisms not related to Arnold diffusion. Nonetheless, it is good to know what it is, as its implications go well beyond classical perturbation theory. Arnold diffusion cannot happen for perfectly integrable systems even in many dimensions, but adding a small perturbation can make the system slightly nonintegrable. Rigorous results are good only for incredibly small perturbations. For example, the perturbation of the orbits of the planets, considered as unperturbed systems one planet at a time orbiting the sun, is of order 0.5×10^{-3}. The cause is largely due to Jupiter. This is huge compared to what the rigorous results can handle, and one of the reasons why we still don't know the ultimate fate of the solar system. However, very careful numerical studies by Laskar, Wisdom, and others have shown that the solar system is chaotic. An error in the position of the earth by 15 feet makes it impossible to predict where the earth would be in its orbit in just over 100 million years.

In two dimensions, the invariant tori intersecting on a Poincaré surface of section are impenetrable. A classical trajectory on a given torus must stay on it, or one inside a given closed torus may never reach it, much less cross it.

In three dimensions (ϕ_1, ϕ_2, ϕ_3) with good actions (I_1, I_2, I_3), the invariant unperturbed system lives on three-dimensional tori embedded in six phase space dimensions. The KAM theorem applies (see sections 3.12 and 4.4), so for small perturbation, phase space is mostly filled with invariant tori. If the trajectory is exactly on such a surviving torus, it is stuck there. But what happens if there is a nearby resonance zone—say, such that $k_1\omega_1 + k_2\omega_2 + k_3\omega_3 = 0$ with the k_i integers? In two dimensions, confirming what we have said, the condition $k_1\omega_1 + k_2\omega_2 = 0$ doesn't do much for wanderlust, since $\omega_i = \partial H/\partial I_1$ and wandering beyond the natural resonance width would require that both actions change in a way that changes the energy, which is forbidden. The resonance condition and requirement to keep energy fixed are two conditions with two freedoms; there is no remaining freedom.

A 3-torus embedded in a six-dimensional phase space is not confining: just off the torus, in what is called the *Arnold web*, all three actions can change, keeping $k_1\omega_1 + k_2\omega_2 + k_3\omega_3 = 0$ *and* $H =$ constant. The myriad of possible actions and integers k_i satisfying these conditions constitute the web. This is still two conditions but now with three freedoms, leaving leeway to move and pass by other, completely different invariant tori. The neighborhood where the wandering is allowed decreases, sometimes dramatically, as $\sum_i |k_i|$ increases, but the wandering is allowed nonetheless. Rigorous

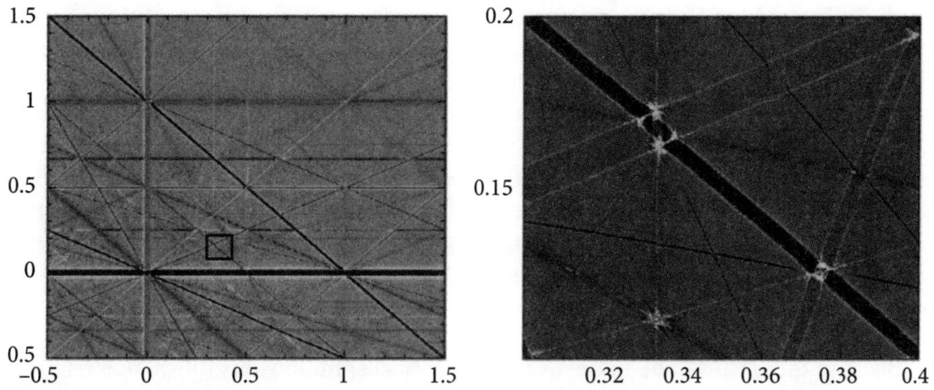

Figure 4.4. Structure of the Arnold web for the Hamiltonian of equation 4.92 according to reference [14]. The region in the small rectangle at the left is shown at the right. The brightness scale is basically time multiplied by the Lyapunov exponent; the brightest regions are chaotic or very unstable, while midrange brightness is due to motion on invariant tori. Some web pathways are dark, with even slower diffusion than on invariant tori.

results showing the diffusion actually happens are hard to come by. Instead, excellent progress has been made detecting the diffusion numerically. This requires "Arnold web detection" and long integration times, as was done in reference [14], which used a clever choice of Hamiltonian

$$H = \frac{I_1^2}{2} + \frac{I_2^2}{2} + I_3 + \epsilon \left(\frac{1}{\cos\phi_1 + \cos\phi_2 + \cos\phi_3 + 4} \right). \tag{4.92}$$

In this system the Arnold web is visible in the two-dimensional plane I_1, I_2, where each point in the plane characterizes a unique torus. All resonances $k_1\omega_1 + k_2\omega_2 + k_3\omega_3 = 0$ are given by straight lines $k_1 I_1 + k_2 I_2 + k_3 = 0$ in the I_1, I_2 plane. The web detection algorithm gave figure 4.4.

4.7 Oscillator Connection

There is a simple equivalence between a classical oscillator system with possibly time-dependent frequencies and couplings, and coupled state quantum amplitudes. The connection is an interesting curiosity, sometimes useful for computations and especially good for harnessing classical intuition about quantum systems.

In its simplest form it can be given as an equivalence between a one-dimensional Schrödinger equation for fixed energy and a parametrically forced classical oscillator:

$$\begin{aligned} \left(\frac{d^2}{dq^2} + 2(E - V(q)) \right) \psi(q) &= 0, \\ \left(\frac{d^2}{dt^2} + \omega^2(t) \right) x(t) &= 0. \end{aligned} \tag{4.93}$$

The simple substitutions $q \leftrightarrow t$, $\psi(q) \leftrightarrow x(t)$, and $\omega^2(t) \leftrightarrow 2(E - V(q))$ produce one form from the other. Note that in the potential tunneling region where $E - V(q) < 0$, the corresponding harmonic oscillator potential is inverted and unstable.

Very delicate maneuvering is required to settle $x(t \to \infty)$ at the top of the unstable barrier, and only if E has certain specific values does $\omega^2(t) = 2(E - V(t))$ accomplish the task. This corresponds to the rarity of normalizable (not exponentially blowing up) eigenfunctions in a potential well. There are solutions of the one-dimensional Schrödinger equation at all energies, except that only a set of measure 0 is normalizable. Similarly, the oscillator is happy to run away to infinity on an inverted, repulsive potential. The gymnastics required for $x(t)$ to settle down at the top of an inverted oscillator for both large positive and large negative time gives insight into just how rare the eigenfunctions are among all solutions. The higher energy quantum eigenfunctions correspond to the classical oscillator remaining attractive for one or more complete oscillations and zero crossings of $x(t)$.

Because there cannot be many different classical simultaneous times, this relationship between wavefunction and oscillator will not extend directly to many quantum coordinates. However, it does extend to many coupled classical oscillators with one time variable, or what is the same thing, many interacting one-dimensional potential energy surfaces. This gets more interesting, because the issue of adiabaticity comes up in both versions. Starting with the coupled classical oscillators, we have

$$\begin{pmatrix} \frac{d^2}{dt^2} + \omega_x^2(t) & \varepsilon(t) \\ \varepsilon(t) & \frac{d^2}{dt^2} + \omega_y^2(t) \end{pmatrix} \begin{pmatrix} x(t) \\ y(t) \end{pmatrix} = \begin{pmatrix} 0 \\ 0 \end{pmatrix}, \tag{4.94}$$

that is equivalent to the evolution of the two-component wavefunction $(\psi_1(q), \psi_2(q))$ in a "curve crossing" situation:

$$\begin{pmatrix} \frac{d^2}{dq^2} + 2(E - V_{11}(q)) & V_{12}(q) \\ V_{21}(q) & \frac{d^2}{dq^2} + 2(E - V_{22}(q)) \end{pmatrix} \begin{pmatrix} \psi_1(q) \\ \psi_2(q) \end{pmatrix} = \begin{pmatrix} 0 \\ 0 \end{pmatrix} \tag{4.95}$$

Clearly, this can be extended to any number coupled oscillators or interacting potential energy surfaces. This equivalence is at the root of the Meyer-Miller method, a "classical analog for electronic degrees of freedom" [15], although they take it much further and make it really useful.

One rather interesting case is $V(q) = \mu \cos(2\pi q/b)$—that is, a sinusoidal periodic potential, which translates to the classical pendulum problem

$$\left(\frac{d^2}{dt^2} + 2(E - \mu \cos(2\pi t/b)) \right) x(t) = 0 \tag{4.96}$$

The standard form of the Mathieu equation is a match: $\ddot{x}(\tau) + [a - 2q \cos(2\tau)]x(\tau) = 0$. This is noted later (equation 19.13), in the context of the electronic spectroscopy of benzophenone, where the "band structure" of the periodic potential becomes the (equivalent) resonance structure of a periodically driven parametric oscillator. The band gaps correspond to unstable classical motion. (See also figure 19.8, later.)

Here, we use the equivalence to gain insight into the adiabatic approximation. The gymnastics required for both $x(t)$ and $y(t)$ to remain finite as $t \to \pm\infty$ is now even more delicate. However, a new element comes in: In the case of the oscillators, under what circumstances will large amplitude motion of $x(t)$ in the far past (with a quiescent $y(t)$) become large amplitude motion of $y(t)$ with a quiescent $x(t)$ in the future? If this

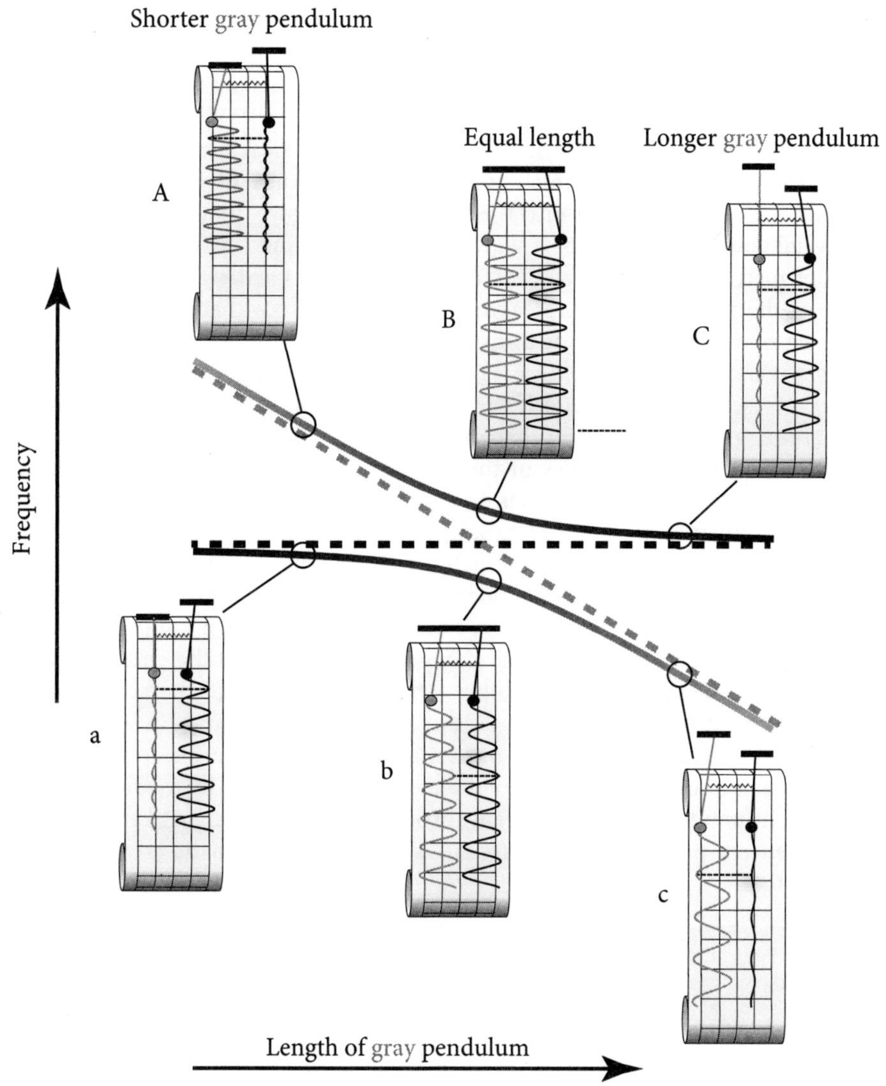

Figure 4.5. Two pendulums, one with a time-dependent length, are coupled by a spring. The modes and their frequencies are indicated at three lengths of the gray pendulum. (The right-hand, black pendulum does not change length; its uncoupled frequency is the horizontal dashed line. The gray dashed line is the frequency of the left uncoupled pendulum as it lengthens from left to right.) The thick solid curves depict the frequencies of the adiabatic modes, obtained by fixing the coupling strength and the length of the pendulums, but including the spring coupling them. The frequencies of oscillation and relative motion of the two pendulums are accurately shown for three lengths of the variable gray pendulum; note that both pendulums oscillate in lock step in each of the modes (A, B, C, a, b, c). The dashed lines connecting the pendulum tracks are guides to the eye for judging the relative phases of oscillation. The shading of the adiabatic curves indicates the character of the oscillation (black = mostly black pendulum motion). Note that by remaining on the upper adiabatic curve, for example, the pendulum energy transfers from the gray to the black oscillator; this is half under way when the pendulums are of equal length. Figure taken from *Why You Hear What You Hear* by the present author, Princeton University Press, 2012.

happens, one is tempted to suppose that something analogous to strong "nonadiabatic coupling" has taken place, but actually it is just the opposite. Equation 4.95 is written in the diabatic representation—that is, the diagonal elements of the matrices represent the uncoupled system.

In the adiabatic limit, assuming a "close encounter" near a time t when $\omega_y(t) = \omega_x(t)$, but $\epsilon(t)$ remains nonvanishing, complete exchange of energy from x to y will occur if the approach to $\omega_y(t) = \omega_x(t)$ is very slow—that is, adiabatic. Large off-diagonal coupling makes for a widely avoided crossing (in the adiabatic picture, where the matrix is diagonalized for every fixed t or fixed q) and allows a faster traverse of the region without breaking adiabaticity. Note that x and y oscillators exchanging energy and amplitude translates to $\psi_2(q)$ becoming large at the expense of $\psi_2(q)$ as q is traversed and $V_{11}(q) = V_{22}(q)$. Inspection of figure 4.5 shows a compete exchange of oscillator amplitude if the traverse is slow and the system remains on the noncrossing upper or lower adiabatic curves, whichever it started on. The mode frequencies determine the values of the adiabatic curves. Solid adiabatic curves are colored dark if the black oscillator has most of the amplitude, and light if the light gray one does. The upper curve labeled (A, B, C) depicts an adiabatic hand-off of pendulum energy from the initially short gray oscillator to the initially long black oscillator. These curves are reliably followed by the modes as length is changed, provided it is changed sufficiently slowly (adiabatically). Note, however, that there is a complete change of character of motion as the pendulums change length, such that the energy flows fully from one individual pendulum to another. This radical change of character before and after the avoided crossing is characteristic of adiabatic motion and corresponds in the quantum analog to a radical change of orbital character, where the "pendulums" are molecular states and the pendulum frequencies are determined by the molecular state energies. If the pendulums were suddenly brought to their starting and ending lengths, the higher energy pendulum would remain so (a so-called diabatic crossing—that is, the dashed lines).

PART II

Quantum and Semiclassical Mechanics

Chapter 5

Aspects of Time-Dependent Quantum Mechanics

5.1 Time-Dependent Perturbation Theory

Suppose we have a Hamiltonian H_0 and we know its eigenvalues E_n^0 and eigenfunctions $|\psi_n^0\rangle$. At $t=0$, the system is in one of its unperturbed eigenstates $|\psi_k^0\rangle$. An applied, possibly time-dependent potential $\lambda V(t)$ comes in at (or anytime after) $t=0$, and will generally induce transitions between the unperturbed eigenstates. The Hamiltonian reads

$$H(t) = H_0 + \lambda V(t), \tag{5.1}$$

and the time-dependent Schrödinger equation is

$$i\hbar\frac{\partial|\psi_k(t)\rangle}{\partial t} = (H_0 + \lambda V(t))\,|\psi_k(t)\rangle. \tag{5.2}$$

The subscript k reminds us that at $t=0$, the system was in the kth unperturbed state. We seek the conditions making the expansion

$$|\psi_k(t)\rangle = |\psi_k^{(0)}(t)\rangle + \lambda|\psi_k^{(1)}(t)\rangle + \lambda^2|\psi_k^{(2)}(t)\rangle + \cdots, \tag{5.3}$$

a solution of the time-dependent Schrödinger equation, order by order in λ. Insertion of equation 5.3 into equation 5.2 followed by collection of powers of λ gives

$$\begin{aligned}
\lambda^0:\ i\hbar|\dot\psi_k^{(0)}(t)\rangle &= H_0|\psi_k^{(0)}(t)\rangle,\\
\lambda^1:\ i\hbar|\dot\psi_k^{(1)}(t)\rangle &= H_0|\psi_k^{(1)}(t)\rangle + V(t)|\psi_k^{(0)}(t)\rangle,\\
\lambda^2:\ i\hbar|\dot\psi_k^{(2)}(t)\rangle &= H_0|\psi_k^{(2)}(t)\rangle + V(t)|\psi_k^{(1)}(t)\rangle,\\
\lambda^n:\ i\hbar|\dot\psi_k^{(n)}(t)\rangle &= H_0|\psi_k^{(n)}(t)\rangle + V(t)|\psi_k^{(n-1)}(t)\rangle.
\end{aligned} \tag{5.4}$$

Writing

$$|\psi_k^{(0)}(t')\rangle = e^{-iH_0t'/\hbar}|\psi_k^{(0)}\rangle = G_0^+(t')|\psi_k^{(0)}\rangle \tag{5.5}$$

allows us to compactly integrate the second of equations 5.4, obtaining

$$|\psi_k^{(1)}(t)\rangle = \frac{-i}{\hbar}\int_{-\infty}^{t} dt'\; G_0^+(t-t')\; V(t')\; G_0^+(t')|\psi_k^{(0)}\rangle; \tag{5.6}$$

this can be directly checked by time differentiation. To first order, and setting $\lambda = 1$, we have (note the brackets in the exponent to denote this is the first-order estimate of the full wavefunction, not just the correction):

$$|\psi_k^{[1]}(t)\rangle = |\psi_k^{(0)}(t)\rangle + |\psi_k^{(1)}(t)\rangle = G_0^+(t)|\psi_k^{(0)}\rangle - \frac{i}{\hbar}\int_{-\infty}^{t} dt'\; G_0^+(t-t')\; V(t')\; G_0^+(t')|\psi_k^{(0)}\rangle. \tag{5.7}$$

The first-order Green function $G_1^+(t)$ is now defined, having the property of taking the initial unperturbed state into the current first-order state,

$$|\psi_k^{[1]}(t)\rangle = G_1^+(t)|\psi_k^{[0]}\rangle. \tag{5.8}$$

Therefore,

$$G_1^+(t) = G_0^+(t) - \frac{i}{\hbar}\int_{-\infty}^{t} dt'\; G_0^+(t-t')\; V(t')\; G_0^+(t'). \tag{5.9}$$

Then $G_2^+(t)$, which takes the initial state through to the solution correct to the second order, is

$$\begin{aligned} G_2^+(t) &= G_0^+(t) - \frac{i}{\hbar}\int_{-\infty}^{t} dt'\; G_0^+(t-t')\; V(t')\; G_1^+(t'), \\ &= G_0^+(t) - \frac{i}{\hbar}\int_{-\infty}^{t} dt'\; G_0^+(t-t')\; V(t')\; G_0^+(t'), \\ &+ \left(-\frac{i}{\hbar}\right)^2 \int_{-\infty}^{t} dt' \int_{-\infty}^{t'} dt''\, G_0^+(t-t')\; V(t')\; G_0^+(t'-t'')\; V(t'') G_0^+(t''). \end{aligned} \tag{5.10}$$

To all orders it is now easy to see by iteration of this equation that

$$\begin{aligned} G^+(t) = G_0^+(t') - \frac{i}{\hbar}\int_{-\infty}^{t} dt'\; G_0^+(t-t')\; V(t')\; G_0^+(t') \\ + \left(-\frac{i}{\hbar}\right)^2 \int_{-\infty}^{t} dt' \int_{-\infty}^{t'} dt''\, G_0^+(t-t')\; V(t')\; G_0^+(t'-t'')\; V(t'') G_0^+(t'') + \cdots. \end{aligned} \tag{5.11}$$

$G^+(t)$ has the property of creating the exact time-dependent wavefunction $|\psi_k(t)\rangle$ starting with the kth zero-order state as an initial condition:

$$|\psi_k(t)\rangle = G^+(t)|\psi_k^{[0]}\rangle = e^{-iHt/\hbar}|\psi_k^{[0]}\rangle. \tag{5.12}$$

Since $G^+(t)$ takes any zero-order state at $t = 0$ to its exact form at $t = t$, and since the zero-order states are a complete set, $G^+(t)$ takes *any* initial state to its time-evolved version under $H = H_0 + V(t')$ at time t.

$G^+(t)$ is the solution of iteration of the following integral equation; its iteration is the perturbation series equation 5.11:

$$G^+(t) = G_0^+(t') - \frac{i}{\hbar}\int_{-\infty}^{t} dt'\; G_0^+(t-t')\; V(t')\; G^+(t'). \tag{5.13}$$

There is no guarantee the series expansion, equation 5.11, has to converge in particular cases. At short times, Δt things should converge; the nth term is of order $(\Delta t)^n$.

There is a nice interpretation of the series 5.11—namely, that the first-order term is a sum (integral) of amplitudes for the system propagating undisturbed by V for a time t', and then being "hit" by V at time t', followed by unperturbed propagation up to time t. The first-order expression sums over all possible times t' to be "hit" once by the potential. These various amplitudes for different times can interfere constructively or destructively. The higher order terms have obvious extensions of the preceding interpretation of the first-order term. One can express this in a fairly obvious diagrammatic series, which is the tip of the iceberg hinting at Feynman diagrams.

5.2 Fermi Golden Rule for Decay and Beyond

We work in the interaction picture, wherein the unperturbed evolution under H_0 is incorporated by writing

$$\psi_I(t) = e^{iH_0t/\hbar}\psi(t);\;\; V_I(t) \equiv e^{iH_0t}V(t)e^{-iH_0t}. \tag{5.14}$$

It is easy to show that $\psi_I(t)$ satisfies

$$i\hbar\frac{\partial\psi_I(t)}{\partial t} = V_I(t)\psi_I(t); \tag{5.15}$$

that is, there is no evolution in $\psi_I(t)$ except that caused by the perturbation $V_I(t)$. We get a perturbation theory of the time-dependent propagation by integrating equation 5.15 with respect to t:

$$\psi_I(t) = \psi_I(0) + \frac{1}{i\hbar}\int_0^t dt'\; V_I(t')\psi_I(t'), \tag{5.16}$$

which is still exact, followed by obtaining a first-order solution $\psi_I^{[1]}(t)$ by replacing $\psi_I(t')$ on the right by its zeroth approximation $\psi_I(0)$, the solution for $V_I(t) = 0$,

$$\psi_I^{[1]}(t) = \psi_I(0) + \frac{1}{i\hbar} \int_0^t dt' \; V_I(t') \; \psi_I(0). \tag{5.17}$$

The second-order correction is given by iteration of this equation—that is,

$$\psi_I^{[2]}(t) = \psi_I(0) + \frac{1}{i\hbar} \int_0^t dt' \; V_I(t') \; \psi_I(0) + \frac{1}{(i\hbar)^2} \int_0^t dt' \int_0^{t'} dt'' \; V_I(t') V_I(t'') \psi_I(0), \tag{5.18}$$

and so on.

Returning to the first-order results, and projecting equation 5.17 from the left by $\langle \psi_{k'} |$ and taking the initial state to be $|\psi_k\rangle$, we get the amplitude for being found in the state $\langle \psi_{k'} |$ at time t,

$$a_{k'k}(t) = \frac{-i}{\hbar} e^{-iE_{k'}t/\hbar} \int_{-\infty}^t dt' \; e^{i(E_{k'} - E_k)t'/\hbar} \; V_{k'k}(t'). \tag{5.19}$$

Then the total probability to be kicked out of the initial state $|\psi_k^0\rangle$ is

$$P_{\not{k}}(t) = \sum_{k' \neq k} |a_{k'k}(t)|^2 = \frac{1}{\hbar^2} \sum_{k' \neq k} \left| \int_{-\infty}^t dt' \; e^{i(E_{k'} - E_k)t'/\hbar} V_{k'k}(t') \right|^2. \tag{5.20}$$

Notice there will be a larger total transition probability if $V_{k'k}(t)$ has components at the Fourier frequency $\omega_{k'k} = (E_{k'} - E_k)/\hbar$.

We now consider an important special case. Suppose we know the system to be in the state $|\psi_k\rangle$ before $t = 0$. The perturbation turns on at $t = 0$ and remains constant thereafter. Then, equation 5.20 becomes

$$\begin{aligned} P_{\not{k}}(t) = \sum_{k' \neq k} |a_{k'k}(t)|^2 &= \frac{1}{\hbar^2} \sum_{k' \neq k} |V_{k'k}|^2 \left| \frac{e^{i\omega_{k'k}t} - 1}{\omega_{k'k}} \right|^2, \\ &= \frac{4}{\hbar^2} \sum_{k' \neq k} |V_{k'k}|^2 \; \frac{\sin^2(\omega_{k'k}t/2)}{\omega_{k'k}^2}, \\ &\approx 4\langle |V_{k'k}|^2 \rangle \int dE \; \rho(E) \; \frac{\sin^2(Et/2\hbar)}{E^2}, \end{aligned} \tag{5.21}$$

assuming the states are dense enough in energy to justify replacing the sum by an integral, and pulling the local average of $|V_{k'k}|^2$ out of the integral—that is, $(\delta k' = 1)$,

$$\sum_{k'} (\cdots) = \sum_{k'} \left(\frac{\delta k'}{\Delta E} \right) \Delta E \; (\cdots) \sim \int dE \; \rho(E)(\cdots). \tag{5.22}$$

Since

$$\int_{-\infty}^{\infty} \sin^2(ax)/x^2\, dx = |a|\pi, \tag{5.23}$$

there results

$$P_{\not k}(t) = \frac{2\pi}{\hbar}\langle |V_{k'k}|^2\rangle \rho(E_k)t. \tag{5.24}$$

The extraction of $|V_{k'k}|^2$ is justified as long as

$$\frac{\sin^2(\omega_{k'k}t/2)}{\omega_{k'k}^2}$$

is sufficiently narrowly peaked (long enough time t) and nonetheless contains sufficiently many E_k's under its main peak (short enough time t). Thus, the time cannot be too long or too short for this to work well. Assuming it does, we can write

$$P_k(t) = 1 - P_{\not k}(t) = 1 - \frac{2\pi}{\hbar}\langle |V_{k'k}|^2\rangle \rho(E_k)\, t \approx e^{-\Gamma t}, \tag{5.25}$$

where

$$\Gamma = \frac{2\pi}{\hbar}\langle |V_{k'k}|^2\rangle \rho(E_k). \tag{5.26}$$

We took the liberty of identifying the two terms we have derived in first-order perturbation theory as belonging to the Taylor series of an exponential. It would take higher order perturbation theory to fully justify this.

The formula for the transition rate $\Gamma = \frac{2\pi}{\hbar}\langle |V_{k'k}|^2\rangle \rho(E_k)$ is called *Fermi's Golden Rule*. Dirac was the first to derive it, but Fermi thought it was so useful he coined the term "golden rule." Any time a state is coupled of a continuum or quasi-continuum, the Golden Rule gives the decay in all but very short and (possibly) very long times.

There is indeed a problem for t too small: the decay of a state must go quadratically as $t \to 0$, as in $P(t) \sim 1 - \alpha t^2$ for some α. The exponential decay is linear in time for short times, of course. To see the necessity of a quadratic short time dependence, we Taylor expand the propagator (assume $V_{kk} = 0$ with no loss of generality),

$$\begin{aligned}\langle \psi_k | e^{-i(H_0+V)t/\hbar} | \psi_k \rangle &\sim \langle \psi_k | (1 - i(H_0 + V)t/\hbar - (H_0 + V)^2 t^2/\hbar^2 + \cdots) | \psi_k \rangle \\ &= 1 - iE_k t/\hbar - (E_k^2 - (V^2)_{kk})t^2/\hbar^2 + \cdots ,\end{aligned} \tag{5.27}$$

and

$$P(t) = \left|1 - iE_k t/\hbar - (E_k^2 - (V^2)_{kk})t^2/\hbar^2\right|^2 = 1 - \alpha t^2 + \beta t^4 + \cdots , \tag{5.28}$$

with $\alpha = (V^2)_{kk}$.

What goes wrong with the Golden Rule derivation at short times? The problem lies with pulling the $|V_{k'k}|^2$ term out of the integral in equation 5.21. We warned that this is not valid unless $\sin^2(\omega_{k'k}t/2)/\omega_{k'k}^2$ is narrowly peaked near $k \approx k'$. This factor becomes extremely broad for small enough t. For any reasonable perturbation the matrix elements $|V_{k'k}|$ must die off if V is smooth and $|E_k\rangle$ and $|E_{k'}\rangle$ become very different for fixed k and large k'.

At very short times we can write

$$\frac{\sin^2(\omega_{k'k}t/2)}{\omega_{k'k}^2} \sim \frac{t^2}{4}, \tag{5.29}$$

but for any finite t, this is not valid for large k', making $\omega_{k'k}t \gg 1$. However, we have noted that the terms

$$\frac{1}{\hbar^2}|V_{k'k}|^2 \, \frac{\sin^2(\omega_{k'k}t/2)}{\omega_{k'k}^2}$$

are very small at large energies (large k') because normally $|V_{k'k}|^2$ will become very small. Using this argument, the approximation 5.29 effectively does not break down, and our more careful analysis, not just ripping the $|V_{k'k}|^2$ term out of the integral, reveals we do have $P_{\not k}(t) \sim t^2$ for very small t, agreeing with the Taylor series argument just given. As t increases further, there comes a time when $\omega_{k'k}t \gg 1$, even though $V_{k'k}$ is not small enough to make the term negligible; then the decay rolls over to $P_{\not k}(t) \sim t$ behavior. All this can be modeled analytically with the following integral, taking a Gaussian fall-off for the matrix elements:

$$I \propto P_{\not k} = \int_{-\infty}^{\infty} e^{-b^2x^2/\hbar^2} \sin^2(tx)/x^2 dx = \frac{\sqrt{\pi}b}{\hbar}\left(e^{-t^2\hbar^2/b^2} - 1\right) + \pi t \,\mathrm{Erf}\left(\frac{\hbar t}{b}\right), \tag{5.30}$$

where Erf is the error function. For $\hbar t/b \ll 1$ this goes as $\sim t^2$, whereas for $\hbar t/b \gg 1$, it goes like t. Specifically, we have (see figure 5.1)

$$I \sim (\sqrt{\pi}\hbar/b)t^2, \text{ short } t,$$

$$I \sim \pi t - \sqrt{\pi b/\hbar}, \text{ large } t.$$

At the "large" end of the timescale, the perturbation theory must break down. The prediction of the initial decay rate $\Gamma = \frac{2\pi}{\hbar}\langle|V|^2\rangle\rho$ will still hold; what happens at longer times is that the exponential decay may not continue. One way this can happen is if the density of states is finite, so that at times longer than $t \sim \hbar/\Delta E = \hbar\rho(E)$, where ΔE is the average spacing between the levels, the decay cannot be exponential. This is because the successive terms in an expansion of the full wavefunction in terms of the full eigenstates "dephase" from one another at this time (that is, they get close to π or more out of phase with their neighbors). A related observation is that at such times $t \sim \hbar\rho(E)$, equation 5.22 is no longer viable.

There are many more possibilities for the application of time-dependent perturbation theory. Exponential decay due to a coupling (just treated) is important, but sometimes the explicit time dependence of $V(t)$ must be incorporated. A very common requirement is that the perturbation oscillates sinusoidally at some frequency rather than remaining constant. This quickly reduces to the previous constant case, however, with an adjustment of the states in the k' spectrum that are resonant. In this way we get photoabsorption, for example.

Another example is an ion flying by an atom. Suppose we assume a straight track for a fast ion, and then the potential felt by the atom is a rising and falling coulomb

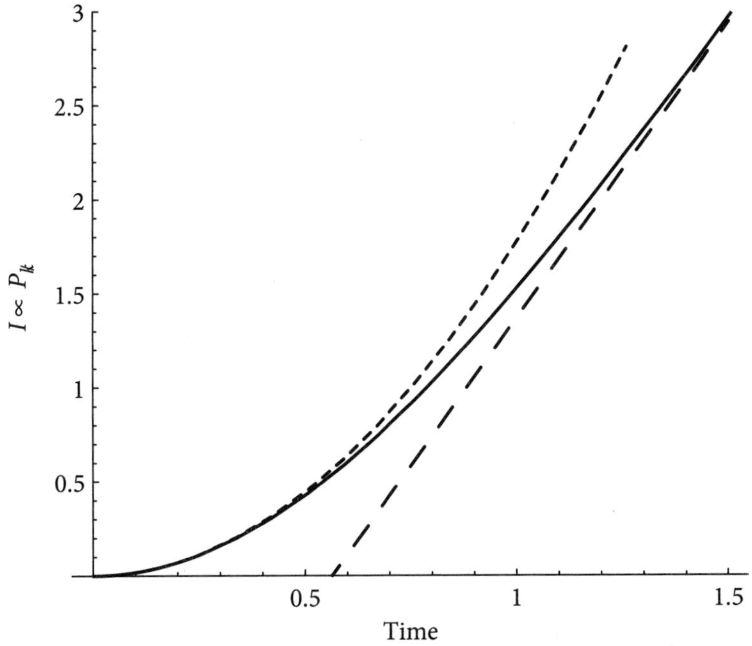

Figure 5.1. Limiting behavior and functional dependence of the integral $I = \sqrt{\pi}(\sqrt{\pi}t\ \mathrm{erf}(t) + e^{-t^2} - 1)$ (equation 5.30 for $b = 1, \hbar = 1$) as a function of t (solid line). Short time limiting behavior is seen in the short dashed line: $I \sim (\sqrt{\pi}\hbar/b)t^2$. Long time behavior (exponential decay) is given by the long dashed line: $I \sim \pi t - \sqrt{\pi}\, b/\hbar$. The transition region is $t \sim b/\hbar = 1$.

potential. The integral 5.20 can be taken over all time to determine the excitation of the atom from state k to k' after the ion has passed.

5.3 System-Bath Interactions, Dephasing, and Depopulation

We now derive some relations (within perturbation theory) involving the decay of a system coupled to a bath. The system has states $|n\rangle$, the bath, $|\boldsymbol{m}\rangle$, and together they have eigenfunctions of an unperturbed Hamiltonian H_0 wherein system and bath do not interact:

$$H_0|n, \boldsymbol{m}\rangle = (E_n + \epsilon_{\boldsymbol{m}})\,|n, \boldsymbol{m}\rangle. \tag{5.31}$$

We take the coupling between system and bath to be V—that is, $H = H_0 + V$.

Suppose the bath starts out in the unperturbed ground state $|n, \boldsymbol{0}\rangle$. From equation (5.11) it evolves as $G^+(t)|n, \boldsymbol{0}\rangle$. The probability amplitude of remaining in the initial state is

$$\langle n, \boldsymbol{0}|G^+(t)|n, \boldsymbol{0}\rangle = e^{-i(E_n+\epsilon_0)t/\hbar}\langle n, \boldsymbol{0}|1 - \frac{i}{\hbar}\int_0^t dt'\ V_I(t')$$

$$-\frac{1}{\hbar^2}\int_0^t dt' \int_0^{t'} dt'' V_I(t')V_I(t'') + \cdots |n, \boldsymbol{0}\rangle. \tag{5.32}$$

We take $\langle n, \mathbf{0}|V_I|n, \mathbf{0}\rangle \equiv 0$; this can always be arranged with a constant shift of the Hamiltonian, and thus V_I, up or down in energy. We thus don't need the first-order term, and now turn attention to the second-order term:

$$-\frac{1}{\hbar^2}\int_0^t dt' \int_0^{t'} dt'' \langle n, \mathbf{0}|V_I(t')V_I(t'')|n, \mathbf{0}\rangle =$$

$$-\frac{1}{\hbar^2}\sum_{\mathbf{m}}\int_0^t dt' \int_0^{t'} dt'' e^{iE_nt'/\hbar} V_{nn}^{\mathbf{0m}} e^{-i(E_n+\epsilon_{\mathbf{m}})(t'-t'')/\hbar} V_{nn}^{\mathbf{m0}} e^{-iE_nt''/\hbar}$$

$$-\frac{1}{\hbar^2}\sum_{\mathbf{m},n'\neq n}\int_0^t dt' \int_0^{t'} dt'' e^{iE_nt'/\hbar} V_{n'n}^{\mathbf{0m}} e^{-i(E_{n'}+\epsilon_{\mathbf{m}})(t'-t'')/\hbar} V_{n'n}^{\mathbf{m0}} e^{-iE_nt''/\hbar}$$

$$\approx -\frac{\pi}{\hbar}\left\langle (V_{nn}^{\mathbf{0m}})^2\right\rangle \rho_\phi\, t - \frac{\pi}{\hbar}\left\langle (V_{n'n}^{\mathbf{0m}})^2\right\rangle \rho_d\, t \equiv -\frac{(\gamma_\phi+\gamma_d)\,t}{\hbar} \equiv -t/T_2, \tag{5.33}$$

where

$$V_{n'n}^{\mathbf{0m}} = \langle n, \mathbf{0}|V|n', \mathbf{m}\rangle,$$

and so on, and where we have used the same procedures as before—for example, equation 5.22. We have split up the contribution to the overall decay constant Γ into two parts, a part coming from "elastic" terms that leave the system state $|n\rangle$ unchanged and another where the system state changes in the intermediate sum. We have correspondingly introduced two different densities of states: One, ρ_ϕ, is the density given that the system state is held fixed. The second, ρ_d, is the total density of states. There is a decay in the probability of $|n, \mathbf{0}\rangle$ even if the system remains in the initial state n; we call $\gamma_\phi/\hbar \equiv 1/T_2'$ the *pure dephasing* constant and T_2' the pure dephasing time; it is due to bath-changing, system state preserving (elastic) interactions.

The whole second-order term (both parts without regard to whether n changes) belongs to an exponential series that under the usual Golden Rule circumstances we can equate to (within second-order perturbation theory)

$$\langle n, \mathbf{0}|e^{-iHt/\hbar}|n, \mathbf{0}\rangle \sim e^{-t/T_2}. \tag{5.34}$$

Next, we examine the decay of the system population irrespective of what the bath is doing, (as opposed to the initial system-bath state decay just investigated):

$$P_n(t) \equiv \sum_{\mathbf{m}} \left|\langle n, \mathbf{m}|e^{-iHt/\hbar}|n, \mathbf{0}\rangle\right|^2,$$

$$= \sum_{\mathbf{m}}\left|\langle n, \mathbf{m}|\left(1 - \frac{i}{\hbar}\int_0^t dt'\, V_I(t') - \frac{1}{\hbar^2}\int_0^t dt' \int_0^{t'} dt''\, V_I(t')V_I(t'') + \cdots\right)|n, \mathbf{0}\rangle\right|^2,$$

$$= \left| \langle n, \mathbf{0}| \left(1 - \frac{i}{\hbar} \int_0^t dt'\, V_I(t') - \frac{1}{\hbar^2} \int_0^t dt' \int_0^{t'} dt''\, V_I(t') V_I(t'') + \cdots \right) |n, \mathbf{0}\rangle \right|^2$$

$$+ \sum_{m \neq 0} \left| \langle n, \boldsymbol{m}| \left(1 - \frac{i}{\hbar} \int_0^t dt'\, V_I(t') - \frac{1}{\hbar^2} \int_0^t dt' \int_0^{t'} dt''\, V_I(t') V_I(t'') + \cdots \right) |n, \mathbf{0}\rangle \right|^2 ,$$

$$= \left| 1 - i \cdot 0 - (\gamma_\phi + \gamma_d)\frac{t}{\hbar} \right|^2 + \sum_{m \neq 0} \left| -\frac{i}{\hbar} \int_0^t e^{i\epsilon_m t'/\hbar} V_{nn}^{\boldsymbol{m}0}\, dt' - \cdots \right|^2 . \tag{5.35}$$

In other language, this is a trace over the bath giving a diagonal element of the system density matrix. Reexpressing the last line, after writing the second term as

$$\begin{aligned}
\sum_{m \neq 0} \left| \frac{-i}{\hbar} \int_0^t e^{i\epsilon_m t'/\hbar} V_{nn}^{0\boldsymbol{m}}\, dt' - \cdots \right|^2 &= \sum_{m \neq 0} \frac{1}{\hbar^2} \int_0^t dt' \int_0^t dt'' e^{i\epsilon_m (t'-t'')/\hbar} V_{nn}^{0\boldsymbol{m}} V_{nn}^{\boldsymbol{m}0} \\
&\to \frac{2\pi}{\hbar} \left\langle (V_{nn}^{0\boldsymbol{m}})^2 \right\rangle \rho_{\boldsymbol{m}} t, \\
&= 2\gamma_\phi t/\hbar,
\end{aligned}$$

we have

$$P_n(t) = 1 - \frac{2}{\hbar}(\gamma_\phi + \gamma_d)\, t + \frac{2}{\hbar}\gamma_\phi t + \mathcal{O}(t^2) = 1 - \frac{2}{\hbar}\gamma_d t + \mathcal{O}(t^2).$$

Thus,

$$P_n(t) = e^{-2\gamma_d t/\hbar} \equiv e^{-t/T_1}, \tag{5.36}$$

where T_1 is the population relaxation time. This last calculation has shown that the population relaxation time is $T_1 = 1/2\gamma_d$. Finally, we have shown the relationship

$$\boxed{\frac{1}{T_2} = \frac{1}{\hbar}(\gamma_d + \gamma_\phi) = \frac{1}{2T_1} + \frac{1}{T_2'}.} \tag{5.37}$$

where the times are defined as

$$\frac{1}{T_1} \equiv 2\frac{\gamma_d}{\hbar} \quad \text{(population relaxation time)} \tag{5.38}$$

$$\frac{1}{T_2'} \equiv \frac{\gamma_\phi}{\hbar} \quad \text{(pure dephasing time)} \tag{5.39}$$

$$\frac{1}{T_2} \equiv \frac{1}{\hbar}(\gamma_\phi + \gamma_d) \quad \text{(state decay time or total dephasing time)} \tag{5.40}$$

One immediate and important consequence of these equations pertains to spectroscopy. Suppose there is a diatomic molecule in a gas of xenon atoms, the latter forming a "bath." Without going into derivations now, the state decays exponentially due to its interaction with a bath, and the absorption spectrum is given by the Fourier transform of the autocorrelation of the evolution—that is,

$$\epsilon(\omega) = \int_{-\infty}^{\infty} \langle \phi(0)|\phi(t)\rangle e^{i\omega t}\, dt \sim \int_{-\infty}^{\infty} e^{-|t|/T_2 + i(\omega-\omega_0)t}\, dt = \frac{2/T_2}{(\omega-\omega_0)^2 + 1/T_2^2}, \tag{5.41}$$

where $|\phi(0)\rangle$ is the initially excited diatomic. A Lorentzian lineshape is observed, by sweeping the incident light frequency through the resonance for exciting the first diatomic vibrational level starting in the ground level. The radiative decay of the vibration is typically very slow. The bath, however, can relax the diatomic vibration in an inelastic collision, causing population decay. It can also collide elastically with the diatomic, each such collision leaving the bath altered. When we see a classic Lorentzian lineshape in absorption, it has a width governed by the exponential decay constant T_2, which in turn is made up of system depopulation and bath dephasing contributions, as per equation 5.37.

This discussion has raised issues related to the subject of "decoherence" (see chapter 25), or loss of system coherence and interference effects due to interactions with a bath. The premise in decoherence is related but still different enough from spectroscopy that it deserves separate treatment. Note that we included a measurement of the "bath" by focusing on the quantity $\langle n, \mathbf{0}|G^+(t)|n, \mathbf{0}\rangle$. This is inherent in spectroscopy, as amplified shortly. But decoherence refers to other experiments in which the bath is ignored altogether; only the system is measured. Thus, after solving for the whole system and bath, the system-only questions are answered after "tracing over the bath variables" in the expression for the full density matrix, leaving only the system variables. Given any interaction of the system and the bath, leading to correlations, or "entanglement," tracing over the bath leaves a system density matrix that mathematically cannot have come from a wavefunction for the system alone, and lacks the coherence such a wavefunction would exhibit.

An absorption spectrum, on the other hand, measures the bath as well as the system. It should not be hard to think of situations where the division of a problem into a "system" and a "bath" is somewhat arbitrary and fungible. The observed spectrum is not aware of our choice. So the spectrum must be a holistic measurement of what we call the system *and* what we call the bath.

For example, consider a large molecule containing a single bond that acts as a "chromophore," or a solid with a "color center" defect. Both of these will absorb light energy more or less locally, and it is tempting to call the chromophore or color center our system. The bath may include many degrees of freedom. This is a basis choice that will affect the magnitudes of T_1 and T_2', but T_2 should remain invariant to our choice, under $1/T_2 = 1/2T_1 + 1/T_2'$. Decoherence is taken up more fully in chapter 25.

5.4 Eigenstates as Fourier Transform of Dynamics

Given the eigenstates $|\psi_n\rangle$ of a Hamiltonian H, the dynamics for an arbitrary initial state $|\varphi(0)\rangle$ is obtained as

$$|\varphi(t)\rangle = e^{-i\hat{H}t/\hbar}|\varphi(0)\rangle = \sum_{n=0}^{\infty} e^{-iE_n t/\hbar}|\psi_n\rangle\langle\psi_n|\varphi(0)\rangle. \tag{5.42}$$

Suppose now that we apply a Fourier transform to both sides of equation 5.42 in order to filter out all but one frequency,

$$\lim_{T\to\infty} \tfrac{1}{2T} \int_{-T}^{T} e^{iE_m t/\hbar} |\varphi(t)\rangle \, dt$$

$$= \sum_{n=0}^{\infty} c_n \lim_{T\to\infty} \frac{1}{2T} \int_{-T}^{T} e^{-i(E_n - E_m)t/\hbar} |\psi_n\rangle \, dt = c_m |\psi_m\rangle, \tag{5.43}$$

where $c_n \equiv \langle \psi_n | \varphi(0) \rangle$. The result of this procedure is to filter out or "project out" just the mth eigenstate, assuming that the energies are nondegenerate.

No matter what approximate method (for example, semiclassical propagation) we use to get $|\varphi(t)\rangle$, approximate eigenstates may be extracted from the dynamics by Fourier transform. Even when we are not able to take T long enough to resolve the eigenenergies from each other, we can still obtain much information about their nature.

We assumed the eigenvalue E_m was known. However, the eigenvalues may be obtained from the dynamics too, since

$$S(E) \equiv \frac{1}{2\pi\hbar} \int_{-\infty}^{\infty} e^{iEt/\hbar} \langle \varphi(0) | \varphi(t) \rangle \, dt,$$

$$= \frac{1}{2\pi\hbar} \sum_{n=0}^{\infty} |c_n|^2 \int_{-\infty}^{\infty} e^{-i(E_n - E)t/\hbar} dt, \tag{5.44}$$

$$= \sum_{n=0}^{\infty} |c_n|^2 \delta(E_n - E). \tag{5.45}$$

The eigenvalues are revealed as peaks in $S(E)$—that is, the spectrum of the state φ. The peaks have a weight $p_n^{\varphi} \equiv |c_n|^2$. The larger this weight, the easier it is to numerically extract the eigenfunction $\psi_n(x)$ by Fourier transform.

A simplification of equation 5.43 occurs for the harmonic oscillator. We can write the 40th eigenstate as a superposition of Gaussians [16–18], since the coherent state (phase space displaced ground state) wavepacket evolution is especially simple and analytic (looking ahead to chapter 10), as figure 5.2 illustrates. It may be shown that the *normalized* harmonic oscillator eigenstate is expressible as

$$\psi_n(q) = \left(\frac{m\omega}{\pi\hbar}\right)^{1/4} \frac{\sqrt{n!\,\omega}}{2\pi} e^{\mu/2} \mu^{(1-n)/2} \int_0^{2\pi} e^{i(n+1/2)\omega t} g(q,t) \, dt, \tag{5.46}$$

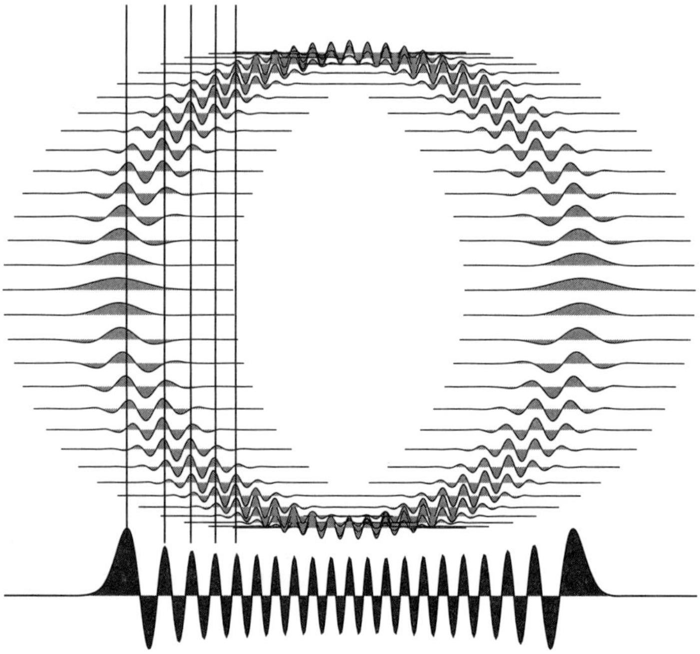

Figure 5.2. Representation of a harmonic oscillator eigenstate for $N = 40$ as a coherent, specifically phased superposition of 60 Gaussian wavepackets. This figure will reappear as the result of a continuous wave (CW) laser excitation; see figure 17.8, later.

where

$$g(q,t) = \exp\left[-\frac{m\omega}{2\hbar}\left(q - \frac{p_0}{m\omega}\sin\omega t\right)^2 + i\frac{p_0}{\hbar}\cos\omega t\left(q - \frac{p_0}{m\omega}\sin\omega t\right) + \frac{i}{2\hbar}\left(\frac{p_0^2}{m\omega}\cos\omega t\sin\omega t - \hbar\omega t\right)\right].$$

Figure 5.2 should be inspected for the interesting way in that the wavepackets conspire to make the eigenstate (using 60 terms in a discrete sum approximating the integral of equation 5.46).

Chapter 6

Stationary Phase Integration

Integration by the method of stationary phase is the core of semiclassical approximation. The subject deserves its own chapter, even though it is fairly elementary in its simplest, one-dimensional form. The extension to many dimensions is straightforward too, but things get more complicated and even murky when going into the subjects of complex stationary phase points (steepest descent) and Stokes phenomena, where the results diverge.

6.1 One Dimension

The integral of a Gaussian is fundamental to the stationary phase idea (here α must have a positive real part):

$$\int_{-\infty}^{\infty} e^{-\alpha x^2}\, dx = \sqrt{\frac{\pi}{\alpha}}. \tag{6.1}$$

The Gaussian integrals of pure imaginary quadratic exponents converge and read, in perfect analogy,

$$\begin{aligned}\int_{-\infty}^{\infty} e^{-i\beta x^2}\, dx &= \sqrt{\frac{\pi}{i\beta}},\\ \int_{-\infty}^{\infty} e^{i\beta x^2}\, dx &= \sqrt{-\frac{\pi}{i\beta}},\end{aligned} \tag{6.2}$$

where β is real positive. These two differ by a phase factor of $\pi/2$, coming from the sign in the exponent. This is intimately related to the so-called Maslov indices, phases in semiclassical integrals that sometimes cause confusion. Here, we see at least in this simple context that the phase choice $\pm\pi/2$ stems merely from the sign of the quadratic form in the exponent.

The main idea of the stationary phase approximation is that integrals of the form $\int f(x)\, e^{i\phi(x)/\hbar}\, dx$ can have exponents that locally look quadratic, near where $i\phi'(x) = 0$ ($\phi(x)$ is stationary, and we know how to do integrals that are quadratic

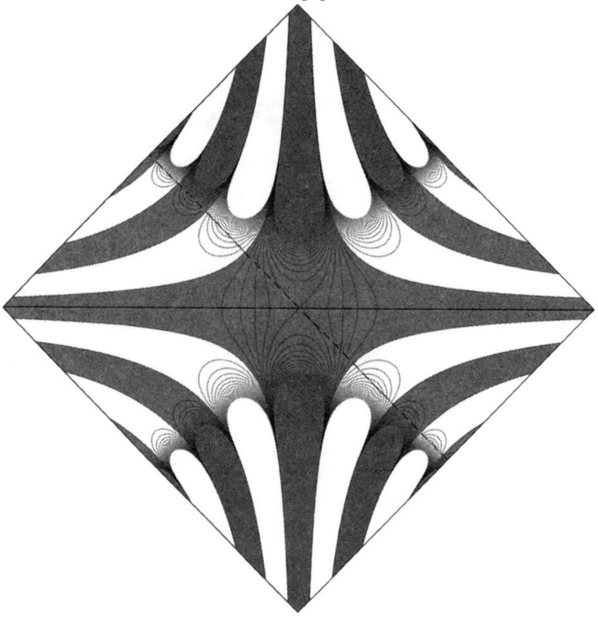

Figure 6.1. Contour plot of Re{exp[$-z^2$]} over the complex z plane. The path of integration of equation 6.1 is along the horizontal real z axis (solid line) passing through the center of the image at $z = 0$. On the other hand, the integrand oscillates wildly in sign for the 135- and 45-degree paths (135 shown dashed) corresponding to the integrals (equation 6.2). Gray regions are positive; white regions are negative. The contours are cut off above and below threshold values, causing the pattern seen in the upper and lower corners.

in the exponent. Details of the integrand away from such points may not matter much, since the exponential may be oscillating wildly and self-canceling, until another point x' with $i\phi'(x') = 0$ is approached, and so on. These points become separate contributions to the estimate of the integral.

The integrals in equation 6.2 are derived from 6.1, supposing α is real positive, by a rotation of the integration path in the complex plane—for example, one that approaches $z = 0$ along the line $z = u(1 - i)/\sqrt{2}$ (figure 6.1). Then

$$\int_{-\infty}^{\infty} e^{-i\alpha z^2}\, dz = \frac{1-i}{\sqrt{2}} \int_{-\infty}^{\infty} e^{-\alpha u^2}\, du = \frac{1-i}{\sqrt{2}}\sqrt{\frac{\pi}{\alpha}} = \sqrt{\frac{\pi}{i\alpha}}. \tag{6.3}$$

COMPLEX STATIONARY PHASE POINTS

The right path can tame integrals with mixed real and imaginary exponents with complex stationary phase points, such as

$$\int_{-\infty}^{\infty} e^{-(i\alpha+\beta)x^2}\, dx = \sqrt{\frac{\pi}{\beta + i\alpha}}, \tag{6.4}$$

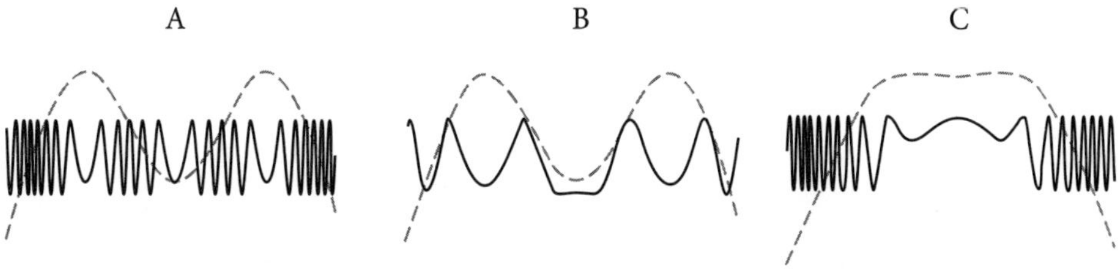

Figure 6.2. One success and two failures of stationary phase. The dashed line corresponds to $\phi(x)$ and the solid line to $\cos(\phi(x)/\hbar)$. In A (success), the three stationary points are well separated, by at least several oscillations in between stationary points. In B, $\hbar$ is larger with no other change, and now the stationary phase approximation fails, although $\phi(x)$ is the same. In C, it fails even for the same value of $\hbar$ as in A, because ϕ is now different and changing too slowly.

for real $\alpha, \beta, \beta > 0$, as pure real Gaussian integrals. Such contour changes can transform nasty oscillatory integrands into tame, nonoscillating real Gaussians. This is called the method of steepest descent—that is, finding a path in the complex plane where the integral is nonoscillating.

Now consider a more general integral of the form

$$I_3 = \int_{-\infty}^{\infty} f(x)\, e^{i\phi(x)/\hbar}\, dx. \tag{6.5}$$

The parameter $\hbar$ is a scale factor that allows us to vary how fast the phase changes. If we plot $\phi(x)/\hbar$, we might get one of the cases shown in figure 6.2: Each max or min in the exponent (dashed line) acts locally like a quadratic near its extrema, allowing us to write

$$I_3 \approx \sum_{\{x_k|\phi'(x_k)=0\}} \int f(x) e^{\frac{i}{\hbar}\phi(x_k)+\frac{i}{2\hbar}\phi''(x_k)(x-x_k)^2}\, dx. \tag{6.6}$$

Then,

$$I_3 \approx \sum_i f(x_k) \sqrt{\frac{2\pi i\hbar}{\phi''(x_k)}} e^{\frac{i}{\hbar}\phi(x_k)} = \sum_i f(x_k) \left|\frac{2\pi\hbar}{\phi''(x_k)}\right|^{\frac{1}{2}} e^{\frac{i}{\hbar}\phi(x_k)-i\pi\nu_k}, \tag{6.7}$$

where we have collected the total phase of the expression in the exponential as a parameter ν_k, taking on the values $\pm 1/4$.

The reason stationary phase works (along the original path on the real x axis) is fast oscillation of the integrand receding from the points of stationary phase x_k on either side, regions where the net integral accumulations are small. This is why the value of $f(x)$ matters most at the stationary points $x = x_k$. Such regions can be "isolated" from each other by intervening fast oscillations, leading to a new stationary phase point at a different x_k. This can certainly fail, as when $\hbar$ is too big, and the regions are not isolated from each other, as in figure 6.2, C. The worst failure comes if $\phi''(x_k) = 0$; this is a singularity in the approximation.

It is also possible to improve on the approximation that f is a constant near each stationary point by writing

$$I_3 \approx \sum_{\{x_k|\phi'(x_k)=0\}} \left[f(x_k) + f'(x_k)(x - x_k) + \frac{1}{2} f''(x_k)(x - x_k)^2 \right]$$
$$\times \int e^{\frac{i}{\hbar}\phi(x_k) + \frac{i}{2\hbar}\phi''(x_k)(x-x_k)^2}\, dx. \tag{6.8}$$

The linear term involving $(x - x_k)$ vanishes when integrated, since the integrand is odd about $x = x_k$. This leaves

$$I_3 \approx \sum_k \left[f(x_k)\sqrt{\frac{2\pi i\hbar}{\phi''(x_k)}} + \sqrt{\frac{\pi}{2}} f'(x_k) \left(\frac{i\hbar}{\phi''(x_k)} \right)^{3/2} \right] e^{\frac{i}{\hbar}\phi(x_k)}. \tag{6.9}$$

PHASOR PICTURE

A very different depiction of an integral well approximated by stationary phase is obtained by dividing up the interval into complex "phasors" (vectors whose length and direction are determined by the real and imaginary parts of $e^{i\phi(q_i)/\hbar}\Delta q$). The phasors are shown as an arrow with magnitude and direction determined by the real and imaginary of the integrand, and their bases are placed at the value of the integral (sum of the phasors) up to that phasor (see figure 6.3):

$$I \equiv \int_{-\infty}^{\infty} e^{i\phi(q)/\hbar}\, dq \approx \sum_i e^{i\phi(q_i)/\hbar}\Delta q. \tag{6.10}$$

STATIONARY PHASE IN SEVERAL DIMENSIONS

We can generalize the idea of stationary phase to many dimensions involving integrals such as

$$I_3 = \int_{-\infty}^{\infty} f(\boldsymbol{q})\, e^{i\phi(\boldsymbol{q})/\hbar}\, d\boldsymbol{q}. \tag{6.11}$$

The stationary phase condition becomes

$$\frac{d\phi(\boldsymbol{q})}{dq_1} = 0, \quad \frac{d\phi(\boldsymbol{q})}{dq_2} = 0, \ \ldots \ \text{ that is, } \nabla\phi(\boldsymbol{q}) = 0, \ \boldsymbol{q} = \boldsymbol{q}_k, \tag{6.12}$$

where $\boldsymbol{q}_k$ is a stationary phase point.

The idea is to approximate the function in the exponent $\phi(\boldsymbol{q})$ with a multivariate quadratic form matched to the second derivatives of $\phi(\boldsymbol{q})$ (the first derivatives being 0) in the vicinity of each stationary phase point $\boldsymbol{q}_k$—that is,

$$\phi(\boldsymbol{q}) \approx \phi(\boldsymbol{q}_k) + (\boldsymbol{q} - \boldsymbol{q}_k) \cdot \boldsymbol{\lambda}_k \cdot (\boldsymbol{q} - \boldsymbol{q}_k)/2. \tag{6.13}$$

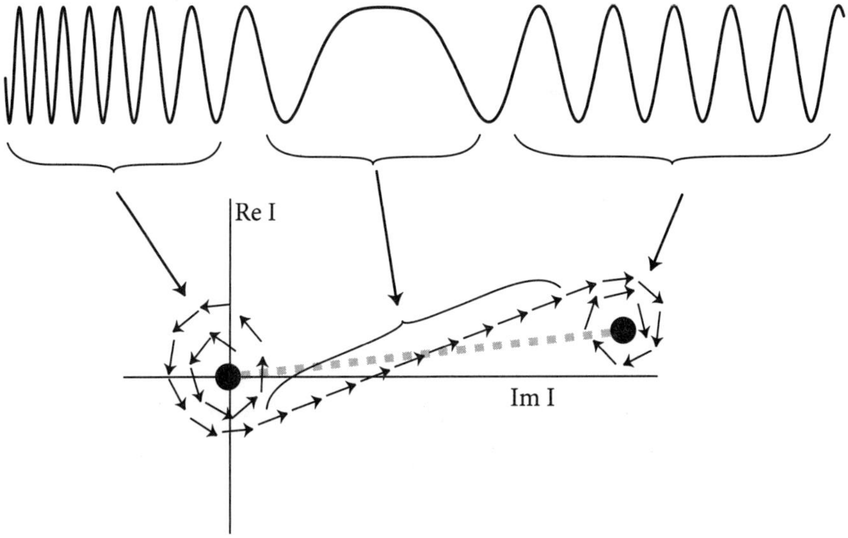

Figure 6.3. Phasor picture of the stationary phase approximation. Only the real part of the integrand is shown at the top. The integral is represented as the limit of a finite numerical sum (see equation 6.3). Vectors (phasors) with fixed length Δq and direction given by the phase of $e^{i\phi(q_i)/\hbar}$ are drawn head to toe, vector adding successive terms in the sum. At first, the result spirals tightly around 0, but as the stationary phase point is approached the net vector addition starts to move away from 0, making steady progress as the phase stabilizes, which can only mean that the many successive phasors remain pointing in the same direction. As the oscillation begins again to the right, the phasor starts to spiral around the final value of the integral. The value of the integral is given by the difference between the starting and ending locations.

The general Gaussian integral, equation A.19 (later), allows us to conclude, in n dimensions with multiple stationary phase points k,

$$I_3 = \int_{-\infty}^{\infty} f(\boldsymbol{q})\, e^{i\phi(\boldsymbol{q})/\hbar}\, d\boldsymbol{q} \approx \sum_k \int_{-\infty}^{\infty} f(\boldsymbol{q}) e^{i(\boldsymbol{q}-\boldsymbol{q}_k)\cdot\boldsymbol{\Lambda}_k\cdot(\boldsymbol{q}-\boldsymbol{q}_k)/\hbar}\, e^{i\phi(\boldsymbol{q}_k)/\hbar} d\boldsymbol{q},$$

$$= (2\pi\hbar)^{\frac{n}{2}} \sum_k f(\boldsymbol{q}_k)|\det[\boldsymbol{\Lambda}_k]|^{-\frac{1}{2}} e^{i\phi(\boldsymbol{q}_k)/\hbar - i\pi\nu_k}. \tag{6.14}$$

A δ FUNCTION

The integral

$$\sqrt{\frac{i\alpha}{\pi}} \int_{-\infty}^{\infty} e^{-i\alpha q^2}\, dq = 1 \tag{6.15}$$

is independent of α. The stationary phase zone becomes ever narrower as $\alpha \to \infty$, and even though the function extends indefinitely, its oscillates symmetrically above and below 0 so rapidly, away from $q = 0$, that any reasonable function $f(q)$ must give vanishing contribution to the integral in that region apart from the $q = 0$

contribution—that is,

$$\lim_{\alpha\to\infty}\sqrt{\frac{i\alpha}{\pi}}\int_{-\infty}^{\infty} f(q)\, e^{-i\alpha(q-q_0)^2}\, dq = f(q_0). \tag{6.16}$$

Then it can be confidently stated that

$$\lim_{\alpha\to\infty}\sqrt{\frac{i\alpha}{\pi}}\, e^{-i\alpha(q-q_0)^2} = \delta(q-q_0). \tag{6.17}$$

6.2 Example: Airy Function and Linear Ramp Potential

Here, we consider the solution to the time-independent Schrödinger equation for the linear ramp potential $V(q) = -\alpha q$—namely,

$$\frac{\hat{p}^2}{2m}\xi_E(q) - (\alpha\,\hat{q} + E)\xi_E(q) = 0, \tag{6.18}$$

that reads, in the coordinate space representation,

$$\frac{-\hbar^2}{2m}\xi_E''(q) - (\alpha\, q\, + E)\xi_E(q) = 0, \tag{6.19}$$

and in momentum space ($q \to i\hbar\,\partial/\partial p$),

$$\frac{p^2}{2m}\bar{\xi}_E(p) - i\alpha\hbar\bar{\xi}_E'(p) = E\;\bar{\xi}_E(p). \tag{6.20}$$

We can reduce the solution at any E to the $E = 0$ case by the shift $q \to q - E/\alpha$—that is, $\xi_0(q) = \xi_E(q - E/\alpha)$. We also make the substitution

$$u = \left(\frac{m\alpha}{\hbar^2}\right)^{1/3} q \equiv \gamma q.$$

Then the Schrödinger equation becomes

$$\frac{1}{2}\phi''(u) + u\phi(u) = 0. \tag{6.21}$$

with $\phi(u) = \xi_0(\gamma^{-1}u)$. In the momentum representation, this becomes

$$\frac{p^2}{2}\phi(p) - i\frac{\partial}{\partial p}\phi(p) = 0. \tag{6.22}$$

The solution is seen to need a cubic in the exponent, and indeed

$$\phi(p) = e^{-ip^3/6} \tag{6.23}$$

is a solution.

The coordinate space wavefunction is a Fourier transform away,

$$\begin{aligned}\xi_0(u) &= \frac{1}{\sqrt{2\pi}} \int_{-\infty}^{\infty} e^{ipu}\phi(p)\, dp, \\ &= \frac{2}{\sqrt{2\pi}} \int_{0}^{\infty} \cos(p^3/6 - pu)\, dp, \\ &= 2^{1/3}\sqrt{2\pi}\, \mathrm{Ai}(-2^{1/3}u), \end{aligned} \tag{6.24}$$

recognizing the second to last line as one definition of an Airy function. Thus

$$\xi_0(q) = \mathrm{Ai}\left[-\left(\frac{2m\alpha}{\hbar^2}\right)^{1/3} q\right]. \tag{6.25}$$

At energy E, the solution is

$$\boxed{\xi_E(q) = \mathrm{Ai}\left[-\left(\frac{2m\alpha}{\hbar^2}\right)^{1/3} (q + E/\alpha)\right].} \tag{6.26}$$

The potential is unbounded and the wavefunction is not normalizable, but the Airy functions obey

$$\int_{-\infty}^{\infty} \mathrm{Ai}(t+x)\mathrm{Ai}(t+y)dt = \delta(x-y); \tag{6.27}$$

this is easy to prove by going back to the momentum space representation. It is convenient and traditional for unbounded wavefunctions to be energy δ-function normalized, so if we define

$$\psi_E(q) = \left(\frac{2m}{\hbar^2\sqrt{\alpha}}\right)^{1/3} \mathrm{Ai}(q + E/\alpha), \tag{6.28}$$

then

$$\int_{-\infty}^{\infty} \psi_E(q)\psi_{E'}(q)dt = \delta(E - E'). \tag{6.29}$$

It is interesting and useful for some applications to know the exact overlap of an Airy function with a general Gaussian wavefunction. This is given by

$$\frac{1}{\sqrt[4]{\pi\alpha}} \int_{-\infty}^{\infty} e^{-\frac{(q-a)^2}{2\alpha} + ip_0(q-a)} \mathrm{Ai}(q)\, dq = (4\pi\alpha)^{1/4}\, e^{\frac{\alpha^3}{12} + \frac{1}{2}\alpha(a + i\alpha p_0) - \frac{\alpha p_0^2}{2}}\, \mathrm{Ai}\left(\alpha^2/4 + i\alpha p_0 + a\right), \tag{6.30}$$

where a is the center of the Gaussian, p_0 is its average momentum, and α may have an imaginary part, giving the Gaussian position-momentum correlation.

The Airy function earns its place in this section if we use the stationary phase approximation to perform the Fourier transform in equation 6.24. It is very instructive to understand this example in detail. The stationary phase points on the integral

$$\psi_E(q) \propto \frac{1}{2\pi} \int_{-\infty}^{\infty} e^{ipq/\hbar} e^{-ip^3/6m\alpha\hbar - iEp/\alpha\hbar}\, dp \tag{6.31}$$

are given by the zeros in the derivative of the exponent:

$$p = \pm\sqrt{2m(E+\alpha q)}. \tag{6.32}$$

The second derivative of the exponent, needed for the prefactor in the stationary phase formula, is just a constant divided by the square root of momentum, yielding for the semiclassical wavefunction

$$\psi_E(q) \sim \frac{c}{\sqrt{p(q)}} \exp\left[\frac{i}{2m\alpha\hbar}[2m(E+\alpha q)]^{3/2} + i\pi/4\right]$$

$$+ \frac{c}{\sqrt{p(q)}} \exp\left[-\frac{i}{2m\alpha\hbar}[2m(E+\alpha q)]^{3/2} - i\pi/4\right],$$

or

$$\boxed{\psi_E(q) \sim \frac{c'}{\sqrt{p(q)}} \cos\left(\int_{E/\alpha}^{q} p(q')\, dq'/\hbar - \pi/4\right).} \tag{6.33}$$

We now analyze the integral schematically and the result numerically (figures 6.4 and 6.5). The area of the action integral $\int p_E(q)\, dq$, taken from the classical turning point into the classically allowed region, plays a crucial role. The two stationary phase points will be badly coalescing if the phase advance from one stationary phase point to the next is less than 2π; this happens if the area enclosed in the shaded region of figure 6.4 is less than Planck's constant h, to the left of the dashed line in figure 6.4. Note that the semiclassical and the exact wavefunction differ visibly to the left of this point, where the area gets even smaller, and agree very well to the right.

Figure 6.5 is a phase space diagram of the Fourier transform integral over momentum, taking the momentum space wavefunction into coordinate space. The kets $|q\rangle$, $|p\rangle$, and $|\psi_E\rangle$ are indicated as "thick" manifolds. The overlaps $\langle q|p\rangle$, $\langle p|\psi_E\rangle$, and $\langle q|\psi_E\rangle$ are shown as the square and diamond-shaped intersections of these manifolds. The areas of these intersections are proportional to the classical probability density $|\langle p|\psi_E\rangle|^2 \sim |\partial^2 S(p, I)/\partial p \partial I|$, $|\langle q|\psi_E\rangle|^2 \sim |\partial^2 S(q, I)/\partial q \partial I|$, and so on. The phase factor $\cos(pq/\hbar - p^3/6m\alpha\hbar + iEp/\alpha\hbar)$ is plotted with the momentum axis vertical and the cosine value g, clearly revealing the two stationary phase points. If q moves to the left, too close to the classical turning point, the stationary phase zones will coalesce, and the stationary phase approximation will break down. In fact, the approximate eigenfunction blows up, with the amplitude rising as $1/|p(q)|^{1/2}$ near the turning point, where $p(q) = 0$. The probability, going as $1/|p(q)|$, rises as does the classical probability, but of course this is badly off compared to the finite quantum result.

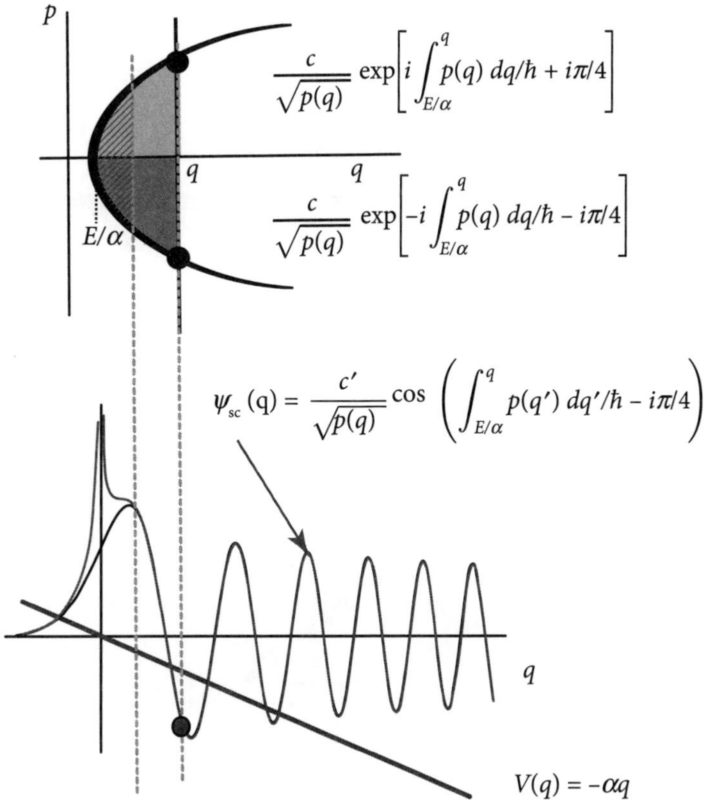

Figure 6.4. (Top) Graphical areas representing the integrals $\pm\int_{E/\alpha}^{q} p(q')dq'$ for the linear ramp potential $V(q) = -\alpha q$ at energy E. (Bottom) Semiclassical Airy function approximation compared to the exact; significant errors develop when the enclosed area (shaded) drops below Planck's constant, h. The semiclassical approximation to the wavefunction in the classically forbidden region, $\psi_{sc}(q) = \frac{c'}{\sqrt{p(q)}}\exp\left(\int_{E/\alpha}^{q} p(q')\,dq'/\hbar\right)$, is also shown at the bottom.

COMMENT ON UNIFORMIZATION

Uniformization is the term used for the repair of semiclassical amplitudes when they fail. Here, we illustrate uniformization in the present semiclassical context of the linear ramp potential. Uniformization often involves improving on a stationary phase integration by performing the integral in some more accurate way—say, by numerical integration or by approximating the integrand by something that can be done analytically.

The Fourier transform over momentum that we just approximated by stationary phase produced the semiclassical approximation, equation 6.33, to the wavefunction of a linear ramp potential. When the stationary phase approximation breaks down, near turning points, an Airy function uniformization can come to the rescue. The Airy function and equation 6.24 is the (in this case, exact) uniformization of the semiclassical stationary phase integral that led to the approximate wavefunction, equation 6.33.

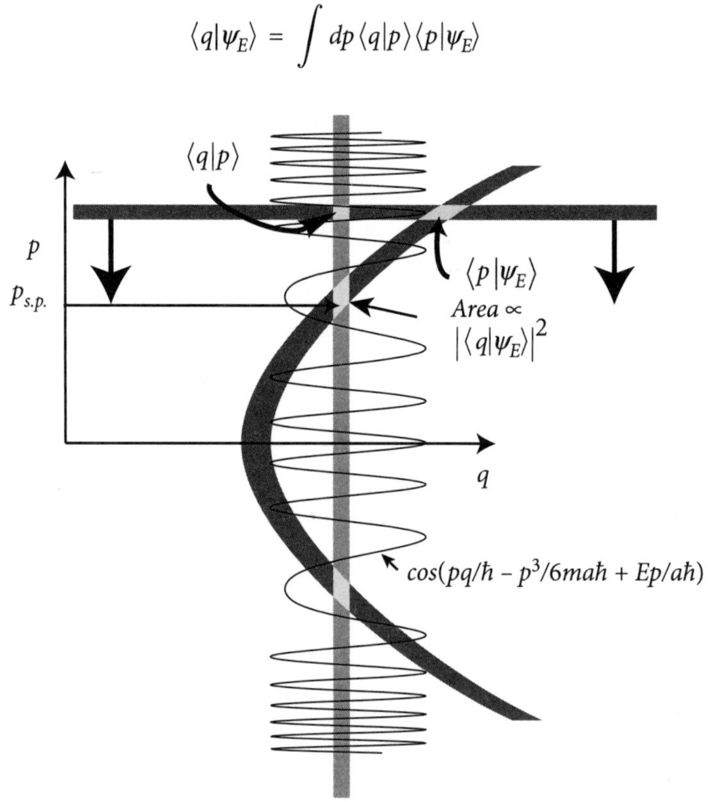

Figure 6.5. Phase space depiction of the momentum integral showing the kets $|q\rangle$, $|p\rangle$, and $|\psi_E\rangle$. Areas of the overlapping regions are proportional to the classical joint probability densities for the semiclassical (that is, stationary phase) versions of $|\langle q|p\rangle|^2$, $|\langle p|\psi_E\rangle|^2$, and $|\langle q|\psi_E\rangle|^2$. The phase factor $\cos(pq/\hbar - p^3/6m\alpha\hbar - iEp/\alpha\hbar)$ is also shown, plotted with the momentum axis vertical and the cosine value horizontal, clearly revealing the two stationary phase points. If q moves to the left, too close to the classical turning point, the stationary phase zones will coalesce, and the stationary phase approximation will break down. The meaning of the intersecting zones came up earlier in section 3.2.

Any smooth, repulsive potential has the same singularity problems as the linear ramp at the turning points. One uniformization procedure is to replace the smooth potential with a straight line approximation to the potential at the turning point, and use the corresponding Airy function as the approximate wavefunction there.

Uniform methods exist even for situations where there are not just two but an infinite number of coalescing stationary phase points in the primitive semiclassical approximation [19]. Several other examples, such as Bessel function uniformization, are well known [20].

6.3 Classically Forbidden (Tunneling) Region

Here, assume $E = 0$, without loss of generality. The stationary phase condition remains $p = \pm\sqrt{2m(E + \alpha q)} = \pm i|\sqrt{2m(E + \alpha q)}|$, but the momentum is imaginary in the classically forbidden region to the left of $q = 0$. Nonetheless, we forge ahead, choosing the branch that decays as it penetrates deeper into the forbidden region. The result is (now dq' is negative)

$$\boxed{\psi_{sc}(q, E) = \frac{c'}{|p(q)|} \exp\left[\int_0^q |p(q')|\, dq'/\hbar\right].} \tag{6.34}$$

Figure 6.4 shows the exact and semiclassical wavefunctions joining nicely as they recede into the tunneling region and away from the turning point.

6.4 Kirchhoff Method

Here, we explore a remarkably useful approach to some scattering problems, one that is essentially quantum mechanical, yet lends itself to semiclassical implementations. When semiclassics based strictly on classical ray paths is insufficient, Kirchhoff's approach can become a type of uniformization that incorporates diffraction (a form of dynamical tunneling; see section 13.1, later) in a natural way.

The basis of the approach is an exact integral expression involving scattering with boundaries present, derived by Kirchhoff in 1883. Kirchhoff also suggested approximations to his formula that are easy to implement and are the essence of what we discuss here. These approximations hark back to Huygen's principle in 1678. (Huygens proposed that every point which a wave reaches becomes a source of an outgoing, forward propagating spherical wave. The sum of these secondary waves determines the form of the wave "downstream" at any subsequent time. He assumed that the secondary waves travelled only in the "forward" direction, and never explained why this is the case.) We focus on two-dimensional scattering problems (as might be useful in quantum semiconductor devices based on AlGaAs or graphene, for example), whereas Kirchhoff was thinking in three dimensions.

The Kirchhoff method is a way of handling reflection and propagation problems by collecting amplitude arriving at a wall or boundary and re-sending it on its way. The boundaries are lines (in 2D) or surfaces (in 3D) consisting of densely packed receivers and repeaters, sending on what they receive, apart from a prefactor that normalizes and phase shifts things correctly. We illustrate this first for "classically allowed" reflection from a mirror, then for two related classically forbidden diffraction problems, and finally for understanding scattering from rough or corrugated surfaces.

As Levine and Schwinger [21] showed, the approximate Kirchhoff method is good only for short wavelengths compared to other scales in the problem. There is no issue with a half-infinite mirror, but there are problems with small mirrors or small apertures. Levine and Schwinger fixed this at considerable effort, but we cannot delve into their methods here. Very small point scatterers, even finitely many of them in some array, may be treated exactly using multiple scattering theory, as in chapter 26.

Within the Kirchhoff approximation, a source at A and a receiver at B communicate via one bounce off a mirror, as shown in figure 6.6 (we do not bother with the direct path). The Kirchhoff receivers and senders lie along the mirror. The amplitude received

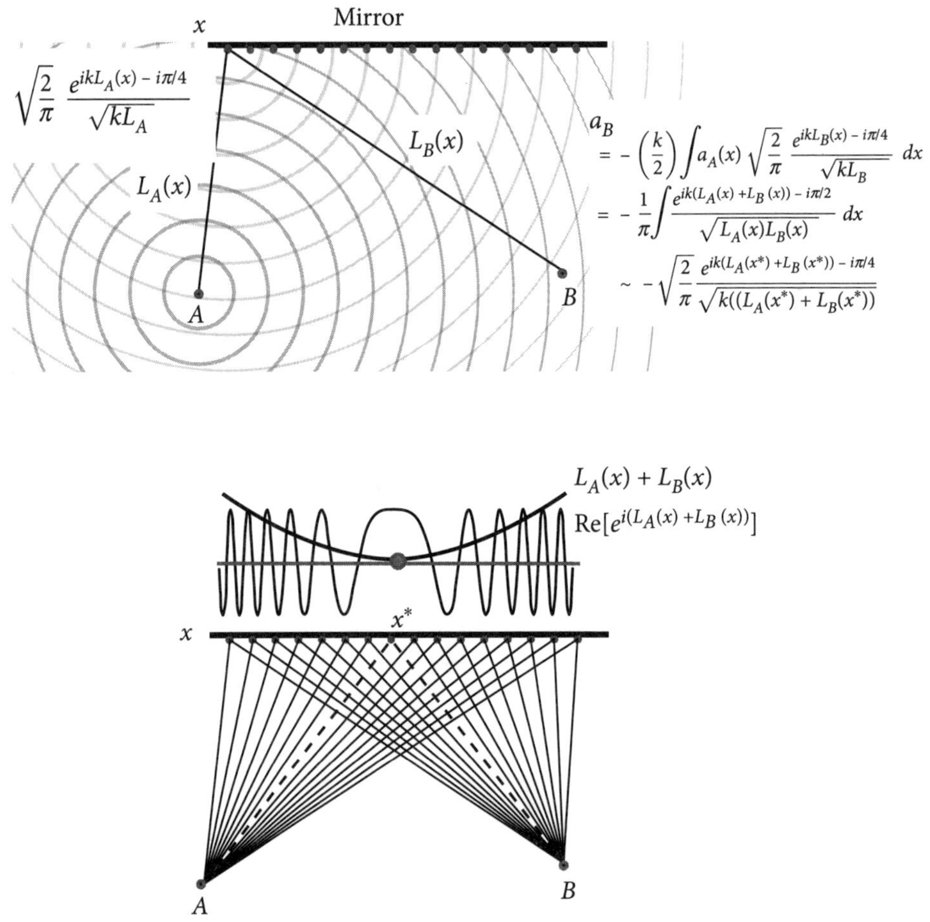

Figure 6.6. Kirchhoff approximation for the case of a source at A reflecting off a mirror, and traveling to a receiver at B. All the paths contribute roughly the same amplitude, but the phase, depending on where the mirror is struck, oscillates rapidly with position except near the specular point x^* (angle of incidence equals angle of reflection), where the phase is stationary. Performing the integral of the amplitude at B from all points on the mirror either exactly or by stationary phase yields a result very close to the exact reflection amplitude, if the mirror is at least several wavelengths long.

at a point x on the mirror, a distance $L_A(x)$ from the sender at A, is

$$a_A(x) = \sqrt{\frac{2}{\pi}} \frac{e^{ikL_A(x) - i\pi/4}}{\sqrt{kL_A(x)}}.$$

The point on the mirror in turn sends out a signal as from a normal point source, proportional to this incoming signal, reaching B with an amplitude (from this one region of length dx around x):

$$a_B(x)\, dx \sim -a_A(x) \left(-\frac{k}{2}\right) \sqrt{\frac{2}{\pi}} \frac{e^{ikL_B(x) - i\pi/4}}{\sqrt{kL_B(x)}}\, dx = -\frac{1}{\pi} \frac{e^{ik(L_A(x) + L_B(x)) - i\pi/2}}{\sqrt{L_A(x)L_B(x)}}\, dx$$

Note the signal leaving the point on the mirror was augmented by a factor $-k/2$ compared to the incoming wave, where k is the wave vector magnitude. This factor can be found by requiring that the incident wave plus the "Kirchhoff wave" for an infinite reflecting wall vanishes everywhere behind the wall.

Now all the points on the mirror are added up, either numerically or by stationary phase integration. The distance *via* the mirror bounce at x is $L_A(x) + L_B(x)$; note that the distance minimizes at the specular point x^*, and the phase $e^{ik(L_A(x)+L_B(x))}$ is stationary there—that is, $d/dx(L_A(x) + L_B(x)) = 0$ at x^*. For typical parameters involving a mirror many wavelengths long and the points A and B many wavelengths away from the mirror, the exact and stationary phase integral are extremely close, and both are very close to the geometric path formula for reflection off a hard wall (that enforces a zero boundary condition and changes the sign of the reflected wave). Stationary phase gives the amplitude a_B at B as

$$a_B = -\frac{1}{\pi}\int dx\, \frac{e^{ik(L_A(x)+L_B(x))-i\pi/2}}{\sqrt{L_A(x)L_B(x)}} \sim -\sqrt{\frac{2}{\pi}}\, \frac{e^{ik(L_A(x^*)+L_B(x^*))-i\pi/4}}{\sqrt{k(L_A(x^*)+L_B(x^*))}}. \tag{6.35}$$

This is just the expression we would use knowing the reflection ought to be specular, with the total distance from A to B given by $L_A(x^*) + L_B(x^*)$. To summarize, a mirror much larger than a wavelength may be considered to be a dense set of point reflectors, each one emitting a signal from A to the receiver at B, but the net signal is very close to that received from the stationary point x^*—that is, the specular reflection.

It is also very instructive to consider the case of a mirror that is not in the right position to image A at B and vice versa. There is no stationary phase location on the mirror, but there are diffractive corrections nonetheless, and the result is that no matter how perfect the (sharp) edge of the mirror, the edges would appear bright to an observer at B if there is a source at A. Assuming that $1/2$ of a full oscillation goes uncancelled, it follows that the brightness of the edge is equivalent to about $1/2$ wavelength of illumination being sprayed out from the edge uniformly in all directions. This example shows that the first mirror to be considered earlier, having a stationary, specular path, still needs to be corrected for diffraction, since the mirror we drew abruptly stops at two ends.

Figure 6.7 reveals the diffraction that results if the mirror ends before the stationary phase point is reached. To be more precise, the diffraction of the end would occur anywhere it ended, but it is minor compared to the specular intensity, if it is present.

The Kirchhoff construction for the region beyond the obstruction, including the dark side diffraction, is shown in figure 6.8. The dark side diffraction is really a continuation of the story of the diffraction from the ends. This is also seen in figure 6.9, and the Kirchhoff treatment is shown in figure 6.8. If the mirror were infinite, the Kirchhoff amplitude is tuned to give 0 on the dark side, but if that infinite integral is truncated at the edge of a finite mirror, the cancellation is not perfect and diffraction shows up behind the mirror. See figure 6.8. This diffractive scattering is easily revealed in a simulation, where a pulse has been arranged to arrive at a finite mirror, as seen in figure 6.9.

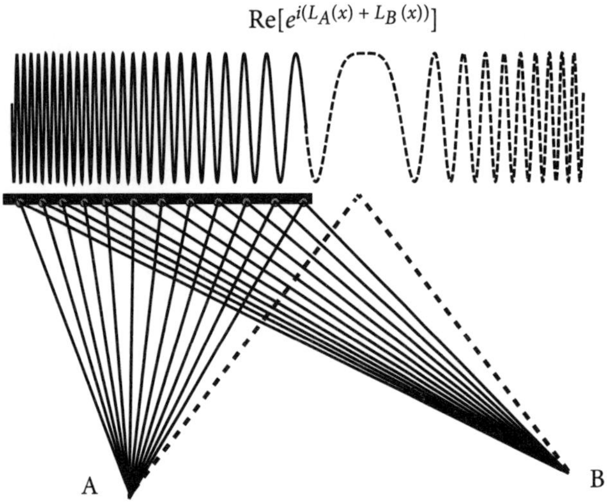

Figure 6.7. If the mirror ends before the stationary phase specular point x^* is reached, the integral does not vanish, but is much smaller, and dominated by the limits of integration at the ends of the mirror. The oscillations of the integrand are not perfectly compensated due to the abrupt end of the integral, and that situation is worse (due to the area of the oscillations) especially at the near end.

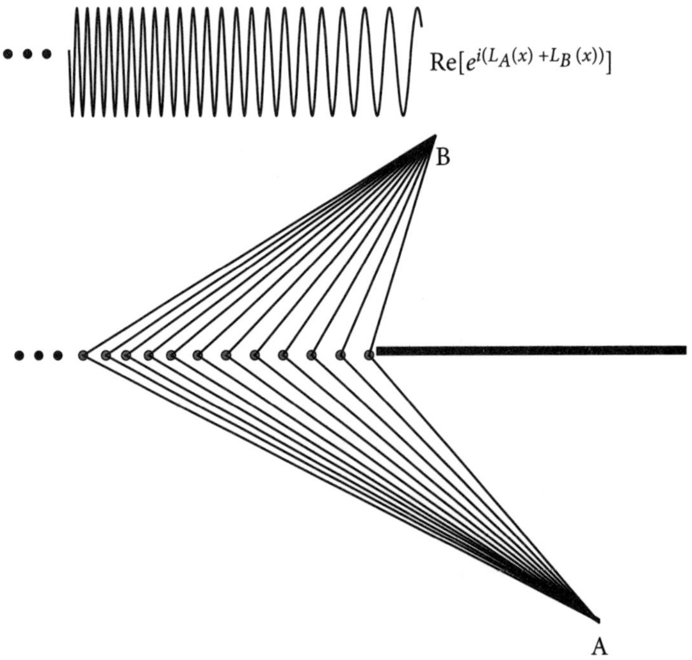

Figure 6.8. Kirchhoff treatment of shadow diffraction, revealing the sudden termination of an integral over a Kirchhoff line extending indefinitely to the left of the barrier wall.

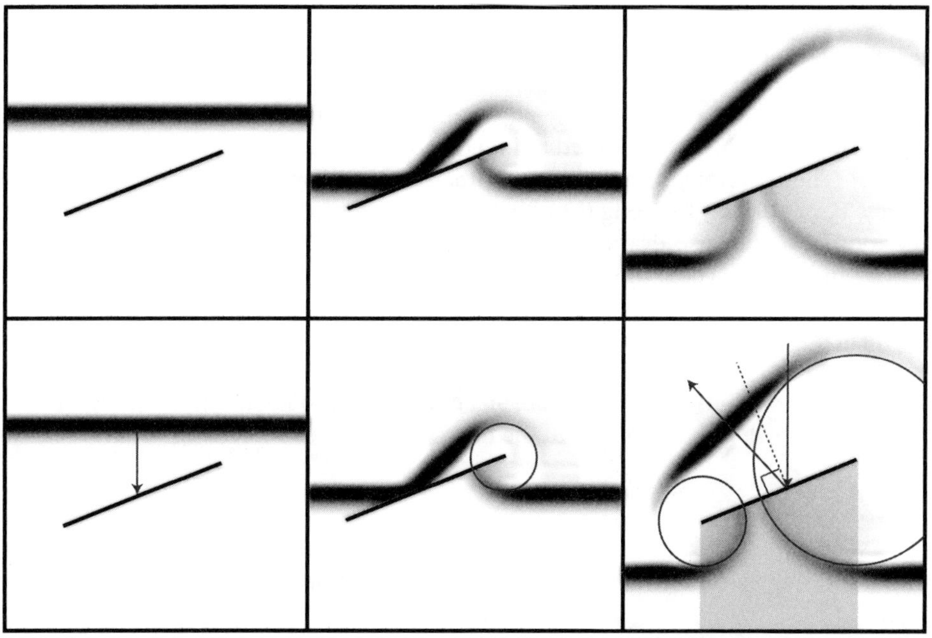

Figure 6.9. A wave pulse arrives at a finite mirror. Circular diffractive waves emanate from both edges, accounting for backscattering and dark-side diffraction. The lower panels are the same sequence, with some marking to guide the eye, including a shaded shadow region. Adapted from *Why You Hear What You Hear*, Eric J. Heller, Princeton University Press, 2012.

6.5 Kirchhoff Method, Rough Surfaces, and Evanescent Waves

NECESSITY OF EVANESCENT WAVES

What happens when a quantum wave of wavelength λ encounters a surface with roughness on a much smaller scale? We suppose that space is free up to the boundary. In optics we are familiar with the fact that apparently perfect mirrors are actually rough on atomic and even much larger scales, up to say a tenth of a wavelength of visible light. Suppose the boundary condition is that the wave must vanish everywhere on the rough surface, when the incoming wave is a single plane wave with wavenumber k and energy $E = \hbar^2 k^2/2m$. Even very close to the surface, the wave $\psi(x, y)$ must satisfy

$$-\nabla^2\psi(x, y) = k^2\psi(x, y). \tag{6.36}$$

Somehow we must be able to use the totality of waves satisfying this equation as a basis for describing the wave near the rough surface. The plane waves of all directions of propagation, $e^{ik(\cos(\theta)x+\sin(\theta)y)}$, for all θ satisfying the equation, but even combinations involving all θ are not capable of describing something much smaller than a wavelength $\lambda = 2\pi/k$. The energy sets k^2, and $k^2 = k_x^2 + k_y^2$.

If the roughness possesses undulations on scales only less than a wavelength, the reflection will then be specular and appear as if no roughness existed several wavelengths and more away from the boundary. Some glass mirror grinding techniques and glass surfaces coming from a melt are almost totally devoid of undulations approaching the wavelength scale of visible light. Then we do not need and in fact cannot use any of the plane waves except the incoming and specularly reflected plane wave to help satisfy the boundary conditions, because if any other plane waves are present they must make an appearance at infinity, belying the specularity of the surface.

So how do we solve the detailed boundary condition on the microscopically rough surface? There are fortunately infinitely many more waves to use, all of the same energy. For a surface flat along the plane $y = 0$ (except for the microscopic roughness deviation), we can use evanescent waves, for $y > 0$—that is, $e^{ik_x x - \kappa_y y}$—with $k^2 = k_x^2 - \kappa_y^2$ and $k_x > k$. These waves also solve equation 6.36. We then have available waves oscillating as fast as needed in x to match the surface roughness, by choice of various κ_y. Smaller scale roughness requires larger κ_y, leading to faster decaying (evanescent) waves above the surface (see figure 23.5).

These observations are embodied in an approach due to Rayleigh, who used a linear combination of evanescent waves of the form we have been discussing—that is,

$$\psi_{evan.}(x, y) = \sum_n a_n e^{-\kappa_n y + ik_x x}, \tag{6.37}$$

with the a_n and κ_n chosen to satisfy the boundary condition at the surface.

Chapter 7

Feynman Path Integral to Van Vleck Propagator

7.1 Semiclassical Program

A bird's-eye view of the program starting with the Schrödinger equation and arriving at the Gutzwiller trace formula through successive stationary phase approximations is depicted in figure 7.1. The time-dependent Van Vleck-Morette-Gutzwiller (VVMG) propagator, or semiclassical Green function, is near the top of the heap. Below the VVMG Green function lie more stationary phase approximations over time and space that must necessarily cause errors in new circumstances.

One of the best reasons for studying classical mechanics is to better understand quantum mechanics. This truth, so crucial in the early days of quantum theory, became almost invisible for decades thereafter. Just look at the textbooks from the 1930s to the 1990s! (And most even today.) Technically, you need the classical Hamiltonian in order to quantize it, but afterward you can nearly dispense with classical mechanics. Some lip service was given to Ehrenfest's theorem in the classic texts, and to the eerily similar appearance of Poisson brackets and quantum commutators (by itself, this is nearly useless). Vague notions of the correspondence principle were used to justify (without much insight, nowadays much enhanced by notions of decoherence) why the world looks so classical to us. Semiclassical WKB theory was usually worked out for one dimension and for a moment a classical trajectory was needed.

One of the artificial roadblocks to development of a semiclassical foundation for quantum mechanics was that classical mechanics is most natural and usually taught in the time domain, where things move, and quantum mechanics has traditionally been taught in the energy domain, where things are frozen. Actually, quantum mechanics is also most natural in the time domain. A new dawn is slowly coming to the teaching of quantum mechanics, brought on mainly by experiments with picosecond, femtosecond, attosecond,... time resolution, and by experiments and theory for many body systems at higher temperatures, where eigenstates are out of the question, and finally by its natural place in semiclassical intuition and approximations.

Schrödinger equation
Feynman path integral propagator

$$i\hbar\frac{\partial\psi(\mathbf{r},t)}{\partial t} = H\psi \quad \longleftrightarrow \quad \int Dr\, e^{iS[r]/\hbar}$$

↓ Stationary phase on the integral over all Feynman paths

Van Vleck-Morette-Gutzwiller semiclassical Green function

$$G(\mathbf{q},\mathbf{q}_0;t) \approx G_{scl}(\mathbf{q},\mathbf{q}_0;t) = \left(\frac{1}{2\pi i\hbar}\right)^{d/2}\sum_j\left|\mathrm{Det}\left(\frac{\partial^2 S_j(\mathbf{q},\mathbf{q}_0;t)}{\partial\mathbf{q}\partial\mathbf{q}_0}\right)\right|^{1/2} \times \exp\left(iS_j(\mathbf{q},\mathbf{q}_0;t)/\hbar - \frac{i\pi\nu_j}{2}\right)$$

↓ Stationary phase on the Fourier transform from time to energy

Van Vleck-Gutzwiller semiclassical energy Green function

$$G_{scl}(\mathbf{r},\mathbf{r}';E) = -\frac{i}{\hbar}(2\pi i\hbar)^{-(d+1)/2''}\sum_{class.\,paths}|\mathcal{D}|^{\frac{1}{2}}\exp[\tfrac{i}{\hbar}S(\mathbf{r},\mathbf{r}',E)]$$

$$\mathcal{D}(\mathbf{r},\mathbf{r}';E) = (-1)^{d+1}\begin{vmatrix}\frac{\partial^2 S}{\partial r\partial r'} & \frac{\partial^2 S}{\partial E\partial r'}\\ \frac{\partial^2 S}{\partial r\partial E} & \frac{\partial^2 S}{\partial E^2}\end{vmatrix}$$

↓ Stationary phase on the trace over all positions

Gutzwiller trace formula for the spectrum

$$\delta g_{scl}(E) = \frac{1}{\hbar\pi}\sum_{po}\frac{T_{ppo}}{\sqrt{|\det(M_{po}-I)|}}\cos(\tfrac{1}{\hbar}S_{po} - \sigma_{po}\tfrac{\pi}{2})$$

Figure 7.1. The hierarchy of semiclassical approximations starting at a fundamental level at the top, revealing the stationary phase approximation needed to arrive at the next level. Classical trajectories are the stationary paths. Interestingly, the last two stationary phase steps can be done in the opposite order, arriving first at a trace formula in the time domain. The author suspects reversing this order would have clarified a lot of problems that arose over the convergence of the energy form of the trace formula. In the energy space trace formula, it was hard to resist summing periodic orbits to infinite order, since once an orbit is discovered, that can be done analytically. The trouble is, this is completely out of the dynamical order, which when followed, seems to be convergent [22].

7.2 Dynamical Postulate of Quantum Dynamics

One of five postulates of quantum mechanics governs the dynamics of wavefunctions—that is, the familiar time-dependent Schrödinger equation

$$i\hbar\frac{\partial\psi(q,t)}{\partial t} = H\psi(q,t). \tag{7.1}$$

We replace this postulate with one that is perfectly equivalent to it, but far superior, as a segue to semiclassical intuition and methods (equation 7.3 below).

Let $\psi(q,0)$ be the wavefunction describing a particle of mass m in a potential $V(q)$ at time $t=0$. At time t, the wavefunction $\psi(q,t)$ depends linearly on $\psi(q,0)$; this can on general grounds be written

$$\psi(q,t) = \int dq'\; G(q,q',t)\,\psi(q',0), \tag{7.2}$$

where $G(q,q',t)$ is the "propagator." Then in the limit $\tau \to 0$, the propagator $G(q,q',\tau)$ is given by

$$\boxed{\lim_{\tau\to 0} G(q,q',\tau) = \langle q|e^{-iH\tau/\hbar}|q'\rangle = \lim_{\tau\to 0}\left(\frac{1}{2\pi i\hbar}\right)^{\frac{1}{2}} \left|\frac{\partial^2 S(q,q',\tau)}{\partial q\,\partial q'}\right|^{\frac{1}{2}} e^{iS(q,q',\tau)/\hbar},} \tag{7.3}$$

where $S(q,q',\tau)$ is the classical action for the path connecting q from q' in a very short time τ, and the vertical bars signify the determinant formed of the mixed second derivatives of the action. The short time limit of the classical action $S(q,q',\tau) = \int_0^\tau L((q,q',t)\,dt$ is easily shown to be

$$\lim_{\tau\to 0} S(q,q',\tau) \approx \frac{1}{2}m\frac{(q-q')^2}{\tau} - V\left(\frac{q+q'}{2}\right)\tau; \tag{7.4}$$

thus the short time propagator is

$$\boxed{\lim_{\tau\to 0} G(q,q';\tau) = \left(\frac{m}{2\pi\hbar i\tau}\right)^{\frac{1}{2}} e^{-\frac{m}{2\hbar i\tau}(q-q')^2 - \frac{i}{\hbar}V\left(\frac{q+q'}{2}\right)\tau}.} \tag{7.5}$$

Either equation 7.3 or equation 7.5 can (and should!) be taken as the founding postulate of quantum mechanics. The time-dependent Schrödinger equation follows quickly from the short time propagator because even if q and q' are not close, the trajectory must travel very fast in the limit $\tau \to 0$; the momentum has no time to change and the average potential is nearly the potential at the average point. Finite times are built of accumulation of much smaller time increments. This is the essence of why this form of the short time propagator is a solid foundation, although mathematicians do not accept these qualitative arguments as any sort of proof.

We insert the short time propagator into equation 7.2 and Taylor expand around $q'=q$: $\psi(q',0) = \psi(q,0) + \psi'(q,0)(q'-q) + 1/2\psi''(q,0)(q'-q)^2$. The expansion works because τ is very small. The function $e^{-\frac{m}{2\hbar i\tau}(q-q')^2}$ oscillates extremely rapidly away from $q' \approx q$, killing any contribution for any reasonably behaved $\psi(q,0)$. Also, it is not important whether the potential is evaluated at the mid-point as stated, or at either end q or q', again because no contribution survives away from $q' \approx q$.

The integral

$$\begin{aligned}
I &= \left(\frac{m}{2\pi\hbar i\tau}\right)^{\frac{1}{2}} \int_{-\infty}^{\infty} dq' e^{-\frac{m}{2\hbar i\tau}(q-q')^2} \psi(q',0), \\
&\approx \left(\frac{m}{2\pi\hbar i\tau}\right)^{\frac{1}{2}} \int_{-\infty}^{\infty} dq' e^{-\frac{m}{2\hbar i\tau}(q-q')^2} [\psi(q,0) + \psi'(q,0)(q'-q) + \tfrac{1}{2}\psi''(q,0)(q'-q)^2], \\
&= \psi(q,0) - i\frac{\hbar^2}{2m}\frac{\partial^2\psi(q,0)}{\partial q^2} \cdot \tau,
\end{aligned} \tag{7.6}$$

is the usual stationary phase integral plus a "correction" term for the variation of $\psi(q,0)$ with q. Thus,

$$\psi(q,\tau) = \left(\frac{m}{2\pi\hbar i\tau}\right)^{\frac{1}{2}} \int_{-\infty}^{\infty} dq' e^{-\frac{m}{2\hbar i\tau}(q-q')^2 - \frac{i}{\hbar}V(\frac{q+q'}{2})\tau} \psi(q',0);$$

$$\boxed{\frac{\psi(q,\tau)-\psi(q,0)}{\tau} = i\frac{\hbar}{2m}\frac{\partial^2\psi(q,0)}{\partial q^2} - \frac{i}{\hbar}V(q)\psi(q,0).} \tag{7.7}$$

This is the short time expansion of $i\hbar\frac{\partial\psi(q,t)}{\partial t} = -\frac{\hbar^2}{2m}\frac{\partial^2\psi(q,0)}{\partial q^2} + V(q)\psi(q,t)$ near $t=0$, thus the equivalence of the dynamical postulate and the usual postulate stating the time-dependent Schrödinger equation is established.

7.3 Feynman Path Integral

We can combine the short time propagators in series to obtain expressions for longer times t. By dividing the interval t up into N approximate propagators for duration t/N, and then letting $N \to \infty$, we reach an exact limit for finite time propagation.

$$\begin{aligned}
G(q,q',t) &= \lim_{N\to\infty} \int \cdots \int dq_1 dq_2 \cdots dq_N\, G(q,q_1,t/N) \\
&\quad \times G(q_1,q_2,t/N)\cdots G(q_N,q',t/N), \\
&= \lim_{N\to\infty} \int \cdots \int dq_1 dq_2 \cdots dq_N \prod_i \left(\frac{e^{iS_i/\hbar}}{A}\right), \\
&= \lim_{N\to\infty} \int \cdots \int \frac{dq_1}{A}\frac{dq_2}{A}\cdots\frac{dq_N}{A} e^{iS_{path}/\hbar},
\end{aligned} \tag{7.8}$$

where

$$A = \left(\frac{mN}{2\pi i\hbar t}\right)^{-\frac{1}{2}}, \tag{7.9}$$

and

$$S_{path} = \sum S_i, \tag{7.10}$$

$$S_i = \frac{m(q_i - q_{i+1})^2}{2\tau} - V(q_{i+1})\tau \tag{7.11}$$

$(\tau = t/N)$.

Only some notation separates equation 7.8 from the Feynman propagator. As $N \to \infty$, the number of paths proliferate, almost all of them arbitrarily jagged, since q_i and q_{i+1} are arbitrary in the course of integrations over coordinates. The expression for the Feynman propagator can be written

$$\boxed{G(q, q', t) = \int e^{i S_{[q(t)]}(q,q',t)/\hbar} D[q(t)],} \tag{7.12}$$

where $[q(t)]$ is the path specified by particular values of the coordinates in the course of integration over the infinitely many paths starting at q' and ending at q in time t.

$$S_{[q(t)]}(q, q', t) = \int L([q(t)])\, dt \tag{7.13}$$

is the integral over the Lagrangian for the path $[q(t)]$.

PRACTICAL MATTERS

As powerful and inspiring as the Feynman path integral seems, it is easy to throw some cold water on it since it represents stringing infinitely many short time propagators together *in the least efficient way possible.* Here's why: Think of $G(q, q', \tau)$ for very short τ as a large matrix $\tilde{G}(\tau)$, labeled by a discretization of q and q' (coming from having approximated the integrals by sums). If the dimension of the matrix is M by M, then $\tilde{G}(q, q', 2\tau)$ is computed by matrix multiplication requiring order M^3 multiplications. Now suppose we want $\tilde{G}(q, q', 3\tau)$. The best way to do this is to use $\tilde{G}(q, q', 2\tau)$ in

$$\tilde{G}(3\tau) = \tilde{G}(\tau)\tilde{G}(2\tau), \tag{7.14}$$

costing another M^3 multiplications. This is the efficient way to do it, because the M terms forming each of the elements of $\tilde{G}(2\tau)$ have been gathered into one number, "forgetting" their parentage so to speak. The Feynman expression does not permit this; each path is individually remembered and enumerated. The path integral way to reach $\tilde{G}(3\tau)$ requires order M^4 multiplications! The Feynman way is to multiply three matrices together with a quadruple "Do loop," whereas of course the efficient way to do it is two sequential triple Do loop calculations. If M = 1000, then one way costs 3×10^6 operations, and the other, 10^9 operations. Obviously, this is only the tip of the iceberg. The multiplications need grow only in proportion to the time, but in the Feynman path integral the number of multiplications grows exponentially in time! For example, for N ($M \times M$) matrices, there are M^{1+N} terms or paths. Supposing two degrees of freedom keeping track of 100 coordinate points in each dimension, with 20 time slices ($N = 20$), needs an astronomical 10^{80} paths. And yet, this problem done efficiently takes seconds on a laptop.

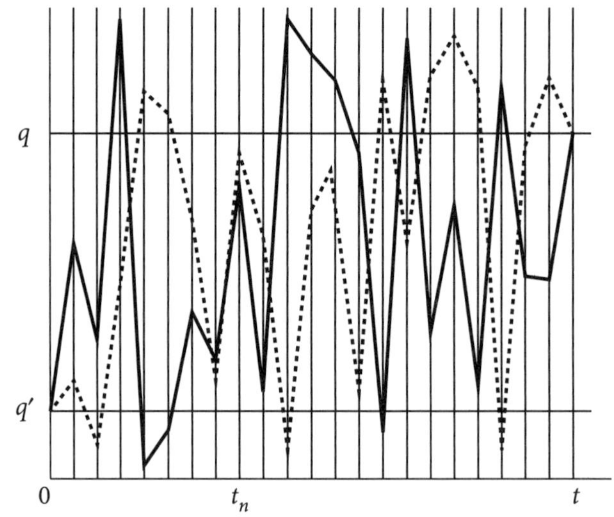

Figure 7.2. Two Feynman paths shown for discretized time, going from q' to q in time t. Actual Feynman paths would be infinitely more jagged fractals, taking infinitesimal time increments.

Things get even better for the non-Feynman approach if one wants to propagate one or a few initial wavefuncitons, as opposed to having the whole propagator. It takes only order M^2 multiplications for one time step of a wavefunction and if the result is stored as a new vector of dimension M, it takes just M^2 multiplications for the next step. So for N steps the bill runs to NM^2 multiplications. Feynman path integration taken literally requires M^{1+N} just for the propagator. Even greater efficiency results when using the "split operator" version of short time propagation, discussed next.

If the propagator is required only at a large time—say, for the prediction of the outcome of a collision—then using $\tilde{G}(4\tau) = \tilde{G}(2\tau)\tilde{G}(2\tau)$, $\tilde{G}(8\tau) = \tilde{G}(4\tau)\tilde{G}(4\tau)$, and so on, one gets one to times $2^K\tau$ in K iterations, but again, this not the Feynman path integral approach.

No one does calculations purely by summing over Feynman paths; they start with the path integral expression and work their way to something calculable through mathematical manipulations. But still, what a wonderful expression it is! Exact (here, nonrelativistic) propagation by a sum over all possible paths, almost all of them impossibly jagged, equally weighted, and given a phase that is the classical action for the (generally nonclassical) paths divided by Plank's constant (see figure 7.2).

SPLIT OPERATOR APPROACH

A very efficient way to do calculations is to repeatedly multiply the short time propagator matrix into the vector representing the evolving wavefunction, as just mentioned. But even this is slow compared to the following method: one writes, for short times τ,

$$\psi(t+\tau) \approx e^{-i\hat{p}^2\tau/2m\hbar} e^{-i\hat{V}\tau/\hbar}\psi(t). \tag{7.15}$$

Defining $\phi(q', t+\tau) \equiv e^{-\frac{i}{\hbar}V(q')\tau}\psi(q', t)$ ($\phi(q', t+\tau)$ is just a direct multiplication); incrementing the wavefunction $\psi(q, t)$ by time τ is seen to require two successive

Fourier transforms:

$$\hat{\psi}(p', t+\tau) = e^{-ip'^2\tau/m\hbar} \int dq'\, e^{ip'q'/\hbar}\phi(q', t+\tau) = e^{-ip'^2\tau/m\hbar}\phi(p', t+\tau),$$
$$(7.16)$$
$$\psi(q, t+\tau) = \frac{1}{2\pi\hbar}\int dp' e^{-ip'q/\hbar}\hat{\psi}(p', t+\tau).$$

Fast Fourier transforms can be used, requiring not M^2 multiplies as before for the matrix times vector operation, but order $M \log M$ multiplications, a huge savings for large M. It is straightforward to show that the error in this approach, compared to exact propagation by the quantum unitary operator $e^{-i\hat{H}\tau/\hbar} = e^{-i\hat{p}^2\tau/2m\hbar - iV(\hat{q})\tau/\hbar}$, is of order τ^2. One can use the symmetrized split-step Fourier method, taking a half time step using either operator, then a full time step with the other, and finally a second half time step again with the first. This yields an error estimate of order τ^3.

7.4 Van Vleck-Morette-Gutzwiller Semiclassical Propagator

Intuitively, it seems that an approximate propagator for longer times is obtained by using equation 7.3 directly for longer times τ, using the classical action over classical paths instead of the jagged Feynman paths, providing a direct path from purely classical mechanics to the quantum propagator! This needs a bit more justification; an extension of this notion will emerge together with a formula that however fundamentally justifies the intuitive hunch.

The idea is to start with the Feynman expression for the exact quantum propagator and use stationary phase over all the intermediate integrations over paths to evaluate it. The end result is the famous semiclassical Van Vleck (1928)–Morette (1951)–Gutzwiller (1970s) formula for the propagator.

We build a propagator for longer times by a succession of short time propagators. The process becomes clear just by combining two short time propagators, using stationary phase to do the required integral:

$$G_{sc}(q, q', 2\tau) = \int dy\, G(q, y, \tau)G(y, q', \tau),$$
$$= \frac{1}{2\pi i\hbar}\int dy \left(\frac{\partial^2 S(q, y, \tau)}{\partial q \partial y}\right)^{\frac{1}{2}} \left(\frac{\partial^2 S(y, q', \tau)}{\partial y \partial q'}\right)^{\frac{1}{2}}$$
$$\times \exp[iS(q, y, \tau)/\hbar + iS(y, q', \tau)/\hbar],$$
$$= \frac{\sqrt{2\pi i\hbar}}{2\pi i\hbar}\frac{\left(\frac{\partial^2 S(q,\bar{y},\tau)}{\partial q\partial\bar{y}}\right)^{\frac{1}{2}}\left(\frac{\partial^2 S(\bar{y},q',\tau)}{\partial\bar{y}\partial q'}\right)^{\frac{1}{2}}}{\left(\frac{\partial^2 S(q,\bar{y},\tau)}{\partial\bar{y}^2} + \frac{\partial^2 S(\bar{y},q',\tau)}{\partial\bar{y}^2}\right)^{\frac{1}{2}}} e^{\frac{i}{\hbar}(S(q,\bar{y},\tau)+S(\bar{y},q',\tau))}, \quad (7.17)$$

where

$$\frac{\partial}{\partial\bar{y}}(S(q, \bar{y}, \tau) + S(\bar{y}, q', \tau)) = -p(q, \bar{y}, \tau) + p(\bar{y}, q', \tau) = 0 \quad (7.18)$$

defines the stationary value of $\bar{y}$. The exponent in equation 7.17 is just what defines the action for a time 2τ (the bar is hereafter omitted over y):

$$S(q, q', 2\tau) = S(q, y, \tau) + S(y, q', \tau), \quad (7.19)$$

with $y \equiv y(q, q')$ through equation 7.18. We still need to simplify the prefactor to reach the desired form, equation 7.3. If we differentiate $-p(q, y, \tau) + p(y, q', \tau) = 0$ with respect to q',

$$\frac{d}{dq'}(p(q, y, \tau) - p(y, q', \tau)) = 0, \tag{7.20}$$

the result is also 0, since equation 7.18 is true at every q'. From this we deduce, since y depends on q',

$$0 = \left(\frac{\partial p(q, y, \tau)}{\partial y} - \frac{\partial p(y, q', \tau)}{\partial y}\right)\frac{dy}{dq'} - \frac{\partial p(y, q', \tau)}{\partial q'}, \tag{7.21}$$

or

$$\begin{aligned}\left(\frac{\partial p(q, y, \tau)}{\partial y} - \frac{\partial p(y, q', \tau)}{\partial y}\right) &= \frac{\partial p(y, q', \tau)}{\partial q'}\frac{dq'}{dy}, \\ &= \left(\frac{\partial^2 S(y, q', \tau)}{\partial y \partial q'}\right)\frac{dq'}{dy};\end{aligned} \tag{7.22}$$

there are cancellations in numerator and denominator in equation 7.17, giving

$$G_{sc}(q, q', 2\tau) = \frac{1}{\sqrt{2\pi i\hbar}}\left(\frac{\partial^2 S(q, y, \tau)}{\partial q \partial y}\frac{dy}{dq'}\right)^{\frac{1}{2}} e^{\frac{i}{\hbar}S(q, q', 2\tau)}. \tag{7.23}$$

Differentiating

$$S(q, q', 2\tau) = S(q, y, \tau) + S(y, q', \tau) \tag{7.24}$$

with respect to q and q', we get, remembering $y = y(q, q')$,

$$\frac{\partial S(q, q', 2\tau)}{\partial q} = \frac{\partial S(q, y, \tau)}{\partial q} + (-p(q, y, \tau) + p(y, q', \tau))\frac{dy}{dq}, \tag{7.25}$$

$$= \frac{\partial S(q, y, \tau)}{\partial q} \tag{7.26}$$

by equation 7.18. Then,

$$\frac{\partial^2 S(q, q', 2\tau)}{\partial q \partial q'} = \frac{\partial^2 S(q, y, \tau)}{\partial q \partial y}\frac{\partial y}{\partial q'}, \tag{7.27}$$

and thus,

$$G_{sc}(q, q', 2\tau) = \frac{1}{\sqrt{2\pi i\hbar}}\left(\frac{\partial^2 S(q, q', 2\tau)}{\partial q \partial q'}\right)^{\frac{1}{2}} e^{\frac{i}{\hbar}S(q, q', 2\tau)}; \tag{7.28}$$

this is what we wanted to show. This looks almost like the postulate, equation 7.5, but there is a crucial and highly simplifying (as well as intuitively satisfying) difference: the action $S(q, q', t)$ is now computed over only classical paths. We have carried along a subscript sc on $G_{sc}(q, q', 2\tau)$ because we used a stationary phase rather than an

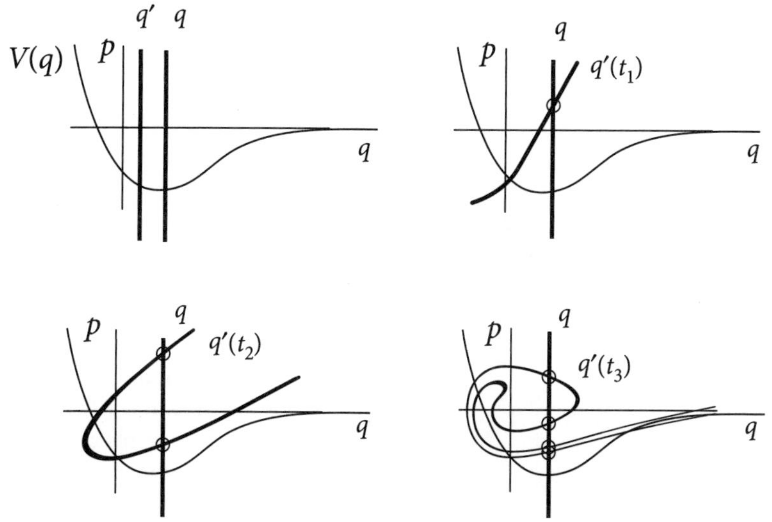

Figure 7.3. For an initially sharp position q', increasingly many paths (circled) reaching a given point q develop as time increases in this example of dynamics in a potential $V(q)$.

exact integral in progressing from time τ to time 2τ. Now that we have contracted the propagators for two time steps of length τ into one of length 2τ, further such contractions go the same way. As time progresses, more than one path from a given q to a final q' may develop, (see figures 7.3 and 7.4, for example). Each path will have its own action S_k. There are also phase factors that come in, coming ultimately from the issue of whether the stationary phase integrations encountered are of the "$\exp[ix^2]$" type or the "$\exp[-ix^2]$" type, so we write, finally,

$$G_{sc}(q, q', t) = \frac{1}{\sqrt{2\pi i \hbar}} \sum_k \left| \frac{\partial^2 S_k(q, q', t)}{\partial q \partial q'} \right|^{\frac{1}{2}} e^{\frac{i}{\hbar} S_k(q,q',t) - \frac{i\pi\nu_k}{2}} \tag{7.29}$$

for a general time t. This is the Van Vleck-Morette-Gutzwiller semiclassical expression for the propagator.

The Van Vleck determinant, a square root prefactor in the sum over classical contributions, must be the square root of a classical probability density, as in the universal formula

$$\text{semiclassical amplitude} \sim \sum_j \sqrt{\text{classical probability density}_j}\; e^{i \text{ classical action}_j/\hbar}, \tag{7.30}$$

and as we have seen already in connection with figure 6.5, earlier. The area of the intersections of the finite width zones for $q'(t)$ and q are proportional to the classical probability density as explained in section 3.2.

It is interesting to examine the semiclassical action for a path that deviates at only one point from a classical path. The phase becomes stationary as the errant point joins the classical path (figure 7.5).

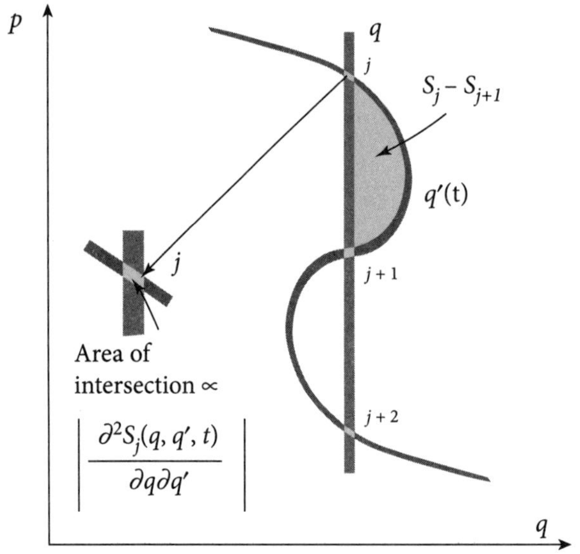

Figure 7.4. For an initially sharp position, q' evolves under a nonlinear Hamiltonian, overlapping the position q at time t. The Van Vleck determinants and actions have a geometrical interpretation shown here and discussed in section 3.2 .

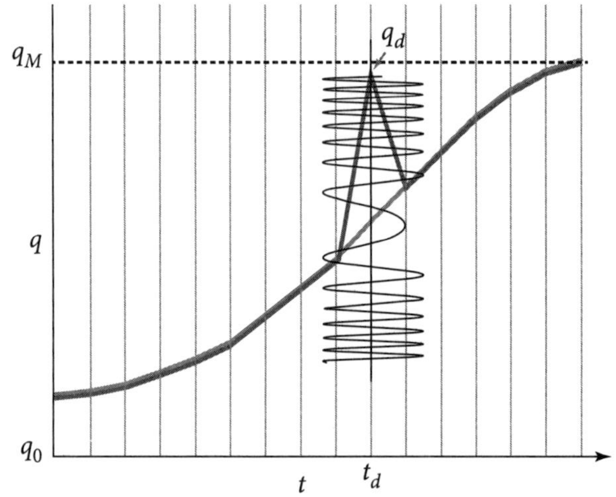

Figure 7.5. The smooth line is a (stationary) classical path $q(t)$, discretized in time. Another path follows this one, except at one time t_d, where it deviates from the classical path. The exponent of the action for the whole path is shown plotted along a vertical axis, the vertical coordinate being the q position of the "jog" that the nonclassical path takes at time t_d. The phase goes stationary as the jog approaches the classical path, as can be seen. $\mathrm{Re}\{\exp[\frac{i}{\hbar}S(q_M, q_{M-1}, \tau) + \cdots + \frac{i}{\hbar}S(q_{d_1}, q_d), \tau) + \frac{i}{\hbar}S(q_d, q_{d-1}\tau) + \cdots + \frac{i}{\hbar}S(q_1, q_0, \tau)\}]$ is plotted as the black oscillating line. The stationary phase point stands out, and corresponds to the new path aligning itself at t_d with the classical path.

7.5 Another Route to the Van Vleck-Morette-Gutzwiller Propagator

A different route to the Van Vleck propagator is instructive, reached by writing the wavefunction as an exponential (no approximation there), as in

$$\psi(q,t) = e^{\frac{i}{\hbar}S(q,t)}, \tag{7.31}$$

where $S(q,t) = -i\hbar \log \psi(q,t)$. Expanding S in a power series in $\hbar$:

$$S(q,t) = S(q,t) + \frac{\hbar}{i} S_1(q,t) - \hbar^2 S_2(q,t) + \ldots, \tag{7.32}$$

we plug this into $i\hbar\, \partial\psi(q,t)/\partial t = H\psi(q,t)$, compare like powers of $\hbar$, and obtain

$$\hbar^0 : \frac{1}{2m}\left(\frac{\partial S(q,t)}{\partial q}\right)^2 + \frac{\partial S(q,t)}{\partial t} + V(q,t) = 0, \tag{7.33}$$

$$\hbar^1 : \frac{\partial S_1(q,t)}{\partial t} + \frac{1}{m}\frac{\partial S}{\partial q}\frac{\partial S_1}{\partial q} + \frac{1}{2m}\frac{\partial^2 S}{\partial q^2} = 0, \tag{7.34}$$

stopping at first order in $\hbar$.

The "boundary condition" here is determined by $\psi(q,0)$, defining $S(q,0)$ of course. Some special $\psi(q,0)$ have standard labels in quantum mechanics, like a momentum state $\psi(q,0) = \exp[ip'q/\hbar] \equiv \langle q\,|\,p'\rangle$, i.e. the ket $|\,p'\rangle$ turning to Dirac notation. The associated boundary condition is that the momentum is $p = p'$ at $t = 0$, irrespective of q'. Quantum mechanically, the conjugate position q' is completely uncertain and uniformly distributed, akin to a classical Lagrangian manifold.

If the position rather than the momentum is initially definite, say $q = q'$ at $t = 0$, then $\psi(q,0) = \delta(q - q') \equiv \langle q|q'\rangle$, with momentum totally indefinite. We can carry over the labels of the boundary conditions to our equivalent exponential form, adding for example the label q' to S

$$S(q,0) \to S(q,q',0) \tag{7.35}$$

in the case of an initial position state.

Equation 7.32 now reads

$$S(q,q',t) = S(q,q',t) + \frac{\hbar}{i} S_1(q,q',t) - \hbar^2 S_2(q,q',t) + \ldots, \tag{7.36}$$

and equation 7.33 now becomes

$$\hbar^0 : \frac{1}{2m}\left(\frac{\partial S(q,q',t)}{\partial q}\right)^2 + \frac{\partial S(q,q',t)}{\partial t} + V(q,t) = 0. \tag{7.37}$$

Now is a good time to mention an important point, which is that all the $\hbar$ expansions anywhere in this book are asymptotic series, and may diverge if taken to high order (or become problematic even sooner), but thank goodness the lowest order terms are usually accurate (except in situations involving breakdown of the stationary phase approximation—that is, coalescence of stationary phase points). The lowest order terms dominate as $\hbar \to 0$, and any stationary phase convergence problems retreat in that limit. Although we are routinely trying to apply semiclassical formulae for fixed, finite $\hbar$, because there lie the applications, one should bear in mind that in the "semiclassical limit" $\hbar \to 0$, the expression equation 7.29 becomes exact for any finite time t.

The order $\hbar^0$ equation is the surviving term in the classical, $\hbar \to 0$ limit. We therefore identify $S(q, q', t)$ as the classical action and equation 7.37 as the Hamilton-Jacobi equation (equation 7.40, later), since

$$\frac{\partial S(q, q', t)}{\partial q} = p, \quad \frac{\partial S(q, q', t)}{\partial q'} = -p', \tag{7.38}$$

and as we learned earlier

$$\frac{\partial S(q, q', t)}{\partial t} = -H(q, q', t). \tag{7.39}$$

Then,

$$\frac{1}{2m}\left(\frac{\partial S(q, q', t)}{\partial q}\right)^2 + \frac{\partial S(q, q', t)}{\partial t} + V(q, t) = \frac{p^2}{2m} + V(q, t) - H = 0. \tag{7.40}$$

To generalize things to include states of the system other than a position eigenstate, we write instead $S(q, \alpha, t)$, where α represents the initial state or boundary condition at $t = 0$.

Turning to S_1, equation 7.34, we try the substitution

$$S_1 = \frac{1}{2}\log\left(\frac{\partial^2 S(q, \alpha, t)}{\partial q \partial \alpha}\right). \tag{7.41}$$

Carrying out the required differentiations in equation 7.34, and multiplying through by $(\partial^2 S(q, \alpha, t)/\partial q \partial \alpha)$, we get

$$\frac{1}{2}\left(\frac{\partial^3 S(q, \alpha, t)}{\partial q \partial \alpha \partial t}\right) + \frac{1}{2m}\left(\frac{\partial S(q, \alpha, t)}{\partial q}\right)\left(\frac{\partial^3 S(q, \alpha, t)}{\partial^2 q \partial \alpha}\right) + \frac{1}{2m}\left(\frac{\partial^2 S(q, \alpha, t)}{\partial q \partial \alpha}\right)\left(\frac{\partial^2 S(q, \alpha, t)}{\partial^2 q}\right) = 0. \tag{7.42}$$

Substituting equations 7.38 and 7.39, we obtain

$$\frac{-1}{2}\left(\frac{\partial^2 H(q, \alpha, t)}{\partial q \partial \alpha}\right) + \frac{1}{2m} p \left(\frac{\partial^2 p}{\partial q \partial \alpha}\right) + \frac{1}{2m}\left(\frac{\partial p}{\partial \alpha}\right)\left(\frac{\partial p}{\partial q}\right). \tag{7.43}$$

But, for example,

$$H(q, \alpha) = p(q, \alpha, t)^2/2m + V(q), \tag{7.44}$$

and

$$\left(\frac{\partial^2 H(q,\alpha,t)}{\partial q \partial \alpha}\right) = \frac{1}{2m}\left(\frac{\partial^2}{\partial q \partial \alpha}\right) p^2,$$
$$= \frac{2}{2m}\left(\frac{\partial p}{\partial q}\right)\left(\frac{\partial p}{\partial \alpha}\right) + \frac{2}{2m} p \left(\frac{\partial^2 p}{\partial q \partial \alpha}\right). \tag{7.45}$$

These terms are just what is needed to cancel the terms in equation 7.43, making the expression 0, as required.

If $\alpha = q'$, we have the semiclassical version of the the Green function. Having verified the substitution equation 7.41, earlier, then with proper normalization the result becomes

$$G_{sc}(q,q',t) = \frac{1}{\sqrt{2\pi i \hbar}}\left(\frac{\partial^2 S(q,q',t)}{\partial q \partial q'}\right)^{\frac{1}{2}} \exp[iS(q,q',t)/\hbar]. \tag{7.46}$$

This is again the Van Vleck semiclassical propagator. The prefactor becomes the "Van Vleck determinant" of the matrix of partial derivatives. There may be more than one classical path connecting q with q' in time t; a separate term with a different action $S_k(q,q',t)$ is required for each path. Maslov phases $\pi\nu_k/2$ connected with focal points come in at longer times; the Maslov index ν_k is the number of such focal points encountered along the path from q to q' at time t.

The full expression becomes

$$G_{sc}(q,q',t) = \sum_k \frac{1}{\sqrt{2\pi i \hbar}}\left|\frac{\partial^2 S_k(q,q',t)}{\partial q \partial q'}\right|^{\frac{1}{2}} \exp[iS_k(q,q',t)/\hbar - i\pi\nu_k/2]. \tag{7.47}$$

This agrees with equation 7.3, obtained by a very different line of reasoning. The Van Vleck propagator in an explicitly multidimensional form reads

$$\begin{aligned} G(\mathbf{q},\mathbf{q}';t) &\approx G_{sc}(\mathbf{q},\mathbf{q}';t) \\ &= \left(\frac{1}{2\pi i\hbar}\right)^{d/2} \sum_k \left|\det\left(\frac{\partial^2 S_k(\mathbf{q},\mathbf{q}';t)}{\partial \mathbf{q}\partial \mathbf{q}'}\right)\right|^{\frac{1}{2}} \\ &\quad \times \exp\left(iS_k(\mathbf{q},\mathbf{q}';t)/\hbar - \frac{i\pi\nu_k}{2}\right); \end{aligned} \tag{7.48}$$

d is the number of degrees of freedom. The Van Vleck determinant seen here is a purely classical object, with the geometrical interpretation given in figures 7.4 and 7.5 for example. It is perhaps almost as important in purely classical mechanics as it is in semiclassical theory, yet it appears in no classical mechanics texts that I and Prof. Wheeler of Reed College are aware of.

The determinant of the matrix of mixed coordinate derivatives is the classical probability for starting at $\mathbf{q}'$ at time $t = 0$ and ending at $\mathbf{q}$ at time t. The mixed coordinate derivatives are elements of the stability matrix; see equation 3.74. We write, again in one-dimensional notation,

$$\left|\frac{\partial^2 S_k(q,q';t)}{\partial q \partial q'}\right|^{\frac{1}{2}} = \left|\frac{\partial p}{\partial q'}\right|^{\frac{1}{2}} = |m_{21}|^{-1/2}. \tag{7.49}$$

The stability matrix records the response of the kth trajectory to small changes in initial conditions. To be complete, the classical action is the time integral of the Lagrangian L along the trajectory, as we already know:

$$S_k(\mathbf{q}, \mathbf{q}'; t) = \int_0^t dt' \, L = \int_0^t dt' \left\{ \mathbf{p}(t') \cdot \dot{\mathbf{q}}(t') - H(\mathbf{p}(t'), \mathbf{q}(t')) \right\}. \tag{7.50}$$

Equation 7.48 was originally derived by Van Vleck in 1928 without the summation over paths or index ν and was accordingly limited to short times.

The Van Vleck propagator leads to an approximate solution of the time-dependent Schrödinger equation, but for years much more work went into the time-independent Schrödinger equation $H\psi = E\psi$ and the energy Green function—that is, the Fourier transform of the Van Vleck propagator, $G_{sc}(\mathbf{q}, \mathbf{q}', E)$. We consider $G_{sc}(\mathbf{q}, \mathbf{q}', E)$ in section 7.6, but remark here that that to arrive at the semiclassical energy Green function, the Fourier transformation integral $G_{sc}(\mathbf{q}, \mathbf{q}', E) = \int_{s.p.} \exp(iEt/\hbar) G_{sc}(\mathbf{q}, \mathbf{q}', t) \, dt$ is done by stationary phase. This makes the energy propagator one more approximation away from exactness than is the Van Vleck time propagator, the stationary phase version of the Feynman propagator.

7.6 Semiclassical Energy Green Function

The energy Green function $G(q, q', E)$ answers the question, "What is the amplitude to reach point q from point q' keeping the energy fixed at E?" The propagator $G(q, q', t)$ keeps time rather than energy fixed.[1] The two Green functions are related by a Fourier transform,

$$G(q, q', E) = \int_{-\infty}^{\infty} dt \, e^{iEt/\hbar} \, G(q, q', t). \tag{7.51}$$

The semiclassical approximation of the energy Green function $G_{sc}(q, q', E)$ is obtained by applying the stationary phase approximation to the time Fourier transform of the time-dependent semiclassical VVMG propagator $G_{sc}(q, q', t)$:

$$G_{sc}(q, q', E) \overset{s.p.}{=} \int_{-\infty}^{\infty} dt \, e^{iEt/\hbar} G_{sc}(q, q', t). \tag{7.52}$$

Two other Green functions are defined similarly, called *retarded* and *advanced*. They are required to vanish unless $t > 0$ or $t < 0$, respectively, obtained by half Fourier transforms over the full time-dependent Green function—that is,

$$G^{+}_{sc}(q, q', E) \overset{s.p.}{=} \int_{0}^{\infty} dt \, e^{iEt/\hbar} G^{+}_{sc}(q, q', t). \tag{7.53}$$

[1] See P. Cvitanović, R. Artuso, R. Mainieri, G. Tanner and G. Vattay, *Chaos: Classical and Quantum*, Chapter 33, *ChaosBook.org* (Niels Bohr Institute, Copenhagen 2012).

The stationary phase points of the integrand in equation 7.52 are given by

$$\frac{\partial S_k(q, q', t)}{\partial t} = -E, \tag{7.54}$$

specifying the corresponding solution times

$$t_{s.p.} = t(q, q', E). \tag{7.55}$$

We expand around such times, writing

$$S_k(q, q', t) \approx S_k(q, q', t_{s.p.}) + \frac{1}{2}\left(\frac{\partial^2 S_k(q, q', t)}{\partial t^2}\right)_{t=t_{s.p.}} (t - t_{s.p.})^2. \tag{7.56}$$

Equation 7.54 implies the Legendre transformation

$$S_k(q, q', E) = S_k(q, q', t_{s.p.}) + E\, t_{s.p.}. \tag{7.57}$$

The prefactor coming from the stationary phase approximation to equation 7.53, together with the prefactor already in the VVMG Green function, makes the combination

$$\left|\frac{\partial^2 S_k(q, q', t)}{\partial q \partial q'}\right|^{\frac{1}{2}} \left(\frac{\partial^2 S_k(q, q', t)}{\partial t^2}\right)^{-\frac{1}{2}}. \tag{7.58}$$

To proceed, consider the $(N+1) \times (N+1)$ determinant D involving derivatives of $S_k(q, q', E)$ abbreviated as S:

$$D(q, q', E) = \begin{pmatrix} \frac{\partial^2 S}{\partial q' \partial q} & \frac{\partial^2 S}{\partial q' \partial E} \\ \frac{\partial^2 S}{\partial q \partial E} & \frac{\partial^2 S}{\partial E^2} \end{pmatrix} = \begin{pmatrix} -\frac{\partial p'}{\partial q} & -\frac{\partial p'}{\partial E} \\ \frac{\partial t}{\partial q} & \frac{\partial t}{\partial E} \end{pmatrix} \equiv \frac{\partial(-p', t)}{\partial(q, E)}. \tag{7.59}$$

From the right-hand side and apart from signs, D is seen to be the Jacobian of the coordinate transformation $(q, E) \to (p', t)$ for fixed q'. Jacobians combine in "chain rule" form, since, for example,

$$\begin{aligned} dxdy &= \left(\frac{\partial(x, y)}{\partial(u, v)}\right) dudv = \left(\frac{\partial(x, y)}{\partial(u, v)}\right)\left(\frac{\partial(u, v)}{\partial(s, q)}\right) dsdq, \\ &= \left(\frac{\partial(x, y)}{\partial(s, q)}\right) dsdq. \end{aligned} \tag{7.60}$$

Thus, by the same token, and since $\det(A \cdot B) = \det A \cdot \det B$,

$$\begin{aligned} \det D &= (-1)^{N+1} \det\left(\frac{\partial(p', t)}{\partial(q, E)}\right) = (-1)^{N+1} \det\left(\frac{\partial(p', t)}{\partial(q, t)}\right) \det\left(\frac{\partial(q, t)}{\partial(q, E)}\right) \\ &= (-1)^{N+1} \det\left(\frac{\partial p'}{\partial q}\right) \det\left(\frac{\partial t}{\partial E}\right) = \det\left[\left(\frac{\partial^2 S}{\partial q \partial q'}\right)\left(\frac{\partial^2 S}{\partial t^2}\right)^{-1}\right] \end{aligned} \tag{7.61}$$

Comparing equations 7.58 and 7.61, we see that the simplification/clarification we are seeking is that the Van Vleck determinant together with the amplitude arising from the time Fourier transform stationary phase integral of equation 7.58 is in fact the determinant of the extended dimensionality matrix D. Allowing for multiple paths k reaching q from q' at energy E,

$$\boxed{G_{sc}(q,q',E) = \frac{1}{i\hbar(2\pi i\hbar)^{(N-1)/2}} \sum_k \left| \det \begin{pmatrix} \frac{\partial^2 S_k}{\partial q' \partial q} & \frac{\partial^2 S_k}{\partial q' \partial E} \\ \frac{\partial^2 S_k}{\partial q \partial E} & \frac{\partial^2 S_k}{\partial E^2} \end{pmatrix} \right|^{\frac{1}{2}} e^{\frac{i}{\hbar} S_k(q,q',E) - \frac{i\pi}{2}\nu_k}.} \tag{7.62}$$

We can cast this in a more useful form by requiring one of the initial and one of the final coordinates to point along the trajectory at the initial and final times, respectively. The other $N-1$ coordinates are perpendicular to the trajectory, so the initial coordinates read $(q'_{\parallel}, q'_{\perp 1}, q'_{\perp 2}, \ldots)$ and final coordinates $(q_{\parallel}, q_{\perp 1}, q_{\perp 2}, \ldots)$. In this special coordinate system, it is evident that the derivatives

$$\frac{\partial S(q,q',t)}{\partial q_{\perp}} = p_{\perp} = 0 \text{ and } \frac{\partial^2 S(q,q',t)}{\partial q_{\parallel} \partial q_{\perp}} = 0 \tag{7.63}$$

must vanish, since the momentum perpendicular to the trajectory vanishes, and that momentum remains 0 as we move along the trajectory. Thus the matrix in the determinant of equation 7.62 is

$$\begin{pmatrix} 0 & 0 & \frac{\partial^2 S}{\partial q'_{\parallel} \partial E} \\ 0 & \frac{\partial^2 S}{\partial q_{\perp} \partial q'_{\perp}} & * \\ \frac{\partial^2 S}{\partial q_{\parallel} \partial E} & * & * \end{pmatrix},$$

where the $*$ indicates a matrix element that does not matter when the determinant is expanded. Since

$$\frac{\partial^2 S}{\partial q_{\parallel} \partial E} = \frac{\partial t}{\partial q_{\parallel}} = \frac{1}{\dot{q}_{\parallel}}, \text{ and } \frac{\partial^2 S}{\partial q'_{\parallel} \partial E} = \frac{\partial t}{\partial q'_{\parallel}} = \frac{1}{\dot{q}'_{\parallel}},$$

thus

$$\left| \det \begin{pmatrix} \frac{\partial^2 S_k}{\partial q' \partial q} & \frac{\partial^2 S_k}{\partial q' \partial E} \\ \frac{\partial^2 S_k}{\partial q \partial E} & \frac{\partial^2 S_k}{\partial E^2} \end{pmatrix} \right| = \left| \frac{1}{\dot{q}_{\parallel}} \frac{1}{\dot{q}'_{\parallel}} \det \left(\frac{\partial^2 S_k}{\partial q_{\perp} \partial q'_{\perp}} \right) \right|, \tag{7.64}$$

and finally

$$\boxed{G_{sc}(q,q',E) = \frac{1}{i\hbar(2\pi i\hbar)^{(N-1)/2}} \sum_k \left| \frac{1}{\dot{q}_{\parallel}} \frac{1}{\dot{q}'_{\parallel}} \det \left(\frac{\partial^2 S_k}{\partial q_{\perp} \partial q'_{\perp}} \right) \right|^{\frac{1}{2}} e^{\frac{i}{\hbar} S_k(q,q',E) - \frac{i\pi}{2}\nu_k}.} \tag{7.65}$$

The mathematical structure of a semiclassical energy eigenstate in coordinate space in one dimension, and its phase space interpretation, is shown in figure 7.6.

$$\langle q|\psi_E\rangle \propto \sum_{k=1}^{2} \left|\frac{\partial^2 S_k(q, E)}{\partial q \partial E}\right|^{1/2} e^{iS_k(q, E)/\hbar + i\alpha_k}$$

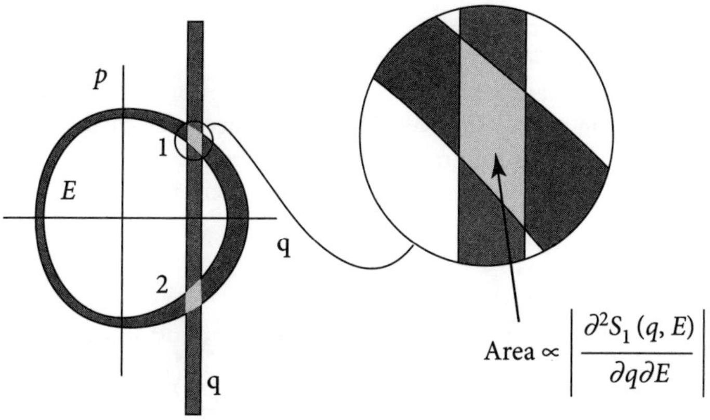

Figure 7.6. Semiclassical formula for and classical interpretation of a bound, semiclassical energy eigenstate evaluated in coordinate space, with the classical joint probability shown. (See sections 3.2 and 6.2.)

7.7 Trace Formula for the Energy Spectrum

For a bound system, we may write the energy Green function as

$$G(\boldsymbol{r}, \boldsymbol{r}', E) = \sum_n \frac{\langle \boldsymbol{r}|\psi_n\rangle\langle\psi_n|\boldsymbol{r}'\rangle}{E - E_n}; \tag{7.66}$$

its trace is

$$\text{Trace}[G(\boldsymbol{r}, \boldsymbol{r}', E)] = \int d^N\boldsymbol{r}\, G(\boldsymbol{r}, \boldsymbol{r}, E) = \sum_n \frac{1}{E - E_n}. \tag{7.67}$$

The famous Gutzwiller "trace formula" seeks to obtain this trace semiclassically within stationary phase, establishing the quantum eigenvalue spectrum having used only classical trajectories. As we show presently, only strictly periodic orbits contribute:

$$\rho(E) = \int_{s.p.} d\boldsymbol{r}\, G_{sc}(\boldsymbol{r}, \boldsymbol{r}, E) = \sum_n \frac{1}{E - E_n} \approx \text{smooth function}$$
$$+\frac{1}{\pi\hbar} \sum_{p,m}{}' \frac{T_p}{||M_p^m - 1||^{\frac{1}{2}}} \cos\left\{n\left[\frac{1}{\hbar}S_p(E) - \frac{\mu_p\pi}{2}\right]\right\}, \tag{7.68}$$

where the sum p is over all distinct periodic orbits (that is, trajectories that repeat themselves), and m is the number of repetitions of the same orbit. M_p^m is a stability matrix element for the mth traverse of the pth periodic orbit, $S_p(E)$ is the classical action for the pth periodic orbit, and μ_p is its Maslov phase. T_p is the base period of the pth orbit. The smooth part is a natural result of the density of energies E_n of the eigenstates. We are interested in the oscillating part, which has the potential to reveal the eigenvalues.

When the stationary phase trace is performed over all positions in the semiclassical energy Green function, it is clear right away that at the very least only closed orbits can contribute to the result, since $\boldsymbol{r} = \boldsymbol{r}'$ and the action and its derivatives are evaluated only there—for example, $S(\boldsymbol{r}, \boldsymbol{r})$. But a closed orbit has to return to the same position $\boldsymbol{r}$ only along some path, perhaps a new one with a new momentum at $\boldsymbol{r}$, and therefore need not be a periodic orbit. However, it presently becomes clear that only periodic orbits appear in the semiclassical trace formula.

The reason is simple: the stationary phase requirement for $S(\boldsymbol{r}, \boldsymbol{r})$,

$$\frac{\partial S(\boldsymbol{r}, \boldsymbol{r})}{\partial \boldsymbol{r}} = 0, \tag{7.69}$$

requires separate differentiation of both $\boldsymbol{r}$ dependencies, and since

$$\boldsymbol{p} = -\frac{\partial S(\boldsymbol{r}, \boldsymbol{r}')}{\partial \boldsymbol{r}} \text{ and } \boldsymbol{p}' = \frac{\partial S(\boldsymbol{r}, \boldsymbol{r}')}{\partial \boldsymbol{r}'}, \tag{7.70}$$

we have, as a condition of stationary phase,

$$\boldsymbol{p} = \boldsymbol{p}', \tag{7.71}$$

where $\boldsymbol{p}'$ is the momentum after a return to the position $\boldsymbol{r}' = \boldsymbol{r}$. Both the position and the momentum are the same upon return, and the contributing orbits to the trace formula are therefore periodic.

Some insight into what the trace formula involves was given in section 3.11, where the *classical* trace of a periodic orbit was presented. Sure enough, the square root of the classical trace from equation 3.108 appears as the prefactor in equation 7.68.

We have not issued a warning that the analysis given here works only for chaotic systems (without further modification anyway), where the periodic orbits are isolated, meaning the orbits do not belong to continuously labeled families. For example, if the orbits were enumerated by the irrationals on the number line, they are isolated, but if they are enumerated by the whole number line they would be nonisolated periodic orbits. The issue of weighing the contribution of a periodic orbit that is part of a family raises new questions we did not address here, but the problem is tractable (see references [23, 24] for a unique and beautiful approach; beware that "closed orbits" in that paper refer to what we now call periodic orbits).

The trace formula has consumed a great deal of ink (or electrons anyway) in its time, due to concerns over its convergence and calculability. Many different schemes have been proposed, including those based on other formalisms altogether. A basic problem is that for a chaotic system, the number of new periodic orbits grows exponentially with time or length of the orbit, resulting in exponentially more work to get linearly more eigenvalues. Even if all the orbits were made readily available, there is evidence that the sum over orbits (after summing over repetitions of the orbit, which is elementary) diverges.

7.8 Closed Orbits for Spectroscopy

In spite of the fame and importance of the trace formula, only the eigenenergies are revealed, but in experiments those energies usually come along with spectral intensities. J. B. Delos and co-workers pointed out that *not* doing the trace and keeping track of closed (coming back to the same position but with a new momentum) but not necessarily periodic orbits can provide accurate estimates of energies *and* intensities [25] for specific initial states relevant to experiments. This is called a local density of states, as opposed to the total density that is the integral over all local densities. The local density is referred to as any appropriate quantum state; local in coordinate space might be appropriate, for example, if an electron comes from a tightly bound orbital, but is photo-ejected into a large Rydberg state. The local density of states for the state ϕ is

$$\rho_\phi(E) = \sum_n |\langle E_n|\phi\rangle|^2 \delta(E_n - E),$$

whereas the total density of states is

$$\sum_\phi \rho_\phi(E) = \sum_n \delta(E_n - E).$$

A 2*s* orbital may indeed look like a point source to a large Rydberg orbital. Delos and co-workers treated closed orbits, giving the intensities for atomic transitions with a magnetic field present. The field was deemed not important within about 50 Å of the nucleus, but the ionic core plus magnetic field at larger distances gives rise to chaotic dynamics, and among all the chaotic paths are closed ones that return to their origin. The retention of more information—that is, the intensities, is clear in hindsight, since information about $G_{sc}(\boldsymbol{r}, \boldsymbol{r}, E)$ is not averaged over all $\boldsymbol{r}$, but rather used at a specific and crucial $\boldsymbol{r}$ for the problem.

Later, we will see that other types of local states—for example, Gaussian wavepackets—are appropriate to molecular systems. To get some perspective on this, we recall that a Green function is the autocorrelation (for a diagonal element) or cross correlation function of a propagated position state—that is,

$$G(\boldsymbol{r}, \boldsymbol{r}', t) = \langle \boldsymbol{r}|e^{-iHt/\hbar}|\boldsymbol{r}'\rangle. \tag{7.72}$$

In this case we require only returning to the same position to contribute.

In the light convergence of the trace formula and exponential proliferation of orbits, the progress with closed orbits has been impressive, reproducing experimental spectra with many peaks accurately in both position and intensity. The exponential proliferation problem does not go away, and certain kinds of spectroscopy may be much more amenable than others.

Other spectroscopies are much more naturally expressed in terms of coherent states. A position state is one limit of a coherent state, in which the position uncertainty has shrunk to 0; there is also the limit in which the free particle momentum uncertainty has shrunk to 0, and everything in between. The coherent state Green function lives at either limit and anywhere in between:

$$G(\alpha, \alpha', t) = \langle \alpha|e^{-iHt/\hbar}|\alpha'\rangle. \tag{7.73}$$

This opens up the situation considerably, in that only close returns in both position and momentum contribute. This is taken up in section 8.3.

7.9 History

Now that we have introduced the Feynman propagator and its various semiclassical descendants, it is easier to discuss the fascinating history of these developments.[2] The Feynman path integral has several antecedents, going back to Norbert Wiener, who introduced the Wiener integral for diffusion and Brownian motion, and was extended by P.A.M. Dirac in a 1933 paper using, like Feynman, the classical Lagrangian to do quantum mechanics. The method was developed by Feynman in his 1942 Princeton PhD thesis "The Principle of Least Action in Quantum Mechanics" and in fuller form in 1948 [26], but he never put much effort into cleaning up some of his ad hoc and intuitive (but correct) ideas, leaving that to others. Perhaps more surprisng is he did not develop the implications of what he had stated were the most important Feynman paths, the stationary action classical paths. That was left to Pauli and Morette independently and simiultaneously, and later Gutzwiller. The crucial Van Vleck determinants give the classical probability densities relevant to the classical manifolds corresponding to the quantum states defining the amplitude. The determinants were apparently reinvented by Pauli for his discussion, and Morette used them with reference to Van Vleck's much earlier paper. Both Pauli[3] and separately Morette [27], who was at the Princeton Institute for Advanced Study at the time, had the essential formula for the Green function, equation 7.48.

Van Vleck's paper was written much earlier, with WKB theory in mind (chapter 12). Van Vleck's point was that the one-dimensional WKB results failed to show the true structure of semiclassical wavefunctions and amplitudes, since in 1D it is too easy to summarize the prefactors without revealing their deeper orgins as derivatives of actions. The issue is embodied in equation 7.30, since the classical probability density can be expressed in mundane ways in one dimension, such as $1/p(q)$ in equation 6.33. It is nonetheless possible to reveal the deeper structure even in 1D—for example, see equation 7.29 or figure 7.5. It is no big leap to see how those cases, written in 1D notation, must generalize to many dimensions (equation 7.48). Still, Van Vleck made this clear only by working in many dimensions, where a real "Van Vleck determinant" is needed.

Martin Gutzwiller completes this short history of major developments [28,29]. First, there is the energy Green function $G(\boldsymbol{r}, \boldsymbol{r}'; E)$ (see equations 7.62 and 7.65) obtained by stationary phase integration on $G(\boldsymbol{r}, \boldsymbol{r}'; t)$. Gutzwiller needed it on his way to the trace formula—that is, the trace of $G(\boldsymbol{r}, \boldsymbol{r}'; E)$ over all coordinate space by stationary phase (equation 7.68). Second, he provided a unified framework for the phase, and third, of course, he gave us the trace formula.

[2] Here I relied partly on the excellent summary given by Prof. Nicholas Wheeler of Reed College, found in chapter 3 of his online book on quantum mechanics.

[3] Pauli Lectures on Physics: volume 6, appendix, English, Dover, 1973.

Chapter 8

Stressing the VVMG Propagator

The Van Vleck-Morette-Gutzwiller (VVMG) semiclassical propagator is the result of stationary phase on the Feynman path integral, as just discussed. It was a long time before anyone tested or tried to use the propagator on nonlinear systems, for three reasons. The first is the necessity of numerical implementation, requiring computers. However, the numerics is no more onerous than running classical trajectories. The second is the traditional preference for the time-independent Schrödinger equation. For many years, eyes and minds were mainly focussed on the energy Green function and the Gutzwiller trace formula. The third is that incorrect arguments were given, suggesting that the VVMG semiclassical propagator would break down rather fast in time for a nonlinear system. It turns out that it breaks down rather slowly, even for many chaotic systems. Its nemesis is hard-core (literally) diffraction.

8.1 Semiclassical Wavepacket Dynamics in an Anharmonic Oscillator

We used the cellular dynamics method [30, 31] (see section 11.3, later), for calculating the Van Vleck-Morette-Gutzwiller (VVMG) semiclassical propagator in a one-dimensional Morse potential acting on an initially narrow wavepacket. The VVMG propagator performed amazingly well (see figure 8.1). This example and the application to a chaotic billiard [22] (see figure 8.6, later) were the first nontrivial tests of which we are aware of the VVMG propagator, the direct descendant of the Feynman path integral by application of the stationary phase approximation, as we have seen.

8.2 Quantum and Semiclassical Revival

Any wavepacket like the one chosen in figure 8.1 can be expanded in terms of eigenstates, evolving as

$$\psi(q,t) = \sum_n a_n e^{-iE_n t/\hbar} \psi_n(q). \tag{8.1}$$

This raises the question, whether at some future time $t \gg 0$ the phase factors $e^{-iE_n t/\hbar}$ will all realign (modulo 2π) with the same relative phase they started with, or very close to it. This would cause the wavepacket to reform, in a so-called quantum revival.

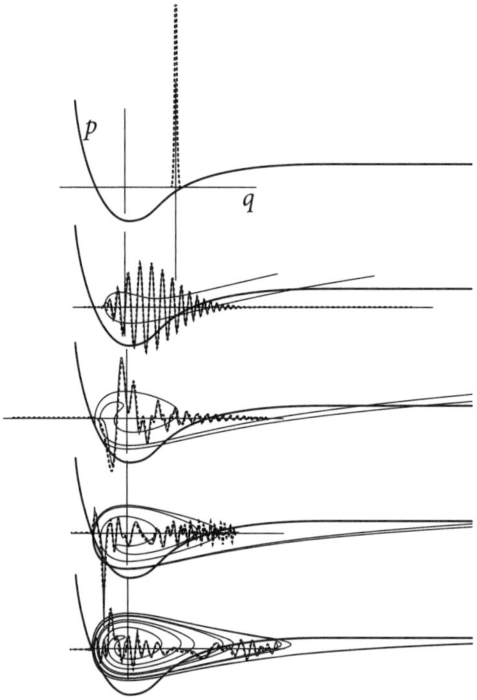

Figure 8.1. The "exact" (fast Fourier transform propagation, dashed line, of the topmost wavefunction as an initial condition) evolving wavefunction compared to the semiclassically calculated time evolution (solid line) in a Morse potential. The evolution of the classical Lagrangian manifold is shown as a thin line.

Indeed this happens, although the times required for a given initial wavepacket to revive can be quite large even for a one-dimensional problem, and ridiculously large in more dimensions, unless there are certain degeneracies present, such as in the case of the hydrogen atom, for example. In a quantum revival, an initially localized wavepacket reforms or "revives" into a compact reincarnation of itself long after it has spread in an unruly fashion over a region restricted only by the potential energy. This is a purely quantum phenomenon, having no classical analog (see figure 8.2).

The challenge is: can the VVMG propagator remain accurate long enough to predict long time revivals?

It is worth noting that in the circumstance shown, the purely classical analog of the quantum revival *never* happens. It is seen in figure 8.1 that the classical distribution continuously spawns more arms of a spiral, essentially uniformly occuping the whole allowed coordinate region at all times beyond the earliest. The cause of the quantum revival is clearly interference between the different arms of the spiral, as discussed shortly.

These are by no means the most adverse circumstances successfully faced by the VVMG propagator, as we will see. It is very instructive to watch what happens near the revival time, long after the time at the bottom of figure 8.1. Nothing new has to happen in the semiclassical procedure, except to compute and add up the ever increasing number of terms in the semiclassical sum, each corresponding to starting near the initial wavepacket location and winding up at a given position. In the case of a one-dimensional anharmonic oscillator, the spiraling seen in figure 8.1 leads at long times to four classical contributions or "arms" of the spiral every complete oscillation of the trajectories with the shortest period. This period is the period of the trajectories in the initial classical manifold having the lowest energy, since the period

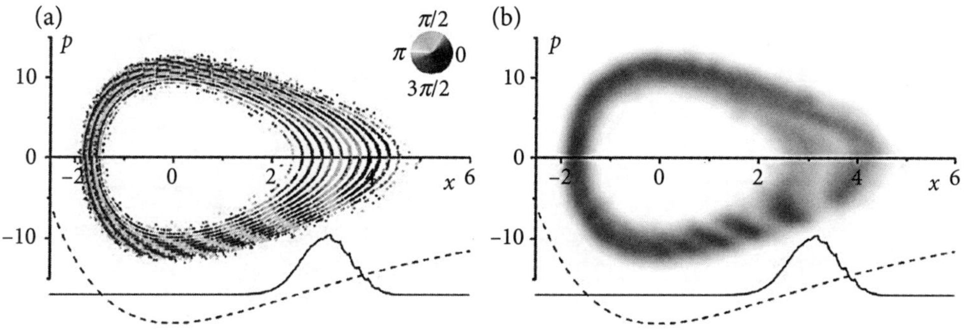

Figure 8.2. A long time quantum revival of a wavepacket captured by purely primitive semiclassical VVMG propagation. An initially vertical classical manifold representing a position state is shown on the left, greatly twisted into a multiple spiral by the nonlinear time evolution. The shades are given by a mapping of the semiclassical phase, including Maslov corrections, onto the color wheel (here reduced to shades of gray). It is seen, on the left and on the right in a blurred version, how the phases cancel when collected vertically, summing over all the branches to give the coordinate space wavefunction, except in the region where the revival is occurring, where the phases add up constructively [32]. The potential is shown by a dashed line, and the semiclassical wavepacket at this time is very close to the exact one, except for the small blips on the right side.

increases monotonically with increasing energy. Every half period two new branches are generated, as can be seen happening at the bottom of figure 8.1.

Now, we try much longer time propagation in the Morse potential, up to times where the quantum revivals can occur. Quantum revival and Anderson localization are members of a small class of subtle interference effects resulting in a quantum distribution radically different from the classical after long time evolution under classically nonlinear evolution [32]. The success of this long time evolution semiclassical VVMG propagated quantum revival is seen in figure 8.2.

8.3 Autocorrelation and Spectrum in a Chaotic System

Next, two much tougher challenges for Van Vleck-Morette-Gutzwiller (VVMG) are faced [22]. In case 3, we launch a wavepacket in the fully classically chaotic stadium billiard in two spatial dimensions, and track the autocorrelation of this wavepacket, Fourier transforming it to get at the spectrum. This latter was done numerically (instead of by stationary phase), using the semiclassical autocorrelation, and represents a uniformization of the spectrum. But the autocorrelation itself was the result of applying the off center guiding method (see section 11.2, later) and is better viewed as a real approximation to the fully primitive semiclassical Generalized Gaussian wavepacket dynamics (GGWPD) method; here, this is the VVMG propagator applied to coherent states (Gaussian wavepackets). The initial wavepacket in the stadium billiard and its early fully quantum evolution (computed for reference only) are shown in figure 8.3.

The autocorrelation function was calculated by adding up all the amplitude, returning on homoclinic orbits, in the manner of the thin (actually it is much thinner

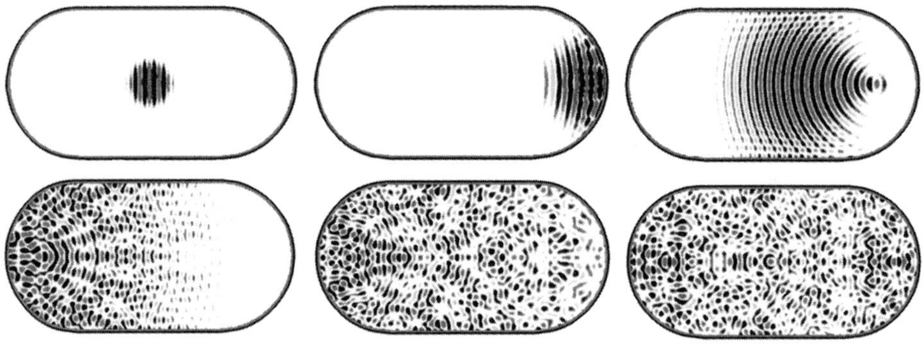

Figure 8.3. Sequential snapshots of the Gaussian launched in the stadium at $t = 0$ (upper left) until just past the first recurrence in the autocorrelation function.

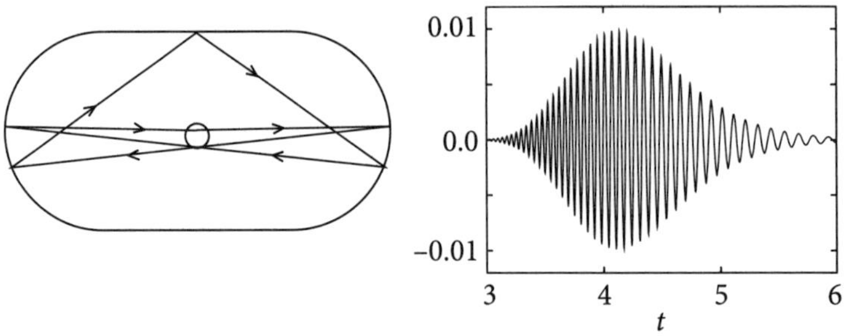

Figure 8.4. A homoclinic classical orbit is shown starting and returning to the vicinity of the initial wavepacket. If followed for longer, this orbit would continue to bounce back and forth horizontally, or nearly so, approaching the perfectly horizontal periodic orbit in the infinite future. A total of 65,000 of these homoclinic orbits were required. The contribution of this particular one is shown in the panel at the right, from the time it first starts until it is no longer important. Note that the short wavelengths arrive first, as the fastest part of the wavefunction makes its way along the orbit [22].

in reality) wavepacket overlapping the initial, circular wavepacket (see figure 11.4, below). Each homoclinic oscillation contributes such a thin wavepacket overlap; the number of them increasing, stretching, and moving toward the central fixed point as time increases. Altogether 65,000 such homomclinic overlaps were used. Each one was obtained by guiding the original, circular Gaussian by an off-center homoclinic orbit starting on the unstable manifold. These homoclinic contributions are now understood to be approximations to the complex GGWPD guided by complex trajectories. The GGWPD makes it clear that all complex initial conditions must be sought that return to the vicinity of the initial wavepacket, if the autocorrelation is being computed. The real, homoclinic orbits are approximations to these. A single homoclinic orbit and its contribution through time to the autocorrelation is shown in figure 8.4.

Figure 8.5 shows a single typical trajectory in the stadium.

Figure 8.6 reveals the exact (black) and semiclassical (gray) autocorrelation function of the initial wavepacket $|\langle\phi(0)|\phi(t)\rangle|$, together with its Fourier transform, giving the spectrum up to a resolution determined by the cutoff time of the transform. That

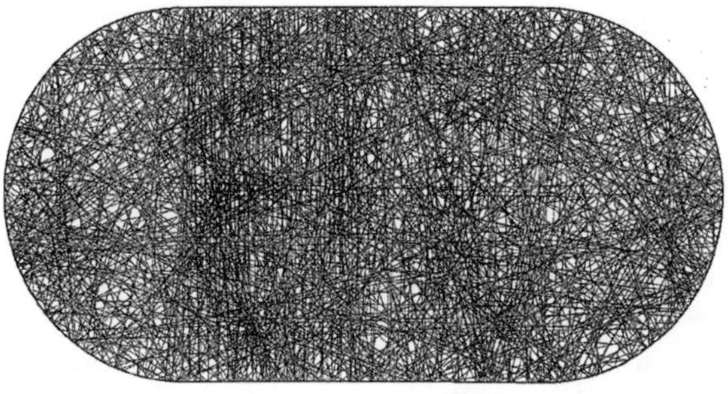

Figure 8.5. The track of a single, typical chaotic trajectory in the stadium billiard.

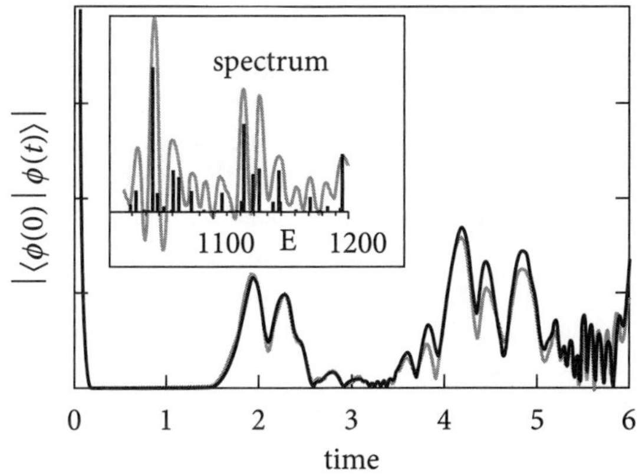

Figure 8.6. A total of 65,000 homoclinic orbits were used to produce the autocorrelation function shown for a wavepacket (see figure 8.4) launched in the Bunimovich stadium billiard. The black line gives the exact quantum mechanical result, and the gray line the semiclassical approximation. As stated in the text, although off-center guided wavepackets (see section 11.2) were used to compute the homoclinic contributions, the method is close to, but not quite the same as, the GGWPD method [22]. The spectrum given by the Fourier transform of the autocorrelation function up to the time computed and shown here, in the inset, up to time equal to six (more than two periods of the orbit), gives the spectrum with the resolution appropriate to that time. The spectrum is shown in light gray in the inset, along with the exact spectrum of the initial wavepacket (black), computed by projecting the eigenstates onto the wavepacket and squaring those amplitudes.

spectrum is shown in gray, and the exact quantum stick spectrum is shown in black in the inset.

8.4 Anderson Localization

Finally, we face perhaps the ultimate challenge testing the VVMG propagator—namely, *Anderson localization*. Or that is to say, Prof. Lev Kaplan faced it, and we recount his results [33].

Anderson localization is a subtle phenomenon, an example of quantum mechanics exploring *less* territory than does classical mechanics. It is easiest to witness what is happening in the case of a bounded random potential in one to three dimensions. By bounding the fluctuations of the potential, we can make Anderson localization appear in a surprising form, namely, that even for energies lying above the highest barrier, so that no classical backscattering occurs on any barrier, all the quantum states are localized—that is, square integrable functions dying off exponentially from some mean position. Classically and quantum mechanically, even backscattering from barriers would naïvely be expected to lead to diffusion and indefinite spreading. Instead, through a subtle quantum interference effect, the wavefunction stops diffusing, defying the classical analog of flying freely over the tops of the barriers. This is called Anderson localization.

Kaplan investigated an Anderson-localized quantum map exhibiting unbounded classical diffusion. As the map is iterated, the initially linear classical manifold he started with exhibits dozens and then thousands of oscillations, all of which semiclassically interfere. In fact, before the Anderson localization sets in, there is a short time of classical-like diffusion, followed by a much longer time as the wavefunction freezes, so to speak. The time is long enough that there is no possibility of numerically accounting for all the branches or contributions to the VVMG propagator, growing exponentially in number. Kaplan was nonetheless able to show that if one *could* numerically follow the trillions of contributions, it would have localized, semiclassically. He did this by looking for convergence of the semiclassical propagator for a long (but not enough to see localization) time T. The propagator is *iterated* (multiplied) to get to $2T, 3T, \ldots$. He noticed that (1) Anderson localization was seen for the iterated propagator, and (2) the result stopped changing if T was taken larger and larger, starting with T large enough. He bounded the "error" in using these iterations instead of the direct semiclassical sum. The evidence was that if all the extremely many semiclassical contributions could have been counted at long times, Anderson-localized solutions would have resulted. Anderson localization can now be seen at least in some cases as destructive interference of classically allowed paths.

Chapter 9

Phase Space Representations of Wavefunctions and Densities

9.1 Wigner Phase Space

Wigner defined a transformation, lifting wavefunctions and density matrices from coordinate or momentum space into a classical-like phase space density, revealing both position and momentum disposition in an intuitive way. The Wigner transformation is

$$\rho_W(q,p) = \left(\frac{1}{\pi\hbar}\right)\int_{-\infty}^{\infty} e^{2ip\cdot s/\hbar}\psi^*(q+s)\psi(q-s)\,ds. \tag{9.1}$$

The overall phase of the wavefunction is lost, but it does not matter in the determination of observables. Quantum densities and operators are lifted into phase space in an analogous way:

$$\rho_W(q,p) = \left(\frac{1}{\pi\hbar}\right)\int_{-\infty}^{\infty} e^{-2ip\cdot s/\hbar}\rho_Q(q+s,q-s)\,ds. \tag{9.2}$$

The Wigner-Weyl transform [34, 35] of an operator is given by

$$A_W(p,q) = 2\int_{-\infty}^{\infty} ds\; e^{2ips/\hbar}\langle q-s|\hat{A}|q+s\rangle. \tag{9.3}$$

All values of physical observables are preserved. Expectation values may be computed as traces over phase space as

$$\langle\hat{A}\rangle = \langle\psi|\hat{A}|\psi\rangle = \mathrm{Tr}[\hat{A}\,|\psi\rangle\langle\psi|] = \mathrm{Tr}[A_W\rho_W] = \int\!\!\int dp\,dq\;A_W(p,q)\rho_W(p,q), \tag{9.4}$$

where Tr denotes the quantum or classical phase space trace, as appropriate. It is easily seen from the definition that the usual coordinate and momentum space probability densities are given by integrating $\rho_W(q,p)$ over the conjugate variable:

$$\frac{1}{h}\int_{-\infty}^{\infty} dp\ \rho_W(q,p) = |\psi(q)|^2, \text{ and}$$
$$\frac{1}{h}\int_{-\infty}^{\infty} dq\ \rho_W(q,p) = |\psi(p)|^2. \tag{9.5}$$

Expectation values of binary operator products are also computable as traces over classical phase space:

$$\mathrm{Tr}[\hat{A}\hat{B}] = \mathrm{Tr}[A_W B_W] = \int dpdq\ A_W(p,q)B_W(p,q). \tag{9.6}$$

All of the equations are easily generalized to many dimensions—for example, the Wigner phase space density becomes

$$\rho_W(\mathbf{q},\mathbf{p}) = \left(\frac{1}{\pi\hbar}\right)^N \int_{-\infty}^{\infty} e^{2i\,\mathbf{p}\cdot\mathbf{s}/\hbar}\phi^*(\mathbf{q}+\mathbf{s})\phi(\mathbf{q}-\mathbf{s})\,d\mathbf{s}. \tag{9.7}$$

How does the Wigner phase space distribution evolve in time? Knowing how the wavefunction evolves according to the Schrödinger equation, we can always choose to calculate the Wigner phase space distribution at each time using the definition. A direct approach starting and staying in Wigner phase space is given in an $\hbar$ expansion:

$$\frac{\partial\rho_t^W(\mathbf{q},\mathbf{p})}{\partial t} = -\{\rho_t^W(\mathbf{q},\mathbf{p}), H(\mathbf{p},\mathbf{q})\} - \frac{\hbar^2}{24}\frac{\partial^3 H}{\partial q^3}\frac{\partial^3\rho_t^W}{\partial p^3} + \mathcal{O}(\hbar^4) + \cdots. \tag{9.8}$$

Among all possible schemes to "lift" wavefunctions into phase space distributions, the Wigner phase space distribution is unique in that the equations governing the time evolution start with the classical Liouville equation in an expansion in $\hbar$.

However, in spite of appearances, the terms of order $\hbar^2$ and higher in equation 9.8 do not "go away" as $\hbar \to 0$; instead, they grow. The reason is that pure state densities ρ_t^W are order $\hbar^{-3}$ in $\partial^3\rho_t^W/\partial p^3$, for example, and order $\hbar^{-5}$ in the next order that involves $\partial^5\rho_t^W/\partial p^5$, and so on. Thermal averaging tames the higher order terms and allows higher powers of $\hbar$ to decrease in magnitude for thermal densities $\rho_T(q+s, q-s)$ with high enough temperature T. This property was behind Wigner's original work *On the Quantum Correction for Thermodynamic Equilibrium* [35]. Corrections to classical thermal chemical reaction rates were calculated, for example.

The preceding equations show that as far as observables are concerned, quantum mechanics can take place in Wigner phase space under this dynamics, using the transformations already given; this is sometimes called the *Wigner equivalent formalism*.

Because the Wigner transformation introduces a quadratic form in the exponent—that is, $2ips/\hbar$, wavefunctions that are also exponentials with quadratic forms, or even sums of such terms, give especially simple results. The wavefunction $\psi(q) = e^{-(2+i)(q-3)^2/2+2i(q-3)} + e^{-(2+2i)(q+3)^2/2} = \psi_1(q) + \psi_2(q)$ is a sum of two Gaussians, one

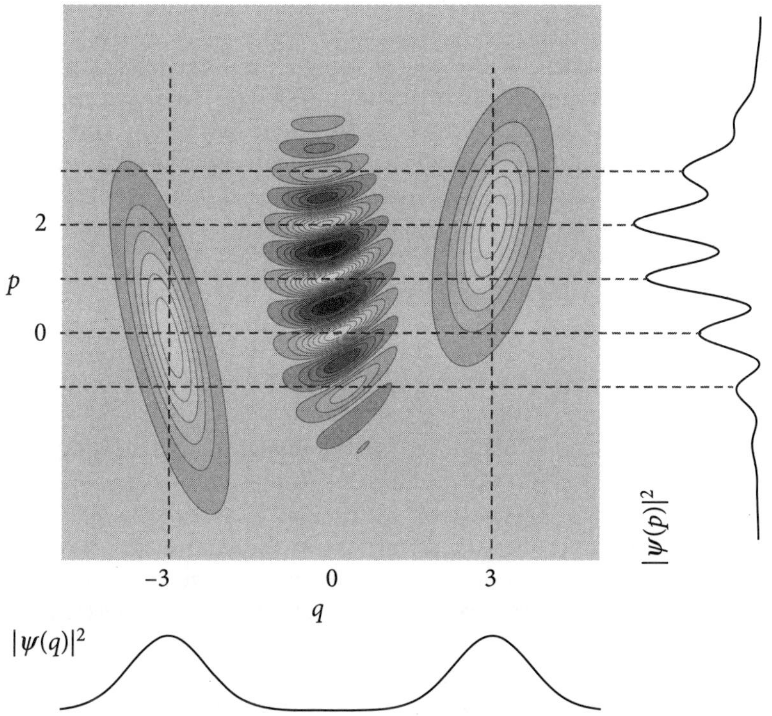

Figure 9.1. The Wigner transform $\rho(p,q) = \rho_1(p,q) + \rho_2(p,q) + \rho_i(p,q)$ of the sum of two Gaussian wavepackets; $\psi(q) = e^{-(2+i)(q-3)^2/2+2i(q-3)} + e^{-(2+2i)(q+3)^2/2} = \psi_1(q) + \psi_2(q)$. Projecting $\rho(p,q)$ vertically—that is, integrating over momentum—results in $|\psi(q)|^2$. The positive and negative portions of the Wigner density lying between the two Gaussian distributions on either side cancel, almost exactly. $\psi_1(q)$ and $\psi_2(q)$ do not interfere perceptibly in position space. On the other hand, projecting the density horizontally—that is, integrating over position—results in $|\psi(p)|^2$; the positive and negative portions of the Wigner interference density $\rho_i(p,q)$ are responsible for the interference oscillations.

centered at $q = -3$, $p = 2$ and another at $q = 3$, $p = 0$. Both have (different) position-momentum correlation. The Wigner transform $\rho(p,q)$ of this sum of two Gaussians has a piece equal to the Wigner transform of each alone, and an interference term, residing between the two (figure 9.1): $\rho(p,q) = \rho_1(p,q) + \rho_2(p,q) + \rho_i(p,q)$. The transform of the sum consists of both positive and negative parts, while $\rho_1(p,q)$ and $\rho_2(p,q)$ are positive definite. Since $|\psi(q)|^2$ is obtained by integrating $\rho(p,q)$ over momentum, it is seen that the positive and negative parts of $\rho_i(p,q)$ very nearly cancel each other in that integral, giving two Gaussian mounds. The momentum space density $|\psi(p)|^2$ is obtained by integrating $\rho(p,q)$ over position, and it is clear from figure 9.1 that it oscillates because of the interference term. The oscillation of the Wigner density of two real Gaussians separated in space by Δq goes like $\cos(2p_x\Delta q/\hbar)$, making it easier for small environmental perturbations to wipe out the oscillations, the farther apart the wavepackets are. The more "macroscopic" the manifestation of coherence, the more easily it gets destroyed.

9.2 Quantum Flux

The Wigner phase space density seems to break the uncertainty principle rules by specifying definite values at sharp points in both position and momentum. Of course, it does not really violate the uncertainty principle: any quantum state is spread over an area, such that $\Delta p^2 = \mathrm{Tr}[p^2 \rho_W(p, q)] = \int p^2 |\psi(p)|^2 \, dp$ and $\Delta q^2 = \mathrm{Tr}[q^2 \rho_t^W(p, q)] = \int q^2 |\psi(q)|^2 \, dq$, so that necessarily $\Delta p \Delta q \geq \hbar/2$. The negative regions in Wigner phase space representations also play a role in holding the line on the uncertainty principle.

We should be just as curious that the quantum flux, defined as

$$\vec{j}(\boldsymbol{q}) = \frac{\hbar}{m} \mathrm{Im}[\psi^*(\boldsymbol{q}) \nabla \psi(\boldsymbol{q})] = \frac{1}{m} \mathrm{Re}[\psi^*(\boldsymbol{q}) \; \hat{\boldsymbol{p}} \; \psi(\boldsymbol{q})], \tag{9.9}$$

apparently gives precise information about the momentum at definite positions $\vec{q}$. Roughly, we can write $\vec{j} \sim \vec{p} \times |\psi^2|$—that is, a magnitude and direction of the (net) momentum is available at any position we choose. The flux seems to get a pass in textbooks and elsewhere on the subject of how to measure it, and how it comes to know so much about both position and momentum of a wavefunction.

How *would* you measure it? Flux is the expectation value of a Hermitian quantum operator and therefore a measurable; the operator is

$$\hat{\boldsymbol{j}}_q = \frac{1}{2m} (|\boldsymbol{q}\rangle\langle \boldsymbol{q}| \hat{\boldsymbol{p}} + \hat{\boldsymbol{p}} |\boldsymbol{q}\rangle\langle \boldsymbol{q}|). \tag{9.10}$$

A careful analysis of this operator [36], by a process that generates the $\boldsymbol{q}$ states as limits of narrow Gaussians, shows that only two of the infinite number of eigenstates of $\hat{\boldsymbol{j}}_q$ have an eigenvalue other than zero, and these two tend to $+\infty$ and $-\infty$, respectively, as the Gaussians become delta functions! Thus, each location where flux is to be measured (at a tiny spot rather than a point, because a true point really is an ill-defined procedure due to the infinities) requires a huge number of trials, almost all of them give 0, and, rarely, a very large, directed value, such that after averaging one arrives the correct textbook flux at that point.

The flux operator is disappointing anyway as a measure of phase space disposition: it is strictly zero everywhere for real wavefunctions. Yet such eigenstates often have nodal structure that strikes the eye and invites probing in phase space.

The flux operator can be very usefully generalized by starting with another phase space distribution—namely, the coherent state or Husimi representation. We introduce that next, and the generalization of the flux that it makes possible.

9.3 Husimi Phase Space and Related Distributions

The Husimi phase space distribution $\rho_\psi(\boldsymbol{q}_0, \boldsymbol{p}_0, \boldsymbol{A}_0)$ of a wavefunction ψ is obtained by projecting it onto Gaussians $G(\boldsymbol{q}; \boldsymbol{q}_0, \boldsymbol{p}_0, \boldsymbol{A}_0)$, varying the position and momentum parameters to obtain a map of the wavefunction's location in phase space [36]

$$\rho_\psi(\boldsymbol{q}_0, \boldsymbol{p}_0, \boldsymbol{A}_0) = |\langle \boldsymbol{q}_0, \boldsymbol{p}_0, \boldsymbol{A}_0 | \psi \rangle|^2 \tag{9.11}$$

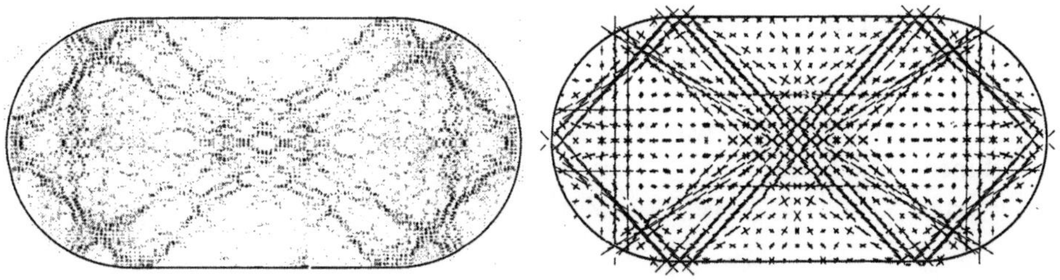

Figure 9.2. Husimi projection of a scarred eigenfunction [37] of the classically chaotic stadium billiard (right), and on the left, a standard contour plot of ψ^2. For the Husimi projection, a grid of positions was used as centers of the Gaussians, and a discrete set of directions of the momentum used, with the resulting projection onto the eigenstate squared plotted as lines whose direction is taken from the momentum, and whose length is proportional to the absolute squared overlap of each Gaussian used. Although this is a fairly transparent example, the Husimi projection clarifies the confused region near the center of the billiard, and shows that the flux along the orbits continues through. This eigenstate is scarred by two periodic orbits: a "figure 8" and a "double triangle" (see chapter 22).

(see equation 10.34), and

$$G(\boldsymbol{q}; \boldsymbol{q}_0, \boldsymbol{p}_0, \boldsymbol{A}_0) = \langle \boldsymbol{q} | \boldsymbol{q}_0, \boldsymbol{p}_0, \boldsymbol{A}_0 \rangle = \left(\frac{\sqrt{2} \det[\mathrm{Im}(\boldsymbol{A}_0)]^{\frac{1}{4}}}{\pi^{\frac{N}{4}}} \right)$$
$$\times \exp\left[\frac{i}{\hbar} \{ (\boldsymbol{q} - \boldsymbol{q}_0) \cdot \boldsymbol{A}_0 \cdot (\boldsymbol{q} - \boldsymbol{q}_0) + \boldsymbol{p}_0 \cdot (\boldsymbol{q} - \boldsymbol{q}_0) \} \right], \tag{9.12}$$

with $\boldsymbol{A}_0$ usually taken to be a diagonal matrix with imaginary elements for the purposes of Husimi plots. The Gaussian is centered at the point $(\boldsymbol{q}_0, \boldsymbol{p}_0)$ in phase space, with a "user defined" width (in coordinate space) controlled by $\boldsymbol{A}_0$. The width is chosen to somehow reveal an optimum picture of the wavefunction; this choice varies considerably according to the wavefunction itself. Of course, $\boldsymbol{A}_0$ may be changed and another map generated with a different perspective.

Writing $\rho_\psi(\boldsymbol{q}_0, \boldsymbol{p}_0, \boldsymbol{A}_0)$ as $|\langle \boldsymbol{q}_0, \boldsymbol{p}_0, \boldsymbol{A}_0 | \psi \rangle|^2 \equiv |\langle G(\boldsymbol{q}_0, \boldsymbol{p}_0, \boldsymbol{A}_0) | \psi \rangle|^2$, it is evident that the Husimi transform can be computed over momentum space representations of the Gaussian and of ψ, and that $\rho_\psi(\boldsymbol{q}_0, \boldsymbol{p}_0, \boldsymbol{A}_0)$ can be written $\mathrm{Tr}[|G\rangle\langle G|\psi\rangle\langle\psi|]$. The latter form makes it clear, upon recalling equation 9.6, that the Husimi distribution $\rho_\psi(\boldsymbol{q}_0, \boldsymbol{p}_0, \boldsymbol{A}_0)$ is a Wigner transform of the wavefunction $|\psi\rangle$ averaged over a Gaussian centered at $(\boldsymbol{q}_0, \boldsymbol{p}_0)$. The Wigner transform can take on positive and negative values, but the Gaussian averaging gives a positive definite result.

The Husimi distribution is positive by definition, but it cannot be used to reconstruct $|\psi(q)|^2$, for example. It lends itself to the intuitive reconstruction of the phase space distribution of eigenstates in two dimensions, by the following trick: Using the semiclassical approximation that $E = (p_x^2 + p_y^2)/2m + V(x, y) = p_0^2/2m + V(x, y)$, a dense set of positions are chosen, and a range of Gaussians differing only in direction of the momentum parameters, centered at each (x_i, y_j) is plotted with short line segments. The lines have a base at (x_i, y_j), a direction the same as the Gaussian momentum, and a length proportional to the magnitude of the Husimi projection. The momentum parameters are $p_{x,k} = p_0 \cos(\theta_k)$, $p_{y,k} = p_0 \sin(\theta_k)$ over an equally spaced range of θ_k on $(0, 2\pi)$. This was done to produce figure 9.2, revealing an image of the four-dimensional Husimi distribution on a three-dimensional slice—that is, locations on the "energy shell" $E = (p_x^2 + p_y^2)/2m + V(x, y)$.

Chapter 10

Gaussian Wavepackets and Linear Dynamics

Before the euro, the 10 Deutsche mark honored Gauss (figure 10.1). In chapter 7, we derived the "gold standard" of semiclassical approximation—namely, the VVMG Green function propagator. In chapter 8, we tested the propagator in various difficult circumstances, and the results were remarkable.

It is clear from equation 7.17 that if the classical action for a system is quadratic, stationary phase is not an approximation, and the exact long time action remains a quadratic form. This gives rise to a number of well-known results for propagation of wavefunctions that can be expressed initially as exponentials of quadratic forms. Clearly the wavefunction remains in quadratic form. We give some of these results in this chapter.

Experiments on atomic and molecular systems are now often conducted in a way that demands that theory be implemented in the time domain. A variety of methods have emerged to handle the time dependent Schrödinger equation, including essentially exact fast Fourier transform (FFT) and other basis set methods. These are, however, restricted to a few degrees of freedom, and sometimes just give "the answer" without much physical insight. The wavepacket approaches we discuss in this chapter are capable of handling many degrees of freedom and giving much insight into the essential physics. Here, we discuss exact solutions of simpler systems like free particles, linear ramp potentials, and the normal or inverted harmonic oscillator. For these systems, the wavepackets have a full set of semiclassical attributes but in fact are exact solutions of the time-dependent Schrödinger equation. We transition into approximate semiclassical uses of wavepackets in more complex situations starting in chapter 11.

Even when experiments are done at fixed energy, there remains a strong motivation to use the time domain for calculations and understanding. For example, calculating high-resolution Raman scattering for a medium-size molecule in the energy domain is next to impossible, owing to the huge number of intermediate states needing to be calculated and summed. Raman scattering is, however, one of many examples where the "essential physics" is over in a short time, even though the data is energy resolved. The time-dependent formulation of Raman scattering (Lee and Heller) [38–40] makes clear the essential (and short-time) information needed to calculate

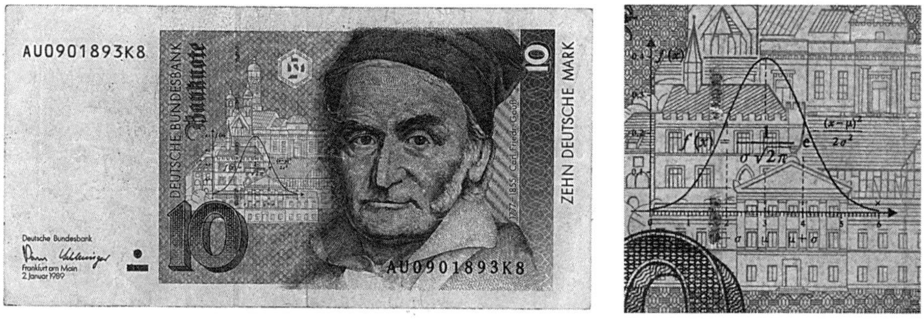

Figure 10.1. Before the euro, there was the 10 Deutsche mark note, not only with a picture of Gauss but also of a Gaussian function.

and understand resonance and preresonance Raman spectra [41]. See sections 17.8 to 17.14.

Scattering theory is another case in point. Collisions are frequently over quickly, yet the usual formulation of particle scattering is in the energy domain. See David Tannor's book, *Introduction to Quantum Mechanics: A Time-dependent Perspective* [1], an excellent introduction to wavepacket formulations of quantum mechanics, and many specific problems, including reactive scattering and many other subjects not treated here.

A general Gaussian wavepacket in one dimension can be written as

$$\psi_1(q) = \left(\frac{a_1}{\pi\hbar}\right)^{1/4} \exp\left[-\frac{a_1 + ib_1}{2\hbar}(q - q_1)^2 + \frac{i}{\hbar}p_1(q - q_1) + \frac{i}{\hbar}\phi_1\right], \tag{10.1}$$

where $(a_1 + ib_1)$ is a complex number with positive real part a_1 and imaginary part ib_1 of either sign. The Gaussian has average position $q_1 = \langle\psi_1|\hat{q}|\psi_1\rangle = \int \psi_1^*(q)\; q\; \psi_1(q)\; dq = \int |\psi_1(q)|^2 q\; dq$, and average momentum $p_1 = \langle\psi_1|\hat{p}|\psi_1\rangle = \int \psi_1^*(q)\left(-i\hbar\frac{\partial}{\partial q}\right)\psi_1(q)\; dq$. The uncertainties in position and momentum are easily calculated:

$$\begin{aligned} \Delta q^2 &= \int \psi_1(q)^*(\hat{q} - q_1)^2 \psi_1(q)\; dq, \\ &= \frac{\hbar}{2a_1}, \end{aligned} \tag{10.2}$$

$$\begin{aligned} \Delta p^2 &= \int \psi_1(q)^*(\hat{p} - p_1)^2 \psi_1(q)\; dq, \\ &= \frac{a_1\hbar}{2} + \frac{b_1^2\hbar}{2a_1}, \end{aligned} \tag{10.3}$$

$$\Delta q \Delta p = \frac{\hbar}{2}\sqrt{1 + \frac{b_1^2}{a_1^2}}. \tag{10.4}$$

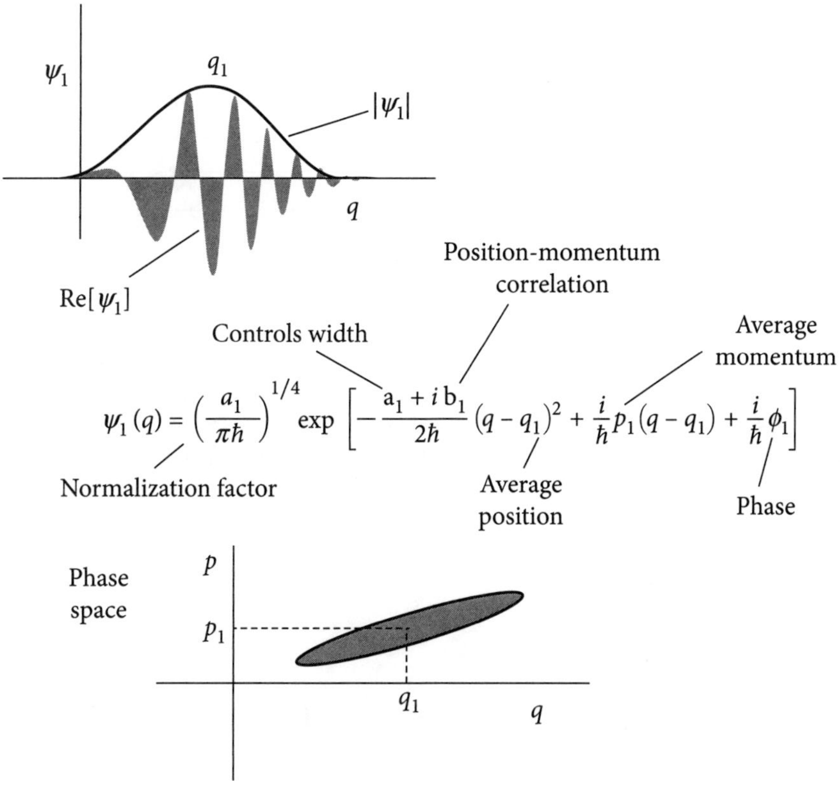

Figure 10.2. Anatomy of a Gaussian. A Gaussian wavepacket is displayed in coordinate space with the role of the parameters $a_1, b_1, q_1, p_1, \phi_1$ shown.

The "anatomy" of the Gaussian is worth discussing. In figure 10.2, we show a normalized, general Gaussian in coordinate space and the aspects that each parameter in the Gaussian controls. The parameters $a_1, b_1, q_1, p_1, \phi_1$ are all taken to be real.

Equation 10.4 shows that the minimum uncertainty of $\hbar/2$ is attained only if the imaginary part b_1 of the "spread" parameter $\alpha_1 = a_1 + ib_1$ vanishes. Note that b_1 controls the "chirp," or position-momentum correlation of the Gaussian. This correlation is seen in figure 10.2, where the local wavelength decreases to the right. One way of understanding the chirp is to notice that the "local momentum," defined as

$$p(q) \sim \hbar \frac{\partial}{\partial q} \mathrm{Im}[\log \psi_1(q)] = p_1 - b_1(q - q_1), \tag{10.5}$$

varies with position. It is the mean momentum p_1 when $q = q_1$, and deviates linearly away from the center according to the sign and magnitude of b_1. (This is not usually the position-momentum correlation axis when the wavepacket is plotted in phase space; see the following).

The reason for failing to be minimum uncertainty if $b_1 \neq 0$ is clear from the phase space schematic at the bottom of figure 10.2. Only if the major axes of the distribution line up with the momentum and position axes can the product of the "shadows," or projections Δp and Δq, be minimized. If $b_1 \neq 0$, there is a rotation, as shown.

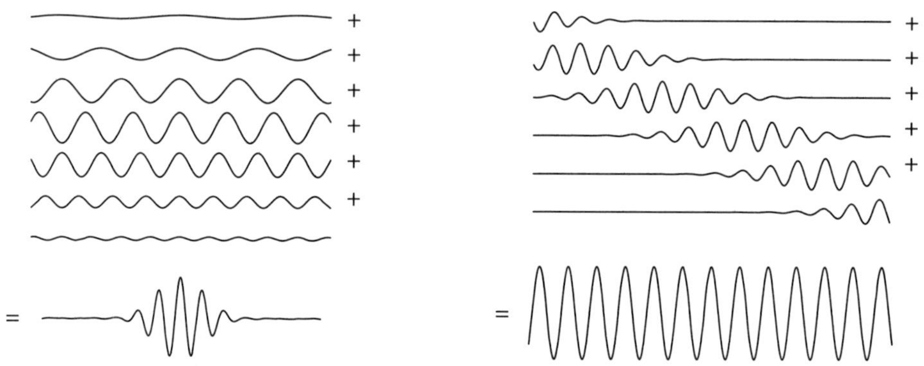

Figure 10.3. (Left) The construction of a Gaussian wavepacket from plane waves. The plane waves above were added directly to produce the near Gaussian wavepacket seen at the bottom. (Right) The construction of plane waves from Gaussian wavepackets.

The momentum representation of the Gaussian is obtained by Fourier transform of $\psi_1(q)$:

$$\psi_1(p) = \int_{-\infty}^{\infty} \langle p|q\rangle\langle q|\psi_1\rangle\, dq = \frac{1}{\sqrt{2\pi\hbar}} \int_{-\infty}^{\infty} e^{-ipq/\hbar}\psi_1(q)\, dq,$$

$$= \left(\frac{a_1}{\pi\hbar\alpha_1^2}\right)^{1/4} \exp\left[-\frac{(p-p_1)^2}{2\hbar(a_1+ib_1)} - \frac{iq_1p}{\hbar} + \frac{i\phi_1}{\hbar}\right]. \tag{10.6}$$

Also

$$\psi_1(q) = \langle q|\psi_1\rangle = \int_{-\infty}^{\infty} \langle q|p\rangle\langle p|\psi_1\rangle\, dp, \tag{10.7}$$

$$= \frac{1}{\sqrt{2\pi\hbar}} \int_{-\infty}^{\infty} e^{ipq/\hbar}\psi_1(p)\, dp. \tag{10.8}$$

From this last perspective the localized wavepacket is written as a superposition of plane waves (momentum eigenstates) $e^{ipq/\hbar}$ with a Gaussian amplitude $\psi_1(p)$. From

$$\hat{p}\, e^{ipq/\hbar} = -i\hbar\frac{\partial}{\partial q}e^{ipq/\hbar} = p\, e^{ipq/\hbar}, \tag{10.9}$$

we see that $e^{ipq/\hbar}$ is a momentum eigenstate with eigenvalue p. In figure 10.3, left, we see seven plane waves approximating a Gaussian wavepacket, as in equation 10.7, approximated as a discrete sum with seven terms,

$$\psi_1(q) \approx \frac{1}{\sqrt{2\pi\hbar}} \sum_{n=1}^{7} e^{ip_n q/\hbar} \psi_1(p_n)\, \delta p_n, \tag{10.10}$$

where $\psi_1(p_n)$ is given in equation 10.6.

Conversely, we can make a plane wave out of Gaussians. This can be done in many ways; the simple relation

$$\sqrt{\frac{a_1}{2\pi\hbar}} \int \exp[-a_1(x-x')^2/2\hbar + ip_0 x/\hbar]\, dx' = \exp(ip_0 x/\hbar) \tag{10.11}$$

is shown in figure 10.3, right, as an approximate discrete sum. This is our first example of decomposing an extended state in terms of localized Gaussian packets.

Very often we require the overlap of two wavefunctions or phase space distributions. Every quantum amplitude $\langle \boldsymbol{\alpha}|\boldsymbol{\beta}\rangle$ is an overlap or "joint probability amplitude" of two certainties, the certainty of being in state $|\boldsymbol{\alpha}\rangle$ and the certainty of being in state $|\boldsymbol{\beta}\rangle$. What are your chances, if definitely in $|\boldsymbol{\alpha}\rangle$, of also belonging to $|\boldsymbol{\beta}\rangle$? Or if definitely in $|\boldsymbol{\beta}\rangle$, of also belonging to $|\boldsymbol{\alpha}\rangle$? The answer in both cases is the same, $|\langle \boldsymbol{\alpha}|\boldsymbol{\beta}\rangle|^2$.

Several factors lead to Gaussian integrals becoming paramount in such amplitudes and probabilities. Of course, if $|\boldsymbol{\alpha}\rangle$ and $|\boldsymbol{\beta}\rangle$ are Gaussian—that is, $\langle \boldsymbol{q}|\boldsymbol{\alpha}\rangle$ is Gaussian, then the N-dimensional Gaussian integral given here is indispensable. It should not be overlooked that a position state like $|\boldsymbol{q}\rangle$ is simply a very narrow Gaussian, and a momentum state like $|\boldsymbol{p}\rangle$ is a Gaussian very broad in position but very narrow in momentum. Finally, if the states $|\boldsymbol{\alpha}\rangle$ and $|\boldsymbol{\beta}\rangle$ have been written as sums of Gaussians, $\langle \boldsymbol{\alpha}|\boldsymbol{\beta}\rangle$ then becomes a sum of Gaussian integrals.

A useful N-dimensional Gaussian integral is

$$\int_{-\infty}^{\infty} e^{-\frac{1}{2}(\boldsymbol{q}-\boldsymbol{a})\cdot\boldsymbol{A}\cdot(\boldsymbol{q}-\boldsymbol{a})+i\boldsymbol{b}\cdot(\boldsymbol{q}-\boldsymbol{a})}\, d\boldsymbol{q} = \frac{(2\pi)^{\frac{N}{2}}}{|\det \boldsymbol{A}|^{\frac{1}{2}}} e^{-\frac{1}{2}\boldsymbol{b}\cdot\boldsymbol{A}^{-1}\cdot\boldsymbol{b}}, \tag{10.12}$$

where $\boldsymbol{A}$ is a (generally complex) symmetric matrix that must have eigenvalues with positive real parts. The vector parameters $\boldsymbol{a}$ and $\boldsymbol{b}$ may be complex.

10.1 Phase Space Pictures of Wavepackets

The Wigner phase space function for the Gaussian equation 10.1 is obtained by carrying out the integral equation 9.1, resulting in

$$\rho^W_{\alpha_1}(q,p) = \frac{1}{\pi\hbar} \exp\left[-\frac{(a_1^2+b_1^2)}{a_1\hbar}(q-q_1)^2 - \frac{1}{a_1\hbar}(p-p_1)^2 - \frac{2b_1}{a_1\hbar}(q-q_1)(p-p_1)\right]. \tag{10.13}$$

The Wigner phase space distribution of a Gaussian thus is itself a Gaussian in phase space. The position-momentum correlation appears as a cross term in the exponent, skewing the Gaussian.

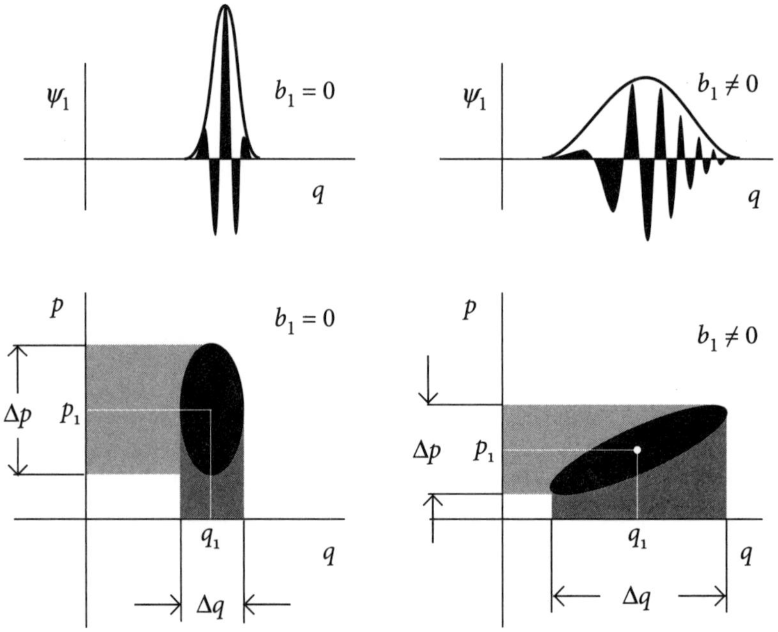

Figure 10.4. Gaussians depicted in coordinate and phase space. The black ellipses are Gaussian densities in the Wigner phase space representation. The wavefunction on the left is of minimum uncertainty $\Delta p \Delta q = \frac{\hbar}{2}$. The position momentum correlation is easily seen in the wavefunction on the right, for which $\Delta p \Delta q > \frac{\hbar}{2}$.

The angle θ formed by the q axis and the major axis of the elliptical shape formed by the Gaussian phase space density is one of the two roots of

$$\tan(2\theta) = -\frac{2b_1}{1 - a_1^2}. \tag{10.14}$$

This is not the same angle as in equation 10.5, unless $a_1 \to 0$—that is, a very wide Gaussian in coordinate space. As a_1 gets larger, position uncertainty diminishes and increasing momentum uncertainty pushes the major axis to lie more vertically, along the momentum axis.

Two cases are seen in figure 10.4, with a schematic diagram of the Wigner phase space density of the Gaussian and the corresponding wavefunction in coordinate space. On the upper left, we see a narrow Gaussian with no position-momentum correlation, and below is its phase space representation. The major axis of the ellipse is parallel to the momentum axis, and the "shadows" on the p and q axes are a minimum, in the sense that

$$\Delta q \Delta p = \frac{\hbar}{2}. \tag{10.15}$$

Minimum uncertainty is achieved only with no position-momentum correlation. On the right is an analogous diagram for $b_1 \neq 0$. The correlation of position with momentum is obvious, with the slower component correlating with smaller q in this case. Here, $\Delta q \Delta p > \frac{\hbar}{2}$.

COHERENT STATES AND GAUSSIAN WAVEPACKETS

The coherent state $|\alpha\rangle$ has a long history[1] and is traditionally defined as the eigenfunction of the harmonic oscillator destruction operator

$$\hat{a} = \frac{1}{\sqrt{2\hbar m\omega}}(m\omega\hat{q} + i\hat{p}),$$
$$\hat{a}|\alpha\rangle = \alpha|\alpha\rangle, \tag{10.16}$$

with eigenvalue

$$\alpha = \frac{1}{\sqrt{2\hbar m\omega}}[m\omega\langle\hat{q}\rangle + i\langle\hat{p}\rangle] = \frac{1}{\sqrt{2\hbar m\omega}}[m\omega q_\alpha + ip_\alpha], \tag{10.17}$$

where q_α and p_α are the expectation values of position and momentum, respectively. The normalized coherent state in coordinate space looks like this:

$$\langle q|\alpha\rangle = \left(\frac{m\omega}{\pi\hbar}\right)^{1/4} \exp\left[-\frac{m\omega}{2\hbar}(q-q_\alpha)^2 + \frac{i}{\hbar}p_\alpha(q-q_\alpha)\right]. \tag{10.18}$$

The coherent states are derived from the ground state of a harmonic oscillator—that is, a Gaussian at rest, the eigenfunction of $\hat{a}|0\rangle = 0$ with eigenvalue $\alpha = 0$. The coherent state $|\alpha\rangle$ with $\hat{a}|\alpha\rangle = \alpha|\alpha\rangle$ is the same Gaussian, but displaced in position by q_α and in momentum by p_α.

The coherent states are complex Gaussians by another name. The set of them for all q_α, p_α is vastly overcomplete. In spite of the overcompleteness, the Gaussians still satisfy a closure relation of the familiar form obtained by summing or integrating over the parameters distinguishing one member of the set from another. For the coherent states, these parameters are q_α and p_α. Thus,

$$\hat{1} = \frac{1}{2\pi\hbar}\iint dq_\alpha\, dp_\alpha\, |\alpha\rangle\langle\alpha|,$$
$$= \frac{1}{\pi}\iint d(\mathrm{Re}\,\alpha)\, d(\mathrm{Im}\,\alpha)\, |\alpha\rangle\langle\alpha|,$$
$$\equiv \frac{1}{\pi}\int d^2\alpha\, |\alpha\rangle\langle\alpha|. \tag{10.19}$$

This may easily be checked by taking the q, q' matrix element of both sides, and verifying that the double integral gives $\delta(q-q')$. (Do the $\int dp_\alpha$ integration first.) Every completeness relation may be viewed as covering all of phase space by summing over successive members of the complete set (see figure 10.5).

Suppose a discrete subset of the coherent states are taken as a basis, ones lying on a rectangular grid with grid spacing such that the area of the enclosed rectangular

[1] Some history of the coherent states: They were first considered by Schrödinger. During World War II, J. Schwinger produced extensive notes on them that developed many of their properties, but these were never published. R. Glauber subsequently used them in his seminal work on quantum optics. For years thereafter, this was the main theater for application of the coherent states. They became known as "Glauber coherent states," but this term is perhaps best applied within the quantum optics context. Many important contributions have been made since by various authors, especially J. Klauder.

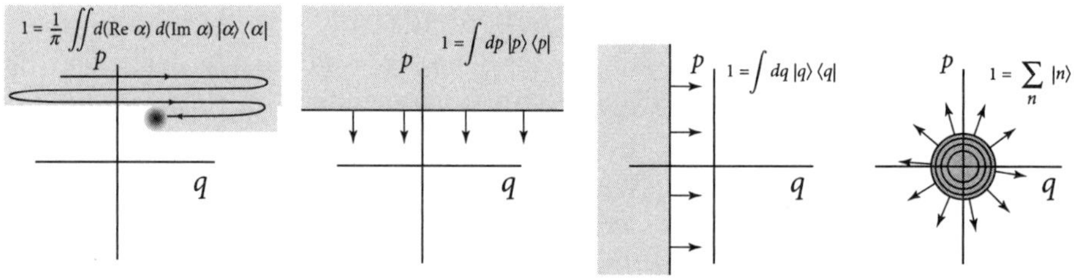

Figure 10.5. Four schematic diagrams showing how different completeness (closure) relations cover the phase plane. The sketch on the right referes to a sum over harmonic oscillator eigenstates $|n\rangle$.

domains is h, the basis extending to infinity in all directions. This basis is countably infinite, nonorthogonal, and complete, but amazingly it is overcomplete by one member even though it has (countably) infinitely many members! It is called a von Neumann basis [42, 43], but unfortunately the experience is that eigenfunctions expanded in this basis converge slowly as the limit of truncation is extended outward in momentum and position. Davis and Heller tried making the grid area locally less than h, with moderate success, and more recently Shimshovitz and Tannor [44] and Halverson and Poirier [45] have proposed variations on the von Neumann lattice idea that are very promising.

10.2 Linear Propagators on Gaussian Wavepackets

Now that some of the properties of complex Gaussians have been discussed, we can look at their dynamics as time-dependent solutions of (possibly time-dependent) Hamiltonian systems (albeit with with linear dynamics—that is, quadratic actions).

We derive these results by direct means, without first constructing the VVMG propagator for the problems, because this will facilitate the subsequent approximate ideas for propagating Gaussians under nonlinear dynamics that are simpler (if more approximate) than first constructing the VVMG propagator.

As Schrödinger realized, the Gaussians are the exact solution for time dependent, at most quadratic potential problems with a standard quadratic kinetic energy. The task is to solve the time-dependent Schrödinger equation

$$i\hbar\frac{\partial\psi(q,t)}{\partial t} = H\psi(q,t) = \left(-\frac{\hbar^2}{2m}\frac{\partial^2}{\partial q^2} + V(q,t)\right)\psi(q,t), \tag{10.20}$$

where the potential is at most quadratic in the coordinate q, but otherwise quite general and even time dependent:

$$V(q,t) = V(q_t,t) + V'(q_t,t)\,(q-q_t) + \frac{1}{2}V''(q_t,t)\,(q-q_t)^2. \tag{10.21}$$

Of course, if the potential is at most quadratic, its second derivative is the same expanded about any point. Even so, we write it as $V''(q_t,t)$ here, as if the point of expansion mattered. We make no error doing this, but the notation becomes useful in the next chapter when we introduce the thawed Gaussian approximation (TGA).

In one dimension, where initially

$$\psi(q,0) = \exp\left[i\frac{A_0}{\hbar}(q-q_0)^2 + \frac{i}{\hbar}p_0(q-q_0) + \frac{i}{\hbar}s_0\right],$$

normalization requires

$$\mathrm{Im}[s_0] = \frac{\hbar}{4}\log\left(\frac{\pi\hbar}{2\,\mathrm{Im}[A_0]}\right).$$

This is a wavepacket that starts out as the general Gaussian equation 10.1, and becomes

$$\psi(q,t) = \exp\left[i\frac{A_t}{\hbar}(q-q_t)^2 + \frac{i}{\hbar}p_t(q-q_t) + \frac{i}{\hbar}s_t\right] \tag{10.22}$$

as time evolves. How do the parameters α_t, q_t, p_t, and ϕ_t depend on time?

Assuming that the potential is at most a (possibly time dependent) quadratic form, the strategy is to plug the Gaussian wavefunction (equation 10.1) into equation 10.20, assuming all the parameters α_t, q_t, p_t, and ϕ_t are time dependent. We compare like powers of $(q - q_t)$, imposing conditions on the parameters α_t, q_t, p_t, and ϕ_t as needed in order to satisfy the dependent Schrödinger equation. The resulting equations are as follows.

Quadratic term:

$$\dot{A}_t = -2\,A_t^2/m - V''(q_t,t)/2, \tag{10.23}$$

Linear term:

$$\begin{aligned}\dot{q}_t &= p_t/m,\\ \dot{p}_t &= -V'(q_t,t),\end{aligned} \tag{10.24}$$

Constant term:

$$\dot{s}_t = -V(q_t,t) + i\,\hbar\,A_t/m + p_t\,\dot{q}_t - p_t^2/2m. \tag{10.25}$$

Small adjustments will bring these expressions into more familiar and useful form. Equations 10.24 are just Hamilton's equations of motion for a trajectory that is the expectation value of the position and momentum of the wavepacket. This connection was first developed in references [16–18], and has been given a group theoretical interpretation in reference [46].

We can tame the nonlinear differential equation and reveal an intimate connection between the equation for A_t and the equations for nearby orbits in classical mechanics for A_t by writing

$$A_t \equiv \frac{1}{2}\frac{P_{zt}}{Z_t}, \tag{10.26}$$

yielding after substitution into equation 10.23

$$\dot{P}_{zt} = -V''(q_t, t)Z_t = -\left(\frac{\partial^2 V(q,t)}{\partial q^2}\right)_{q_t} Z_t,$$
$$\dot{Z}_t = \frac{P_{zt}}{m}. \tag{10.27}$$

Since one variable has been replaced by two, there is more than one pair (Z_t, P_{zt}) corresponding to a given A_t. (There is a kind of gauge structure tied up in this.) At the initial time $t = 0$, one is free to choose any (Z_0, P_{z0}) corresponding to the given initial A_0, with no effect on the physics. Equations 10.27 are just the (linear) equations of motion of a harmonic oscillator with (time dependent) force constant $V''(q_t, t)$, if Z_t is the position of the oscillator and P_{zt} is the momentum.

This can be put in matrix form:

$$\frac{d}{dt}\begin{pmatrix} P_{Z_t} \\ Z_t \end{pmatrix} = \begin{pmatrix} 0 & -V''(q_t,t) \\ m^{-1} & 0 \end{pmatrix}\begin{pmatrix} P_{Z_t} \\ Z_t \end{pmatrix} \equiv K(t)\begin{pmatrix} P_{Z_t} \\ Z_t \end{pmatrix}. \tag{10.28}$$

This is a *linear* differential equation for P_{Z_t} and Z_t. If K is time independent—that is, the potential is constant in time—then it is integrated to give

$$\begin{pmatrix} P_{Z_t} \\ Z_t \end{pmatrix} = M(t)\begin{pmatrix} P_{Z_0} \\ 1 \end{pmatrix}, \tag{10.29}$$

where

$$M(t) = \exp[Kt]. \tag{10.30}$$

The boundary condition $Z_0 = 1$ has been used. $Z_0 = 1$ forces the boundary condition $P_{Z_0} = 2A_0$. If K is time dependent, $M(t)$ cannot be integrated so cleanly as in equation 10.30, and usually must be found by integrating the equations of motion (10.28) numerically.

To proceed with the remaining parameters, we set

$$L_t = -V(q_t, t) + p_t\dot{q}_t - p_t^2/2m,$$
$$= p_t\dot{q}_t - E_t, \tag{10.31}$$

where $E_t = p_t^2/2m + V(q_t, t)$ is the energy, and $L_t = p_t\dot{q}_t - E_t$ is the Lagrangian for the trajectory p_t, q_t. Now we can write equation 10.25 as

$$\dot{s}_t = i\hbar\frac{1}{2m}\frac{P_{zt}}{Z_t} + L_t,$$
$$= i\hbar\frac{1}{2}\frac{\dot{Z}_t}{Z_t} + L_t, \tag{10.32}$$

which can be integrated to give

$$s_t = s_0 + i\,\hbar\frac{1}{2}\mathrm{Tr}\left[\log\left(Z_t\right)\right] + S_t,$$

$$S_t = \int_0^t L_t\,dt. \tag{10.33}$$

The trace in this last equation is superfluous in one dimension, but it shows how the phase-normalization factor s_t is obtained in the N-dimensional case.

In multidimensional form, the normalized Gaussian wavepacket is given by

$$\begin{aligned}\psi_t(\boldsymbol{q}) &\equiv \psi_t(\boldsymbol{q}_t, \boldsymbol{p}_t; \boldsymbol{q}) \\ &= \exp\left[\frac{i}{\hbar}\{(\boldsymbol{q}-\boldsymbol{q}_t)\cdot \boldsymbol{A}_t\cdot(\boldsymbol{q}-\boldsymbol{q}_t) + \boldsymbol{p}_t\cdot(\boldsymbol{q}-\boldsymbol{q}_t) + s_t\}\right],\end{aligned} \tag{10.34}$$

where $\boldsymbol{A}_t$ is an N×N-dimensional matrix for N coordinates, and $\boldsymbol{q}$, $\boldsymbol{q}_t$, $\boldsymbol{p}_t$ are N-dimensional vectors that obey

$$\frac{d}{dt}\boldsymbol{q}_t = \nabla_p H, \tag{10.35}$$

$$\frac{d}{dt}\boldsymbol{p}_t = -\nabla_q H, \tag{10.36}$$

$$\boldsymbol{A}_t = \frac{1}{2}\boldsymbol{P}_Z\cdot \boldsymbol{Z}^{-1}, \tag{10.37}$$

$$\frac{d}{dt}\begin{pmatrix}\boldsymbol{P}_Z\\ \boldsymbol{Z}\end{pmatrix} = \begin{pmatrix}\boldsymbol{0} & -\boldsymbol{V}''(t)\\ \boldsymbol{m}^{-1} & \boldsymbol{0}\end{pmatrix}\begin{pmatrix}\boldsymbol{P}_Z\\ \boldsymbol{Z}\end{pmatrix}, \tag{10.38}$$

$$\dot{s}_t = L_t + \frac{i\hbar}{2}\mathrm{Tr}[\dot{\boldsymbol{Z}}\cdot \boldsymbol{Z}^{-1}], \tag{10.39}$$

or

$$s_t = s_0 + S_t + \frac{i\hbar}{2}\mathrm{Tr}[\ln \boldsymbol{Z}]. \tag{10.40}$$

$\boldsymbol{V}''$ and $\boldsymbol{m}^{-1}$ are N-dimensional matrices of mixed second derivatives of the Hamiltonian with respect to position and momentum coordinates, respectively. That is,

$$\left[\boldsymbol{V}''\right]_{ij} = \frac{\partial^2 H}{\partial q_i \partial q_j}. \tag{10.41}$$

Also

$$\begin{pmatrix}\boldsymbol{P}_{Z_t}\\ \boldsymbol{Z}_t\end{pmatrix} = \boldsymbol{M}(t)\begin{pmatrix}\boldsymbol{P}_{Z_0}\\ 1\end{pmatrix}, \tag{10.42}$$

where from equation 10.38 and equation 10.42, we have

$$\frac{d\boldsymbol{M}(t)}{dt} = \begin{pmatrix}\boldsymbol{0} & -\boldsymbol{V}''(t)\\ \boldsymbol{m}^{-1} & \boldsymbol{0}\end{pmatrix}\boldsymbol{M}(t), \tag{10.43}$$

and the stability matrix,

$$\boldsymbol{M}(t) = \begin{pmatrix} \boldsymbol{m}_{11} & \boldsymbol{m}_{12} \\ \boldsymbol{m}_{21} & \boldsymbol{m}_{22} \end{pmatrix}, \tag{10.44}$$

where $\boldsymbol{m}_{11} = \partial \boldsymbol{p}_t/\partial \boldsymbol{p}_0|_{\boldsymbol{q}_0}$, $\boldsymbol{m}_{12} = \partial \boldsymbol{p}_t/\partial \boldsymbol{q}_0|_{\boldsymbol{p}_0}$, $\boldsymbol{m}_{21} = \partial \boldsymbol{q}_t/\partial \boldsymbol{p}_0|_{\boldsymbol{q}_0}$, and $\boldsymbol{m}_{22} = \partial \boldsymbol{q}_t/\partial \boldsymbol{q}_0|_{\boldsymbol{q}_0}$.

SPECIAL CASES OF LINEAR DYNAMICS

We discuss some important special cases of wavepacket motion on time-independent potentials at most quadratic in q. Free particle dynamics, motion on a linear ramp potential $V(q) = \beta q$, and harmonic motion $V(q) = \pm\frac{1}{2}m\omega^2 q^2$ are examples of linear dynamics, where the current positions and momenta are linear functions of the initial positions and momenta.

FREE PARTICLE

For a free particle with Hamiltonian $H = p^2/2m$, we have

$$q_t = q_0 + \frac{p_0}{m}t, \tag{10.45}$$

$$p_t = p_0. \tag{10.46}$$

This is already tantamount to solving the stability equations, but since we have a time-independent Hamiltonian we can use equation 10.47:

$$M(t) = \exp[Kt],$$

$$M = \exp\left[\begin{pmatrix} 0 & 0 \\ 1/m & 0 \end{pmatrix} t\right] = \begin{pmatrix} 1 & 0 \\ t/m & 1 \end{pmatrix}. \tag{10.47}$$

Thus,

$$P_Z = P_{Z_0}, \; Z = 1 + P_{Z_0}t/m, \tag{10.48}$$

so that A_t is

$$A_t = \frac{A_0}{1 + 2A_0 t/m}. \tag{10.49}$$

The classical action integral becomes

$$s_t = s_0 + \frac{p_0^2}{2m}t + \frac{i\hbar}{2}\mathrm{Tr}\left[\ln(1 + 2A_0\, t/m)\right]. \tag{10.50}$$

Every parameter in the Gaussian, equation 10.22, is now determined.

Figure 10.6 shows the time evolution of the real part of a free Gaussian wavepacket. At first, the wavelength is shorter to the left of the wavepacket, corresponding to faster motion. This position-momentum correlation causes a narrowing of the wavepacket with time, as the faster part catches up with the middle, and the slower part on the right is caught. The wavepacket reaches minimum size and momentarily becomes a minimum uncertainty wavepacket with $b_{1t} = 0$. Following this, the position-momentum correlation reverses, with the faster part (shorter wavelength) forever after

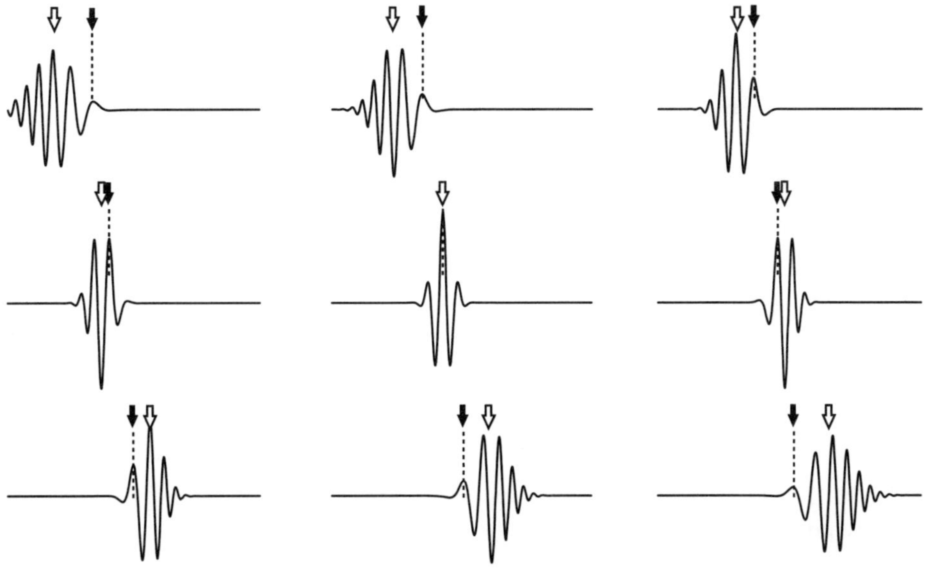

Figure 10.6. The real part of a Gaussian wavepacket propagating in a constant potential. Note the position-momentum correlation, changing during the motion. The wavepacket at first gets narrower, then broader. The phase velocity is less than the group velocity. In a broad wavepacket, these differ by a factor of 2.

leading. The wavepacket spreads indefinitely thereafter. Figure 10.6 also shows the distinction between the *phase velocity* and the *group velocity* of the wavepacket. The latter is the velocity of the probability density of the wavepacket; we already know this is p_t/m. The phase velocity is slower for the time-dependent Schrödinger equation. The black arrow points to a local maximum—that is, it is a place of constant phase of the wave from frame to frame in figure 10.6. The open arrow shows the motion of the center of the wavepacket.

Figure 10.7 shows a Gaussian wavepacket with initial position-momentum correlation for a constant potential. The correlation is revealed by a skewed ellipse in phase space. This ellipse is a filled contour of the corresponding Wigner phase space distribution. Since the center of the distribution has a positive momentum, it moves (on the average) to the right. There is a shear in the phase space distribution corresponding to trajectories with higher momentum traveling faster. This causes an initial narrowing of the wavepacket in coordinate space, since for this wavepacket the faster particles started out in the rear. After the minimum uncertainty wavepacket is reached, the faster trajectories continue and give the reverse position-momentum correlation, with the faster trajectories in the lead. The dashed lines show the contours of constant energy. Since the energy of each trajectory remains constant, the distribution is confined by the same contours at all times. The phase space distribution has area h at all times.

We now consider the free particle Green function. The Green function is related to an initial wavepacket with vanishing initial width. In phase space, the "wavefunction" $\delta(q-q')$ is an infinite vertical line. Because the potential is a quadratic form, the phase space distribution evolves classically, with p_t and q_t given by the linear equation 10.45. This motion is a shear that rotates and stretches the line, becoming $p(q) = m(q-q')/t$

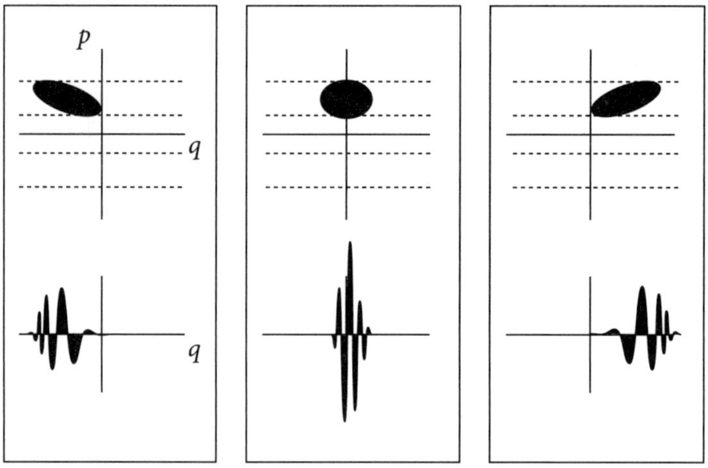

Figure 10.7. Phase space and coordinate space pictures of a Gaussian evolving on a constant potential.

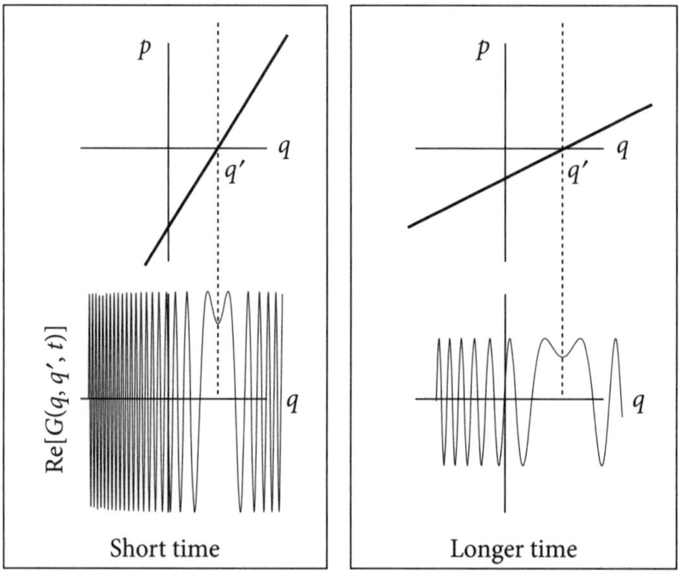

Figure 10.8. Phase space and coordinate space pictures of an initially very narrow Gaussian (or coordinate state, located at the dashed line) evolving on a constant potential at two times >0; this is the evolution of the Green function $G(q, q', t)$.

for $t > 0$. The tilting line in figure 10.8 corresponds to classical trajectories getting to q in time t with velocity $(q - q')/t$. The Green function for $t > 0$ is an undamped function whose oscillations are more rapid away from the initial position q', in accord with the local deBroglie wavelength given by $\lambda(q, t) = h\, t/|(q - q')|m$.

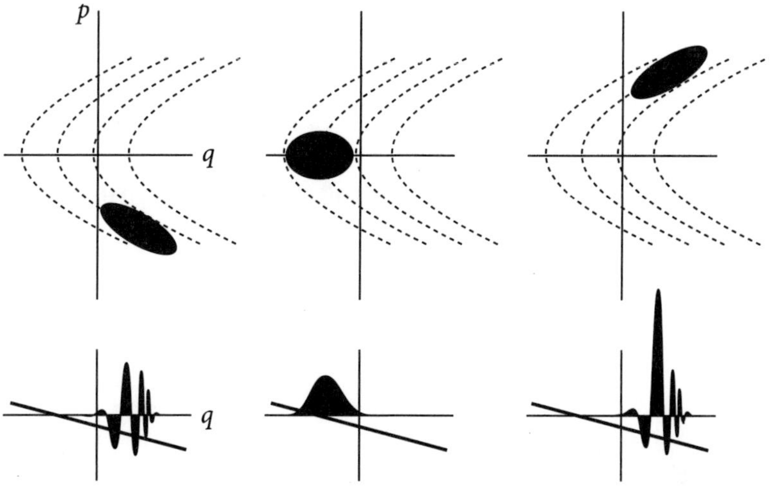

Figure 10.9. Phase space and coordinate space pictures of a Gaussian evolving on a linear potential $V = -\beta q$.

LINEAR RAMP POTENTIAL

For a particle on a linear ramp with Hamiltonian $H = p^2/2m - \beta q$, we have

$$\begin{aligned} p_t &= p_0 + \beta t, \\ q_t &= q_0 + \frac{p_0}{m} t + \frac{\beta}{2m} t^2, \\ s_t &= s_0 + \frac{1}{2} i\hbar \ln\left[1 + \frac{P_{Z_0} \cdot t}{m}\right] + \frac{p_0^2}{2m} t - \beta q_0 t + \frac{\beta p_0}{m} t^2 + \frac{\beta^2}{3m} t^3. \end{aligned} \tag{10.51}$$

Note that since again $V'' = 0$, the equations for P_Z and Z are unaffected as compared to the constant potential (free particle). Thus, A_t is again given simply by equation 10.49. Viewing in phase space, the result is shown in figure 10.9. The center of the wavepacket in position and momentum is now governed by the constant acceleration of the ramp. The stability equations for the linear and the constant potential are identical, since the second derivative vanishes for both. Thus, the *shape* of the phase space distribution for both problems is the same at equal times, but the location in phase space and a plot of the real part of the wavefunction is quite different, because q_t and p_t are different. Again, the contours of constant energy bound the phase space density.

LINEAR POTENTIAL GREEN FUNCTION

Next consider the Van Vleck propagator for a linear potential energy $V(q) = \alpha q$ in one dimension, and its Fourier transform energy form. The free particle case is contained in the results for $\alpha = 0$ (as it is for $\omega = 0$ for the harmonic oscillator). The Hamiltonian for the problem is $H = p^2/2m + \alpha q$. The coordinate q evolves as $q(t') = q' + p't'/m - \alpha t'^2/2m$, where q' and p' are the initial coordinate and momentum at time $t' = 0$. We want $q(t) = q$—that is, the coordinate reaches the point q at time t, that can be used to solve for p' in terms of q, q', t:

$$S(q, q', t) = \int_0^t L(t')\, dt' = \frac{m(q-q')^2}{2t} - \frac{\alpha}{2}(q+q')t - \frac{\alpha^2 t^3}{24m}. \tag{10.52}$$

As promised, this is a quadratic form in q and q'. Note that if $\alpha = 0$ the familiar action for the free particle is recovered:

$$\begin{aligned} G(q, q', t) = G_{sc}(q, q', t) &= \frac{1}{\sqrt{2\pi i\hbar}} \left| \frac{\partial^2 S(q, q', t)}{\partial q \partial q'} \right|^{\frac{1}{2}} \exp[iS(q, q', t)/\hbar] \\ &= \sqrt{\frac{m}{2\pi i\hbar t}}\, e^{i\frac{m(q-q')^2}{2\hbar t} - i\frac{\alpha}{2\hbar}(q+q')t - i\frac{\alpha^2 t^3}{24\hbar m}}. \end{aligned} \tag{10.53}$$

HARMONIC OSCILLATOR

For a harmonic oscillator with Hamiltonian $H = \frac{1}{2m}p^2 + \frac{1}{2}m\omega^2 q^2$, it is a standard result, easily verified, that

$$\begin{aligned} p_t &= p_0 \cos(\omega t) - m\omega q_0 \sin(\omega t), \\ q_t &= q_0 \cos(\omega t) + (p_0/m\omega) \sin(\omega t) \end{aligned} \tag{10.54}$$

solve the classical equations of motion for the guiding trajectory $(q_t,\ p_t)$. The 2×2 matrix M is given by

$$\begin{aligned} M = \exp[Kt] &= \exp\left[\begin{pmatrix} 0 & -m\omega^2 \\ 1/m & 0 \end{pmatrix} t\right], \\ &= \begin{pmatrix} \cos(\omega t) & -m\omega \sin(\omega t) \\ \sin(\omega t)/m\omega & \cos(\omega t) \end{pmatrix}, \end{aligned} \tag{10.55}$$

while the phase is

$$s_t = s_0 + \frac{1}{2}\left[p_t q_t - p_0 q_0\right] + \frac{i\hbar}{2}\mathrm{Tr}\left[\log Z\right]. \tag{10.56}$$

This specifies all the parameters in the Gaussian. From this we calculate A_t as

$$A_t = \frac{1}{2}\frac{2A_0 \cos(\omega t) - m\omega \sin(\omega t)}{(2A_0/m\omega)\sin(\omega t) + \cos(\omega t)}. \tag{10.57}$$

It is easily seen that A_t oscillates periodically with twice the frequency of the oscillator, unless $A_0 = im\omega/2$. The displacement by q_0 is accomplished by substituting $q - q_0$ for q in the ground state, and the boost is done by multiplying the ground state by $\exp[ip_0(q - q_0)/\hbar]$. In the special case that $A_0 = im\omega/2$,

$$s_t = s_0 + \frac{1}{2}[p_t q_t - p_0 q_0 - \omega t]. \tag{10.58}$$

If the displacement and boost are zero, then $q_0 = p_0 = q_t = p_t = 0$, $s_t = s_0 - \frac{1}{2}\omega t$, $A_t = A_0$. If the Gaussian wavepacket starts out as very wide compared to the ground state, it becomes narrower after one-quarter of a period, wider again after one-half

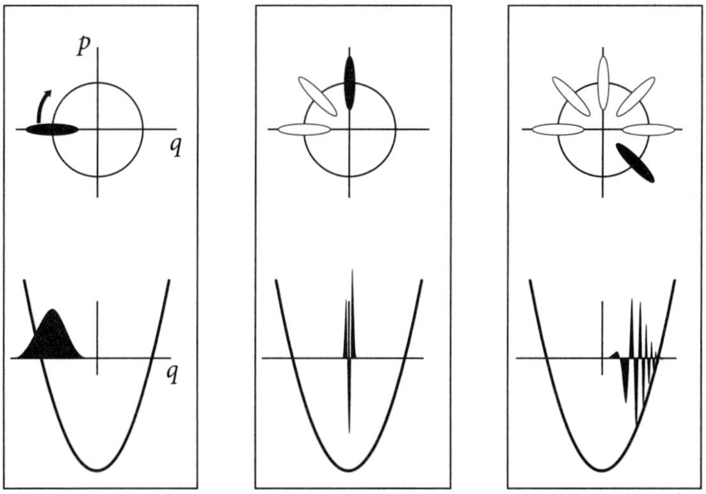

Figure 10.10. Phase space and coordinate space pictures of a Gaussian evolving on a harmonic potential. This Gaussian is initially much broader than the "coherent state"—that is, the ground state of the oscillator. Note the wavepacket becomes narrow after 1/4 of a period and again at 3/4 of a period, and so on.

vibrational period, and so on, as seen in equation 10.57. Between these mileposts, it develops position-momentum correlation, as evidenced by a complex A_t. The behavior will stand out more clearly when we look at the phase space picture of the dynamics.

Figure 10.10 shows an initial wavepacket on the left that is much wider than the coherent state of the oscillator. The coherent state would be circular in this diagram, since the classical energy contours are circular. All the trajectories travel in circles around the origin in phase space with an angular velocity that is independent of energy. This means that the ellipse rigidly rotates around the origin as shown. From the position and orientation of the ellipse at any moment, we can sketch the wavefunction as is seen in each panel. Since the dynamics is linear, the classical dynamics of the initial Wigner density is identical to the quantum evolution of the Wigner density. As stated earlier in connection with equation 10.57, we see the width of the wavepacket oscillating at twice the frequency of the classical oscillator.

HARMONIC OSCILLATOR GREEN FUNCTION

The action for the harmonic oscillator of mass m and frequency ω was given in equation 2.30 as

$$S(q_t, q_0, t) = \frac{1}{2} m\omega \cot \omega t \left[-\, q_0^2 + \frac{2\, q_0 q_t}{\cos \omega t} - \, q_t^2 \right].$$

The corresponding VVMG (and exact) Green function is

$$G(q, q', t) = \left(\frac{m\omega}{2\pi i \hbar \sin(\omega t)} \right)^{1/2} \exp \left[-\frac{m\omega[(q^2 + q'^2)\cos(\omega t) - 2qq']}{2i\hbar \sin(\omega t)} \right]. \quad (10.59)$$

The energy form of the propagator, $G_{sc}(q, q', E)$ can be inferred from equations 7.62 and 7.64. In one dimension, the only coordinate is along the trajectory of course, so we have

$$G(q,q',E) \sim \frac{1}{i\hbar(2\pi i\hbar)^{(N-1)/2}} \frac{1}{|\dot{q}\,\dot{q}'|^{\frac{1}{2}}} e^{\frac{i}{\hbar}S(q,q',E)-\frac{i\pi}{2}\nu}, \tag{10.60}$$

where the action $S(q,q',E) = \int_q^{q'} p_E(q'')\, dq''$. Note that the semiclassical energy Green function blows up at the classical turning points of energy E, if either $\dot{q}=0$ or $\dot{q}'=0$. These are a manifestation of the breakdown of the stationary phase approximation due to the coalescence of stationary phase points in time, if either q or q' is near a turning point where $|\dot{q}| \sim 0$ or $|\dot{q}'| \sim 0$.

UNSTABLE HARMONIC OSCILLATOR

If the potential is a quadratic *barrier* rather than a well, we have a pure imaginary frequency, $\omega \to i\omega$, $V(q) \to -\frac{1}{2}m\omega^2 q^2$. In equation 10.54, we get $\cos(\omega t) \to \cosh(\omega t)$, $\sin(\omega t) \to i\sinh(\omega t)$, giving

$$\begin{aligned} p_t &= p_0 \cosh(\omega t) + m\omega q_0 \sinh(\omega t), \\ q_t &= q_0 \cosh(\omega t) + (p_0/m\omega)\sinh(\omega t), \\ P_Z &= P_{Z_0}\cosh(\omega t) + m\omega \sinh(\omega t), \\ Z &= \cosh(\omega t) + (P_{Z_0}/m\omega)\sinh(\omega t). \end{aligned} \tag{10.61}$$

The phase term carries over directly,

$$s_t = s_0 + \frac{1}{2}\left[p_t q_t - p_0 q_0\right] + \frac{i\hbar}{2}\mathrm{Tr}\left[\log Z\right], \tag{10.62}$$

with equation 10.61 giving p_t, and so on. It follows that

$$A_t = \frac{1}{2}\frac{2A_0\cosh(\omega t) + m\omega\sinh(\omega t)}{(2A_0/m\omega)\sinh(\omega t) + \cosh(\omega t)}. \tag{10.63}$$

This example is a prototype of the important subject of unstable motion near a periodic orbit (the point $q = p = 0$ here). It is only a slight extension to add another, noninteracting stable degree of freedom to the unstable harmonic oscillator, making a two degree of freedom system with an unstable periodic orbit. We will do this in Chapter 22, in a study of scarring caused by periodic orbits.

The wavefunction of the inverted harmonic oscillator is shown in figure 10.11 at three times, along with the corresponding phase space density. The exponential escape of trajectories on the inverted barrier leads to very rapid spreading of the wavepacket along the *unstable* axis.

The stability matrix M is

$$M = \begin{pmatrix} \cosh(\omega t) & m\omega\sinh(\omega t) \\ \sinh(\omega t)/m\omega & \cosh(\omega t) \end{pmatrix},$$

with $\det M = 1$, $\mathrm{Tr} M = 2\cosh(\omega t) > 2$. The eigenvalues of M are $e^{\omega t}$ and $e^{-\omega t}$.

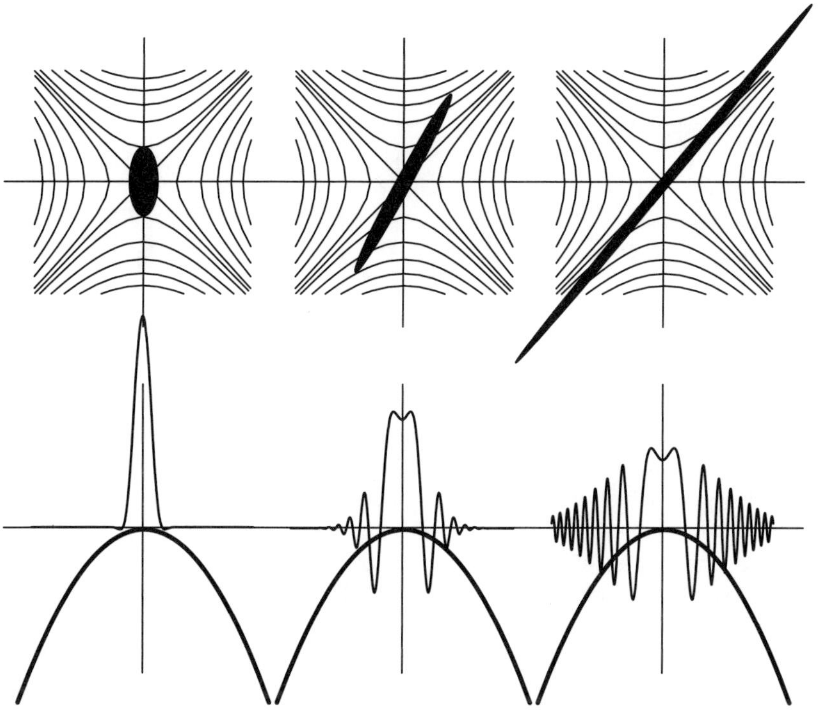

Figure 10.11. Phase space and coordinate space pictures of a Gaussian evolving on an inverted harmonic oscillator.

10.3 Linear Dynamics Applied to More General Wavefunctions

We have treated linear dynamics (quadratic action functions) acting on wavefunctions that were exponentials of quadratic forms. Everything is analytic and exact. Before proceeding to full-blown nonlinear dynamical time evolution in the next chapter, we consider linear dynamics acting on more complicated wavefunctions. This is more than an exercise; it can be easily applied in practice. For example, suppose an initial Gaussian wavefunction is subjected to a nonlinear Hamiltonian but only for a finite time. Perhaps the wavefunction passes through a lens and then back into free space. The propagator becomes quadratic form again and the dynamics is linear, but the initial data or wavefunction is not any variant of an exponential quadratic form. We can still write

$$\psi(x) = e^{iS(x)/\hbar},$$

where $S(x)$ is in general complex but not a quadratic form.

For example, suppose $S(x) = \alpha \exp[-x^2/\beta]$. If we use $p(x) = dS(x)/dx$, then $p(x) = -2\alpha/\beta\, x \exp[-x^2/\beta]$. Classically, this is initial data $p(x_0)$; under free particle time evolution it takes on the form $(x_0, p(x_0)) \to (x_t(x_0, p(x_0)), p_t(x_0, p(x_0))) = (x_0 + p(x_0)t, p(x_0))$. This example is displayed in figure 10.12, and plays an important role in the quantum mechanics of cusps and the structure of branched flow, to be treated in chapter 24. The exact free particle time evolution of $\psi(x) = \exp[iS(x)/\hbar]$; $S(x) = \alpha \exp[-x^2/\beta]$ is obtained by performing the integral involving a free propagator on

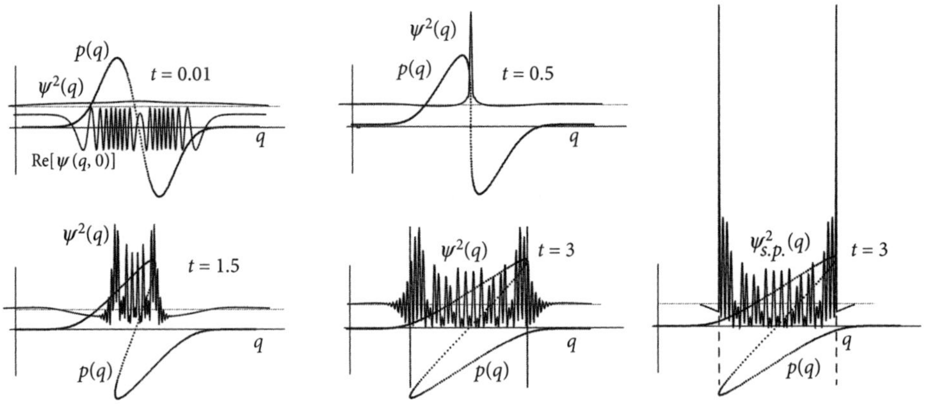

Figure 10.12. Surprisingly complex wavefunction evolution results from the free particle propagation of a wavefunction that is not represented initially by a linear manifold. Semiclassically, the initial wavefunction here possesses nonlinear position momentum correlation, given by the dotted line showing the corresponding classical manifold and its density at the upper left. (The proximity of dots give the density.) The real part of the initial wavefunction is shown as a solid line there. The time evolution of this classical manifold, and of the initial wavefunction, in a constant potential, is shown at $t = 0.5$, $t = 1.5$, and $t = 3$. The vertical projection of the classical manifold onto the coordinate axis corresponds to a singularity in the classical density, and a spike in the quantum mechanical wavefunction, where the manifold is vertical. At later times the classical manifolds fold on themselves, causing interference of three branches in the central region. Vertical singularities in the lower-right panel mark the turning points corresponding to vertical sections of the classical manifold. The quantum mechanical wavefunction is seen to oscillate to the right and left beyond this, due to tunneling contributions interfering with the classically allowed amplitude beyond the classically singular points. At the lower right, the result of real valued stationary phase integration is shown at $t = 3$, to be compared with its exact neighbor to its left. The stationary phase evaluation shows the usual maladies at classical turning points, with no tunneling beyond them (the classical branch is present but its interference with the tunneling contribution is missing). This figure is also related to work on branched flow; see chapter 24 [19].

the initial wavefunction; this may easily fail to be analytic, but could be performed numerically. It may yield to stationary phase evaluation. In the case of free particle propagation, the real part of the initial wavefunction is shown as a solid line at the upper left. Subsequent frames show increasing time evolution. The cusp or singularity is reached already in the second frame at $t = 0.5$. All the wavefunctions shown are the numerically determined "exact" ones, except at the lower right, where the result of stationary phase evaluation is shown at $t = 3$, revealing its usual maladies of singularities at classical turning points and no tunneling beyond them (the third branch is present but the interference of it with the tunneling contribution is of course missing).

Chapter 11

Gaussian Wavepacket Methods for Nonlinear Time Evolution

If the potential is smooth but not quadratic, we can make an approximation by supposing the wavepacket is narrow enough and the potential smooth enough that we can expand the potential quadratically around the (moving) center of the wavepacket q_t. This again results in a potential of the form seen in equation 10.21, where the parameters are obtained by expanding the potential around the changing position q_t of the moving wavepacket. This is called the *thawed Gaussian approximation* (TGA) [16–18].

11.1 Thawed Gaussian Approximation

We know that a Gaussian wavepacket in a quadratic or lower order potential (that is, under the influence of a possibly time-dependent potential most quadratic in its coordinates) has an analytic solution. Is it promising to try to propagate wavepackets semiclassically in anharmonic potentials?

Figure 11.1 gives a visual answer: if the center of a moving wavepacket is used as the center of Taylor expansion of the potential to second order in the coordinate, and the resulting quadratic potential is used rather than the actual potential, the wavepacket evolution is known analytically in the next instant. In that instant the center of the wavepacket will have moved. This is not a problem, if we insist the center of the wavepacket is always used as the place around which to Taylor expand the potential. This will ensure that the largest amplitudes in the wavepacket are given the most accurate approximation to the potential.

A time-dependent quadratic potential thus results from a continuous reexpansion of an anharmonic potential about the moving center of the wavepacket. The resulting approximate action is still a quadratic form, and an initial Gaussian wavepacket will remain Gaussian under its influence. We call this the "thawed" Gaussian wavepacket approximation, or TGA [16–18]. It is very important to recognize that there can be a different effective potential for different Gaussian packets on the same fixed potential, and thus if we have broken up an extended state in terms of many Gaussian pieces, we have a different propagator for each one. This is not as dangerous as it sounds and in fact is a powerful idea that resurfaces often.

The errors clearly will be largest in the wings of the wavepacket where the quadratic expansion of the potential may be failing; this will be worse if the wavepacket is

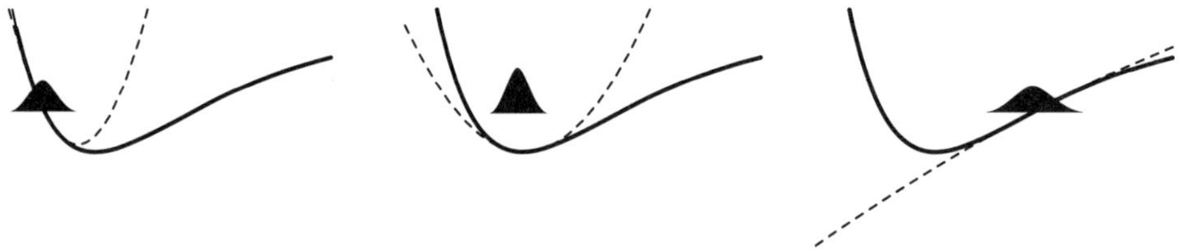

Figure 11.1. Motion of a wavepacket on an anharmonic potential energy surface is approximated by locally expanding the potential to second order about the instantaneous center of the wavepacket. This gives an effective, time-dependent harmonic potential; the wavepacket remains Gaussian; the equations of motion are just those given earlier (equations near 10.35).

broad, and/or if the anharmonicity is large. Since wavepackets tend to spread, the accumulation of errors limits the approximation to relatively short times in many cases. Figure 11.1 shows the main idea. The TGA leads to simple approximations for absorption and Raman spectra [40, 47], the latter is intrinsically short time in nature (if not it passes over to what is called resonance fluorescence). The TGA also yields relatively accurate descriptions of intrinsically short time photodissociation [47] and scattering [48]. See chapter 19.

We have already derived all the equations we need (section 10.2), having now approximated the anharmonic problem to a tailor-made (different for each new wavepacket), time-dependent quadratic potential.

We state without proof that in the limit $\hbar \to 0$, the TGA becomes exact for any finite time, if the potential is smooth. The TGA is therefore a semiclassical method, although not necessarily most accurate for a given finite $\hbar$. The exactness as $\hbar \to 0$ is clear if we note that the wavepacket width can be taken to scale as $\sqrt{\hbar}$ for example, making the local quadratic approximation arbitrarily accurate as $\hbar \to 0$. If the dynamics is unstable, the wavepacket will spread exponentially fast, so that we only extend the accurate time domain logarithmically in $\hbar$ as $\hbar \to 0$. Nonetheless, small $\hbar$ eventually wins out for any finite time.

A hard wall can also be treated exactly, by the method of images borrowed from optics. Even curved walls may be treated accurately, by arranging the image wavepacket, emerging from behind the wall, to have different parameters than the incoming wavepacket it replaces. The parameters are chosen to make the superposition of the incoming and emerging wavepackets nearly vanish on the curved wall; this includes specularly reflecting p_t at the moment the centers of the incoming and emerging wavepackets coincide.

If the initial wavefunction is not a single Gaussian, we can write it as a superposition of such Gaussians, and propagate each one as if alone, banking on the linearity of the Schrödinger equation. The example of the expansion of a plane wave in terms of Gaussians is given in figure 10.3.

11.2 Generalized Gaussian Wavepacket Dynamics

Generalized Gaussian wavepacket dynamics (GGWPD) began with the realization that an infinity of complex pairs of initial positions and momenta are mathematically identical to the usual real choices q_0, p_0 [49–52].

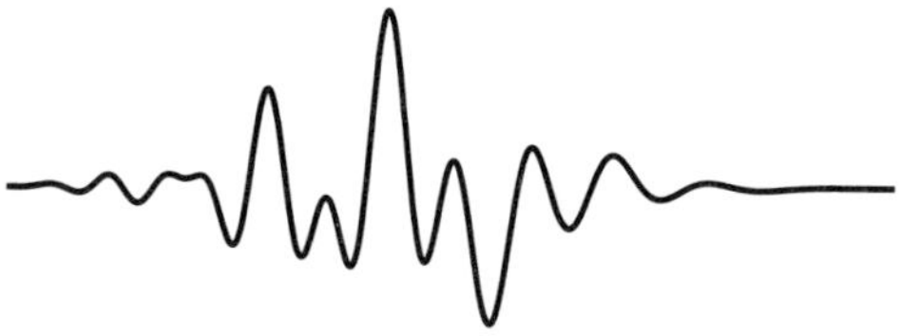

Figure 11.2. The real part of the sum of two Gaussians.

The central idea in references [51, 52] is that in spite of the fact that the complex guiding centers are just another way of writing the same initial (and final) Gaussian, for $t > 0$ they travel classically to places the real guiding centers do not. This means we can look for complex initial conditions, leading directly to certain complex final conditions, both sets of conditions belonging to Gaussians unchanged from the original. The workhorse is thawed Gaussian dynamics as before, but now with complex $\tilde{q}_t, \tilde{p}_t$. This analytic continuation of real Gaussian wavepacket dynamics attains the full extension of the Van Vleck-Morette-Gutzwiller (VVMG) semiclassical Green function to coherent states. Weissman [50] considered coherent state to coherent state semiclassical generalization of the VVMG propagator. Here and in references [51, 52], we derive these results in a simpler way, and extend his results somewhat by using quite general $A = P_z/2Z$ in the initial and final states, allowing general coherent state to coherent state, coherent state-to-position state, and even the VVMG position state to position state limits in one expression.

The generalization of an ordinary Gaussian with real valued position and momentum parameters to complex parameters is straightforward. Consider the ket Gaussian $\psi_\alpha(q; q_0, p_0) = \langle q|\alpha_0; q_0, p_0\rangle$

$$\psi_\alpha(q; q_0, p_0) = e^{-\alpha_0(q-q_0)^2/\hbar + i p_0(q-q_0)/\hbar + c_0/\hbar}, \tag{11.1}$$

with

$$c_0 = \ln\left[\left(\frac{2\mathrm{Re}(\alpha_0)}{\pi}\right)^{1/4}\right]. \tag{11.2}$$

The "equivalence class" of initial complex Gaussian centers, defined by $\psi(q; \tilde{q}_0, \tilde{p}_0) = \psi_{\alpha_0}(q; q_0, p_0)$ giving the same Gaussian is given by

$$\begin{aligned} 2\alpha_0\tilde{q}_0 + i\tilde{p}_0 &= 2\alpha_0 q_0 + i p_0, \\ \tilde{c} &= c_0 + \alpha_0(\tilde{q}_0^2 - q_0^2) + i(\tilde{p}_0\tilde{q}_0 - p_0 q_0), \\ e^{-\alpha_0(q-\tilde{q}_t)^2/\hbar + i\tilde{p}_0(q-\tilde{q}_0)/\hbar + \tilde{c}_0/\hbar} &= e^{-\alpha_0(q-q_0)^2/\hbar + i p_0(q-q_0)/\hbar + c_0/\hbar}. \end{aligned} \tag{11.3}$$

$\tilde{q}_0, \tilde{p}_0$ are to be the initial conditions for new guiding trajectories. Starting with real valued positions and momenta, changing the position q_0 to the (possibly complex) value $\tilde{q}_0$, the momentum p_0 to $\tilde{p}_0$ must also change according to the first equation, in order to keep the Gaussian unchanged. The second equation then gives the necessary change of the normalization c. A new set of complex initial conditions results and a new $\tilde{c}$, completely equivalent to the original Gaussian, at $t = 0$. Each new choice of complex initial $\tilde{q}_0$ gives a unique partner $\tilde{p}_0$ that becomes a new thawed Gaussian under time evolution

$$\psi(q; \tilde{q}_t, \tilde{p}_t) = e^{-\tilde{\alpha}_t(q-\tilde{q}_t)^2/\hbar + i\tilde{p}_t(q-\tilde{q}_t)/\hbar + \tilde{c}_t/\hbar}. \tag{11.4}$$

The corresponding equations for a bra Gaussian are

$$\begin{aligned} 2\alpha_0^*\tilde{q}_0 - i\tilde{p}_0 &= 2\alpha_0^* q_0 - i p_0, \\ \tilde{c}^* &= c_0^* + \alpha_0^*(\tilde{q}_0^2 - q_0^2) - i(\tilde{p}_0\tilde{q}_0 - p_0 q_0). \end{aligned} \tag{11.5}$$

There are two new parameters per degree of freedom, since if a real initial position q_0 is changed to an arbitrary complex one, a specific complex initial momentum follows.

A subtle but crucial shift has taken place from the TGA. Rather than propagate the initial Gaussian according to its real classical initial conditions, making use of whatever it becomes, we look for positions on the initial Gaussians' complex manifold that classically evolve to positions on the final Gaussians' complex manifold. This is the exact analog of the position-to-position VVMG Green function, which looks for initial real momentum at the initial real position connecting classically with final real momentum at the final real position. Gaussian wavepacket dynamics has thus been upgraded to the status of the "gold standard" VVMG Green function: the coherent state extension of VVMG, which we call generalized Gaussian wavepacket dynamics (GGWPD).

The transformations can be reversed, so that starting with a complex Gaussian with complex guiding centers, we can find the equivalent real guiding centers. The equations are easily derived for the ket manifold, with $\tilde{q}_0 = \tilde{q}_0^r + i\tilde{q}_0^i$ and so on, as

$$\begin{aligned} q_0 &= \tilde{q}_0^r - \frac{1}{2\alpha^r}(2\alpha^i\tilde{q}_0^i + i\tilde{p}_0^i), \\ p_0 &= \tilde{p}_0^r + \frac{\alpha^i}{\alpha^r}\tilde{p}_0^i + 2\tilde{q}_0^i\left(\alpha^r + \frac{\alpha^{i2}}{\alpha^r}\right), \\ c_0 &= \alpha(q_0^2 - \tilde{q}_0^2) + i(q_0 p_0 - \tilde{q}_0\tilde{p}_0) + \tilde{c}_0, \end{aligned} \tag{11.6}$$

and for the bra manifold

$$\begin{aligned} q_0 &= \tilde{q}_0^r + \frac{1}{2\alpha^r}(2\alpha^i\tilde{q}_0^i + i\tilde{p}_0^i), \\ p_0 &= \tilde{p}_0^r - \frac{\alpha^i}{\alpha^r}\tilde{p}_0^i - 2\tilde{q}_0^i\left(\alpha^r + \frac{\alpha^{i2}}{\alpha^r}\right), \\ c_0^* &= \alpha^*(q_0^2 - \tilde{q}_0^2) - i(q_0 p_0 - \tilde{q}_0\tilde{p}_0) + \tilde{c}_0^*. \end{aligned} \tag{11.7}$$

How are we going to use this new freedom? Suppose we want to know the time evolution

$$\langle\psi(q_1, p_1)|e^{-iHt/\hbar}|\psi(q_0, p_0)\rangle, \tag{11.8}$$

where $\psi(q_{0,1}, p_{0,1})\rangle$ are Gaussians with real guiding centers. Before, using real thawed Gaussian propagation, we would have (q_t, p_t) as the guiding orbit, leading to a single time-dependent Gaussian that we evaluate at q' at time t, with possibly inaccurate results in an anharmonic system after some time evolution.

Paths are sought in complex phase space that start at $\tilde{q}_0, \tilde{p}_0$ in the ket manifold M_0 and end after time t at $\tilde{q}_1, \tilde{p}_1$ in the bra manifold M_1. The equations of motion are just like those for real valued trajectories, including the equations for P_{Z_t}, Z_t, except everything is complex. This sounds like a bothersome "root search"; however, it is easily avoided if many final state amplitudes M_f are needed, as in reconstructing the

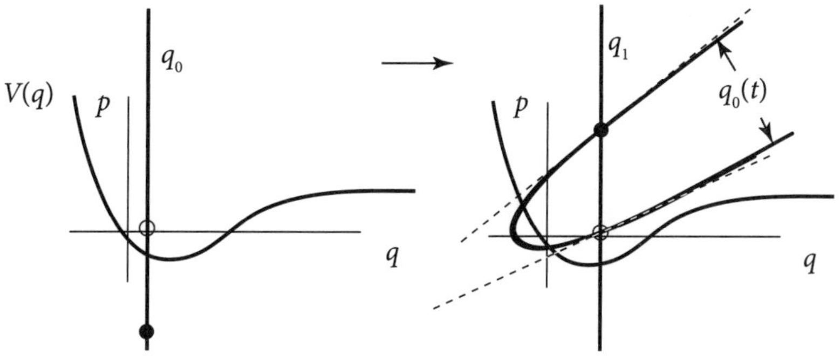

Figure 11.3. For an initially sharp position q_0, here two real classical paths reach q_1 at time *t*. GGWPD would employ the same two paths in this case, as solutions to equations 11.6 and 11.7.

whole propagated wavefunction. Then every initial complex Gaussian belonging to M_0 counts, except for those that are discarded, as follows: Some trajectories leading to runaway and divergent results are discarded following a procedure going back to Stokes (trajectories "beyond the Stokes lines"). Stokes lines (G. G. Stokes, 1847 and 1858) are slightly shrouded in mystery. Nonetheless, when the asymptotic quantity you are interested in starts to blow up beyond some boundary in the complex plane, you have found your Stokes line in practice. It is beyond the scope of this book to discuss *why* the divergence occurs, but the instructions are: if it blows up, discard it.

There are other details of the procedure, and we refer the interested reader to references [51–54].

Returning to initial and final position states—that is, the semiclassical VVMG Green function, we can now view it as a limit of the coherent state procedure just outlined, as the initial and final Gaussian become narrow. (See figure 11.3.) If $\alpha_0 \to \infty$, the first of the equations 11.3, $2\alpha_0\tilde{q}_0 + i\,\tilde{p}_0 = 2\alpha_0 q_0 + i p_0$, requires $\tilde{q}_0 = q_0$ and therefore it would seem $\tilde{p}_0 = p_0$. However, suppose we divide the equation through by $\sqrt{\alpha_0}$ as $\alpha_0 \to \infty$. Now, we still require $\tilde{q}_0 = q_0$, but $\tilde{p}_0$ can be any finite value. This corresponds to the fact that an infinitely narrow wavepacket has infinite momentum uncertainty, and the initial momentum cannot matter. But it *does* matter if the wavepacket is propagated in time semiclassically, using the real q_0 and new $\tilde{p}_0$ as initial conditions.

Consider the example in figure 11.3, a real position ket at q_0 propagating for a time t on an anharmonic potential. We seek the amplitude $\langle q_1|q_0(t)\rangle$. The two final time solutions to the GGWPD problem are the dashed lines in figure 11.3. Both are terrible solutions to the correct evolution in the anharmonic potential, *except* where they are needed—that is, the intersection points with the final state $|q\rangle$. Each solution came from a different initial $p_0 = p_a$, $p_0 = p_b$ (open circle and closed circle), and both end up where they need to be after time t. Each contribution also gives exactly the VVMG Green function amplitude.

One might ask why go to this trouble when one attains the accuracy of the Van Vleck propagator, and presumably no better. The answer is that the theory is written directly in terms of smooth Gaussian wavepackets. These are often very close to physical initial and final conditions. There is no need to integrate over initial positions with Gaussian weighting, for example. Autocorrelation functions may involve just one Gaussian initial and final state.

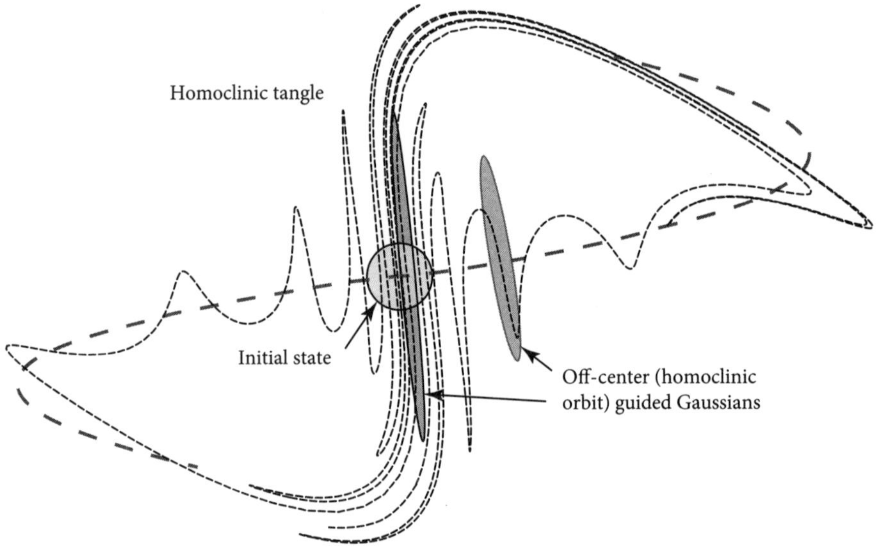

Figure 11.4. Returning (homoclinic) pieces of stretched and folded phase space domains of a chaotic system are approximately linear in the region of overlap with the initial Gaussian (circle). A "thawed" Gaussian wavepacket is shown returning and overlapping the initial Gaussian, centered on (and guided by) a primary homoclinic orbit (primary because it is returning on the first pass), stretching along the unstable direction and doing a good job representing one homoclinic contribution to the overlap at this time. An approaching wavepacket is seen guided by another primary homoclinic orbit, not yet overlapping the initial state. Each homoclinic intersection with the stable manifold (dashed line) carries a similar guided wavepacket. For the purposes of this autocorrelation (overlap of the initial Gaussian with the time evolved wavefunction), the initial wavepacket is decomposed into many thawed Gaussians, each guided by a different homoclinic orbit. See reference [22].

The problem of dealing with complex trajectories, some of which run away, has kept GGWPD from the mainstream until recently. But GGWPD has been greatly simplified, linking off-center guiding with GGWPD and showing how to replace the complex dynamics with real trajectories [54, 55].

OFF-CENTER GUIDING

What if a given initial Gaussian is propagated by thawed Gaussian dynamics, but using a different real center than the natural one (q_0, p_0)? This is very reminiscent of using different complex centers in the generalized Gaussian wavepacket dynamics just discussed, and the intimate connections between the two have been worked out [54, 55].

The off-center guiding allows treatment of very chaotic systems for rather long times. The stretching and folding of phase space that happens on a short timescale in strongly chaotic systems dramatically increases the number centers used for guiding. But the stretching can be put to use, in the following way [56–59]: In computing an autocorrelation function $\langle\phi|\phi(t)\rangle$, for example, we require knowledge only of the amplitude near the region of phase space spanned by the initial state $|\phi\rangle$. In such a restricted region, the returning amplitude may consist only of quasi-linear homoclinic oscillations, easily approximated by a linearized transformation, one for each homoclinic "branch." Each linearization corresponds to expansion of the initial wavepacket's motion not about its center, but about a homoclinic trajectory starting on an unstable manifold within the domain of the initial wavepacket, returning to the stable manifold, approaching the periodic orbit. So-called primary homoclinic orbits join the stable manifold on the first pass so to speak; secondary homoclinic orbits join it on the second pass after missing the periodic orbit on the first try; and so on.

This is illustrated for an unstable chaotic system generating a homoclinic tangle in figure 11.4. Increasingly, many off-center guided wavepackets contribute to the overlap with the initial state (autocorrelation) as time develops. In this figure, only two of them have been populated with wavepackets stretched by the local stability, for clarity.

THE UNIFICATION OF OFF-CENTER GUIDING AND GGWPD

Wavepackets that are guided by classical trajectories not necessarily starting at their center (average position and momentum) may be used to guide them to represent paths or processes where the center does not go, as just discussed. What justifies this procedure?

The complex Gaussian GGWPD solutions are rigorously motivated as the generalization of the VVMG propagator to coherent states, and more generally, arbitrary Gaussian wavepackets. But they are inconveniently guided by complex trajectories that may cross Stokes lines that have to be discarded by hand.

It was noticed that the most important and well behaved of these complex Gaussians were in fact guided by complex trajectories that were nearly real. The idea arose that completely real, off-center guiding trajectories might be used to extrapolate to the nearby complex ones, without having to actually "run" them [22]. The rigor and presumably higher accuracy of the GGWPD would then be recovered without the hassle, so to speak.

Van Voorhis and the author [53] used this idea to get accurate photodissociation spectra of CO_2 that included several periodic orbits arising from the vicinity of the initial wavepacket.

Tomsovic and co-workers [60] independently extended the off-center theory and made it much more systematic and useful, allowing only real trajectories to be used once more. They made the method quite general, so that it may well become the "best" semiclassical use of wavepackets. We quote from reference [60], stating their essential advance very well:

> The GGWPD method has been in existence for more than 25 years, but has not developed into a widely used and practical technique. The barriers to its direct implementation are considerable. Most important are the difficulties of analytically continuing classical dynamics in a complex domain where both position and momenta are complex. It requires high-dimensional root searches in multiple degrees of freedom with quantities exhibiting highly complicated functional behaviors. It also requires determining whether a particular saddle should be kept or dropped. By recognizing that real classical dynamics imprints itself on the complex dynamics, it is possible to develop indirect root search methods that are vastly easier to implement and avoid several of the pitfalls that a direct root search method would have to confront.

The possibility of computing or knowing eigenstates fades drastically as the number of degrees of freedom grows. Fortunately, in most situations eigenstates are not needed, not measured, and not even measurable in practice. Instead, aggregate properties of eigenstates are needed: for example, eigenstates are summed over in expressions for Raman scattering, or correlation functions of various sorts, a big hint that one should not have to first calculate eigenstates individually. In contrast, time-dependent solutions to the Schrödinger equation are relatively easy to obtain for short times, and often the sums over states, correlation functions (and thus statistical quantities) can be expressed instead in terms of the dynamics; the eigenstates are never needed. This suggests the utility and power of using the time-dependent VVMG propagator and GGWPD, which is easier to calculate (using classical trajectory input) and especially accurate for short times, but often surprisingly accurate for long times too—even if the dynamics is chaotic.

11.3 Initial Value Representation

As Prof. William Miller showed, there is an entirely parallel universe of semiclassical expressions, a "semiclassical algebra," for quantum amplitudes, with all the identities and Dirac notation applying in the semiclassical world just as in the exact quantum world, *as long as all integrals relating semiclassical quantities are done by stationary phase* [61]. If they are done analytically (rarely possible) or numerically instead, then the resulting expression is called "uniformized" and does not fit into the Dirac algebra of semiclassical amplitudes. Uniformized amplitudes are normally more accurate, even dramatically so, than their "primitive semiclassical" counterparts; we shall give examples of this later.

In 1970 Miller introduced the *initial value representation* (IVR) [62, 63], wherein integrals that would have been done by stationary phase integration (to remain within the nonuniformized, semiclassical algebra), are done numerically instead. This is the main idea of most uniformizations, but Miller introduced a crucial variant, by making an advantageous change of variables, *writing the integral as a sum over all initial momenta at fixed initial positions—that is, an integral over initial values of the trajectories.* If the integrals are done numerically, the result is a uniformization, and the elimination of a root search for the stationary phase points. The root search should never have been necessary, however, if one needs to find the dynamics of $\psi_{n_1}(q_1)$, since every q_1 and every p_1 of relevance to $\psi_{n_1}(q_1)$ presumably goes someplace needed for the determination of $\psi_{n_1}(q_1, t)$. The IVR streamlines this truism and uniformizes the result in the process. This means, for example, that some classically forbidden amplitudes will now have finite results because they are expressed as an integral done accurately, and other amplitudes will be finite that would have been singular.

IVR GREEN FUNCTION

We apply the IVR idea first to the Green function, writing

$$\begin{aligned} G^{sc}(q, q_0; t) &= \left(\frac{1}{2\pi i\hbar}\right)^{\frac{1}{2}} \sum_k \left|\left(\frac{\partial^2 S_k(q_0, q)}{\partial q_0 \partial q}\right)\right|^{\frac{1}{2}} \exp[i S_k(q_0, q)/\hbar - i\nu\pi/2], \\ &= \left(\frac{1}{2\pi i\hbar}\right) \int dp_0 \left|\left(\frac{\partial q_t}{\partial p_0}\right)_{q_0}\right|^{\frac{1}{2}} \delta(q - q_t(q_0, p_0)) \exp[i S(q_0, p_0)/\hbar \\ &\quad - i\nu\pi/2]. \end{aligned} \tag{11.9}$$

In spite of the integral over initial momenta, the Green function has not been uniformized (the δ function has seen to that), but rather we have simply rewritten it. We take the action S to be a function of q_0, p_0; $q_t(q_0, p_0)$ is the position of the trajectory that leads from the initial conditions $q = q_0$, $p = p_0$.

The expression 11.9 is intriguing, as it constructs the VVMG Green function in terms of position δ functions $\delta(q - q_t(q_0, p_0))$ moving classically. The functions are not spreading or constrained by the uncertainty principle. Each one is weighted by its own time-dependent amplitude and phase. We will soon give a convenient method for calculating it.

The IVR Green function equation 11.9 applied to extended initial and final states quickly yields Miller's IVR equations:

$$\begin{aligned}
&\int dq_2 \int dq_1 \psi^*_{n_2}(q_2) G^{sc}(q_2, q_1; t) \psi_{n_1}(q_1) \\
&= \int dp_1 \int dq_2 \int dq_1 \psi^*_{n_2}(q_2) \left|\frac{\partial q_2}{\partial p_1}\right|^{1/2} \delta(q_2 - q_t(q_1, p_1)) \psi_{n_1}(q_1) e^{iS(q_2,p_1)/\hbar - i\nu\pi/2}, \\
&= \int dp_1 \int dq_1 \left|\frac{\partial q_t}{\partial p_1}\right|^{1/2} e^{iS(q_2,p_1)/\hbar - i\nu\pi/2} \psi^*_{n_2}(q_t) \psi_{n_1}(q_1). \qquad (11.10)
\end{aligned}$$

Now there are uniformizations at work if the integrals are done numerically.

CELLULAR EVALUATION OF THE IVR GREEN FUNCTION

The VVMG seems daunting to evaluate: for each q and q' we have to find all the trajectories connecting these coordinates, sum their contributions, and then repeat everything for each new time t. But if we follow essentially all the trajectories (using some interpolation between explicitly sampled trajectories) leading from q_0 at $t = 0$, then we cannot fail to find all the final positions q that can be reached, and furthermore we can just increment the time in the usual way using all the previous information (that is, run trajectories to get the next time's data without starting over). Following all the trajectories sounds like a lot of work. The IVR-cellular dynamics method, presented here, makes this much more tractable.

IVR-cellular dynamics enabled the first detailed evaluation of the VVMG propagator [30] in a nonlinear potential. We can write the semiclassical VVMG time-dependent Green function as an integral over the initial values of momentum—that is, equation 11.9.

CELLULAR IMPLEMENTATION OF IVR

Some way is needed to simplify and organize the calculation of the IVR amplitudes. Since integrals that were once done by stationary phase are being done explicitly, the integrands are bound to be oscillatory. Various methods have been proposed [63], including some that do not retain the interference of alternative classical paths. Here we give an approach that has the advantage of running classical wavepackets unrestricted by the uncertainty principle.

We can see directly from equation 11.9 that the VVMG Green function is calculated semiclassically by summing over weighted classical trajectories represented as moving Dirac delta functions, launched from the same place $q = q_0$, but with all possible momenta p_0. Each point along the line $q = q_0$ is a separate trajectory for different p_0 that carries a weight $\left|\partial q_t/\partial p_0\right|^{\frac{1}{2}}_{q_0} dp_0$ and a phase $S(q_0, p_0)/\hbar - \nu\pi/2$. The direct implementation of this form of the Green function is still foreboding, since roots still need to be found for each q and q_0.

We form a nearly constant function (as nearly as we please) as a sum of Gaussians as

$$1 \approx \eta \sum_n \exp[-\beta(y - na)^2]. \tag{11.11}$$

A careful choice of the widths of the Gaussians (one wants them as narrow as practicable, since they will represent a range of initial conditions) achieves a sum that is nearly constant as a function of y with a reasonable spacing parameter a. The factor η normalizes the sum in the sense that it hovers very near 1. In effect, this will allow *classical* Gaussian wavepackets to be used to calculate the VVMG propagator. Applying equation 11.9, we have

$$G^{sc}(q, q_0; t) = \left(\tfrac{\eta^2}{2\pi i\hbar}\right)^{\frac{1}{2}} \sum_m \int dp_0 \left|\tfrac{\partial q_t}{\partial p_0}\right|_{q_0}^{\frac{1}{2}} \delta(q - q_t(q_0, p_0)) e^{-\beta(p_0-p_m)^2} e^{iS(q_0,p_0)/\hbar - i\nu\pi/2}. \tag{11.12}$$

If the exponential factor β is large enough, the mth term will not have much contribution except near $p_0 \approx p_m$. This allows us to expand the action $S(q_0, p_0)$ about $p_0 = p_m$, which we do to second order. Also, we expand q about $q_t(q_0, p_m)$ to first order, consistent with the second-order expansion of S. The net result of these expansions will be to make $G^{sc}(q, q_0; t)$ simply a sum of Gaussians, quadratic q and q_0. One Gaussian follows each initial trajectory with initial position q_0 but different momentum p_m. It is clear there are easy extensions of this idea to include a range of initial positions as well.

The Maslov phase is $-i\nu\pi/2$. The integer ν increases by one whenever the classical trajectory connecting q_0 to q encounters a focal point or caustic, where $\partial^2 S(q, q_0)/\partial q \partial q_0$ diverges. This happens individually for each cell; the phase is retarded for the cell by $\pi/2$ whenever $(\partial q_t/\partial p_0)|_{q_0}$ changes sign.

We perform the integral over p_0, which imposes $q = q_t$, solving analytically for the root value p_0 for each m, namely,

$$p_0 = p_m + (1/m_{21})(q - q_t). \tag{11.13}$$

The expansion for S plus the quadratic term, in the exponent of the mth term is

$$\begin{aligned} &\frac{i}{\hbar}\left[S(q_0, p_m) + (m_{21}\, p_t(q_0, p_m))(p_0 - p_m) + \frac{1}{2} m_{21}\, m_{11}(p_0 - p_m)^2\right] \\ &\quad - \beta(p_0 - p_m)^2 - i\nu\pi/2. \end{aligned} \tag{11.14}$$

(See section 3.8 for definitions of the stability matrix elements m_{21} and so on.) With equation 11.13, this becomes a pure quadratic form in q and q_0, the arguments of the Green function. Thus the full semiclassical Green function is given as a sum of Gaussians quadratic in q and q_0:

$$G^{sc}(q, q_0, t) = \sum_m g_m(q, q_0). \tag{11.15}$$

Fast oscillation of exponents is not a problem, because it is within a quadratic form and the integrals will be analytic, though perhaps close to 0. As β is made larger and more terms in the sum over m are taken, the method converges rigorously on the

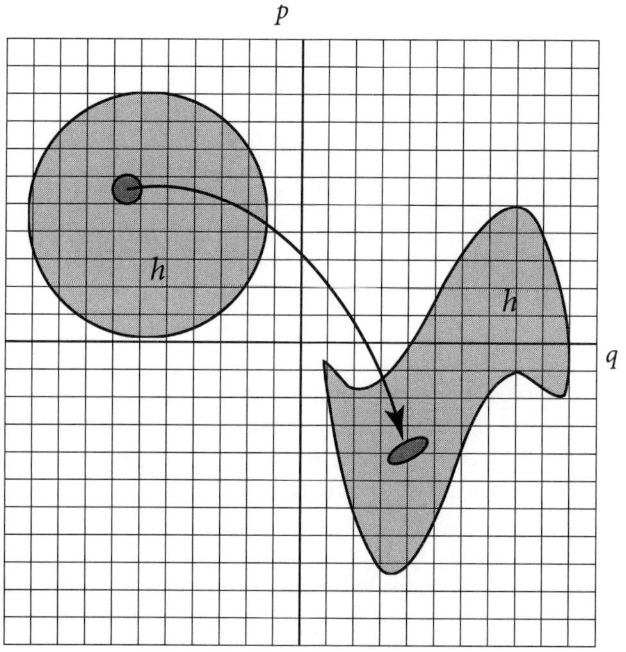

Figure 11.5. A region of size *h*, Planck's constant, evolves nonlinearly. It is decomposed into much smaller classical regions that evolve locally nearly linearly.

construction of the exact semiclassical time dependent Green function (see figure 11.5). For any finite time there is always a linearizable regime around each trajectory so that it is unnecessary to calculate a more dense set of trajectories than the linearizable domains dictate. In a sense, we have beaten the uncertainty principle, since we are not tying the linearizable domains to Planck's constant. A grid of initial conditions for the classical trajectories determines all the parameters; there are no trajectories to search for. It may be argued that by running the initial grid of trajectories, we have determined all the relevant dynamics anyway. This is true, but if the entire wavefunction is needed it is difficult to see how we could propagate it with less than this knowledge. Further, by judicious use of the linearizable domains we have made global use of point-wise trajectory information.

Cellular dynamics is a way of computing the VVMG. We displayed four tests of the VVMG propagator, of increasing challenge, in chapter 8: (1) an anharmonic oscillator for short and intermediate times, (2) the same oscillator for times so long that "quantum revivals" take place, (3) autcorrelation and spectra in a classically chaotic billiard, and (4) Anderson localization captured by semiclassical Van Vleck propagation. The first two were done within the cellular approach.

FROZEN GAUSSIANS; HHKK METHOD

An even simpler idea than the TGA is the *frozen Gaussian approximation* (FGA) [64]. As the name suggests, in this approach one does not let individual wavepackets spread. For a single wavepacket, this is forbidden. The hope, however, was that if every initial state (even a Gaussian) is expanded in terms of many frozen Gaussians, each with a different classical initial condition, then collectively there might be a proper representation of spreading, and so on. Also, there is no need to follow the stability matrix, a serious computational overhead for many Gaussians in many degrees of freedom. This approach is unfortunately not exact even as $\hbar \to 0$. Surprisingly, Herman and Kluk [65, 66] found a way to make the sum not only semiclassically exact but also importantly a uniformization. Unfortunately, it comes at the cost of reintroducing and calculating the stability matrices. Kay [67] made clear the basis of the method and extended it. The Herman-Kluk (HK) or Heller-Herman-Kluk-Kay (HHKK) method is used often, usually in a Monte-Carlo sampling over initial frozen Gaussians. There are many impressive examples of the application and extensions of the HHKK method. An interesting and instructive example is "Spectra of Harmonium in a Magnetic Field Using an Initial Value Representation of the Semiclassical Propagator," by F. Grossmann and T. Kramer [68].

Considering that typically many thousands of frozen Gaussians are needed in HK and there is no a priori selection of the most useful among them, a method was introduced to greatly reduce the number of Gaussians needed [69], taking advantage of the structure of phase space as it evolves nonlinearly.

Manoloupolis developed a hybrid of the Herman approach and cellular dynamics [70,71], used to do nontrivial quantum dynamics in 15 degrees of freedom. Kay [72] has given another globally uniform expression inspired by similar notions, wherein time-independent wavefunctions are written as integrals over Gaussian wavepackets.

DEGRADATION OVER TIME

As was clear when we considered the semiclassical version of the Airy function, the coordinate space semiclassical wavefunction is inaccurate and even singular as the area of intersection between the parabolic constant energy track of the linear ramp potential and the vertical coordinate space manifold becomes less than Planck's constant (see section 6.2). Under integrable dynamics, much less chaotic dynamics, the classical manifold representing an initial position develops convolutions and hairpins that would seem to cause a devastating proliferation of singularity problems like those just mentioned. Fortunately, this happens much more slowly than one might imagine, and may not even be noticeable if the semiclassical propagator is being applied to an initial wavefunction with some extent in coordinate space, or projected onto one, or both. This is all the more surprising in chaotic systems, where the hairpin turns in the classical manifolds are proliferating exponentially in number [58]. In the worst-case scenario, evaluating an initial coordinate space state propagated semiclassically as a function of final position, the fraction of "clean" contributions of area greater than Planck's constant decays as $F(t) = \exp[-a\lambda\hbar^{1/3}t]$, where a is a system-dependent parameter, and λ is the Lyapunov exponent characteristic of the system.

Against this exponential decay, even moderately extended initial or final states do wonders to smooth out singularities, for both integrable and chaotic systems.

11.4 Variational Wavepacket Methods

The notion of solving the time-dependent Schrödinger equation variationally by introducing a time-dependent basis of semiclassical wavepackets was introduced in 1976 [73]. The time-dependent Dirac-Frenkel-McLachlin [74,75] variational principle was used, minimizing the functional

$$I = \int d\tau \left| H\psi(t) - i\hbar \frac{\partial \psi(t)}{\partial t} \right|^2, \qquad (11.16)$$

where $d\tau$ represents integration over all space, which results in the condition

$$\mathrm{Re} \int \delta\theta^*(H\psi(t) - i\hbar\theta)d\tau = 0 \qquad (11.17)$$

for variational guesses to $\theta = \partial\psi(t)/\partial t$.

Some suggestions for variational wavepacket treatments of nonadiabatic corrections to Born-Oppenheimer theory, particlarly at surface crossings, were made in reference [73]. Early on, Metiu and co-workers developed more ideas for extending the Gaussian wavepacket methods to surface crossing situations and have over the years made many innovative applications of wavepacket ideas [76–79]. The work by Metiu and co-workers, using wavepacket bases and time-dependent variational principles [80], contained many promising ideas presented and some tried explicitly in following papers, including some spawning ideas like those mentioned next. Modern multiprocessor computers suggest that some of these ideas could now be profitably revisited.

Time-dependent variational methods were applied to Hartree theory [74, 81], wherein many dimensional time-dependent trial solutions were written as direct products of lower dimensional wavefunctions. The work was extended to multiconfigurational Hartree products by Cederbaum and co-workers starting in 1990 [82] with many following papers and extensions.

Variational wavepacket methods were taken to a high art by Mandelshtam and co-workers, obtaining accurate quantum thermodynamic functions for Ne_{38} clusters [83].

Another systematic approach is that of Coalson and Karplus [84, 85], in which the extensions to Gaussian wavepacket dynamics in terms of expansions involving Hermite polynomials multiplying Gaussians were developed. The equations of the Hermite polynomials extending the frozen Gaussian approximation (FGA) were developed in reference [84,85].

11.5 Wavepacket Spawning Methods

In the author's opinion, the most powerful and innovative use of wavepackets in difficult, many-body circumstances has been Martinez's ab initio multiple spawning approach. An early pioneer with similar ideas was H. Metiu [80]. Wavepackets are generated as needed to describe an ever more complex wavefunction, on multiple Born-Oppenheimer potential energy surfaces, giving some of the best insights yet into what really happens at the ubiquitous molecular Born-Oppenheimer violating degeneracies, which are the essential step in many chemical processes. Early work is found in references [86, 87]. These articles are the tip of the iceberg of a large body of work capable of handling the bifurcation of wavepackets upon encountering anharmonic potential surface crossings or narrows avoidances, and worse, conical intersections. The work is well informed by high-level electronic structure theory. State of the art multiprocessor technologies, including graphics processors, have been put to use.

11.6 Maslov and Geometric Phase

The often bothersome, sometimes mysterious, but always essential Maslov phase, (the integer ν_j in equation 7.48) doesn't appear explicitly in the earlier wavepacket derivation of the TGA. Accordingly, as noted by Littlejohn [88], the TGA has in it the seeds of a very simple method to find Maslov indices. The trick is to keep track of the phase

$$\frac{i\hbar}{2}\mathrm{Tr}[\ln \boldsymbol{Z}]$$

in equation 10.40; sudden jumps of the overall phase are to be avoided, yet Tr[ln $\boldsymbol{Z}$] will jump suddenly in phase if left to its own devices. The addition of a phase correction to cancel a jump of $\pm\pi/2$ in the phase of the trace keeps the phase advancing smoothly and gives the Maslov correction automatically.

Child [89] has drawn attention to an interesting geometric phase that arises in Gaussian wavepacket dynamics with a nonadiabatic time-periodic Hamiltonian that happens to give a periodic wavepacket solution (except for the phase). The phase is automatically included in the equations already given, but it is important to point out a geometric (Berry [90]) phase when present, as Child has done.

One of the great advantages of working with wavepacket representations is the automatic uniformization that they provide. This is not to say that all possible caustic singularities are avoided, but some of the most common kinds are ameliorated, much like oil on troubled waters. These advantages were pointed out in the early wavepacket work [16–18], and are easy to understand in a phase space picture. Take for example the semiclassical coordinate space representation of an eigenstate of an oscillator. In the traditional position space, there is a well-known caustic singularity at the classical turning point, with the wavefunction diverging as the inverse velocity. The tangency of the two delta function manifolds $\delta(q - q_0)$ and $\delta(E - H(p, q))$ seen in figure 6.4, earlier, when q_0 is at the turning point, leads to the singularity; a Gaussian amplitude, however, has no such singularity. There is no such singularity in the semiclassical coherent state representation.

Chapter 12

WKB Methods

12.1 Derivation of the WKB Wavefunction

The Wentzel-Kramers-Brillouin (WKB) methods go back to 1926, the year quantum mechanics was invented by Schrödinger. WKB methods are semiclassical approximations to time-independent quantum mechanics. It is a testament to the level of familiarity with classical mechanics that such approximate solutions were put forth (three times over) in the first year of quantum mechanics. They were derived directly in the energy domain, although they are obtainable by stationary phase Fourier transform from the time domain.

In parallel with the time-dependent derivation of the VVMG propagator, we write an energy eigenstate $\psi_E(q)$ as

$$\psi_E(q) = e^{iS(q,E)/\hbar}. \tag{12.1}$$

We expand the exponent in powers of $\hbar$ (here we go further in $\hbar$ than we really need just to exercise Mathematica a bit):

$$S(q,E) = S_0[q,E] + \frac{\hbar}{i}S_1[q,E] + \left(\frac{\hbar}{i}\right)^2 S_2[q,E] + \left(\frac{\hbar}{i}\right)^3 S_3[q,E]. \tag{12.2}$$

Inserting this into the time-independent Schrödinger equation

$$-\frac{\hbar^2}{2m}\frac{d^2\psi_E(q)}{dq^2} + (V(q) - E)\psi_E(q) = 0 \tag{12.3}$$

and comparing powers in $\hbar$, we get (here in Mathematica code, to third order in the exponential $\hbar$):

$$\psi[q] := \mathrm{Exp}\left[\frac{i}{\hbar}\left(S_0[q,E] + \frac{\hbar}{i}S_1[q,E] + \left(\frac{\hbar}{i}\right)^2 S_2[q,E] + \left(\frac{\hbar}{i}\right)^3 S_3[q,E]\right)\right], \tag{12.4}$$

$$\mathrm{Collect}\left[\mathrm{Simplify}\left[E + \frac{\hbar^2}{2m}\partial_{\{q,2\}}\psi[q]/\psi[q] - V[q]\right], \hbar\right], \tag{12.5}$$

$$
\begin{aligned}
&E - V[q] - \left(\frac{S_0^{(1,0)}[q, E]^2}{2m}\right) \\
&\quad + \hbar\left(\frac{i S_0^{(1,0)}[q, E] S_1^{(1,0)}[q, E]}{m} + \frac{i S_0^{(2,0)}[q, E]}{2m}\right) \\
&\quad + \hbar^2\left(\frac{S_1^{(1,0)}[q, E]^2}{2m} + \frac{S_0^{(1,0)}[q, E] S_2^{(1,0)}[q, E]}{m} + \frac{S_1^{(2,0)}[q, E]}{2m}\right) \\
&\quad + \hbar^3\left(-\frac{i S_1^{(1,0)}[q, E] S_2^{(1,0)}[q, E]}{m} - \frac{i S_0^{(1,0)}[q, E] S_3^{(1,0)}[q, E]}{m} - \frac{i S_2^{(2,0)}[q, E]}{2m}\right) \\
&\quad + \hbar^4\left(-\frac{S_2^{(1,0)}[q, E]^2}{2m} - \frac{S_1^{(1,0)}[q, E] S_3^{(1,0)}[q, E]}{m} - \frac{S_3^{(2,0)}[q, E]}{2m}\right) \\
&\quad + \frac{i\hbar^5 S_2^{(1,0)}[q, E] S_3^{(1,0)}[q, E]}{m} \\
&\quad + \frac{\hbar^6 S_3^{(1,0)}[q, E]^2}{2m}.
\end{aligned}
\tag{12.6}
$$

We will not need all this complexity, but the structure of the equations is displayed to show how one can solve them successively, by starting with the $\hbar^0$ term, and moving up, getting one new quantity in terms of older known ones at each new level in the hierarchy.

When the $\hbar^0$ term is set to 0, we have the Hamilton-Jacobi equation:

$$\frac{1}{2m}\left(\frac{\partial S_0(q, E)}{\partial q}\right)^2 + V(q) - E = 0. \tag{12.7}$$

In the one-dimensional case under consideration, we have

$$\frac{dS_0(q, E)}{dq} = \pm\sqrt{2m(E - V(q))} = \pm p_E(q) \tag{12.8}$$

and

$$S_0(q, E) = \pm \int p_E(q')\, dq'. \tag{12.9}$$

The order $\hbar$ terms are zero if

$$\frac{\partial S_0(q, E)}{\partial q}\frac{\partial S_1(q, E)}{\partial q} + \frac{1}{2}\frac{\partial^2 S_0(q, E)}{\partial q^2} = p(q, E)\frac{\partial S_1(q, E)}{\partial q} + \frac{1}{2}\frac{\partial p(q, E)}{\partial q} = 0. \tag{12.10}$$

Inspection or integration shows that

$$S_1(q, E) = -\frac{1}{2}\log|p(q, E)| + \text{const.} \tag{12.11}$$

Stopping at this stage gives an approximate so-called WKB wavefunction:

$$\psi_E^{WKB}(q) \sim \frac{1}{\sqrt{|p(q, E)|}} e^{\frac{\pm i}{\hbar} \int^q p(q', E)\, dq'}. \tag{12.12}$$

Just as in the time-dependent semiclassical case, there may be more than one term in $\psi_E^{WKB}(q)$, corresponding to distinct ways of "being at q with energy E," although we show just one here. In a one-dimensional well, for example, there are two terms corresponding to heading left or right; see the $\pm$ present in equation 12.8, earlier.

We write equation 12.12 in a form making it easier to see how the analysis generalizes to more than one dimension:

$$\psi_E^{WKB}(q) \sim \sum_k \left(\frac{\partial^2 S_k(q, E)}{\partial q \partial E} \right)^{\frac{1}{2}} e^{\pm \frac{i}{\hbar} S_k(q,E)}. \tag{12.13}$$

Note in equation 12.12 the prefactor $|p(q, E)|^{-1/2}$, proportional to the square root of the classical probability density of "being at q, and with energy E." (This classical probability density goes as 1/velocity). This is a very common (in fact universal) pattern in semiclassical approximations—that is,

$$\psi^{semiclassical}(q) = \sum_n \left[\text{classical probability density}(q)_n\right]^{\frac{1}{2}} e^{i\ (\text{classical action}(q)_n)/\hbar}, \tag{12.14}$$

where the sum is over all the ways (if there are more than one) of being at q with energy E.

12.2 Tunneling and Connection Formulas

We have so far tacitly assumed that the position q is in a classically allowed region, where $p(q, E) = \pm\sqrt{2m(E - V(q))}$ is real—that is, $V(q) \leq E$. However, except for the problem mentioned earlier near the turning point $V(q) = E$, there is no reason not to trust the $\hbar$ expansion when $p(q, E)$ is imaginary. In some sense the meaning of imaginary momenta is immaterial—mathematically, the expansion is working as long as the error terms are small. In the forbidden regions,

$$\psi_E^{WKB}(q) \sim \frac{1}{\sqrt{|p(q, E)|}} e^{\pm \frac{1}{\hbar} \int^q |p(q', E)|\, dq'}. \tag{12.15}$$

Again in one dimension there are two solutions, one exponentially increasing and one decreasing. This is not different from the exact solution of Schrödinger's equation in such regions. The problem is one of boundary conditions: what solutions are acceptable given the physical requirements? In the case of bound states in confining potentials, we reject the solutions that increase without bound in either direction, since that would make the eigenstate badly unnormalizable.

We now have the building blocks to make WKB eigenfunction approximations—namely, the "local" solutions equation 12.12 and equation 12.15, but we have to sew them together to solve the boundary conditions correctly. In the case of a potential well, this will not even be possible except at certain energies, leading to a semiclassical estimate of the eigenvalues. The process of sewing the solutions together turns out to be

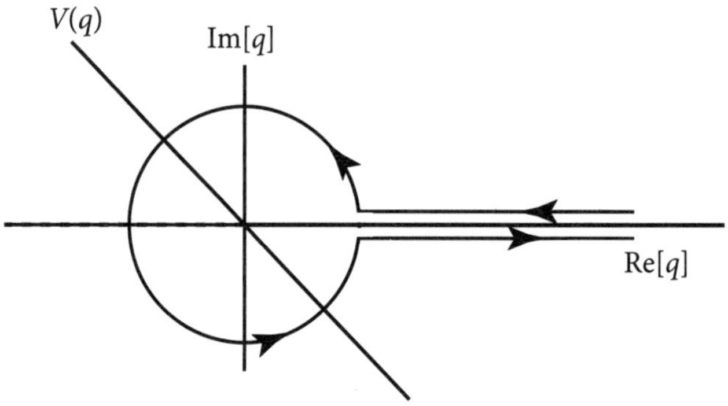

Figure 12.1. Contour in complex plane taken to examine the phase shift at a turning point.

a little subtle, and most books beg off here and quote the result. There is a phase shift involved; we should not be surprised since we have already seen them arise at focal points (see discussion around equation 7.47). Like the classical turning point, this is a singular point in coordinate space. Passing through such points always leads to a phase shift of some sort.

To justify the phase shift, we use an argument of Landau: stay well away from the singular point $V(q) = E$ by circumnavigating it in the complex q plane. Consider a smooth potential barrier decreasing to the right; in fact, we might just as well take it to be a linear potential (any curvature in the potential is unimportant since we are interested here only in the "turning point" region $V \approx E$). On the right, we start with the wave "incoming" on the linear potential from right to left—namely,

$$\begin{aligned}\psi(q) &\sim \frac{1}{\sqrt{p(q)}} e^{-i\int^q p(q')\,dq'/\hbar} \\ &= \frac{1}{[2m(E - V(q))]^{1/4}} e^{-i\int^q \left[2m\left(E - V(q')\right)\right]^{\frac{1}{2}}\,dq'/\hbar} \qquad (12.16)\end{aligned}$$

(the integral taken from some reference point a to q). The idea is to go into the forbidden region, circumnavigate the branch point, visit real but forbidden q on the way, and keep going in the same (counterclockwise) direction back to real allowed q, and see what emerges as a result of continuously following the original wave as a function of complex q (see figure 12.1). We take $V(q) = -\beta' q$, $\beta' > 0$. Letting $z = E/\beta + q$, the wavefunction becomes

$$\psi(q) \sim \frac{1}{(\beta z)^{1/4}} e^{-i\int^z (\beta z')^{\frac{1}{2}}\,dz'/\hbar}, \qquad (12.17)$$

where $\beta = 2m\beta'$. Imagine heading around a circle of radius 1 starting at $z = 1$ as shown. Then

$$z' = e^{i\theta'}; \quad \theta' = 0 \to 2\pi, \qquad (12.18)$$

and

$$(\beta z')^{\frac{1}{2}} = (\beta)^{\frac{1}{2}} \to (\beta e^{2i\pi})^{\frac{1}{2}} \to -(\beta)^{\frac{1}{2}} \qquad (12.19)$$

(that is, the integrand changes sign), and

$$\frac{1}{(\beta z)^{1/4}} \to e^{-i\pi/2}\frac{1}{(\beta z)^{1/4}}, \tag{12.20}$$

so what goes in as

$$\psi(q) \sim \frac{1}{(\beta z)^{1/4}} e^{-i\int^z (\beta z')^{\frac{1}{2}}\, dz'/\hbar} \tag{12.21}$$

comes out as

$$\psi(q) \sim \frac{e^{-i\pi/2}}{(\beta z)^{1/4}} e^{i\int^z (\beta z')^{\frac{1}{2}}\, dz'/\hbar}. \tag{12.22}$$

Note that this wave is indeed heading to the right. This tells us that there is a $-\pi/2$ phase shift upon encountering the turning point. In other words, burrowing into the potential barrier and coming out again means you bounce off (reversed momentum) and get a phase shift of $-\pi/2$.

If we stop halfway round the circle—that is, after circumnavigation by π, we arrive on the real q axis in the forbidden region. We have

$$\begin{aligned}\psi =\sim & \frac{1}{(\beta z e^{i\pi})^{1/4}} e^{-i\int_{q_L}^z (\beta z' e^{i\pi})^{\frac{1}{2}}\, dz'/\hbar} \\ = & \frac{1}{(\beta z e^{i\pi})^{1/4}} e^{\int_{q_L}^z (\beta z')^{\frac{1}{2}}\, dz'/\hbar},\end{aligned} \tag{12.23}$$

where q_L is the left turning point; this we choose (arbitrarily) as a reference point. Equation 12.23 is exactly what we want in the forbidden region, since the solution is exponentially *decreasing* to the left.

We can clean up the phases a little by starting with a phase of $i\pi/4$:

$$\psi(q) \sim \frac{e^{i\pi/4}}{(\beta z)^{1/4}} e^{-i\int_{q_L}^z (\beta z')^{\frac{1}{2}}\, dz'/\hbar} \quad \text{(incoming)}, \tag{12.24}$$

then in the forbidden region we have a real wavefunction

$$\begin{aligned}\psi(q) \sim & \frac{e^{i\pi/4}}{(\beta z e^{i\pi})^{1/4}} e^{-i\int_{q_L}^z (\beta z' e^{i\pi})^{\frac{1}{2}}\, dz'/\hbar} \\ = & \frac{1}{(\beta z)^{1/4}} e^{\int_{q_L}^z (\beta z')^{\frac{1}{2}}\, dz'/\hbar}\end{aligned} \tag{12.25}$$

that emerges (reflects) as

$$\psi(q) \sim \frac{e^{-i\pi/4}}{(\beta z)^{1/4}} e^{i\int_{q_L}^z (\beta z')^{\frac{1}{2}}\, dz'/\hbar}. \tag{12.26}$$

In terms of more general potentials we would have, in the forbidden region,

$$\psi(q) \sim \frac{1}{\sqrt{|p(q, E)|}} e^{\int_{q_L}^q |p(q', E)|\, dq'/\hbar}. \tag{12.27}$$

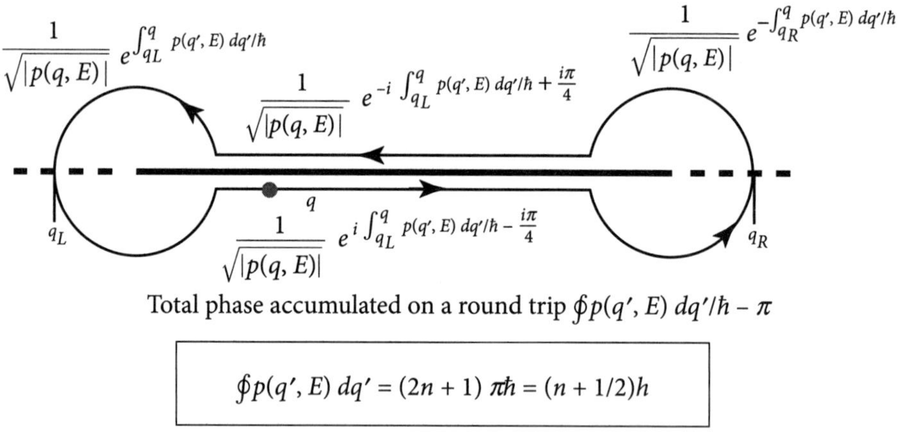

Figure 12.2. Quantization based on in-phase requirement of round-trips along the dog-bone path.

WKB OR BOHR-SOMMERFELD EIGENVALUES

In the classically allowed region, we have

$$\psi(q) \sim \frac{e^{-i\pi/4}}{(\beta z)^{1/4}} e^{i\int_{q_L}^{z} (\beta z')^{\frac{1}{2}}\,dz'/\hbar} + \frac{e^{i\pi/4}}{(\beta z)^{1/4}} e^{-i\int_{q_L}^{z} (\beta z')^{\frac{1}{2}}\,dz'/\hbar}$$

$$\sim \frac{1}{(\beta z)^{1/4}} \cos\left(\int_{q_L}^{z} (\beta z')^{\frac{1}{2}}\,dz'/\hbar - \frac{\pi}{4}\right), \tag{12.28}$$

or in terms of classically allowed q and for more general potentials,

$$\psi(q) \sim \frac{1}{\sqrt{|p(q,E)|}} \cos\left(\int_{q_L}^{q} p(q',E)\,dq'/\hbar - \frac{\pi}{4}\right), \tag{12.29}$$

where q_L is the left-hand turning point (that is, for a potential with a negative slope).

The analysis for the right-hand turning point is of course nearly the same; again there is a phase shift of $-\pi/2$. The quantization rule follows from asking that the accumulated phase be a multiple of 2π. A round-trip must give a phase change of an integer multiple of 2π (see figure 12.2):

$$\oint p(q', E_n)\,dq' = (n+1/2)h, \tag{12.30}$$

where the (discrete) classical energies where this is true are labeled E_n, and are the WKB estimates of the quantized energies.

In 1916–1917 there had been (more heuristically) the Bohr-Sommerfeld quantization rules, for integrable systems of 2 degrees of freedom— namely,

$$\oint_{c_j} \mathbf{p}\cdot d\mathbf{q} = n_j h, \tag{12.31}$$

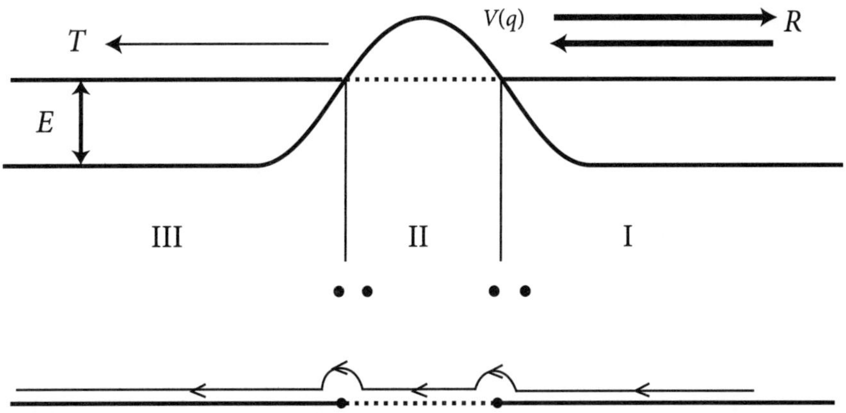

Figure 12.3. Tunneling through a potential barrier in one dimension. For the path shown for complex q, the plane wave incident from the right is analytically continued into the forbidden region (dashed) and emerges in the allowed region on the other side of the barrier.

where the paths labeled by c_j are topologically distinct. (Giving for example the hydrogen atom classical quantizing orbits two quantum numbers rather than the one circular "Bohr orbit" quantum number that Bohr had guessed.) This is N topologically independent integrals on the N-torus of an integrable system, each with its own action quantum number n_j or $n_j + 1/2$. It is important that the momenta and the positions can be generalized coordinates, but Sommerfeld had in mind the extension of Bohr theory to elliptical orbits, thus bringing in angular momentum.

12.3 WKB Through-Barrier Penetration

We suppose a particle approaches a barrier traveling from right to left, with energy insufficient to penetrate the barrier. We expect some amplitude will certainly reflect, and we anticipate some will tunnel, too. Let the reflection amplitude be R, so in region I (figure 12.3) to the right of the barrier (and far enough from it that the potential is zero)

$$\psi_I(q) = \frac{1}{\sqrt{p_0}} e^{-ip_0q/\hbar} + R\,\frac{1}{\sqrt{p_0}} e^{ip_0q/\hbar}, \tag{12.32}$$

where p_0 is the asymptotic momentum at energy E—that is, $p_0 = \sqrt{2mE}$. In region III, we have far to the left of the barrier

$$\psi_{III}(q) = T\,\frac{1}{\sqrt{p_0}} e^{-ip_0q/\hbar}, \tag{12.33}$$

where T is the transmission amplitude. In region I, the WKB wave is of the form (we assume unit reflection amplitude; our goal here is to get the transmission amplitude)

$$\psi_I(q) = \frac{1}{\sqrt{|p(q)|}} e^{-i\int_{q_A}^{q} p(q')\,dq'/\hbar + i\frac{\pi}{4}} + \frac{1}{\sqrt{|p(q)|}} e^{\int_{q_A}^{q} p\,dq'/\hbar - i\frac{\pi}{4}}. \tag{12.34}$$

In the tunneling region II, we have

$$\psi_{II}(q) = \frac{1}{\sqrt{|p|}} e^{-\int_{q_A}^{q} |p(q')|\, dq'/\hbar}. \tag{12.35}$$

Finally, in region III on the other side of the barrier we have

$$\begin{aligned}
\psi_{III}(q) &= \frac{1}{\sqrt{|p|}} e^{-i\int_{q_A}^{q} p(q')\, dq'/\hbar - \frac{i\pi}{4}}, \\
&= \frac{1}{\sqrt{|p(q)|}} e^{\int_{q_A}^{q_B} |p(q')|\, dq'/\hbar}\, e^{-i\int_{q_B}^{q} p(q')\, dq'/\hbar - \frac{i\pi}{4}}, \\
&\equiv T \frac{1}{\sqrt{|p(q)|}}\, e^{-i\int_{q_B}^{q} p(q')\, dq'/\hbar - \frac{i\pi}{4}},
\end{aligned} \tag{12.36}$$

so that $R = 1$ and

$$T = e^{\int_{q_A}^{q_B} |p(q')|\, dq'/\hbar - i\pi/2} = e^{-\int_{q_B}^{q_A} |p(q')|\, dq'/\hbar - i\pi/2}. \tag{12.37}$$

Of course, $R = 1$ is inconsistent with the finite transmission, but all this theory is asymptotic ($\hbar \to 0$) and in that limit the error is exponentially small. This is also the case for thick barriers and finite $\hbar$.

12.4 Errors and Quantum Potential

A good way to understand the errors in the WKB method is to plug the WKB wavefunction into the Schrödinger equation and see what emerges as the violation of $H\psi = E\psi$. Starting with

$$\psi^{WKB}(q) := \mathrm{Exp}\left[\frac{i}{\hbar}\left(S_0[q, E] + \frac{\hbar}{i} S_1[q, E]\right)\right] \tag{12.38}$$

and using Mathematica, we find

$$\begin{aligned}
&\mathrm{Collect}[\mathrm{Simplify}[E\ \psi[q] + \frac{\hbar^2}{2m}\partial_{\{q,2\}}\psi[q] - V[q]\psi[q]], \hbar] \\
&\qquad = \hbar^2 \left(\frac{S_1^{(1,0)}[q, E]^2}{2m} + \frac{S_1^{(2,0)}[q, E]}{2m}\right)\psi[q].
\end{aligned} \tag{12.39}$$

With $S_1 = -1/2 \log |p|$, we obtain

$$\begin{aligned}
(H - E)\psi^{WKB}(q) &= \frac{\hbar^2}{2m}\left(\frac{3p'^2}{4p^2} - \frac{p''}{2p}\right)\psi^{WKB}(q) \\
&= \frac{\hbar^2}{32m}\left(\frac{5V'(q)^2 + 4\,[2m(E - V(q))]\,V''(q)}{[2m(E - V(q))]^2}\right)\psi^{WKB}(q), \\
&\equiv -V^{WKB}(q)\psi^{WKB}(q).
\end{aligned} \tag{12.40}$$

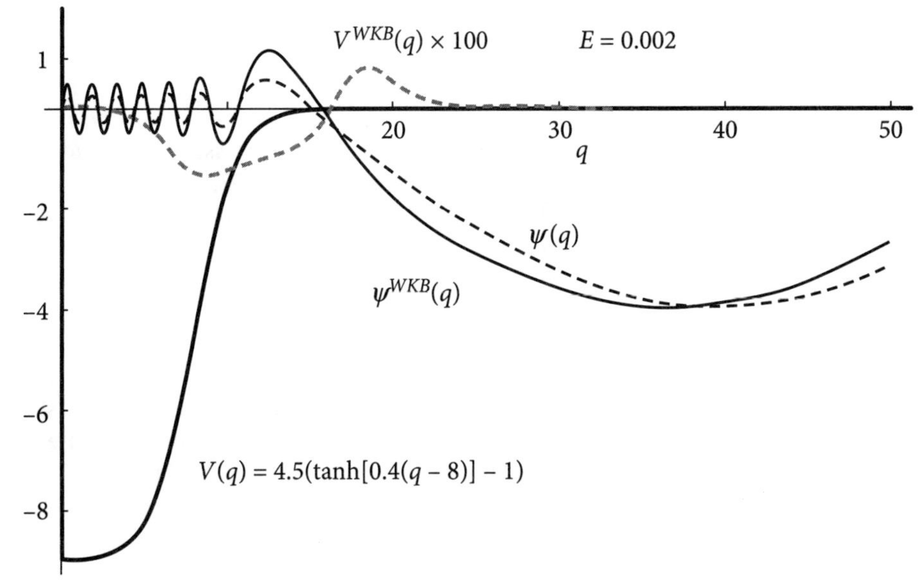

Figure 12.4. The potential $V(q)$ (thick solid line), the WKB (solid line) and exact (dashed line) wavefunctions for $V(q)$, the potential V^{WKB}, magnified by 100, (dashed line) for the energy $E = 0.002$ above the flat ramp at $E = 0$. The error in the WKB wavefunction and the beginnings of quantum reflection are evident.

where we have used

$$p' = \frac{dp}{dt}\frac{dt}{dq} = -\frac{V'(q)m}{p}. \tag{12.41}$$

The term on the right in equation 12.40 is a violation of the Schrödinger equation, in the form of an energy-dependent potential. The WKB wavefunction exactly solves the problem

$$(T + V(q) + V^{WKB}(q) - E)\psi^{WKB}(q) = 0. \tag{12.42}$$

This is interesting because the potential $V^{WKB}(q)$ "turns off" classically forbidden events, since the WKB wavefunction is built on classically allowed ones. Specifically, diffractive backscattering of a wave traveling with energy above a barrier is completely suppressed for the potential $V(q) + V^{WKB}(q)$. Could this concept be generalized? Is there a more general way to get fully quantum (not semiclassical) solutions to a modified Hamiltonian that has no dynamical tunneling or diffraction of any sort?

12.5 Quantum Reflection

Certainly, the errors in the WKB method have everything to do with the size and shape of the potential term V^{WKB}. It is clear that things are desperate when $E - V(q) = 0$, i.e. at classical turning points. A tug-of-war results though if $E - V(q)$ is small positive definite, but also V' and V'' are small. This happens at low energies near a ramp potential such as $V(q) = a(\tanh(b\ q) - 1)$ (figure 12.4), bringing up a new phenomenon called *quantum reflection*. In figure 12.4, the exact and WKB wavefunctions are normalized to be the same amplitude to the right of the ramp, but the exact wavefunction is somewhat smaller amplitude than the WKB function to the left, after traversing the ramp downhill (we discuss boundary conditions more carefully shortly). Indeed, this is a trend: the WKB wavefunction is guided by classical motion on the exact potential and has no difficulty going down the ramp no matter how slowly it approaches from the right. But compared to this classical amplitude, the exact quantum result is smaller at this low energy of $E = 0.002$, and will fail to go down the ramp altogether as $E \to 0$. Instead, it reflects off the start of the downhill ride, where the "start" moves farther to the right as $E \to 0$. The penetration of the ramp region can be shown to drop as a coefficient times $k \sim \sqrt{E}$, where the coefficient is smaller for sharper onsets and steepness of the ramp. The k dependence is exactly in line with the known "threshold law" for exdothermic reactive scattering in 3D. Exothermic implies a downhill potential, and in three dimensions only the s-wave (angular momentum $\ell = 0$) component of the incoming wave matters as $E \to 0$, since a radial potential barrier $\sim \ell(\ell+1)/r^2$ bars entry for $\ell = 1, 2, \ldots$. Thus the scattering is s-wave and amounts to a barrier-less 1D problem in the radial coordinate on $(0, \infty)$. Exothermic reaction cross sections diminish as k as $k \to 0$. The exception is the long-range Coulomb potential, which decreases so uniformly that there is no quantum reflection from it.

The wave vanishes at $q = 0$ in the potential shown in figure 12.4. There are bound states in the well with $E < 0$; these do not much affect the quantum reflection, *unless* there is a bound state *exactly* at threshold, $E = 0$, in which case there is a spectacular resonant build up of amplitude beyond the ramp. Then quantum reflection is defeated by resonance. This resonance may be viewed as a result of a very small portion of the incident wave penetrating past the ramp, which accelerates and—here is the key—returns from the repulsive the wall in phase with the next arriving portion. A wavepacket at such low energy is essentially arriving for a very long time, giving the resonance time to build up amplitude beyond the ramp.

In figure 12.4 the potential $V(q)$ (thick solid line), $V^{WKB} \times 100$ (dashed), the exact (dashed line), and WKB (solid line) wave functions at an energy $E = .002$ are plotted.

Quantum reflection is apparent in figure 12.5, where the exact wave function is shown for boundary condition $\psi = 0$ at the left. The plot was varied to span the range of the wavefunction, so the potential seems to change, but is always $V(q) = 2\tanh(q/5) - 2$, with $H = p^2/2 + V(q)$. Crucially, the wavefunction is normally smaller when the potential dips, because, semiclassically at least, the wavefunction amplitude goes as $1/\sqrt{p(q)}$. ψ^{WKB} is not shown, but its oscillation range is shown as dashed lines. At low energies, $E = 0.0001$, the contrast between inner and outer wavefunction amplitudes is dramatic compared to the semiclassical dashed lines, showing a great suppression of the quantum wavefunction inside compared to beyond the ramp. If the energy is $E = 0.04$, already the quantum reflection is nearly absent and both inner and outer parts of the exact solutions agree with the WKB amplitudes.

The absolute magnitude of the "quantum potential" $V^{WKB}(q)$ is not the figure of merit; rather, it is the magnitude of $V^{WKB}(q)$ compared to the kinetic energy; this ratio is

$$\begin{aligned} E(q) &= V^{WKB}(q)/(p^2/2m) \\ &= \frac{\hbar^2}{32m}\left(\frac{5V'(q)^2 + 4\,[2m\,(E - V(q))]\,V''(q)}{[2m\,(E - V(q))]^3}\right), \end{aligned} \tag{12.43}$$

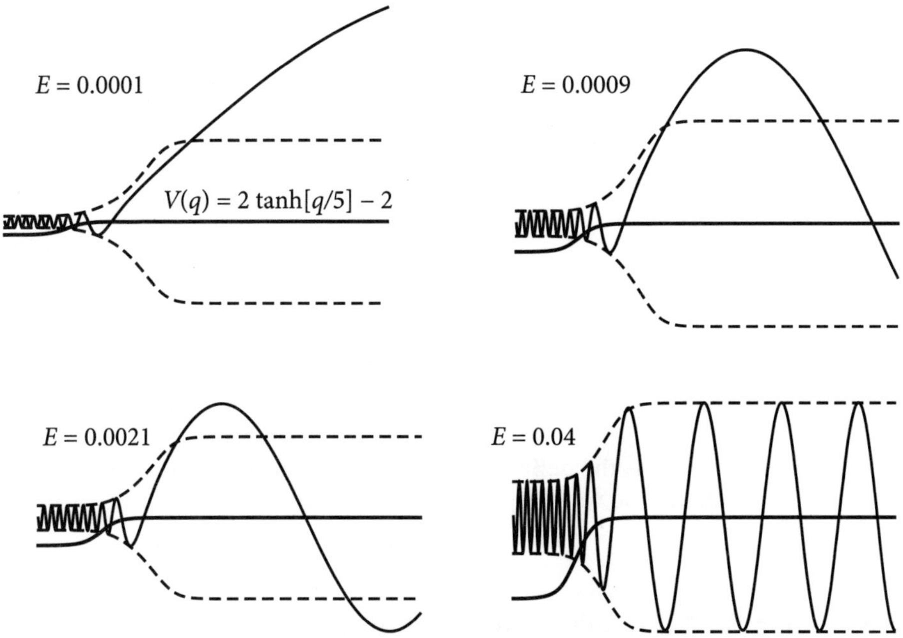

Figure 12.5. The dashed line shows the range of the WKB wavefunction amplitude, which is a plot of $1/\sqrt{p(q)}$. Starting at low energy at the upper left, and continuing to higher energy at the lower right, the suppression of the wavefunction to the left of the ramp compared to its amplitude to the right goes from dramatic at lowest energy to nearly no suppression compared with the semiclassical prediction at higher energy, unambiguously illustrating the quantum reflection/suppression phenomenon. The potential is the same in each case but the plotting range was changed from frame to frame to capture the wavefunctions.

and we need $E(q) \ll 1$ for agreement between the quantum and semiclassical WKB wavefunction.

There is a simple rule for the validity of WKB wavefunctions in the presence of a changing potential—namely, that the wavelength should not change significantly over the distance of one wavelength, lest the notion of a local wavelength become meaningless. This means

$$\left| \frac{d\lambda(q)/2\pi}{dq} \right| \sim \hbar \left| \frac{p'(q)}{2p(q)^2} \right| < 1. \tag{12.44}$$

We can get the same result from our analysis if we take only the first term in equation 12.40. This is justified because this term is positive definite and the second term can be shown to be roughly the same size. Thus, take

$$\frac{\hbar^2}{2m}\left(\frac{3p'^2}{4p^2}\right), \tag{12.45}$$

divide it by the kinetic energy, obtaining

$$\hbar^2(\frac{p'}{p^2})^2 \ll 1, \tag{12.46}$$

or what is the same thing

$$\hbar\frac{p'}{p^2} = \frac{m\hbar|V'(q)|}{p^3} \ll 1. \tag{12.47}$$

This criterion can be written

$$\left|\frac{d(\lambda/2\pi)}{dq}\right| \ll 1, \tag{12.48}$$

where $\lambda(q) = 2\pi\hbar/p(q)$ is the deBroglie wavelength. That is, the criterion for WKB to be accurate is again: "the wavelength should not change much over the scale of a wavelength."

12.6 Double Well Tunneling and Splitting

The important problem of splitting of energy levels in a double potential well is a worthy application of WKB techniques. Some of the techniques we use hint at ways of approaching multiple wells in several dimensions, but we focus on one-dimensional tunneling here.

If the wells on the left and right are identical, and an infinite impenetrable wall lies between, the energy levels must come in degenerate pairs. If the barrier becomes barely penetrable, the degeneracy is lifted but the levels come in closely spaced pairs. Let ψ_α and ψ_β be such a pair of tunnel split eigenstates, with energies $E_\beta > E_\alpha$. Since the Hamiltonian is left unchanged upon taking $q \to -q$, the eigenstates of H must transform as even or odd with respect to this interchange. The lower energy state ψ_α will have even parity ($\psi_\alpha(-q) = \psi_\alpha(q)$, with $q = 0$ being the midpoint of the double well). ψ_β is slightly higher in energy since it has an additional node at $q = 0$, $\psi_\beta(-q) = -\psi_\beta(q)$. Nodes "cost" or imply higher energy, but if the wavefunction is of low amplitude, the energy cost is lower, as is easy to show. The wave is of very low amplitude in the tunneling region.) The energies of the states are E_α and $E_\beta > E_\alpha$.

Define the localized, nonstationary states ψ_L and ψ_R as (see figure 12.6)

$$\psi_L = \frac{\psi_\alpha + \psi_\beta}{\sqrt{2}}; \ \psi_R = \frac{\psi_\alpha - \psi_\beta}{\sqrt{2}}; \text{ also}$$

$$\psi_\alpha = \frac{\psi_L + \psi_R}{\sqrt{2}}; \ \psi_\beta = \frac{\psi_L - \psi_R}{\sqrt{2}}.$$

Such a localized a state $\psi_L(q)$ satisfies Schrödinger's equation in the left well at energy E_L except in the barrier region and out to the other side, where it remains small. Likewise, there is a right-localized state that has energy E_R. For a symmetric double well, $E_L = E_R$.

Suppose we didn't know the exact eigenstates and energies, and all we have is approximate $\psi_L(q)$ and $\psi_R(q)$ by WKB methods. How do we find the tunnel splitting?

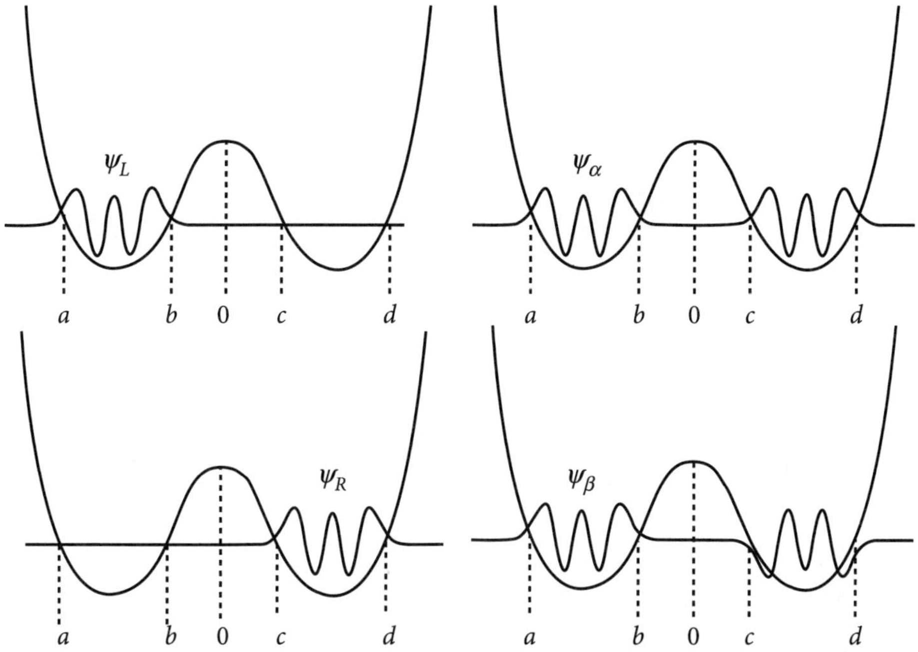

Figure 12.6. Diagram showing a tunneling pair of eigenstates, the symmetric state ψ_α on the upper right, and the anti-symmetric state ψ_β on the lower right, for a symmetric double well. The localized linear combinations of these eigenstates, ψ_L and ψ_R are shown on the left.

We have

$$\psi_L(q) = \frac{\eta}{\sqrt{|p|}} \cos\left(\int^q p\, dq - \frac{\pi}{4}\right) \tag{12.49}$$

with η fixed by

$$1 = \int_{q_a}^{q_b} |\psi_L|^2\, dq = \eta^2 \int_{q_a}^{q_b} \frac{dq}{|p|} \cos^2\left(\int^q p\, dq - \frac{\pi}{4}\right)$$

$$\approx \frac{\eta^2}{2m} \int_{q_a}^{q_b} \frac{dq}{\frac{dq}{dt}} = \frac{\eta^2}{2m} \int_0^{\tau_L/2} dt = \frac{\eta^2}{2m} \frac{\tau_L}{2},$$

where we have used the fact that $\langle\cos^2\rangle \sim 1/2$ if cosine oscillates fast enough (true as $\hbar \to 0$) and where τ_L is the classical period of one vibration in the left well at the energy $(E_\alpha + E_\beta)/2$. This gives, for the normalization constant η,

$$\eta = \left(\frac{2m\omega_L}{\pi}\right)^{\frac{1}{2}}. \tag{12.50}$$

To get the tunnel splittng, we use Herring's formula, derived next. With

$$H\psi_\alpha = E_\alpha \psi_\alpha; \;\; H\psi_\beta = E_\beta \psi_\beta, \tag{12.51}$$

we integrate only over the left half to form (for real ψ_α, ψ_β),

$$I = \int_{-\infty}^{0} \left(\psi_\alpha(q) \left[-\frac{\hbar^2}{2m}\frac{d^2}{dq^2} + V(q) \right] \psi_\beta(q) - \psi_\beta(q) \left[-\frac{\hbar^2}{2m}\frac{d^2}{dq^2} + V(q) \right] \psi_\alpha(q) \right) dq,$$

$$= (E_\beta - E_\alpha) \int_{-\infty}^{0} \psi_\alpha(q)\, \psi_\beta(q)\, dq = -\frac{\hbar^2}{2m} \left[\psi_\alpha \psi_\beta' - \psi_\beta \psi_\alpha' \right] \Big|_{-\infty}^{0};$$

the last equality follows by choosing each of two ways to evaluate the preceding line.

Then, we have

$$(E_\beta - E_\alpha) \int_{-\infty}^{0} \psi_\alpha(q)\, \psi_\beta(q)\, dq = \frac{\hbar^2}{2m} \left(\psi_\alpha \psi_\beta' - \psi_\beta \psi_\alpha' \right) \Big|_{q=0}; \tag{12.52}$$

this is Herring's formula. It is exact.

For a symmetric well tunnel doublet, to a very good approximation,

$$\int_{-\infty}^{0} \psi_\alpha(q)\, \psi_\beta(q)\, dq = \frac{1}{2}. \tag{12.53}$$

Also as is easily seen,

$$\left(\psi_\alpha \psi_\beta' - \psi_\beta \psi_\alpha' \right) \Big|_{q=0} = \left(\psi_L \psi_R' - \psi_R \psi_L' \right) \Big|_{q=0}; \tag{12.54}$$

thus,

$$(E_\beta - E_\alpha) = \frac{2\hbar^2}{2m} \left(\psi_L \psi_R' - \psi_R \psi_L' \right) \Big|_{q=0}. \tag{12.55}$$

Now we substitute semiclassical estimates of the wavefunctions $\psi_L(0)$ and $\psi_R(0)$. From section 12.3, we know

$$\psi_L(0) = \frac{1}{2} \left(\frac{2m\omega_L}{\pi} \right)^{\frac{1}{2}} \frac{1}{\sqrt{|p|}} e^{-\int_{q_b}^{0} |p|\, dq/\hbar}, \tag{12.56}$$

$$\psi_R(0) = \frac{1}{2} \left(\frac{2m\omega_L}{\pi} \right)^{\frac{1}{2}} \frac{1}{\sqrt{|p|}} e^{\int_{q_c}^{0} |p|\, dq/\hbar}. \tag{12.57}$$

The factors of $1/2$ in front of these expressions arise from the splitting of cos into

$$\cos() = \frac{1}{2}(e^{i()} + e^{-i()}),$$

together with the requirement that only the left-moving exponential is taken into the tunnel region from the right hand well, and so on. (The other blows up and is not used; this is tied up with the Stokes phenomenon mentioned earlier.) When these wavefunctions are substituted into Herring's formula, the derivatives acting on $1/\sqrt{|p|}$ cancel, giving

$$\Delta E_{\alpha\beta} = (E_\beta - E_\alpha) = \left(\frac{\hbar\omega_L}{\pi}\right) e^{-\int_{q_b}^{q_c} |p|\, dq/\hbar}, \tag{12.58}$$

the sought-after formula for the energy splitting. Starting in the left well, the overlap evolves in time as

$$|\langle\psi_L|\psi_L(t)\rangle|^2 = \tfrac{1}{2}\left(1 + \cos(\Delta E_{\alpha\beta} t/\hbar)\right), \tag{12.59}$$

as the wave sloshes back and forth with frequency $\Delta E_{\alpha\beta}/\hbar$.

If we had viewed the calculation as a 2×2 matrix diagonalization, with two states of equal energy coupled by a matrix element V_{LR}—that is,

$$M = \begin{pmatrix} E & V_{LR} \\ V_{LR} & E \end{pmatrix}, \tag{12.60}$$

the splitting of the eigenvalues $E + V_{LR}$ and $E - V_{LR}$ would be $2V_{LR}$. Thus, we have found that the potential coupling between the left- and right-hand states is

$$V_{LR} = \left(\frac{\hbar\omega}{2\pi}\right) e^{-\int_{q_b}^{q_c} |p|\, dq/\hbar}. \tag{12.61}$$

The physical interpretation of V_{LR} is illuminating. The tunnel splitting is proportional to $\hbar$, but there is much stronger $\hbar$ dependence operating in the exponential. As $\hbar \to 0$, the tunnel splitting equation 12.58 dies faster than any power law. This is characteristic of all "classically forbidden" processes: they shut down like $e^{-|S|/\hbar}$ as $\hbar \to 0$. (In several or more dimensions, more tunneling pathways open up using multiple smaller hops in phase space as $\hbar \to 0$, and even though each primitive hop goes like $e^{-|S_j|/\hbar}$, the action differences S_j are typically getting smaller. The overall expected diminishment of the rate of tunneling with $\hbar$ in such complex systems has not been determined to the author's knowledge, but an extremely good place to start is "Resonance-assisted Tunneling," Brodier, Schlagheck, and Ullmo, reference [91].)

The factor ω in V_{LR} makes sense if you think of it as the "number of attempts to get through the barrier per second." The wavefunctions ψ_L and ψ_R are each proportional to $\sqrt{\omega}_L = \sqrt{\omega}_R$; interaction between them is thus enhanced by a factor of $\omega_L = \omega_R = \omega$.

Chapter 13

Tunneling

13.1 Tunneling below and above a Barrier

There is much to be learned about systems possessing single potential barriers, now that we understand a bit about WKB methods. Three such problems are (1) the symmetric double well, the wells formed by a central barrier in the potential (just treated earlier); (2) the unbounded particle (free except for a localized barrier) in 1D; and (3) a particle on a ring lying perpendicular to a gravity potential field, with a bump causing a potential barrier. All three problems are interesting for particles approaching with energy above or below the top of the barrier. For the unbounded particle, both above-barrier reflection and below-barrier penetration problems are interesting. On a ring, tunneling under the barrier is a little (but not totally) moot, since the particle can get to the other side of the barrier classically by going around the other way. Tunneling would be a small correction, but working out the tunneling correction to the below barrier eigenvalues would be interesting. Reflection above the barrier on the ring will become a key paradigm of dynamical tunneling.

For the double well, reflection above the barrier is of diminished importance, because it is a small correction to the particle reversing direction classically at the bounding walls. Inclusion of the above-barrier reflection would slightly shift the eigenvalues, but we bypass this opportunity.

The real reason we study these problems is that they retain their essence even in several, perhaps many, degrees of freedom.

BACKSCATTERING DIFFRACTION IN 1D UNBOUNDED MOTION

Backscattering diffraction is a key paradigm for this book, as it forms a "type case" for dynamical tunneling even in many degrees of freedom. Dynamical tunneling is a form of diffraction, or it *is* diffraction. The word *diffraction* has been used loosely for years, including (erroneously) quantum or wave processes that are classically allowed. Using the term *dynamical tunneling* is more precise. *Classically forbidden* (a term that includes dynamical tunneling) processes were discussed in connection with collisions in W. H. Miller and T. F. George [92] and other papers, and is another altogether appropriate term.

Dynamical tunneling is defined as any process that happens quantum mechanically but is forbidden classically, without the intervention of a potential barrier that must be penetrated. This concept widens the horizon of tunneling considerably. We will show, for example, that photo-exciting a molecule from the ground vibrational state $n = 0$ directly to, say, $n = 4$ is dynamical tunneling. Backscattering diffraction (waves returning from a barrier that at the same energy classical particles would have surmounted and continued on their way) is dynamical tunneling [93].

We now examine the backscattering for a wave flying above (in the sense of its energy) a potential barrier in one dimension, from the perspectives of perturbation theory, distorted wave perturbation theory, semiclassical dynamical tunneling integrals, and the exact result. We first consider one encounter with a barrier in an unbounded domain, but later get closer to important applications by considering a barrier affecting a particle on a ring, with periodic boundary conditions.

ABOVE-BARRIER REFLECTION: FIRST-ORDER PERTURBATION THEORY

For transitions between degenerate scattering states, first-order perturbation theory takes the form of the Born approximation, discussed now. We are concerned specifically with a transition starting with a right-trending plane wave and want to know how much of it reflects, at an energy above a local barrier maximum, to become a left-trending plane wave of the same energy. The Born approximation is derived starting with the Lippmann-Schwinger equation for the exact scattering wavefunction in the presence of the perturbation—namely,

$$\psi = \psi_0 + G_0^+ V \psi,$$

where $G_0^+ = (E - H_0 + i\epsilon)^{-1}$ and ϵ is an infinitesimal positive constant. ψ_0 satisfies $H_0 \psi_0 = E \psi_0$. The Lippmann-Schwinger equation, an integral equation for ψ, is checked by multiplying it by $(E - H_0)$ and recovering the Schrödinger equation. The Born approximation is obtained by evaluating the first term in the iteration of the Lippmann-Schwinger equation to give the Born series:

$$\psi = \psi_0 + G_0^+ V \psi_0 + G_0^+ V G_0^+ V \psi_0 + \cdots .$$

In one dimension, Green function $G_0^+(q, q', E) = m/ik \exp[ik|q - q'|]$. Then, to the left of the potential where the backscattered wave would be found,

$$\psi(q) \approx e^{ikq} + \frac{m}{ik} e^{-ikq} \int dq' e^{2ikq'} V(q'). \tag{13.1}$$

The wave has probability 1 of having come in from the left, so the probability of heading back—that is, reflecting, is the reflection coefficient

$$\mathcal{R} = \frac{m^2}{\hbar^2 p^2} \left| \int_{-\infty}^{\infty} V(x) e^{\frac{2i}{\hbar} p x} dx \right|^2 ; \tag{13.2}$$

this is the Born perturbation theory result [94]. The formula is valid under certain "smallness" conditions of the potential:

$$\left| \frac{a\ V_0}{\hbar v} \right| \ll 1,$$

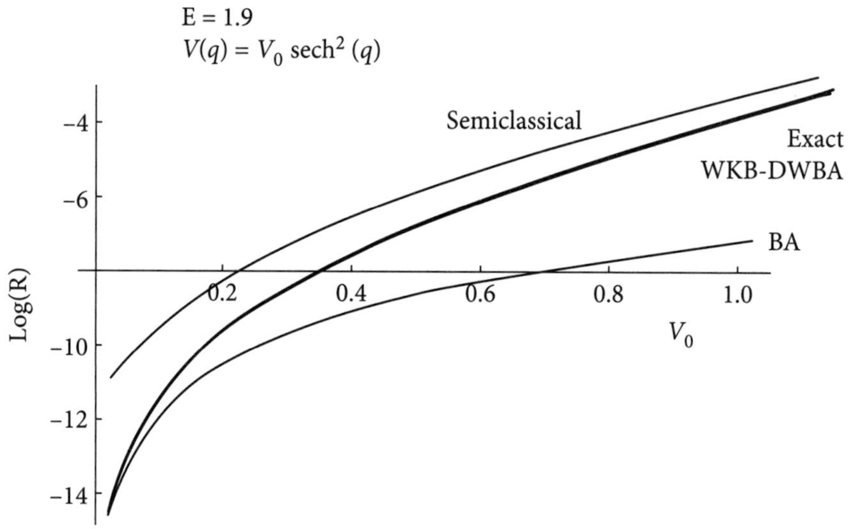

Figure 13.1. Natural logarithm of reflection probabilities for the barrier V_0 sech$^2(q)$, energy E $= 1.9$, for a range of V_0. As expected, the Born approximation is good only for very low barrier heights, while the semiclassical result is slightly disappointing, although expected to break down at low barrier heights. The distorted wave perturbation theory using the WKB solutions and the perturbation V^{WKB} essentially lies on top of the exact answer, for the whole range of potential strengths shown.

where $v = p/m$ is the classical velocity of the particle, V_0 is the barrier height, and a is the width of the potential. The scattering potential height V_0 must also be weak compared to the kinetic energy of the incoming particle [94]. This is the ordinary perturbation "Born approximation" reflection probability; see figure 13.1.

ABOVE-BARRIER REFLECTION: DISTORTED WAVE PERTURBATION THEORY USING WKB SOLUTIONS

The goal of distorted wave perturbation theory is to use a wave that (1) has the correct boundary condition, and (2) is a known solution closer to the exact answer than the undistorted wave (figure 13.2, right). The distorted wave solves a potential $V'(x)$ exactly; the difference $V(x) - V'(x)$ becomes the new "distorted wave" perturbation, smaller than it was for the undistorted wave. We have such an approximate solution available: the WKB wavefunction. The WKB wave distorts in the presence of the barrier, but it doesn't reflect from it. $-V^{WKB}$ becomes the perturbation used to estimate the reflection. We don't even have to construct V^{WKB} to get the perturbation matrix element (see equation 12.40). First, the left- and right-trending WKB waves are orthogonal:

$$\int_0^{2\pi} \psi_{WKB}^{-,*}(q)\, H_{WKB}\, \psi_{WKB}^{+}(q)\, dq = E \int_0^{2\pi} \psi_{WKB}^{-,*}(q)\, \psi_{WKB}^{+}(q)\, dq = 0.$$

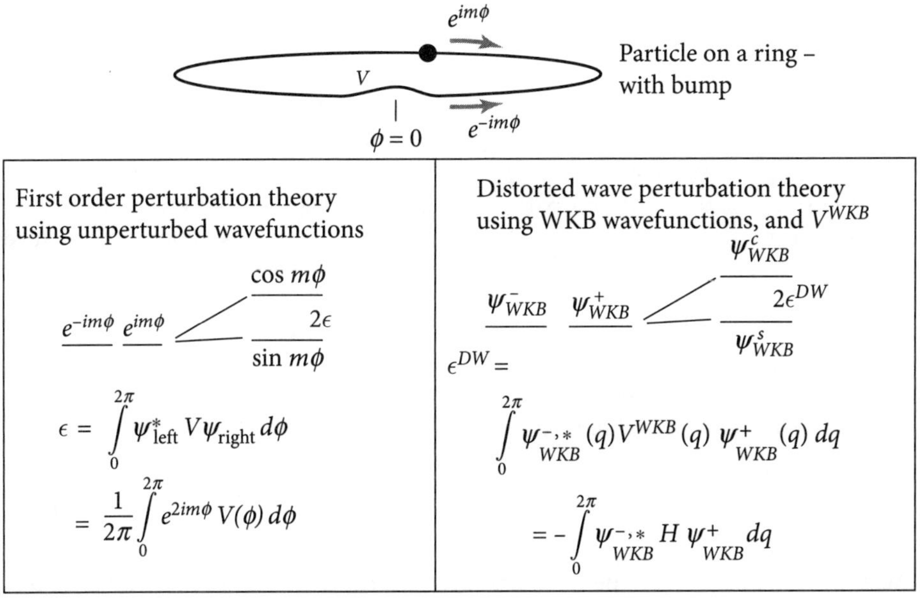

Figure 13.2. Bead on a frictionless ring, in a gravity field perpendicular to the ring so that the barrier represents a potential hill to climb. In the left box, the first-order perturbative solution is sketched out, also called the Born approximation in scattering theory. A more robust plan, for larger potentials and energies closer to, but above the top of the barrier, is to start from the WKB solutions; these are still degenerate and rotate only clockwise or counterclockwise, since the classical orbits guiding the WKB solutions are not reflected by the barriers for energies above the barrier tops. However, the WKB wave *is* distorted by the barrier. The perturbation needed is the difference between the exact potential, and the potential $\psi^{\pm}_{WKB}$. This as we have learned earlier is $V^{WKB}(q)$, equation 12.40. This "distorted wave" perturbation theory is sketched at the bottom right.

Then

$$\mathcal{R}^{DW} = \frac{m^2}{\hbar^2 p^2} \int_0^{2\pi} \psi^{-,*}_{WKB}(q) \, V^{WKB}(q) \, \psi^{+}_{WKB}(q) \, dq,$$

$$= \frac{m^2}{\hbar^2 p^2} \int_0^{2\pi} \psi^{-,*}_{WKB}(q) \, (H_{WKB} - H) \, \psi^{+}_{WKB}(q) \, dq,$$

$$= -\frac{m^2}{\hbar^2 p^2} \int_0^{2\pi} \psi^{-,*}_{WKB}(q) \, H \, \psi^{+}_{WKB}(q) \, dq. \tag{13.3}$$

This is the distorted wave reflection probability, completing the idea set out in equation 13.1; see figure 13.1. The distorted wave is a much more robust approach compared to the bare plane wave, good for stronger potentials and energies closer to, but above the top of the barrier.

SEMICLASSICAL ABOVE-BARRIER REFLECTION

Finally (apart from "exact" numerical solutions), we come to the semiclassical dynamical tunneling treatment, very different from perturbation theory.

This is justified as follows [95], starting with Feyman path integrals in momentum space, implying a summation over all paths $p[\tau]$ starting with momentum p and ending at time t with momentum $-p$. $S(p[\tau])$ is the classical action $\int_0^t \mathcal{L}(p[\tau])d\tau$ along the (not usually classical) path $p[\tau]$; $\mathcal{L}$ is the Lagrangian. We perform both the time integral and the path integral by stationary phase. As we know, this reduces the problem to a vastly smaller sum only over paths satisfying the classical equations of motion. For energies greater than the maximum potential there are no classical reflected paths for real initial conditions and real time. We can however analytically continue the integrand to the lower half complex time plane, picking up contributions from complex stationary phase points (that is, by steepest descent). The semiclassical reflection coefficient derived in this way for the barrier along a line is [95]:

$$\mathcal{R} = \frac{e^{-\frac{2}{\hbar}\Im \int_{-p_o}^{p_o} q(p')dp'}}{1 + e^{-\frac{2}{\hbar}\Im \int_{-p_o}^{p_o} q(p')dp'}}, \tag{13.4}$$

where p_o is the classical value for momentum at the center of the potential, $q(p)$ is the classical solution for the position in terms of momentum, and $\Im$ stands for the imaginary part. For the ring (see the following), we may use

$$\epsilon^{SC} = \left(\frac{1}{2\pi}\right) e^{-\frac{2}{\hbar}\Im \int_{-p_o}^{p_o} q(p')dp'}. \tag{13.5}$$

For the case of the linear potential with barrier $V(q) = V_0 \operatorname{sech}^2(q)$ and energy $E = 1.9$, all four methods (including the numerical result) are compared in figure 13.1.

ABOVE-BARRIER REFLECTION: BARRIER ON A RING

Adding a barrier is an interesting variation on the usual particle on a ring. With no barrier, there are degenerate pairs of unperturbed solutions $\psi_{m\pm} = 1/\sqrt{2\pi}\exp[\pm i m\phi]$ propagating clockwise and counterclockwise. With a barrier present, initial clockwise rotation of an unperturbed wave will experience backscattering, but this is not quite the same thing as a single encounter with a barrier in one dimension. The backscattered amplitude on a ring can soon come around and be rescattered in the original direction. It is better to look at this as an unperturbed degenerate two state problem and compute the splitting of the degeneracy due to the barrier by perturbation theory. Apart from normalization, this is the same integral we have just presented, except now it figures in the off diagonal terms of a 2×2 degenerate perturbation theory matrix. We suppose the barrier is everywhere positive but small; this will simplify things below. The diagonal elements of the perturbation theory matrix are therefore equal and slightly greater than zero. The eigenstates will be equal amplitude linear combinations of the two unperturbed circulating states, making one of them $\cos(m\phi)$ and the other $\sin(m\phi)$. Knowing the barrier is positive and supposing it to be symmetric, centered at $\phi = 0$, we can be sure that the lower energy solution will be $\psi_1(\phi) = \sqrt{2}/\sqrt{2\pi}\sin(\phi)$. Since its node is in the middle of the potential barrier, the expectation value of the energy will be lower than its partner, the cosine solution, $\psi_2(\phi) = \sqrt{2}/\sqrt{2\pi}\cos(\phi)$, which has an anti-node there. The splitting of the two states will be twice the perturbation matrix element of the barrier taken between

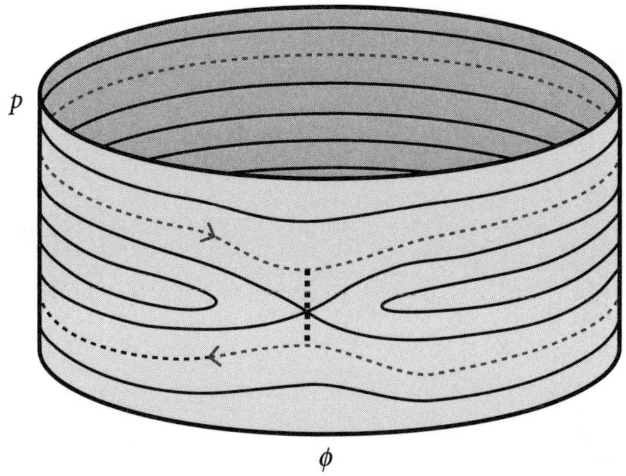

Figure 13.3. A cylindrical phase portrait of the particle or bead on a ring with a barrier in a gravity field normal to the ring. The semiclassical dynamics tunneling calculation is justified in the text and in reference [95].

the left and right circulating states. As with the case of the one-dimensional infinite line, the strength of the backscattering or here the splitting depends on the Fourier transform of the potential at the backscatter momentum change (see figure 13.2). The cylindrical phase portrait in figure 13.3 shows the classical momentum diminishing near the barrier; there is an associated separatrix with stable and unstable manifolds. The classical paths in phase space over the barrier at a given energy are shown dashed. A thicker vertical dashed line crosses the separatrix showing a tunneling path, very analogous to tunneling under the barrier, except the integral needed will involve the "vertical" path $\int q(p)\, dp$ for complex $q(p)$ and real momentum, rather than the "horizontal" barrier tunneling path $\int p(q)\, dq$ for complex $p(q)$ and real position.

ROLE OF COHERENCE IN TUNNELING; RESONANT TUNNELING

It is very instructive to consider another example of tunneling, this time starting from a well, tunneling through a barrier just like before, but now out into "free space"rather than into another mirror image well. Later, we will embellish this free space to include other degrees of freedom, and even decoherence.

The situations we have in mind are illustrated in figure 13.4. At the top, we see two potentials, case A, a symmetric double well, and case B, another barrier well that is the same as case A in the middle and on the left but free on the right. It seems right intuitively that the "rate limiting step" to escape from the well at the left is tunneling through the barrier, if the wavefunction is confined there initially. Thus the time to empty out the well on the left should be approximately the same in the two cases. This is in fact very far from being the case, and coherence (or lack of it) is to blame for the difference. Understanding this will permit insight into tunneling in many body systems.

In the lower part of figure 13.4, a series of double well potentials is depicted. The stretching and extension of the right hand well, if continued, leads continuously from case A to case B. A full understanding should include what happens in the transition

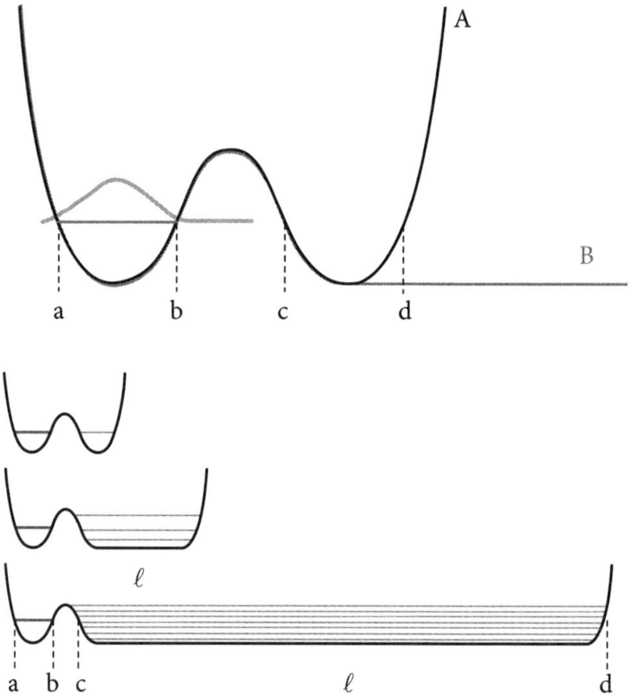

Figure 13.4. (Top) A symmetric double well, case A. Case B follows case A but becomes flat beginning at the bottom of the right-hand well. (Bottom) a succession of ever wider right-hand wells that transition from the symmetric case A to the free case B. Light gray horizontal lines depict energy levels within the right-hand well; the energy of the initial state in the left well is shown also.

from the symmetric double well to the double well that is much larger on the right hand side than the left.

We follow the program in the lower part of figure 13.4, starting with the symmetric case A, and trending to the limit of a very wide right hand well, case B. We need to alter the symmetric double well problem and make it work for an asymmetric well. This is not too difficult. We already know that in the symmetric double well problem the coupling matrix element V_{LR} is given by equation 12.61. The left well and right well wavefunctions are unit normalized, and have a factor $\sqrt{\omega_L}$ in their normalization. This implies that the analogous formula for the two-state coupling for an asymmetric well pair must be

$$V_{LR} = \frac{\hbar\sqrt{\omega_L \omega_R}}{2\pi} e^{-\int_b^c |p|\, dq/\hbar}, \tag{13.6}$$

where ω_R is the classical frequency in the right-hand well, becoming small as the well widens. The coupling V_{LR} may be insufficient to significantly couple left and right states for most intermediate right hand well sizes, depending on the energy gap between the left well energy and the nearest energies of the right-hand well. But if the well continues to grow, the left-hand state ψ_L is guaranteed to couple significantly to many right-hand well states, for the following reason: the matrix element V_{LR} drops off, since $\omega_R \sim 1/\ell$, where $\ell = d - c$ is the width of the well. The right-hand well is essentially a particle in a box, and the energy levels go as $E_n \sim n^2\pi^2\hbar^2/2m\ell^2$, so the level spacing decreases faster than the drop off in coupling, $1/\ell^2$ as opposed to $\sqrt{1/\ell}$. For a large enough right-hand well, the level spacing will be much less than the coupling, and many right-hand well states will be coupled to a single left-hand well state.

We may therefore apply Fermi's Golden Rule when the density of states in the right-hand well grows large enough. The density of states is most usefully expressed here in terms of the Bohr correspondence rule, stating

$$\Delta E = E_{n+1} - E_n \approx \hbar\omega_{class}, \tag{13.7}$$

where ω_{class} is the classical frequency of motion. (For example, here $E_{n+1} - E_n \approx n\pi^2\hbar^2/mL^2$; this is easily seen to be equal to $\hbar\omega_{class}$.) Thus, the density of states is $\rho = 1/\hbar\omega_{class}$. Then according to the Golden Rule,

$$\begin{aligned}\Gamma &= \frac{2\pi}{\hbar} < V_{RL}^2 > \rho = \left(\frac{2\pi}{\hbar}\right)\left(\frac{\hbar^2\omega_L\omega_R}{(2\pi)^2}\right)\left(\frac{1}{\hbar\omega_R}\right)e^{-2\int_{q_b}^{q_c}|p|\,dq/\hbar},\\ &= \frac{\omega_L}{2\pi}e^{-2\int_{q_b}^{q_c}|p|\,dq/\hbar},\end{aligned} \tag{13.8}$$

and the decay of the state proceeds as

$$|\langle\psi_L|\psi_L(t)\rangle|^2 = e^{-\Gamma t}, \tag{13.9}$$

independent of the width of the right-hand well, once it gets large enough. "Large enough" is when the decay is over before any amplitude can return to the barrier.

The semiclassical formulas for the energy splitting in the double well and the decay rate through a barrier are both well known. The same tunneling integral appears in both, as intuitively expected. What is apparently not noticed very often is the factor of 2 that sneaks into the exponent of the barrier decay problem (equation 13.8). This means that whereas the time to empty the left-hand well in the symmetric double well goes as

$$\tau_{\mathrm{dw}} \sim e^{\int_{q_b}^{q_c}|p|\,dq/\hbar}/\omega_L,$$

the half life in the well for the decay to freedom through the same barrier goes as

$$\tau_{barrier} \sim e^{2\int_{q_b}^{q_c}|p|\,dq/\hbar}/\omega_L.$$

The huge difference in escape rates (see figure 13.5) can be attributed to coherence. Suppose after some very short but not infinitesimal time τ amplitude ϵ appears on the right, as in figure 13.6. After another time increment τ, another amplitude ϵ adds to what is on the right (we can neglect the small diminution of amplitude on the left as an order ϵ^2 correction). The first increment ϵ has evolved on the right in time τ. This does nothing (except a phase factor, the same one, importantly, accruing on the left) in the case of the symmetric double well, so the new increment adds coherently and constructively to the first, making amplitude 2ϵ on the right and probability $4\epsilon^2$. Thus, the probability is growing quadratically in time on the right. In the case of a flat

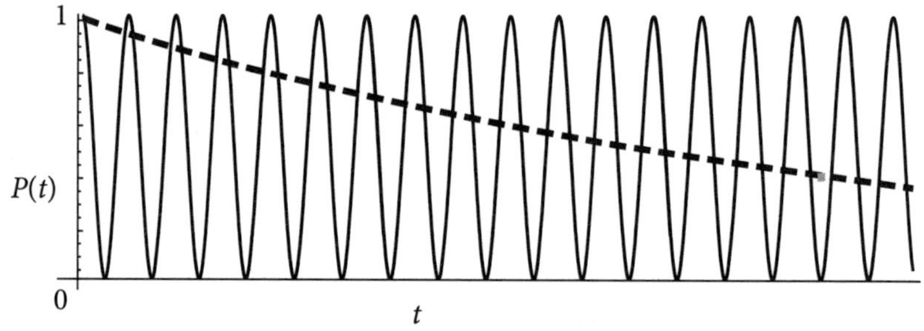

Figure 13.5. Time dependence of the survival probability in a symmetric double well (solid line) compared to the survival probability tunneling from the same well through the same barrier to a flat potential (dashed line).

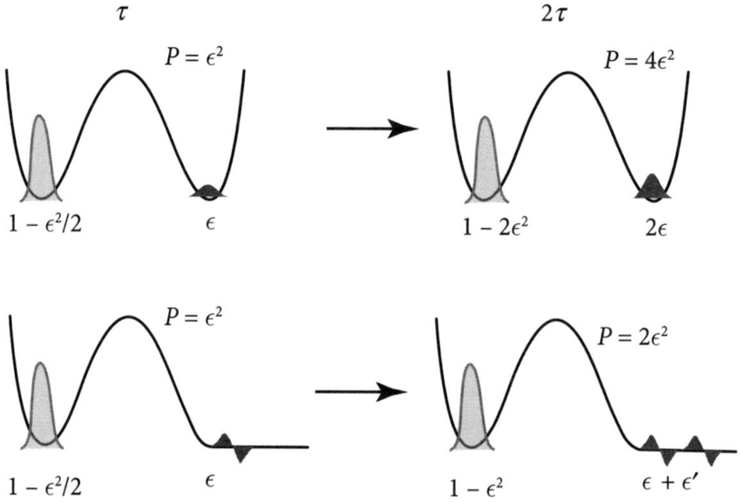

Figure 13.6. In the symmetric double well at the top, a small portion of amplitude ϵ appearing on the right after time τ adds constructively to the next portion at time 2τ, so then the amplitude is 2ϵ and the probability of being on the right is $4\epsilon^2$. This is the beginning of quadratic growth on the right. In the case of the free right-hand side, at the bottom, the first amplitude contribution ϵ moves away and is not available to interfere constructively with the amplitude ϵ arriving behind it, meaning the probably is just $2\epsilon^2$, setting up linear growth with time.

potential, the initial increment moves away and fails to overlap the next increment to arrive, meaning the probability on the right is just $2\epsilon^2$, growing only linearly in time.

SPECTRUM AND DECAY

If the decay leaving the left-hand well is essentially complete before any amplitude can reflect and return to the barrier region, then $\tau_{\text{barrier}} \ll 2\pi/\omega_R$. The exponential decay $\exp[-\Gamma t]$ implies a Lorentzian line width (full width at half maximum) of $2\Gamma = 2/\tau_{\text{barrier}}$. The level spacing is therefore $\hbar\omega_R \ll \Gamma$; i.e. there are many levels within the line width of the spectrum, as seen in figure 13.7. The density of states on the right

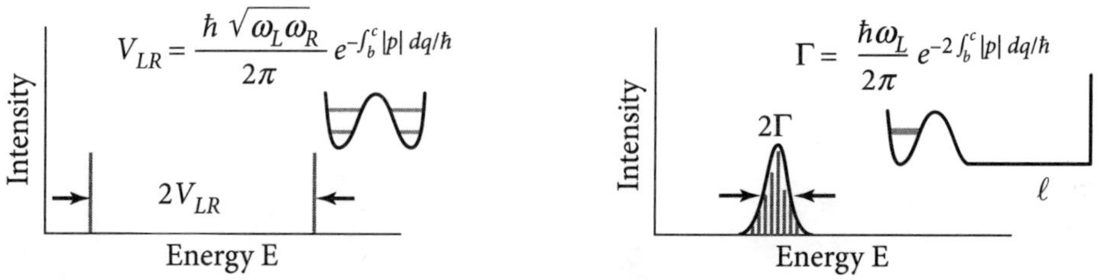

Figure 13.7. The spectrum $\epsilon(E)$ is the time Fourier transform of $\langle\phi(0)|\phi(t)\rangle$, $\epsilon(E) = \sum_n |\langle\phi(0)|\phi_n\rangle|^2\delta(E - E_n)$, where $\phi(0)$ is the initial state confined to the small well on the left, ϕ_n are the eigenstates, and E_n are the eigenenergies of the entire potential (in the case of a double well, there are just two eigenenergies in the region shown, separated by $2V_{RL} \sim \exp[-\int |p(q)|dq/\hbar])$. The line width of the spectrum Γ for a wide right-hand well (but the same left-hand well and barrier, right) becomes $\Gamma \sim \exp[-2\int |p(q)|dq/\hbar]$ if the decay is virtually complete before the first wall reflection arrives back at the barrier, much narrower than the splitting of the symmetric double well. The continuous spectrum if there is no wall at the right—that is, it is infinitely far away, is shown as an envelope of the discrete spectrum.

is high enough that the Fermi Golden Rule may be applied; the energy width of the left-hand state due to decay has become greater than the energy spacing of levels on the right.

TUNNELING AND DECOHERENCE

The preceding examples are a strong hint as to the fate of tunneling in many degrees of freedom, where the idea of "decoherence" might apply. The rule is simple: if the tunneling from a given state is not to an isolated, degenerate state, but rather to a dense quasi-continuum of states, it *may* revert to exponential decay and barrier tunneling, unless some remnants of coherence or constructive interference remain. It really doesn't matter whether we follow all the degrees of freedom explicitly, or trace over (integrate over) most of them. All that matters is whether the amplitude delivered across the barrier "lingers" there at that energy (and for how long if it does).

Clearly, there are intermediate cases, as the following situation shows. We add one more degree of freedom, call it y (representing perhaps many in fact), that switches from attractive in the left well to repulsive (with variable strength of repulsion) on the right. In the x degree of freedom, we suppose a symmetric double well (some tuning of the x well will be necessary to bring the resonance there in line with the resonance in the left hand well, since the y zero point energy on the left is absent on the right). As amplitude arrives on the right, it does not all leave immediately. Depending on the strength of the repulsion, there is more or less lingering on the right, setting up intermediate scenarios that combine exponential and sinusoidal time dependence (see figure 13.8).

The tunneling we have considered in this chapter is a piece of a larger story of classically forbidden processes. Essentially, whenever ordinary classical mechanics refuses to go somewhere or do something that does happen quantum mechanically, we call the quantum process *classically forbidden*. A semiclassical approach to such problems is to try to find analytically continued (where one or more of the classical variables become complex) classical trajectories that behave in the desired fashion; then there will be

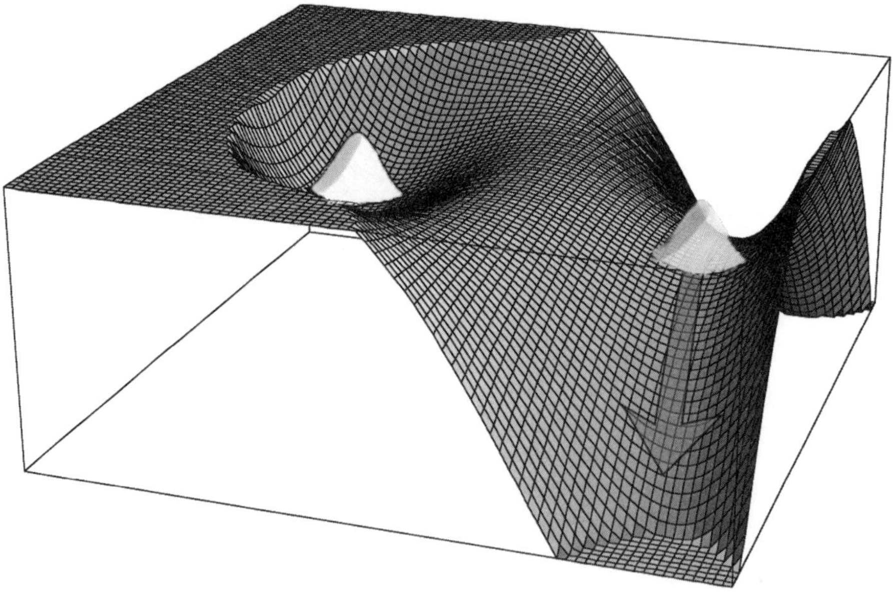

Figure 13.8. A two-dimensional barrier-saddle potential that can be tuned to generate partially resonant decay. Because of an energy shift arising from the downhill saddle, the attractive part of the potential on the right needs to be adjusted to bring the left-side tunneling state into degeneracy with the right-side resonance state, since the *y* zero point energy on the left is absent on the right. That done, the coherent part of the decay diminishes as the repulsive part of the saddle is made steeper.

an associated complex classical action and a concomitant exponential damping. Or, in some cases a perturbation treatment may be applied, where the perturbation is a term in the Hamiltonian that enables the tunneling. When the classical forbiddance is not caused by potential energy barriers and instead is dynamical in nature, we call it *dynamical tunneling* [93]. It is difficult to say who did the first dynamical tunneling calculation, but one could guess it was Huygens (because diffraction is a form of dynamical tunneling). The first molecular dynamical tunneling calculation, on H_2O, was probably by Lawton and Child [96]. An excellent, often detailed reference for many of the parts of this book is Child, "Semiclassical Methods in Molecular Scattering and Spectroscopy" [97].

ANOTHER ASPECT OF RESONANT TUNNELING

The distinction between what is quantum tunneling and what is classically allowed need not be perfectly sharp. Suppose there is a small hole in a wall, where trajectories may leak through at a certain rate. The quantum rate of escape through this hole can be, smaller, or much larger. When it is larger, the situation still qualifies as quantum tunneling.

This is just what happens in the Westervelt resonator (figure 13.9). Such a resonator has a source of wave amplitude (at the top of figure), illuminating two very small holes in the horizontal wall that are barely visible in the figure. On the left, there is an empty chamber (no resonator) with absorptive walls that simply lets any flux emerging from the whole escape. On the right, where we see far more amplitude, there is a Westervelt

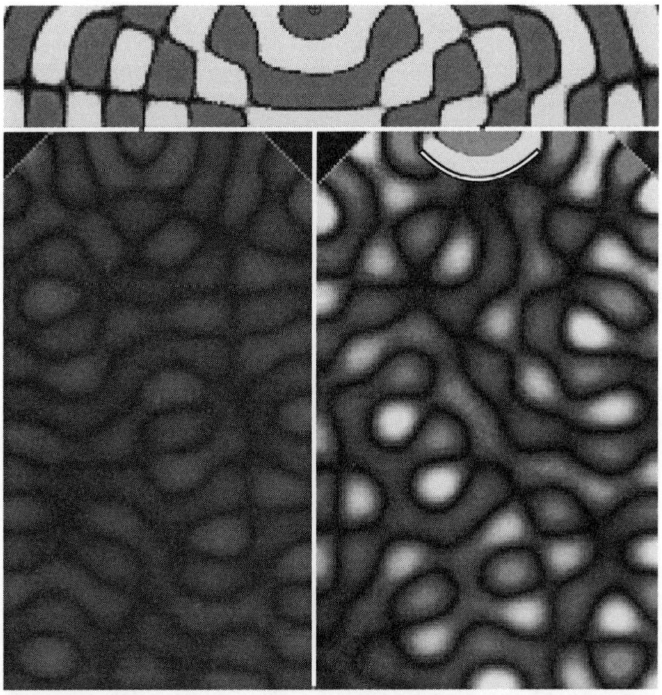

Figure 13.9. A source at the top illuminates two identical small holes through the horizontal wall, below which there are identical chambers. The walls are absorptive on both sides and the bottom (but not the top wall, which is 100% reflective). On the right, this is also a hard semicircular mirror seen just below the hole. At the energy shown, there is a resonance buildup between the wall and the mirror, as evidenced by the enhanced amplitude everywhere in that chamber. This resonance enhances the flux of quantum amplitude leaking through the hole, because fresh amplitude emerging from the hole finds itself immersed in constructively interfering amplitude, enhancing the probability of emerging. On resonance, much more quantum flux gets through the hole than it does classically. This is a case of quantum resonant tunneling, even though there is a dribble of trajectories in the analogous classical system. This is a snapshot of the waves propagating from the source and through the holes and finally to absorbing walls; it is not a standing wave.

resonator, forming a cavity with a reflecting mirror near the hole, but otherwise the same chamber. Classically, the mirror would only lower the flux of particles through the hole, by causing some trajectories to return back through it. But at this energy quantum mechanically, there is a resonance buildup between the semicircular mirror and the wall, causing large amplitude waves in-phase with amplitude escaping through the hole coming from the source. The influx is reinforced, adding constructively to the amplitude inside the resonator. The flux through the hole is well above the quantum rate without the resonator and well above the classical rate as well. Resonant tunneling is taking place, and even though there is a dribble of flux in the classical system, dynamical tunneling is dominating.

13.2 Rigid Asymmetric Top—Showcase of Dynamical Tunneling

The classical problem of a rigid asymmetric top spinning in free space applies widely, although a truly "rigid" body is at best an approximation in real systems, and at worst not even close to the truth.[1]

The best treatment of and insight into the classical, quantum, and semiclassical problem of a rigid asymmetric top is given by Harter and Patterson [98], and subsequent work by Harter. See also the important work by Colwell, Handy, and Miller [99], which is notable for its uniformization of the primitive semiclassical values, correcting for barrier tunneling and reflection above the barrier. The Harter-Patterson method of tunneling on the rotational energy surface (defined shortly) is however a more elegant way to proceed.

Classically, the rotation energy is given by the sum (where x, y, and z are principal axes of inertia in the body frame)

$$E = \frac{J_x^2}{2I_x} + \frac{J_y^2}{2I_y} + \frac{J_z^2}{2I_z} \equiv AJ_x^2 + BJ_y^2 + CJ_z^2, \tag{13.10}$$

where I_x is the moment of inertia about the principal x axis, and J_x is the angular momentum about the x axis, and so on. A, B, C are the rotational constants.

We begin with the definition of Harter's rotational energy surface, a construct that is of great utility. The rotational energy surface is a radial plot of the classical rotational energy as a function of the direction of the classical angular momentum $\boldsymbol{J}$ in the body frame, for fixed $|\boldsymbol{J}|$. Essentially, one "skewers" the body through the center of mass with a rod at some angle with respect to the principal axes, and spins up the body around this axis until angular momentum magnitude $|\boldsymbol{J}|$ is reached. The energy required for this spin-up is used as the length of a vector from the center of mass to the rotational energy (RE) surface, at the angle of the skewer. An RE surface and much more are shown for a triatomic, rigid water-like molecule in figure 13.10. We will here just scratch the surface of what Harter and co-workers have done to analyze the spectra not only of the rigid rotor, but also deformable, nonrigid systems like the SF_6 molecule, through a powerful group theoretical formalism leading to great insight about the hierarchy of spectral clusters seen in experiments.

The ubiquitous separatrix has shown up again, two of them in fact, with their stable and unstable axes joined (smoothly, because this is an integrable system) and symmetrically on opposite sides of the RE surface. The unstable fixed points correspond to the difficulty in spinning asymmetric tops near their unstable axis of rotation.

Tunneling across the separatrices is a big part of the quantum dynamics and the spectrum of the rigid asymmetric top. *Classically, it corresponds to an impossible end-for-end flipping, as seen in figure 13.10.* As with a barrier in 1D, there are two types of classically forbidden paths across the separatrix. For the barrier, there is under the barrier potential tunneling and over the barrier reflection or dynamical tunneling. Here, both paths across the separatrix may be viewed as dynamical tunneling. We cannot go into full detail, except to refer to the source [98] and to quote major results. Formulas for the tunnel splitting should look familiar. For the most asymmetric top,

[1] An example of a spinning body that is rigid for practical purposes is seen in the NASA video at https://youtu.be/1n-HMSCDYtM, and some that are nonrigid, quickly finding their large moment of inertia spin axis due to internal friction (see story surrounding figure 13.12, later) are seen at https://youtu.be/BPMjcN-sBJ4.

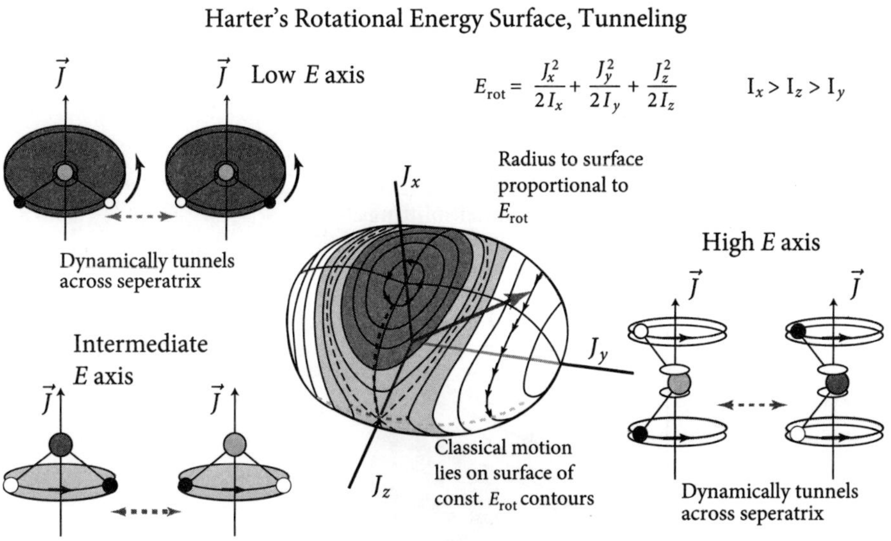

Figure 13.10. The rotational energy surface for A = 0.2, B = 0.4, and C = 0.6 and angular momentum J = 10. The distance of the rotational energy surface from the origin is given by the energy of rotation $E = \frac{J_x^2}{2I_x} + \frac{J_y^2}{2I_y} + \frac{J_z^2}{2I_z}$ for spinning up the rotor to $J = 10$ about the vector with components (J_x, J_y, J_z) and $J_x^2 + J_y^2 + J_z^2 = 100$. Representative motion is shown for an object (a triatomic molecule) with these rotational constants and spinning near the three principal moment of inertia axes. If the system is started with a broken symmetry, by using a linear combination of tunneling doublets, with the light atom leading the rotation as shown in the upper left, dynamical tunneling will lead to the classically impossible situation (with no torques and fixed $\vec{J}$) of the dark atom becoming the leading atom, cycling back and forth between these extremes as with any two state tunneling problem. The same can be said about the cycling of which atom is on top at the lower right, starting with one of them definitely on top. The lower-left case is more complicated, since there is a classical instability flipping the central atom up and down relative to the side atoms, unless it is started exactly at its fixed point (essentially impossible in practice). Moreover, there is no classical difference between the two cases shown; one becomes the other in half of a classical tumble. The tunneling can be dramatic only for initiation some distance from the unstable fixed point, and is hard to draw. The dynamical tunneling then is facile, traveling only a short distance across the separatrix.

B = 0.4, with A = 0.2 and C = 0.6, the tunneling integral is

$$S_n = \frac{e^{-|\theta|}}{T}, \tag{13.11}$$

where T is the classical period on the level curves connected by tunneling, and

$$\theta = i \int_{path} J_\gamma(E_n)\, d\gamma, \tag{13.12}$$

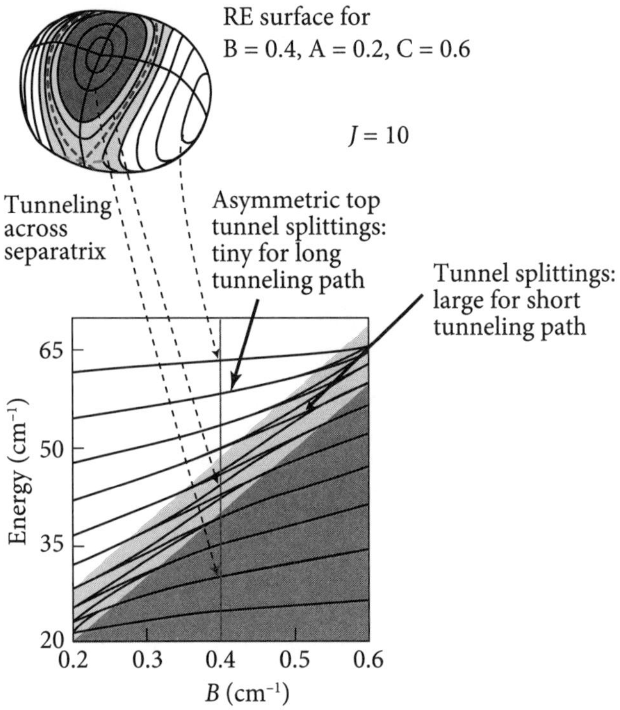

Figure 13.11. Asymmetric top energy levels as a function of the intermediate rotational constant B, for A = 0.2, C = 0.6. Except in the light strip (corresponding to motion near the separatrix), the levels are narrowly split doublets, so close together that they appear to be a single line. All the splittings are attributable to dynamical tunneling as shown in figure 13.10.

where

$$J_\gamma = \sqrt{\frac{J^2(A\cos^2\gamma + B\sin^2\gamma) - E)}{[A\cos^2\gamma + B\sin^2\gamma - C]}},$$

and γ is the usual (body fixed) Euler angle. This accurately gives the splitting energy, 4.44×10^{-1} cm^{-1} quantum mechanically, and 4.8×10^{-1} cm^{-1} semiclassically, for the sixth cluster of close eigenvalues (see figure 13.11).

SEMIRIGID TOPS

The year 1958 marked the United States' first successful satellite launch into orbit, *Explorer 1.* As the press conference shown in figure 13.12 took place, the satellite was busily tumbling out of control. James van Allen, seen in the middle of the photograph, soon realized that the intermittent signal from the satellite was due to its unplanned tumbling. Fortunately, enough antennas were bristling from the satellite that it still gave much useful data, resulting in discovery of the van Allen radiation belts. The satellite was set to rotate around its long axis. The tumbling began because internal friction caused by initially slight wobbling (and probably hastened by the whipping antennas) is converted to heat, lowering the rotational energy, but leaving

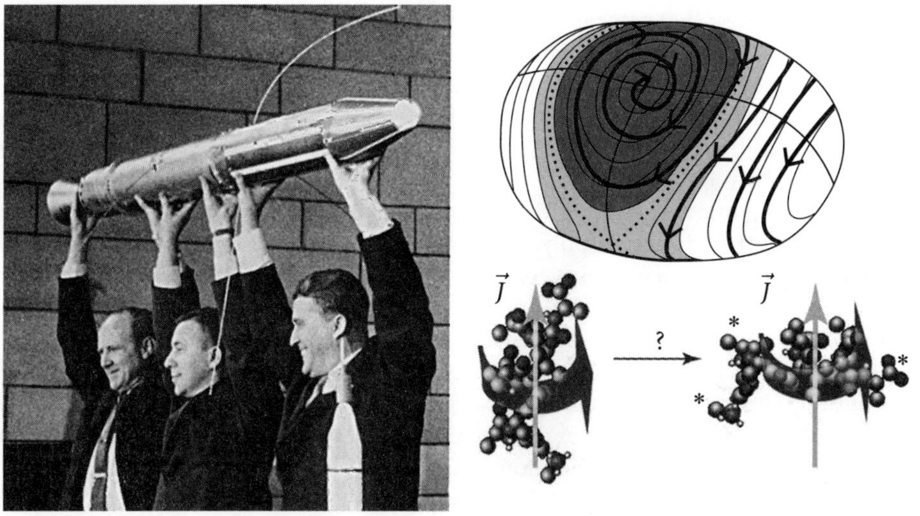

Figure 13.12. A NASA-JPL blunder early in the US space program. Satellites are not rigid bodies, especially with floppy whip antennas attached. The *Explorer 1* satellite was the first launched successfully by the United States. Seen in the left panel are James van Allen (center), William Pickering (left), and Werner von Braun, with a full-size model of the satellite, just after it was successfully orbited in 1958. As this press conference took place, the satellite was busily tumbling out of control. Fortunately, enough antennas were bristling from the satellite that it still gave much useful data, resulting in discovery of the van Allen radiation belts. (The author thanks Prof. William Harter for pointing out the existence and the physics of this story.) In the right panel at the top, we see a dissipative path on a rotational energy surface. Without dissipation, the trajectory would stay on the level curves shown as thin black closed curves lying on the surface. The dissipative path (thick black line) lies on the rotational energy surface because it applies to fixed $\vec{J}$, which was the case for the satellite in space with no external torques. At the lower right, the question is posed: how big does a molecule have to be dump energy from rotation to vibration? We suggest that as size increases, the first mechanism by which this relaxation can take place is dynamical tunneling. The reason is that the classical resonance relaxation pathways may not exist, and the molecule's motion may be partially quasi-periodic. To be able to dynamically tunnel, the molecule has to have a certain threshold density of states, in order to have states corresponding to vibration + tumbling degenerate with the initial quantum level corresponding to spinning around a low moment of inertial axis.

the total space-fixed angular momentum fixed. The process accelerates as the wobbling grows worse. The satellite must rotate around ever lower rotational energy axes, until it bottoms out in end-over-end tumbling rotation, which has the lowest possible rotational energy for the given angular momentum. See reference [100].

Figure 13.12, right, reveals the fascinating and important consequences of some flexure and friction in a rotor. Note the track of declining energy on the rotational energy surface. The whole surface has the same fixed angular momentum, but the radius from the origin is given by the rotational energy, which is not fixed. The same relaxation of rotational energy for fixed angular momentum renders Saturn's rings flat, and makes the solar system lie close to the plane of the ecliptic. Collisions of primordial bodies generate heat that gets radiated away, lowering energy. The angular momentum is fixed though, and the result is a disk-like object on the path to minimizing its rotational energy for a given value of angular momentum. Collisions must eventually stop as the bodies adopt distinct orbits, and finally further rotational energy loss ceases. It must, because there is a minimum amount of rotational energy required to maintain the fixed angular momentum.

An intriguing question is: does the spontaneous dumping of rotational energy into internal vibrations happen in polyatomic molecules, or other very small things like nanoparticles or microscopic dust grains? These after all are not rigid rotors either. When does "friction" play a role? For small systems friction amounts to transfer of energy to vibrational modes so numerous that the pure rotations are essentially never reached again, because of entropy or phase space arguments. The density of states in the flat spin is huge, compared to what was spinning around the axis of lowest moment of inertia. Much of the rotational energy, initially trapped in a rather simple motion, has been put into myriads of vibrations, making available much more phase space—essentially in a drive toward increasing entropy.

It is not certain yet, but as size increases it is likely that this process will begin to happen by quantum dynamical tunneling before it can happen classically for the same molecular Hamiltonian. The classical relaxation on the Born-Oppenheimer potential energy surface for small molecules may be inhibited by lack of classical resonances, since rotation is so slow compared to molecular vibration, unless the molecule gets quite large and floppy. Before that, accidental near-degeneracies lead to quantum tunneling between states of different rotational energy that cannot communicate classically (the states have the same total energy and angular momentum, however). The tendency of moderate sized polyatomic molecules to spend almost no time spinning around their low moment of inertia axis, even if launched that way, heralds the onset of statistical, even pseudo-thermodynamic behavior.

13.3 Resonance and Dynamical Tunneling in Multidimensional Systems

Here, we rely on some of what we learned about action-angle variables, integrable systems, and use the 1D paradigm of "above-barrier reflection" we studied for guidance. Some of the chapter 4 discussion on classical resonance is carried over to quantum systems, including dynamical tunneling, induced by classical resonances. The classical resonances enable quantum tunneling to dramatically different sets of actions with no corresponding classical activity.

A typical scenario is shown in figure 13.13; on the left is a Poincaré surface of section, revealing resonance zones lying between relatively unperturbed regions. The resonance zones are themselves mostly integrable; however, a canonical transformation is required to find the new constant actions inside them. This problem has received considerable attention from a semiclassical perspective [91, 101–105].

The fundamental "primitive" semiclassical way to quantize an integrable system of several degrees of freedom (with a phase space sliced by invariant tori) is the extension of the Bohr quantization rule, equation 12.30, to quantize each of the integrals

$$\oint \vec{p} \cdot d\vec{q} = \left(n + \frac{1}{2}\right) h \tag{13.13}$$

around each topologically independent path.

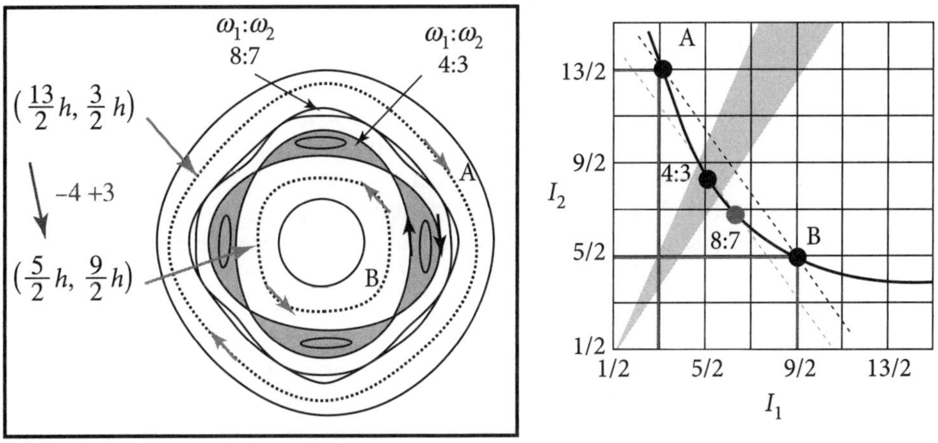

Figure 13.13. A thick curve for $E(I_1, I_2) =$ constant is shown at the right, passing though two semiclassically quantized action pairs, A $= (I_2, I_1) = (13/2h, 3/2h)$ and B $= (I_2, I_1) = (5/2h, 9/2h)$. These sets of actions are related by an increase of $\Delta I_2 = 4h$ and a decline of $\Delta I_1 = -3h$, a 4:3 exchange of action (and quantum numbers). This primitive semiclassical pair is "accidentally" degenerate: there are two pairs of actions satisfying the Einstein-Brillouin-Keller (EBK) quantization condition—that is both actions I_1 and I_2 are (integer + 1/2) h at the same classical energy. A full quantum calculation shows the degeneracy is lifted, indicating dynamical tunneling between the distinct classical motions corresponding to the two sets of quantized actions, shown as closed dashed curves on the Poincaré surface of section on the left. The mean value theorem (see text) guarantees that a 4:3 resonance zone must lie somewhere between the quantized actions on the black curve. The splitting can also be calculated semiclassically, by performing the action integral connecting the tori, or more conveniently, extracting the term responsible for the resonance and treating it as a perturbation. The relevance of the dynamical tunneling above-barrier reflection model problems (section 13.1) is evident.

Figure 13.13 could belong to a Hamiltonian such as

$$H(\boldsymbol{I}, \boldsymbol{\theta}) = \omega_1^0 I_1 + \beta_1 I_1^2 + \omega_2^0 I_2 + \beta_2 I_2^2 + \lambda I_1 I_2 \; v(\theta_1, \theta_2), \tag{13.14}$$

$$\equiv H_0(I_1, I_2) + V(I_1, I_2, \theta_1, \theta_2), \tag{13.15}$$

where $v(\theta_1, \theta_2)$ is periodic in the angle variables θ_i. The base frequencies are given by the parameters ω_i^0 and the diagonal anharmonicity is controlled by β_i. The potential has the expansion

$$v(\theta_1, \theta_2) = \sum_p \sum_{n_1, n_2}{}' v_{n_1, n_2}^{(p)} \exp[ip\,(n_1\theta_1 - n_2\theta_2)], \tag{13.16}$$

where (n_1, n_2) are relatively prime. The unperturbed frequencies are

$$\omega_i = \frac{\partial H_0(I_1, I_2)}{\partial I_i}. \tag{13.17}$$

A thick line giving the location of $E(I_1, I_2) =$ constant is shown at the right in figure 13.13. It is seen passing though two semiclassically quantized action pairs, $(I_2, I_1) = (5/2h, 9/2h)$ and $(13/2h, 3/2h)$. These two sets of actions are related by an increase of $\Delta I_2 = 4h$ and a decline of $\Delta I_1 = -3h$, a 4:3 exchange of quantum numbers.

Two different sets of actions that EBK quantize at the same classical energy can be called "accidentally" degenerate semiclassically. The frequencies are a function of the actions. The fact that the thick constant energy curve on the right passes through the dots at $\mathrm{B} = (5/2h, 9/2h)$ and $\mathrm{A} = (13/2h, 3/2h)$ with a 4:3 resonance in between is a consequence of the mean value theorem (given a planar arc between two endpoints, there is at least one point at which the tangent to the arc is parallel to the secant through its endpoints). Then, the perturbation term involving $\cos(3\theta_1 - 4\theta_2)$ is slow and can have profound effects of the dynamics: a 4:3 resonance zone is activated (see figure 13.13). For reasonable perturbations $\eta V(\boldsymbol{I}, \boldsymbol{\theta})$, the higher Fourier components $v^{(p)}_{n_1,n_2}$ will generally fall off rapidly with p, so for example the $\omega_1 : \omega_2 = 8{:}6$ resonance zone is expected to be smaller. Dashed lines in figure 13.13, left, lie on two classically nonresonant tori (sliced by the surface of section) that are both action quantized at the same energy—that is, both carry integer + 1/2 units of Planck's constant inside their boundaries at this energy. This is rare in such a low-dimensional system, and represents something that can be arranged if a parameter is varied in H_0, causing the semiclassically quantized energy levels of H_0 to cross—that is, $H = E(13/2h, 3/2h) = E(5/2h, 9/2h)$ at that value of the parameter. Figure 13.13, left, shows the topology and consequences of this tunneling. Notice that the frequency ratios $\omega_1 : \omega_2$ are not special at these quantizing actions, with $\omega_1/\omega_2 < 4{:}3$ at B and $\omega_1/\omega_2 > 4{:}3$ at A. Now, $dE = 0 = (\partial H/\partial I_1)\ \delta I_1 + (\partial H/\partial I_2)\ \delta I_2 = \omega_1 \delta I_1 + \omega_2 \delta I_2$ and by the mean value theorem $\delta I_2/\delta I_1$ must reach $= -4/3$ at one point at least, meaning the frequency ratio must reach 4:3. Clearly, infinitely many other rational ratios ω_1/ω_2 will be attained between A and B, most giving extremely narrow resonances and involving large integers, but only the 4:3 resonance matters for connecting the specified actions, at least if the connection is direct (for indirect connection, see [91]).

We see from the gray wedge that the 4:3 resonance zone applies over both higher and lower ranges of action, with the resonance zone growing in width as the perturbation term typically becomes stronger. The action-angle variables that apply to A and B do not apply inside the 4:3 resonance, and instead there is a set of linked unstable and stable manifolds, fixed points, and islands, as seen in figure 13.13.

The phase space structure is like that of a particle on a ring, but with four symmetrically placed barriers instead of one. The circulation of successive intersections with the surface of section is indicated with arrows in figure 13.13, left. The motion is isomorphic with above-barrier reflection. Therefore, we have the choice of perturbation treatment of the tunnel coupling of the $\mathrm{B} = (5/2h, 9/2h)$ and $\mathrm{A} = (13/2h, 3/2h)$ tori, or alternatively an action integral, connecting the quantized tori across the seperatrix, the same choice we made in connection with above-barrier reflection in section 13.1. The tunneling interaction will cause a splitting of the primitive degenerate levels. If the parameter mentioned above is varied, we will notice an avoided crossing as a function of that parameter, whereas without considering the dynamical tunneling there is a crossing of levels at the degeneracy point.

Suppose two states with quantum numbers (1,5) and (4,1) are involved in an avoided crossing. The resonance zone appears between the tori involved in the tunneling for the reason seen in figure 13.13. A 4:3 resonance mediates the tunneling between the A torus with actions (1+1/2, 5+1/2) = (3/2,11/2) and the B torus with actions (9/2, 3/2). The heavy line delineates the level curve of constant energy in action space; note that $E_A = E_B$. By the mean value theorem, somewhere between points A and B must lie actions where the slope S of the constant energy curve is equal to the slope of the line connecting the two sets of actions.

$$-S = \frac{\omega_1}{\omega_2} = -\frac{\partial H/\partial I_1}{\partial H/\partial I_2} = -\left.\frac{\partial I_2}{\partial I_1}\right|_E; \tag{13.18}$$

this is denoted by 4:3 and a black dot. At this point on the level curve, the tangent is parallel to the line drawn between A and B.

$S = -4/3$ corresponds to the zone in classical phase space where the resonance (4:3 in this example) exists. In this way we can always find an appropriate resonance zone that will connect two tori by over-the-barrier dynamical tunneling.

Even if the quantum numbers of the states involved in the avoided crossing are very different, say differing by n quanta in action I_1 and m quanta in I_2, there is in general an n:m classical resonance island chain lying between the tori corresponding to the states that can be the agent of the quantum tunneling. In a generic coupled but nearly integrable phase space, an island chain will exist for every rational winding number. In this sense classical phase space structures are the cause even of very narrow avoided crossings between states of very different character. We caution, however, that the direct coupling may be weaker than indirect paths involving intermediating states. See reference [91] and later.

CALCULATION OF THE TUNNELING INTERACTION

In the case of a 2:2 resonance, canonical transformation to new action-angle coordinates $(J_1, J_2, \phi_1, \phi_2)$ is accomplished via the generator

$$F(\boldsymbol{J}, \boldsymbol{\theta}) = J_1(\theta_1 - \theta_2) + J_2(\theta_1 + \theta_2), \tag{13.19}$$

giving

$$\begin{aligned} \phi_1 &= (\theta_1 - \theta_2) \text{ and } \phi_2 = (\theta_1 + \theta_2), \\ J_1 &= \frac{1}{2}(I_1 - I_2) \text{ and } J_2 = \frac{1}{2}(I_1 + I_2). \end{aligned} \tag{13.20}$$

The Hamiltonian in the new coordinates is (taking only a pair of perturbation terms, $v_{1:1}^{(\pm 2)}$) (see equation 13.16)

$$\begin{aligned} H &= \omega^- J_1 + \omega^+ J_2 + \beta^+ J_1^2 + \beta^+ J_2^2 + 2\beta^- J_1 J_2 + 2\lambda(J_2^2 - J_1^2)\, v_{1:1}^{(2)} \cos(4\phi_1) \\ &\equiv H_0(J_1, J_2) + 2\lambda(J_2^2 - J_1^2)\, v_{1:1}^{(2)} \cos(2\phi_1), \end{aligned} \tag{13.21}$$

where $\omega^{\pm} = (\omega_1^0 \pm \omega_2^0)/2$ and $\beta^{\pm} = (\beta_1 \pm \beta_2)$. It is clear that J_2 is a constant of the motion because ϕ_2 is absent in H. Now suppose the tori labeled A and B have exactly the same energy at the primitive (EBK) level—that is, $H(I_1^A, I_2^A) = H(I_1^B, I_2^B)$ where $I_1^A = (n_1^A + \alpha_1^A)h$, and so on, where n_1^A is an integer and α_1^A arises from the well-known Maslov phase corrections.

There are several ways to calculate the tunneling between the tori caused by the intervening resonance zone. They all begin with a resonance analysis as just described. In two closely related approaches, one gets the tunneling by stopping short of the fully semiclassical analysis, substituting a little quantum mechanics into the one dimensional effective Hamiltonian produced by the resonance analysis. This produces a "uniform" approximation; in essence one is doing full quantum mechanics on a classically pre-processed Hamiltonian. One approach quantizes the (J_1, ϕ_1) Hamiltonian for fixed J_2 with the Ansatz $J_1 \to -i\hbar d/d\phi_1 + \hbar$ [102, 106]. The resulting Schrödinger equation is then solved numerically, or analytically if possible. The pendulum-like Hamiltonian in equation 13.21 contains the nonclassical below barrier tunneling (and above barrier reflection). A closely related scheme uses the Heisenberg (matrix) formulation of quantum mechanics by finding matrix elements of the resonant interaction term and diagonalizing. Bohr correspondence is invoked in the following way: the resonant term depends on both actions and angles of the nonresonant part of the Hamiltonian, as in equation 13.14. The actions appearing in the resonant term are set to their mean values $(\bar{J}_1, \bar{J}_2)$ and the off diagonal tunneling matrix element becomes

$$\langle J_1, J_2 | H | J_1', J_2' \rangle = \frac{1}{(2\pi)^2} \int\int e^{i\, p(n_1\theta_1 - n_2\theta_2)} V(\bar{J}_1, \bar{J}_2, \theta_1, \theta_2)\, d\theta_1\, d\theta_2$$
$$\equiv V(\bar{J}_1, \bar{J}_2)^{(p)}_{n1,n2}, \tag{13.22}$$

with $\bar{J}_1 = (J_1 + J_1')/2$, and so on. The method is uniform since the matrix of EBK diagonal energies and off-diagonal couplings is numerically diagonalized. An example of the use of a procedure much like that just described is found in Roberts and Jaffe [103]. The key point is the reduction of the problem to an isolated pendulum-like Hamiltonian in one degree of freedom. Even in the presence of other pertubations of different order, the isolation of one resonance may suffice, justified by the notion of averaging over the others.

AVOIDED CROSSING OWING TO DYNAMICAL TUNNELING

Figure 13.14 shows a succession of three values of a parameter λ in a 2D anharmonic Hamiltonian. The states are degenerate in zero order for $\lambda = 0.5$. This is a clear case of an avoided crossing due to dynamical tunneling: the classical analog states are evident on either side of the figure, and the mixing of them by tunneling interactions is obvious in the middle pair. The middle pair of eigenstates is not degenerate but split by a small amount, at the avoid crossing. A transfer of spectral intensity occurs if one of the two states carries most or all of the oscillator strength (transition intensity) before the crossing.

Dynamical tunneling interactions and their consequent avoided crossings cause mixing of states that had been living on classical structures before the mixing. An example is given by the Hamiltonian

$$H = \frac{p_x^2}{2} + \frac{p_y^2}{2} + \frac{1}{2}\omega_x^2 x^2 + \frac{1}{2}\omega_y^2 y^2 + \gamma x y^2, \tag{13.23}$$

where ω_x is used as the variable parameter. As ω_x is increased, all the energy levels rise, but obvious collisions are taking place (see figure 13.15). We suppose that one unmixed state carries most of the oscillator strength in a spectroscopic transition. Labeling the oscillator strength as dark, we see an admixture of shades develop as a function of ω_x. There is a corresponding dilution of a single spectral line into three separate lines. If ω_x continues to increase however, the dark character of one of the curves returns and a single line is seen again in the spectrum.

This example heralds a much wider and more general class of problems, where many more than two or three states may be involved simultaneously. A nonstationary state, which happens by its nature to carry the oscillator (optical transition moment)

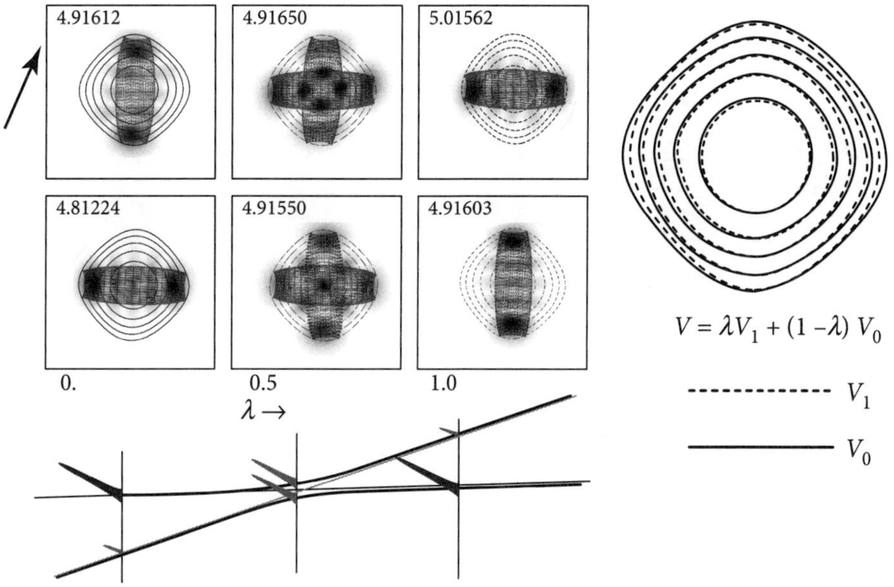

Figure 13.14. An avoided crossing due to dynamical tunneling. The potential is varied in a way that "squeezes" the left-right trending, integrable trajectory much more than the up-down one, raising the left-right primitive (uncorrected for tunneling) semiclassical energy almost linearly as λ increases from 0 to 1, and leaving the up-down primitive semiclassical energy almost untouched. The evolution of the energies with λ is shown at the bottom, along with a spectrum supposing the transition moment (arrow) is aligned with the mostly up-down direction. When the tunneling interaction becomes comparable to the zeroth order splitting, the zero order states mix and the level curves avoid each other by the tunnel splitting, becoming a 50-50 mix of the unperturbed states when the potential has a symmetry making the two semiclassical states degenerate, at $\lambda = 0.5$. The two extreme potentials, for $\lambda = 0$ and $\lambda = 1$, are shown at the right. The classical orbits corresponding to the primitive semiclassical (not tunneling corrected) states are shown in the six panels at the upper left. Also shown in the same panels are the lower and higher energy eigenstate probability densities at $\lambda = 0$, 0.5, and 1.0. At the avoided crossing ($\lambda = 0.5$), the exact eigenstates are seen to be mixtures (tunneling pairs) of the two types of pure classical motion.

strength, is coupled by interactions (be it tunneling, or classically allowed mode mixing, etc.) into other states. These are in turn coupled to more states, etc. The result is a fractionated spectrum of essentially unassignable lines. This has important experimental implications, since not all zero order (pretunneling) states are equal when it comes to coupling to the outside world. If tunneling interactions mix a state carrying a strong dipole into many others nearby in energy, then a strong solitary spectral line may be shared by many nearby eigenstates, all of them having some of the character of the original state; see figure 13.16.

Several resonances of higher order may conspire to make tunneling jumps shorter and the overall tunneling across the main separatrix much more facile than direct tunneling (figure 13.17). This *resonance assisted* tunneling mechanism may be more common than direct processes.

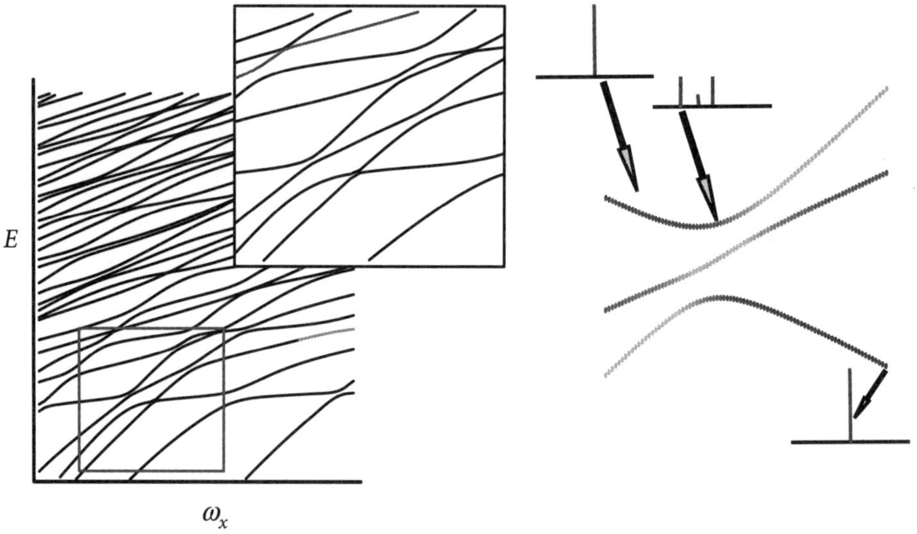

Figure 13.15. (Left) Energy levels as a function of ω_x for the Hamiltonian, equation 13.23. (Right) Avoided crossing involving three states as a function of a parameter, in this case causing dilution of dipole oscillator strength belonging to one of them seen as dark on the left, before the mixing began (dipole strength indicated by relative darkness of the lines). At the far right, the mixing has again become negligible and the state with strong oscillator strength has recovered its character.

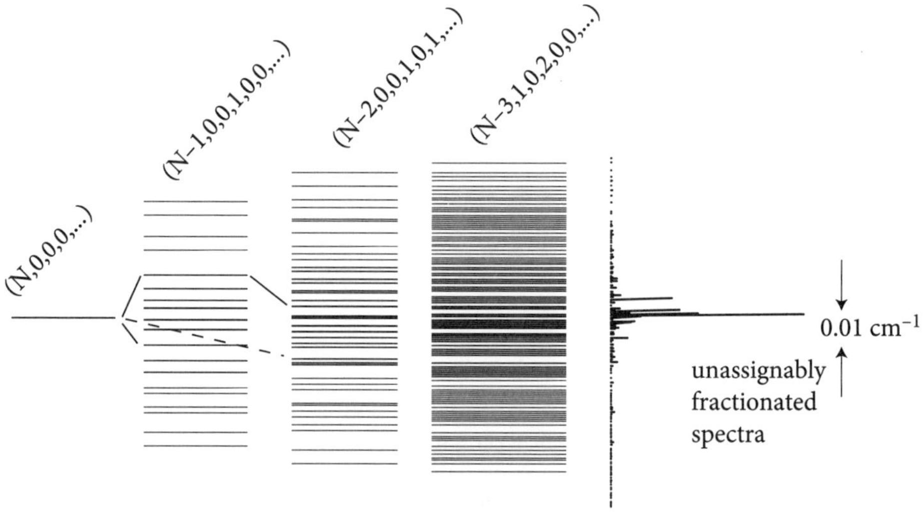

Figure 13.16. A hierarchy of tunneling interactions starting from a state with large oscillator strength leads finally to a fractionated spectrum of narrow bandwidth, the narrowness reflecting the weakness of the tunneling interactions. This kind of spectral fractionation is not common in two degree of freedom systems, since their density of states is typically insufficient. It should, however, be commonplace in systems with a few degrees of freedom, at moderate energies where classical chaos is not strong or dominant but the densities of states are high.

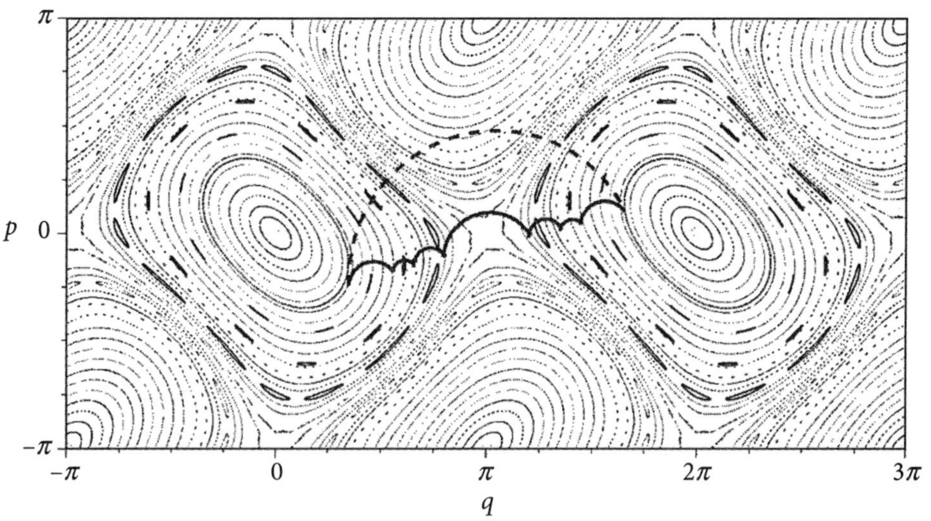

Figure 13.17. Resonance-assisted tunneling. Here we see the possibility of several resonances of higher order conspiring to make tunneling jumps shorter and the overall tunneling across the main separatrix much more facile. The example comes from the kicked Harper model. The dashed line is shown for reference; it is a direct path giving much weaker tunneling than the indirect, multihop one. Reprinted with permission from Elsevier from the work cited in reference [91].

CHAOS- AND RESONANCE-ASSISTED TUNNELING

When two energy levels closely avoid each other as a function of a parameter in the presence of other levels, it is tempting to think, and implied by the above discussions, that those levels are directly connected to each other, most likely by tunneling. However, the presence of other levels much further away may dominate the coupling and tunneling of the pair under focus [91, 107]. In fact, the presence of classical chaos nearby in phase space, or underlying classical resonances may be primarily responsible for the strength of the coupling between two seemingly isolated states.

At the simplest level, this can be seen by considering a 3×3 matrix problem, shown in two forms in figure 13.18.

CLASSICALLY FORBIDDEN QUANTUM RADIATIVE TRANSITIONS

Suppose we apply weak monochromatic radiation of energy $h\nu$ to an anharmonic oscillator in its ground state, resonant (in the quantum sense) with a high lying eigenstate of the oscillator with $h\nu \gg \hbar\omega$, where ω is the local frequency of the oscillator near its minimum. We assign a harmonic oscillator of angular frequency $2\pi\nu$ to the field, as in quantum electrodynamics, and treat it as a degree of freedom coupled to the anharmonic potential. This spectroscopic *overtone transition*, as it is called, is classically forbidden. The frequency of the radiation is much too high and nonresonant to pump much energy in or out of the oscillator classically. Essentially, the oscillator just shivers at frequency $2\pi\nu$ near the bottom of the well. However the transition is enabled by quantum dynamical tunneling. See figure 13.19.

$$\begin{pmatrix} 1 & -0.01 & 0 \\ -0.01 & 0.5x+1 & 0 \\ 0 & 0 & 1.75 \end{pmatrix} \qquad \begin{pmatrix} 1 & -0.01 & 0.15 \\ -0.01 & 0.5x+1 & 0.15 \\ 0.15 & 0.15 & 1.75 \end{pmatrix}$$

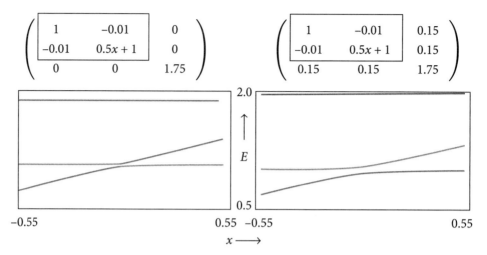

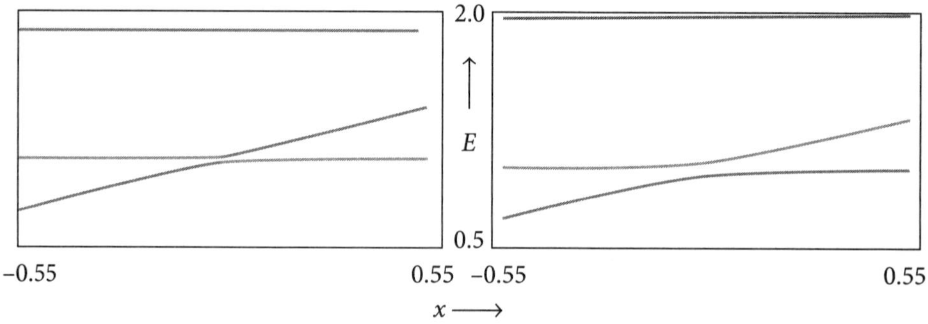

Figure 13.18. On the left, two levels are coupled by a constant 0.01, as a parameter is changed and their unperturbed energies intersect and the eigenenergies avoid by a gap of 0.02. The third state, above, is uncoupled to the lower two. On the right, the coupling is the same, except that the third state, lying quite far away, is coupled by a constant 0.15 to each lower state. It is seen that the avoided gap on the right-hand side is now quite a bit larger. Both of the lower states have acquired a component in the upper one, causing their mutual coupling to be much larger than the previous 0.01 direct coupling. In the current context, the upper state could belong to a chaotic zone, or alternately be related by a classical resonance to each of the first two. The figure and indeed this topic shows that narrow avoided crossings are by no means universally governed by direct coupling. The minimum gap with the coupling to the third state included is now 0.075. See "Resonance-assisted Tunneling," by Brodier, Schlagheck, and Ullmo, reference [91].

Suppose there are four quanta in the field mode, and zero in the oscillator initially. This state is by design almost degenerate with three quanta in the field and four in the oscillator. Although this implies the existence of a 4:1 resonance in the classical phase space (see below), the zone lies between the initial and final invariant tori, well out of the way. Nonetheless, it mediates the 4:3 quantum exchange.

In figure 13.19, we sketch a two-dimensional oscillator potential illustrating the low frequency mostly x-direction (horizontal) motion of the anharmonic potential and the much tighter high-frequency y motion of the field oscillator. Eigenstates corresponding to the initial and final state as described earlier are shown. Classically, the initial motion with all four quanta in the field mode and ground state energy in the oscillator is indicated by the up-down arrow, and the final state motion with three quanta in the field and four in the anharmonic oscillator is indicated by the right-left arrow. These two motions do not communicate classically, but quantum mechanically there is a nonvanishing matrix element connecting them. Figure 13.19, right, displays the phase space view of this, with the convention that a $y = 0$ (field oscillator) surface of section is constructed, in which x and p_x are plotted every time the trajectory penetrates $y = 0$ with $p_y > 0$. Resonance islands appear corresponding to the 4:1 resonance and higher order resonances.

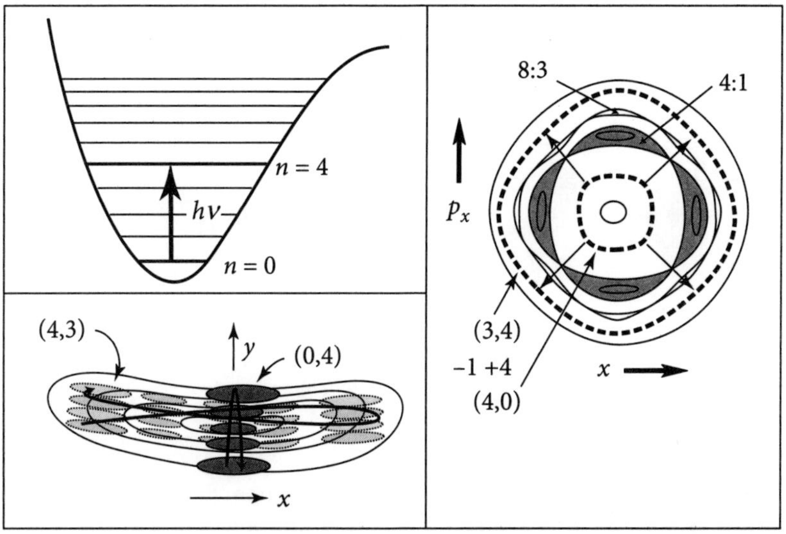

Figure 13.19. (Top left) The energy representation of a one-photon, multiquantum bond oscillator transition. (Bottom left) A coordinate space picture, illustrating the potential (properly treating the quantized electromagnetic field as a harmonic oscillator). The wavefunctions of the field oscillator (vertical coordinate), the anharmonic bond oscillator (horizontal coordinate) are shown for the (0,4) initial state and the (4,3) final state. The qualitative classical motion corresponding to the ground and excited final states is shown. (Right) Poincaré surface of section for the field oscillator-anharmonic oscillator problem, showing resonance zones in phase space corresponding to the 4:1 resonance. The initial torus with 4 quanta in the field oscillator and 0 in the diatomic bond oscillator, (4,0), is almost undistorted from what it would be without the coupling to the field. It corresponds to slow bond oscillation and fast but weak "jitter" of the bond motion due to the field. The jitter does nothing on average, and the classical motion is confined to the torus, corresponding to the classically forbidden nature of the transition.

The relevance of the above-barrier reflection problem to dynamical tunneling is clear: the local phase space structure near the islands is the same as the above-barrier problem. This means that we can use perturbation theory or distorted wave perturbation theory to determine the tunneling interaction between nearly degenerate states with different actions, in this case (4,0) and (3,4), just as we did for the barrier reflection problem.

This discussion applies to many classically non-resonant single photon processes, such as photoionization of helium or the photoelectric effect for electrons on a metal surface. Indeed, the usual textbook discussion of the photoelectric effect, often cited as an example of the need for quantum mechanics, is an example of dynamical tunneling: the photoemission process fails classically under the experimental circumstances of weak, classically non-resonant light, yet it is allowed quantum mechanically.

TUNNELING IN THE TIME DOMAIN

Tunneling is a matter of quantum mechanical amplitude finding its way from one kind of classical motion to another of fixed energy, when the motions do not communicate classically. The quantum mechanical amplitude finds its way from the

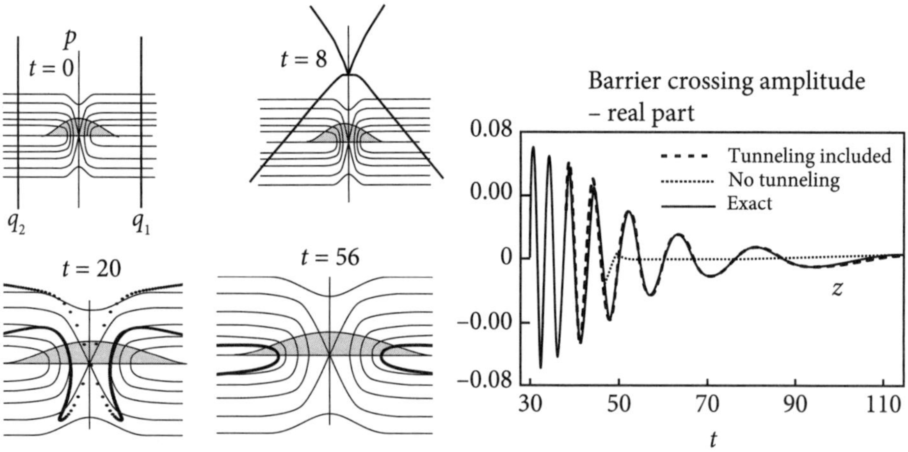

Figure 13.20. How does a time-dependent wavefunction get across a barrier, from one position to another on different sides? Here, the amplitude is split into forward propagation from the left position q_2 for time $t/2$ and backward propagation from the right position q_1 for the same time. This choice yields right-left symmetry, which is more intuitive, and is equivalent to the propagation only from one side for the full time. For early times, at $t = 8$ for example, the amplitude is dominated by trajectories that have enough energy to cross over the barrier. Already by time $t = 20$, the classically allowed contributions have thinned out enormously in the barrier vicinity, and by $t = 56$ their contribution is truly infinitesimal, with none visible in the image. In contrast, the trajectories on either side contributing to under-barrier tunneling are quite dense. At $t = 56$ the Green function for transfer across the barrier is entirely dominated by tunneling contributions under the barrier. The times reported are the forward-only times. Forward-backward times are half of these.

vicinity of one type of motion to the vicinity of the other in the time domain too, but the telltale evidence is revealed also in the energy domain.

A good example is the off-diagonal time domain Green function $G(q_2, q_1; t)$ in the vicinity of a potential barrier [108]. q_2 lies to the left of the barrier, and q_1 lies to the right (figure 13.20). Of course, there are classically allowed paths that connect these two points, corresponding to trajectories with sufficient energy to surmount the barrier; these are included in the manifold that corresponds to specific initial position, since this manifold has infinite momentum uncertainty. Because of the exponential instability near the barrier top however, the classical trajectories that are connecting the two points by passing over the barrier thin out exponentially fast as time progresses.

The tunneling under the barrier must also be considered. This amplitude is also exponentially small. A "battle of the exponentials" is set up: which exponential dominates, the small amplitude from over the barrier trajectories or the small amplitude for tunneling under the barrier? It turns out that tunneling paths start to dominate over classically allowed over-the-barrier paths after some time has elapsed. The situation is depicted in figure 13.20, where inspection of the situation at $t = 56$ reveals that the answer just given is indeed compelling.

Chapter 14

Bixon-Jortner and Fano-Anderson Models

We cannot bypass a chance to discuss two approaches to decay and lineshape problems that should be in everyone's toolbox. The premise here is that a known starting state finds itself coupled to a dense (Bixon-Jortner) or continuous (Fano-Anderson) set of states of nearby energy. The coupling could arise from many types of problems, and might be calculated semiclassically in some circumstances. An example of the physical context was discussed in connection with resonant tunneling, in section 13.1.

The Fano-Anderson model [109] was introduced in 1961. It involved the coupling of an isolated state to a continuum of states. It is almost exactly solvable, but for a one-dimensional integral that is typically not analytic. The Fano-Anderson model led to simple ways to view line width and line shift through the so-called self-energy, with direct application to spectroscopy and many body physics. Anderson applied it in condensed matter physics, and Fano to atomic physics.

Subsequently and independently, Bixon and Jortner [110] introduced a closely related model, involving a state coupled to a discretized quasi-continuum of states, and used it to represent radiationless transitions in molecules.

14.1 Bixon-Jortner Model

We begin with the discrete, Bixon-Jortner version. In figure 14.1, left, there are N states $|\psi_n\rangle$, $n = 1, \ldots, N$ spaced perhaps unevenly, coupled to the lone state $|\psi_0\rangle$ (the first N states have been "prediagonalized," but a new state $n = 0$ is introduced, usually in the middle of the pack somewhere) at some energy E_0. The matrix eigenvalue equation for the model is

$$\begin{pmatrix} E_1 - E_k & 0 & 0 & \cdots & 0 & h_{10} \\ 0 & E_2 - E_k & 0 & & & h_{20} \\ 0 & 0 & E_3 - E_k & & & h_{30} \\ 0 & & & \ddots & & \vdots \\ \vdots & & & & E_N - E_k & h_{N0} \\ h_{01} & h_{02} & h_{03} & \cdots & h_{0N} & E_0 - E_k \end{pmatrix} \begin{pmatrix} a_{1k} \\ a_{2k} \\ a_{3k} \\ \vdots \\ a_{Nk} \\ a_{0k} \end{pmatrix} = \begin{pmatrix} 0 \\ 0 \\ 0 \\ \vdots \\ 0 \\ 0 \end{pmatrix}, \tag{14.1}$$

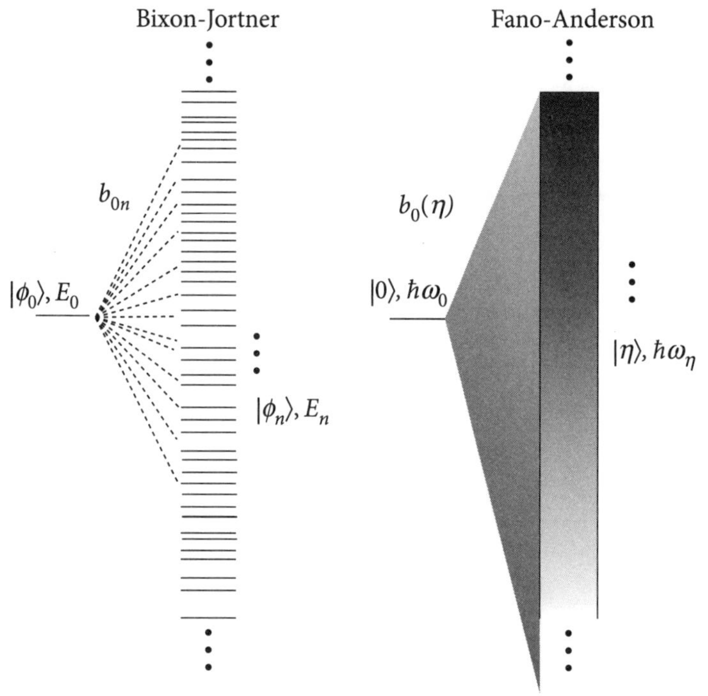

Figure 14.1. Bixon-Jortner (left) and Fano-Anderson (right) models involving lone states coupled to a quasi-continuum, and a true continuum, respectively. The quasi-continuum acts like a continuum when the coupling b_{0n} is larger than the distance between nearest neighbors, for times shorter than the inverse of the level spacing.

where the E_k are the sought-after new eigenvalues of the $N+1$ state problem, and the new eigenstates ψ_k read, in terms of the basis ϕ_n,

$$\psi_k = a_{0k}\phi_0 + \sum_{n=1}^{N} a_{nk}\phi_n. \tag{14.2}$$

For the first n rows we have

$$\begin{aligned}(E_n - E_k)a_{nk} + h_{n0}a_{0k} &= 0\text{—that is,}\\ a_{nk} &= -\frac{h_{n0}a_{0k}}{(E_n - E_k)}.\end{aligned} \tag{14.3}$$

Then on the last row we find

$$\sum_{n=1}^{N} h_{0n}a_{nk} + (E_0 - E_k)a_{0k} = -\sum_{n=1}^{N} \frac{h_{0n}^2 a_{0k}}{E_n - E_k} + (E_0 - E_k)a_{0k} = 0, \tag{14.4}$$

that is,

$$\sum_{n=1}^{N} \frac{h_{0n}^2}{(E_n - E_k)(E_0 - E_k)} = 1. \tag{14.5}$$

The system can almost be solved analytically, except the solutions for equation 14.5 need to be found by a numerical root search for the E_k. Then the eigenvalues E_k can be substituted into equation 14.3 to quickly find the coefficients a_{nk}. Normalization requires

$$|a_{0k}|^2 + \sum_{n=1}^{N} |a_{nk}|^2 = |a_{0k}|^2 + \sum_{n=1}^{N} \frac{|a_{0k}|^2}{(E_n - E_k)^2}$$

$$= |a_{0k}|^2 \left(1 + \sum_{n=1}^{N} \frac{1}{(E_n - E_k)^2}\right) = 1, \qquad (14.6)$$

and everything is settled.

The time evolution of the probability amplitude that the system, starting in the state ψ_0 with probability 1, will remain in ψ_0 (that is, the autocorrelation; the Fourier transform of the autocorrelation is the spectrum of ψ_0) is

$$a(t) = \langle \psi_0 | e^{-iHt/\hbar} | \psi_0 \rangle = \sum_{k=1}^{N+1} |a_{0k}|^2 e^{-iE_k t/\hbar}. \qquad (14.7)$$

The E_k will assiduously avoid coming too close to E_n or E_0, lest a single term become too large in equation 14.5. Even a zero-order level n with $E_n = E_0$ is not particularly special when it comes to its participation in any of the new eigenstates. The results of a Bixon-Jortner calculation are shown in figure 14.2. The vertical scale is energy, with the initial state $|\phi_0\rangle$ shown at the left. The couplings h_{0n}^2 to the unperturbed states $|\psi_k\rangle$ are shown as the thickness of the lines, with the energies E_k correctly given. Then the final eigenvalues E_k are shown, with the spectra $\epsilon_k = |\langle \phi_0 | \psi_k \rangle|^2$ shown at the right as the length of the lines.

14.2 Fano-Anderson Model

The Fano-Anderson model is more useful when true continua are present, and also nearly analytic for yielding formulas (save for an integral that must be done, usually numerically) for the energy shifts and linewidth. Following Longhi [111], we take the lone state $|0\rangle$ to be coupled to a continuum of states $|\eta\rangle$, where neither the coupling nor the density of the continuum states has to be constant. (See figure 14.1, right.) Now,

$$H_0 = \hbar\omega_0 |0\rangle\langle 0| + \hbar \int d\eta \; \omega_\eta \; |\eta\rangle\langle\eta| \qquad (14.8)$$

is the noninteracting Hamiltonian, η is the index of the continuum, and

$$V = \hbar \int (b_0(\eta)|0\rangle\langle\eta| + b_0(\eta)^* |\eta\rangle\langle 0|) \, d\eta. \qquad (14.9)$$

The normalization of the basis states is $\langle 0|0\rangle = 1$, $\langle \eta|\eta'\rangle = \delta(\eta - \eta')$. We expand the wavefunction that evolves from $|0\rangle$ as $e^{-iHt/\hbar}|0\rangle = c_0(t)|0\rangle + \int d\eta \; c_\eta(t)|\eta\rangle$; the

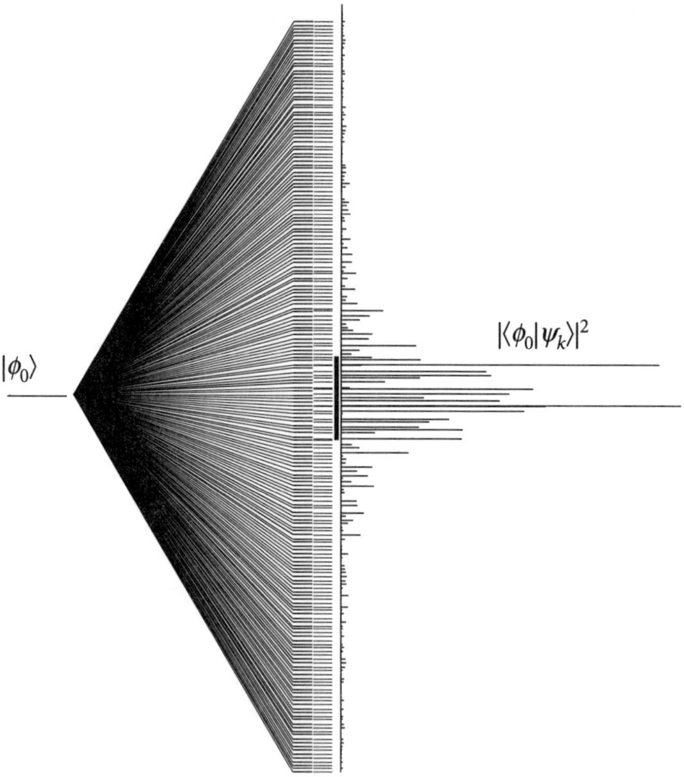

Figure 14.2. Details of the results of a Bixon-Jortner calculation. The predicted Golden Rule width of the band, $\Gamma = 2\pi \langle h_{0n}^2 \rangle \rho$, is shown as a dark band.

coefficients $c_0(t)$ and $c_\eta(t)$ satisfy

$$\begin{aligned} i\,\dot{c}_0(t) &= \omega_0\, c_0(t) + \int d\eta\; b_0(\eta) c_\eta(t), \\ i\,\dot{c}_\eta(t) &= \omega_\eta\, c_\eta(t) + b_0(\eta)^* c_0(t). \end{aligned} \tag{14.10}$$

We Laplace transform $c_0(t)$, $c_\eta(t)$:

$$c_{0,\eta}(s) = \int_0^\infty e^{-st} c_{0,\eta}(t)\, dt, \tag{14.11}$$

and from equations 14.10

$$\begin{aligned} isc_0(s) &= 1 + \omega_0 c_0(s) + \int d\eta\; b_0(\eta) c_\eta(s), \\ isc_\eta(s) &= \omega_\eta c_\eta(s) + b_0(\eta)^* c_0(s), \text{ that is,} \\ c_\eta(s) &= \frac{b_0(\eta)^* c_0(s)}{is - \omega_\eta}. \end{aligned} \tag{14.12}$$

Then

$$isc_0(s) - \omega_0 c_0(s) = 1 + \int \frac{|b_0(\eta)|^2}{is - \omega_\eta} c_0(s),$$
$$c_0(s) = \frac{1}{is - \omega_0 - \Sigma(s)}, \tag{14.13}$$

where

$$\Sigma(s) = \int d\eta \frac{|b_0(\eta)|^2}{is - \omega_\eta} = \int_{\omega_1}^{\omega_2} d\omega' \frac{\rho(\omega')|b_0(\omega')|^2}{is - \omega'}. \tag{14.14}$$

By inverse Laplace transform,

$$c_0(t) = \frac{1}{2\pi i} \int_{0-i\infty}^{0+i\infty} \frac{ds\, e^{st}}{is - \omega_0 - \Sigma(s)}. \tag{14.15}$$

$\Sigma(s)$ is the "self-energy," a shift from the original energy of the unperturbed state given by equation 14.14.

Suppose the state $|0\rangle$, without the coupling to the continuum, carries all the transition strength—that is, from some other state there is coupling to $|0\rangle$ but not to any $|\eta\rangle$. The coupling of $|0\rangle$ to the continuum will broaden it into a band, corresponding to a finite life time. The band center need not be at ω_0, but can be shifted by the real part of the self-energy. This shift is easy to understand: interaction between any two states pushes them apart. The shift is greater the nearer the two states are in energy, and the stronger their interaction. The continuum facing $|0\rangle$ has states above and below it in energy, pushing in opposite directions. The real part of the integral for the self energy in equation 14.14 can be seen measuring the relative influences of strength, density of interacting states, and their proximity. The self-energy is complex. Writing (careful to pick up the pole after the change of variable starting with the inverse Laplace transform)

$$\Sigma(s = -i\omega \pm 0^+) = \Delta(\omega) \mp i\pi\rho(\omega)\langle|b_0(\omega)|^2\rangle, \tag{14.16}$$

the real part of the self-energy is

$$\Delta(\omega) = \mathcal{P} \int_{\omega_1}^{\omega_2} d\omega' \frac{\rho(\omega')|b_0(\omega')|^2}{\omega - \omega'}. \tag{14.17}$$

The imaginary part, controlling the width of the band, is seen to be equivalent to the Fermi Golden Rule formula for the decay rate,

$$\Gamma = \frac{2\pi}{\hbar}\langle v^2\rangle\rho. \tag{14.18}$$

Both the Fano-Anderson and the Bixon-Jortner models are important and useful in their respective domains.

Chapter 15

Power of Degeneracy

Degenerate perturbation theory is a standard topic in many books on quantum mechanics. One wants to know linear combinations of the degenerate or near-degenerate set of states that don't change as a given perturbation is turned on. This is the same as finding the eigenfunctions of the full Hamiltonian within the subspace of the degenerate set. Because they are degenerate, or quasi-degenerate, ordinary perturbation theory fails. Even an infinitesimal perturbation drastically mixes a degenerate basis. If the states are only approximately degenerate but the perturbation is larger than the level spacing of the states, again ordinary perturbation theory fails.

The usual procedure is to brute-force diagonalize the full Hamiltonian in the degenerate basis set of functions. This can be unsatisfactory for two reasons: (1) the degeneracy may be huge, making the diagonalization intractable, and (2) even if tractable, physical insight may be hidden by the brute force approach.

Fortunately, the right linear combinations of degenerate states can sometimes be found based on physical principles or even intuition. If the trial combinations do not change much when the perturbation is turned on, the game has been won, and is tantamount to finding the eigenstates (or at least some of them, for example the one having the lowest energy with the perturbation turned on) without diagonalizing a matrix.

15.1 Generator States

Is there a systematic way to create linear combinations of degenerate or near-degenerate states that approach the exact solutions (within the degenerate or quasi-degenerate set)? One procedure is through projection onto *generator states*. These are functions $|\gamma\rangle$ that possess some desired property but are themselves not eigenstates. From the set of degenerate or near-degenerate states $|E, n\rangle$, we form the linear combination

$$|E, \gamma\rangle = \sum_n |E, n\rangle\langle E, n|\gamma\rangle. \tag{15.1}$$

If the set $|E, n\rangle$ were complete, $|\gamma\rangle$ would simply be reproduced. Instead, $|\gamma\rangle$ spans much more of Hilbert space than encompassed by the degenerate manifold. The incomplete set of states with coefficients $\langle E, n|\gamma\rangle$ "tries" to reproduce the state $|\gamma\rangle$

within the limitations of the basis. If the generator state has classical-like aspects—for example, a wavepacket with a well-defined (within the uncertainty principle) position and momentum—the degenerate or quasi-degenerate combination will take on those classical properties, passing strongly through the targeted position and momentum while remaining an eigenstate of H_0 (if the set was exactly degenerate). The classical property might be chosen to avoid a perturbation localized somewhere in coordinate space, or even to avoid another particle. This will be the case for the waveguide with a protrusion treated next, for the Luukko scars later, and for a guess at an eigenstate for two repelling bodies in a common confining potential.

WAVEGUIDE WITH OBSTRUCTION

Consider a two-dimensional waveguide with parallel walls a distance a apart, except for a protrusion in a small zone along one side. A wave incoming from the left, of energy $E = \hbar^2 k_n^2/2m + n^2\pi^2\hbar^2/(2ma^2)$,

$$\psi_n(x, y) = \sqrt{\frac{m}{\hbar k_n}} e^{ik_n x} \sin(n\pi y/a), \tag{15.2}$$

without the protrusion, becomes

$$\psi_n(x, y) = \sqrt{\frac{m}{\hbar k_n}} e^{ik_n x} \sin(n\pi y/a) + \sum_{n'} \sqrt{\frac{m}{\hbar k_{n'}}} R_{nn'} e^{-ik_{n'} x} \sin(n'\pi y/a), \tag{15.3}$$

well to the left of the protrusion, with $k_n = \sqrt{2mE/\hbar^2 - n^2\pi^2/a^2}$ real, and

$$\psi_n(x, y) = \sum_{n'} \sqrt{\frac{m}{\hbar k_{n'}}} T_{nn'} e^{ik_{n'} x} \sin(n'\pi y/a), \tag{15.4}$$

well to the right of the protrusion. These forms hold everywhere except close to the protrusion, where evanescent waves dwell that are not included in the asymptotic forms earlier. This set of waves is exactly degenerate. The number of propagating "channels," or distinct transverse modes, is N, where $k_N = \sqrt{2mE/\hbar^2 - N^2\pi^2/a^2}$ is the last real valued k—that is, k_{N+1} is imaginary. The incoming flux $\vec{j}$ lies strictly along x—that is, $j_y = 0$; however, $j_x(x, y)$ depends on y. We integrate over y to pick up all the right-trending incoming flux to the left of the barrier; this is

$$j_{in} = \frac{\hbar}{2mi} \int_0^a \{\psi_n^* \frac{\partial \psi_n}{\partial x} - \psi_n \frac{\partial \psi_n^*}{\partial x}\} dy = 1.$$

The reflected, left-trending wave well to the left of the barrier is $j_{refl} = |R_{nn'}|^2$. On the right of the protrusion, the flux is $j_{trans} = \sum_{n'} |T_{nn'}|^2$. Since probability is not lost, we must have

$$1 = \sum_{n'} |R_{nn'}|^2 + \sum_{n'} |T_{nn'}|^2, \tag{15.5}$$

that is, flux either is reflected or else it propagates beyond the protrusion.

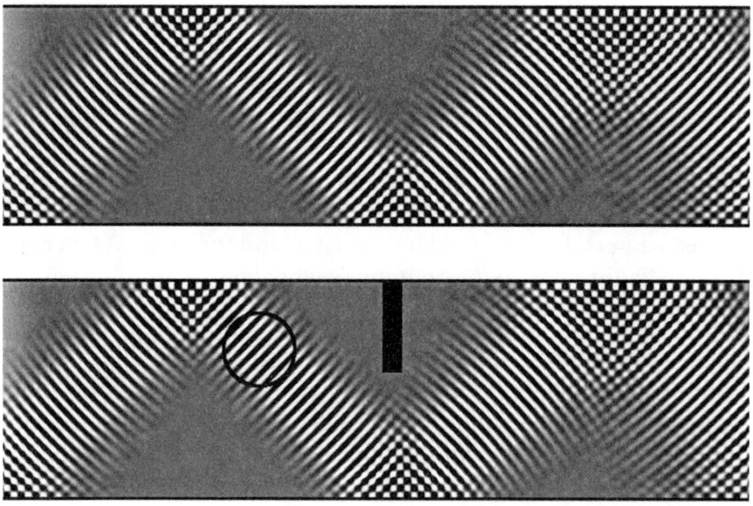

Figure 15.1. A portion of a quantum eigenstate of energy E in an infinite hard wall waveguide. It has an obvious classical analog, but is nothing more than a linear combination of the usual transverse propagating modes of the form $\psi_n(x,y) = \exp[ik_n x] \sin(n\pi y/L)$, with $E = \hbar^2 k_n^2/2m + n^2\pi^2\hbar^2/2mL^2$. The coefficients of the combination are determined by overlaps of the transverse modes with a "generator state" (shown schematically in the lower panel as the circle; this is the location of the Gaussian generator state with momentum along the direction of the flow there). A barrier is also shown in the lower panel; the mode shown would transmit beyond the barrier with very nearly unit probability (nearly zero reflection, corresponding to an eigenvalue of the transmission matrix near 1), which is not true of any of the usual transverse modes.

Since all the states we have been discussing are degenerate, we are free to make any new linear combinations we want, whose structure and appearance will differ from the "standard" set described earlier—that is, $\psi_n(x, y) = \sqrt{m/\hbar k_{n'}} e^{-ik_n x} \sin(n\pi y/a)$. For example, $\psi(x, y) = e^{-ik_7 x} \sin(7\pi y/a) + e^{-ik_10 x} \sin(10\pi y/a)$ is a perfectly good solution of the Schrödinger equation if there are 10 or more channels open, where $\hbar^2 k_n^2/2m + E_n = E$, and E_n is the transverse energy.

However, we have a special purpose in mind: what linear combinations of these degenerate states would backscatter minimally off the protrusion and approach unit transmission, and which ones would backscatter maximally? (See figure 15.1.).

Earlier, we mentioned the "brute-force" method of diagonalizing a matrix to achieve a desired property. Here, suppose we diagonalize the T matrix previously determined in the standard basis (this is just for argument's sake, not the generating state approach). We perform a unitary transformation on the channels—that is, the R and T matrices, after which equation 15.5 still holds in the new basis that we label by η. The eigenvalues will go from largest to smallest, and we see from equation 15.5 that the maximum possible diagonal element of T is 1, the minimum is 0, and the same is true of R. The eigenvectors ψ_η with $T_{\eta,\eta} \sim 1$ (if they exist) have near unit flux, somehow bypassing the protrusion, and those with $T_{\eta,\eta} \sim 0$, if they exist, reflect almost totally off the protrusion. Again, this is the brute-force method.

The much more efficient and intuitive way to find states with near unit transmission is to use a generating state, seen as the circle (representing a Gaussian with momentum

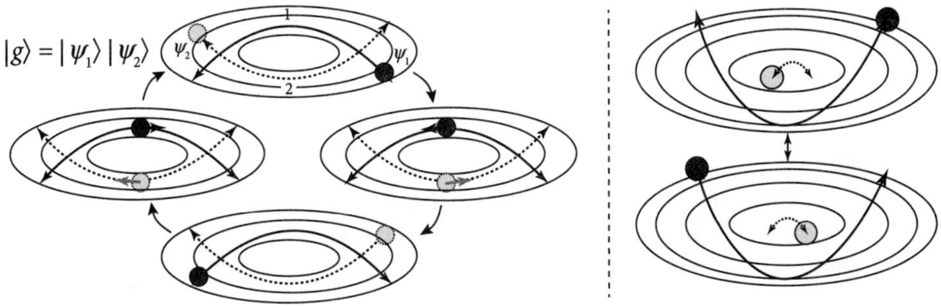

Figure 15.2. Generating state (direct product of black and gray Gaussians) and its dynamics for two particles in a 2:1 degenerate two-dimensional harmonic oscillator. (Left and right) Note that the particles never approach each other very closely in their motion, and the generated states have the same property. The states generated left and right can be taken to have the same energy. The generators are used to give expansion coefficients of all the two-particle degenerate states in a given two-particle manifold. A basis function for Fermions or bosons consists of the sum of two such two-particle states, with the particles exchanged in their orbits and an appropriate sign between terms in the sum.

shown) in figure 15.1. It arranges a superposition of degenerate states by equation 15.1 that almost entirely avoids the obstruction. The degenerate combination is hardly disturbed (bottom) when a barrier is introduced. Every one of the original waveguide modes $\psi_n(x, y)$ above with definite transverse quantum number n are significantly backscattered by the presence of the barrier.

GENERATOR STATES: TWO INTERACTING PARTICLES IN A POTENTIAL

In figure 15.2, we show a scheme to create a reasonably good two-particle eigenstate (four degrees of freedom total) by using the generating state shown at the top. The particles are initiated as symmetrical reflections of each other—that is, two wavepackets, one for each particle, temporarily at rest at the extreme of their motion. As the successive panels show, the wavepackets avoid each other over their whole orbit. Thus, an eigenstate (of the noninteracting particles in the 2:1 harmonic potential shown) created by this generator state by Fourier transform of the dynamics (that is not necessary to perform—see the following) will be minimally perturbed when a repulsive short-range interaction between the particles is turned on. Although we do not need to perform a time-dependent analysis to implement the generators, it is good to point out that the generator wavefunction is equivalent to

$$\psi(\vec{r}_1, \vec{r}_2) = \int_0^\tau e^{iEt}\psi_1(\vec{r}_1, t)\psi_2(\vec{r}_2, t)\, dt, \tag{15.6}$$

where E is the energy of the degenerate manifold, and τ is the classical period of the symmetrically related periodic orbits 1 and 2. Equation 15.6 generates a correlated, entangled basis function, whereas the direct product basis functions are not individually entangled.

A state built out of degenerate states can lead to choreographed motion residing in a single highly correlated function, a linear combination of all the uncorrelated, direct product degenerate basis states. For the four coordinates in figure 15.2, we can write

$$\psi(\vec{x}_1, \vec{x}_2) = \sum_{\substack{K=0,M \\ m_1=0,M-K \\ m_2=0,K}} C_{K,m_1,m_2} \, |M-K-m_1, 2m_1\rangle|K-m_2, 2m_2\rangle. \tag{15.7}$$

The unperturbed energy of each state in the sum is $E = 3/2\hbar\omega + \hbar\omega(M-K-m_1) + 2\hbar\omega/2m_1 + \hbar\omega(K-m_2) + 2\hbar\omega/2m_2 = \hbar\omega(M+3/2)$, independent of K, m_1, or m_2. The potential both particles experience neglecting their interaction is $V(x,y) + 1/2\omega^2\, y^2 + +1/8\omega^2\, x^2$.

The distinct comparative advantage of the guess produced by the generating state, assuming repulsive interactions between the particles, is that the original direct product basis functions can't avoid close contact between the particles, apart from any exchange hole at the intersection of the two orbitals. Introducing the repulsion has a first-order effect on every direct product basis function and requires a standard degenerate perturbation theory diagonalization of a matrix in this basis with the perturbation, resulting in linear combinations presumably coming close to states produced by good generating states.

HYDROGEN ATOM

Hydrogen atom eigenfunctions as usually given come in degenerate manifolds of eigenfunctions,

$$\psi_{n\ell m}(r,\vartheta,\varphi) = \sqrt{\left(\frac{2}{na_0}\right)^3 \frac{(n-\ell-1)!}{2n(n+\ell)!}}\, e^{-\rho/2}\rho^\ell L^{2\ell+1}_{n-\ell-1}(\rho) Y^m_\ell(\vartheta,\varphi), \tag{15.8}$$

where $\rho = 2r/na_0$, a_0 is the Bohr radius, $L^{2\ell+1}_{n-\ell-1}(\rho)$ is a generalized Laguerre polynomial, and $Y^m_\ell(\vartheta,\varphi)$ is a spherical harmonic, where ℓ is the angular momentum, $\ell = 0, 1, \cdots n-1$, and m is degenerate at the nth eigenvalue, E_n. In figure 15.3, four linear combinations of states are given with coefficients given by projections on a generating wavepacket and parameters given in the caption.

QUASI-DEGENERACY

When a manifold of states is not degenerate, but the energy splittings within it are small compared to the perturbation to be applied, the generator state methods may still be useful. For example, in the carbon atom, the $2s$ and $2p_x$ orbitals are not degenerate (within the Hartree-Fock self-consistent mean field approximation) and yet the $\psi_{2s} + \psi_{2p_x}$ "sp" hybrid is a much better starting point than either orbital alone when a diatomic bond is to be formed with another atom.

This problem is one of the oldest in quantum chemistry and is not traditionally formulated in terms of a generator state. Any generator chosen is subject to "chemical intuition" as to what is best, although variational methods may be brought to bear. Clearly, in this case it should be some sort of lobe straddling a line connecting the nuclei. The mixture of $2s$ and $2p_x$ would not come out 50-50 in general, nor should it be.

LUUKKO SCARS

Recently, Esa Räsänen and Perttu Luukko discovered that a two-dimensional r^5 potential perturbed by smooth random potential bumps possesses many deeply scarred

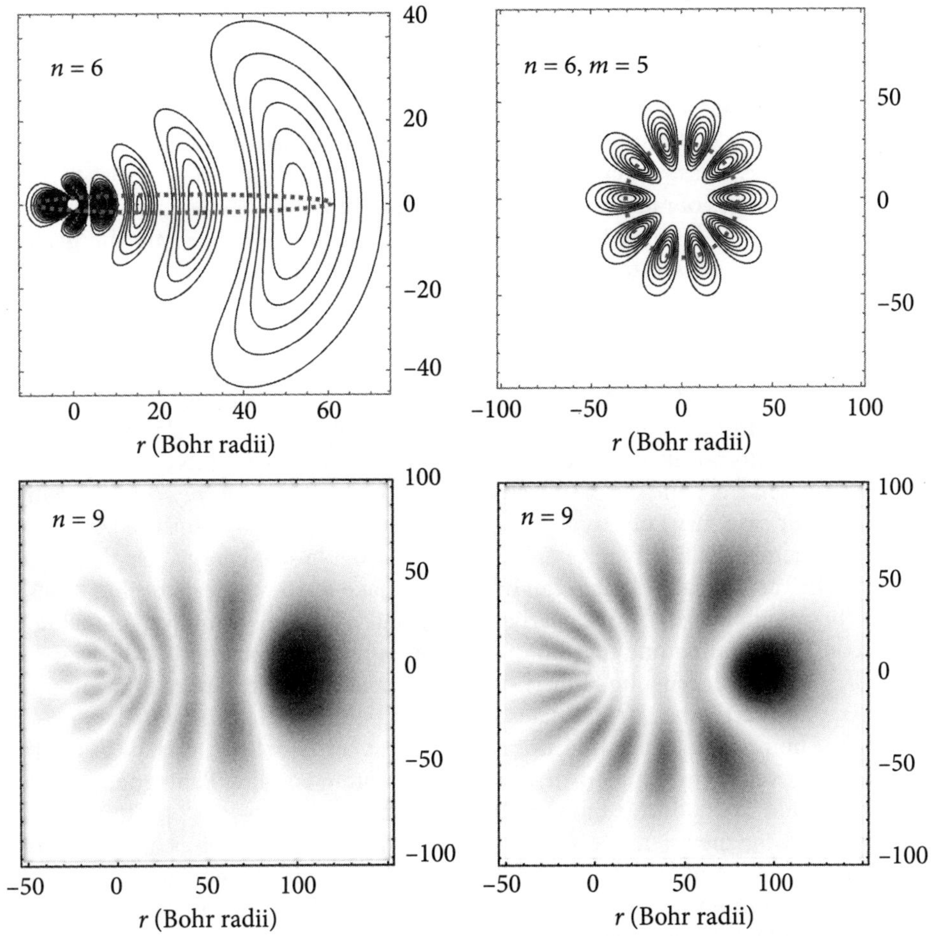

Figure 15.3. (Top left) An eigenfunction $\psi_6^2(r, \theta, \phi)$ of the $n = 6$ manifold of the hydrogen atom, corresponding to the generating wavefunction $\exp[-0.01(r-50)^2 - 6\theta^2 - 6\phi^2]$. It is plotted in the $z = 0$ plane. The axes are labeled in units of a_0. The wavefunction maximizes on zero angular momentum $\ell = 0, m = 0$ and corresponds to a direct collision with the nucleus from a position at rest at the end of the ellipse shown. The nucleus is shown as a white circle imposed on the dense contours at $r = 0$. Only states with $n = 6$ were used in the construction of this wavefunction, and thus it is an energy eigenfunction but not an angular momentum eigenfunction. It is also confined in the θ direction, large only near the $z = 0$ plane, much like the confinement seen here in ϕ. (Top right) The function $\psi_{6,5,5}^2(r, \theta, \phi)$ in the same plane. This is the only classical-like (rather than a superposition becoming classical-like) orbit among the usual representations of the $n = 6$ manifold of degenerate states that are taken to be eigenfunctions of the $\hat{L}^2$ and $\hat{L}_z$ operators in addition to H. (Bottom left and right) $n = 9$ degenerate superpositions using generating states with low (left) and higher (right) expectation values of angular momentum.

eigenstates (see chapter 22) [112]. Especially prominent among them are five-pointed pentagram scars. These arise from a 5:2 classical resonance (that is, pentagram shaped orbits), built into this potential. The resonance zone can be found at any energy, since this is a self-similar "scaling potential"—that is, the motion of any trajectory at one energy has a double at any other energy by scaling distance and time. The classical resonance causes quasi-degenerate groups of unperturbed quantum states to form under 5:2 exchanges of radial and angular quantum numbers or classical action—that is, $|n, \ell\rangle$, $|n-2, \ell+5\rangle$, $|n-4\ell+10\rangle$, ... have approximately the same energy. The random potential, chosen to be smooth on the scale of the local wavelength, couples these quasi-degenerate states and, perhaps surprisingly, creates some strongly scarred eigenstates. Why do the scars form? Some are surprisingly localized to the classical 5:2 orbits.

In chapter 22, the driving principle behind the classical-like periodic orbit form of "scars" in the midst of chaos is the time it takes to "fall off" shorter classical unstable (all periodic orbits are all unstable in a chaotic system) periodic orbits embedded, with zero measure, in a chaotic sea of trajectories. The driving force for the scar localization in the present case turns out to be the variational principle: what linear combination of the degenerate (or nearly degenerate) states has the lowest energy? Or the highest energy? The variational principle can be used to make the best linear combinations of a basis (best in the sense of most nearly an eigenstate of the full Hamiltonian including the barriers) by finding one that minimizes the expectation value of the energy. In a finite dimensional degenerate Hilbert space, the same principle applies to maximizing the energy. Since the basis eigenstates for the r^5 potential ride more or less equally over all the bumps (see figure 15.4, upper left), linear combinations must be found that instead ride over high bumps or inverted bumps preferentially. Pentagonal classical orbits that ride over positive bumps can maximize their energy; the positions of the bumps determine the orientation of the pentagrams, which otherwise could have their arms at any angle of rigid rotation of the pentagram. Even if all the bumps are positive, energy can be minimized by avoiding them. An example of this is seen at the lower right in figure 15.4. (Negative bumps are just as effective, with a reversal of whether the high- or low-energy states ride over them and avoid them. This is exactly what is seen in the most clearly scarred eigenstates. Combinations of positive and negative bumps also beckon states that avoid the negative and seek the positive bumps, and vice versa.) It is no surprise then that as energy increases and higher quasi-degenerate manifolds are involved, the scarred states (*scar* meaning the eigenstates have a partial concentration to a periodic orbit, even in the presence of classical chaos; see chapter 22) are found multiple times in almost the same orientation—avoiding or riding on top of the same bumps. For large enough energy difference, pentagonal scars will change absolute size and may need to find a new orientation the maximizes a strategy using new bumps.

Figure 15.5 shows a few examples of the eigenfunction amplitudes. The 5:2 resonance means that most classical trajectories precess in coordinate space, as they cover the two dimensional 5:2 phase space torus embedded in the four dimensional phase space. They trace out pentagonal shapes that almost close on themselves. However, there is a "line of fixed points" (see Poincaré-Birkhoff fixed-point theorem, section 3.7) that is a continuous set of closed pentagonal orbits, differing only in their fixed rotation angle about the center of the potential. It is not clear as of this writing what role the Poincaré-Birkhoff fixed-point theorem plays in forming these scars.

We have not made use of generator states in connection with Luukko scars. If we use a Gaussian as a generator, pointing its momentum along a classical pentagonal

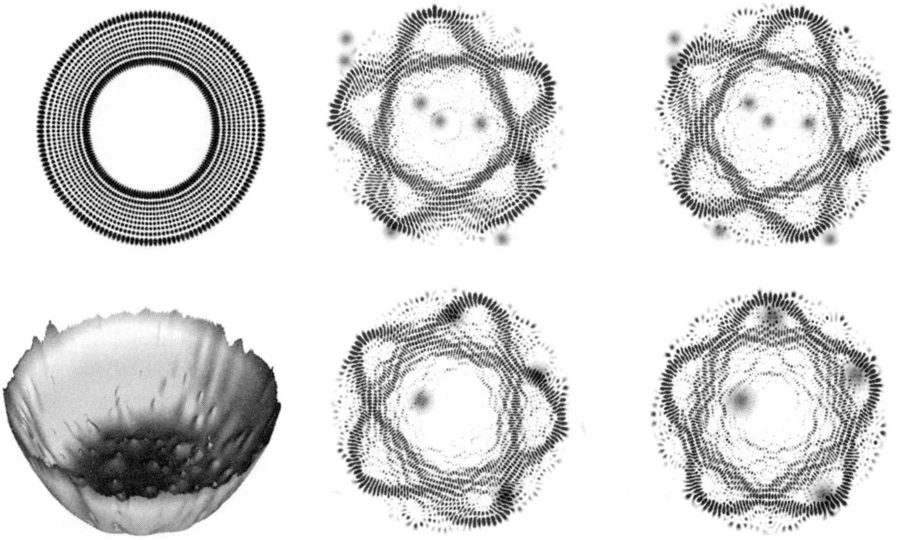

Figure 15.4. The scarred eigenstates in this image are a result of the variational principle, here for the maximal perturbed energy (and on the lower right, for the minimal perturbed energy) in a sum of quasi-degenerate basis states. One of the unperturbed basis states of the r^5 potential is shown at the upper left. Below it, the r^5 potential is shown including a typical random set of barriers. To the right are four examples of eigenfunctions with just a few barriers added to the potential. Inspection of these images reveals that full eigenfunctions find a solution maximizing (or minimizing) their energy: they are made of near-degenerate states and concentrated so that their overlap with the barriers gives an energy contribution as large as possible or as small as possible, while still maintaining a pentagonal orbit structure consistent with a 5:2 resonance (five radial oscillations for every two 2π circulations of the origin). With positive bumps some states maximize their variational energy by orienting the pentagonal orbits to lie on top of as many bumps as possible, and minimize their variational energy by orienting the pentagonal orbits to avoid as many as possible. This way they "win" the competition to become optimal linear combinations of unperturbed states in the variational sense. Three of the examples shown run over the top of the barriers, to maximize their energy. On the lower right, a scarred state has avoided barriers and minimized its variational energy, which is just as variationally rewarding.

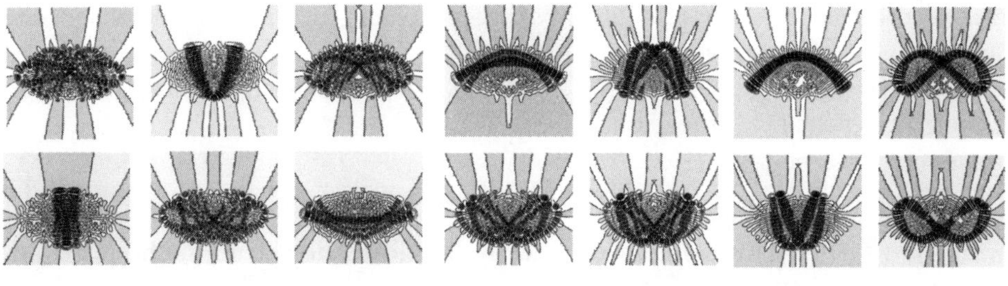

Figure 15.5. Contour maps of some of the eigenstates of a slightly perturbed 2:1 degenerate harmonic oscillator, all formed by diagonalizing a smooth bumpy potential in the exactly degenerate harmonic basis $|N, 0\rangle$, $|N-1,2\rangle$, $\ldots$, $|0,2N\rangle$.

periodic orbit, giving it approximately the right average energy, it will generate a pentagonal localized state as a linear combination of the quasi-degenerate states. This could be done so as to maximize the pentagonal orbit over positive potential bumps for example, coming close to an eigenstate with only intuition as a guide to choosing the generator state.

15.2 Degeneracy and Gauge Transformations

In electromagnetic theory, gauge transformations change vector potentials without changing the physical quantities like electric fields calculated from them. For example, for a single electron in two dimensions in a perpendicular magnetic field, different gauges generate different linear combinations of degenerate states as single-particle eigenstates. In this way, gauge transformations can play a role in creating states resistant to change when a given perturbation is turned on. Generators of the sort mentioned earlier play a role analogous to gauges.

Suppose the 2D Hamiltonian with vector potential $\boldsymbol{A}$ is

$$H = \frac{1}{2m}\left(\boldsymbol{p} - \frac{e}{c}\boldsymbol{A}\right)^2, \tag{15.9}$$

with

$$\boldsymbol{B} = \nabla \times \boldsymbol{A}. \tag{15.10}$$

In the Landau gauge

$$\hat{\boldsymbol{A}} = \begin{pmatrix} 0 \\ Bx \end{pmatrix}, \tag{15.11}$$

the Hamiltonian reads

$$H = \frac{\hat{p}_x^2}{2m} + \frac{1}{2m}\left(\hat{p}_y - \frac{eB\hat{x}}{c}\right)^2. \tag{15.12}$$

The natural eigenstates are

$$\Psi_{k_y,n}(x, y) = e^{ik_y y}\phi_n(x - \hbar k_y/m\omega_c), \tag{15.13}$$

where $\phi_n(x)$ is the nth state of the harmonic oscillator with mass m and frequency $\omega_c = eB/mc$, displaced by an amount $\hbar k_y/m\omega_c$ controlled by k_y, the wave vector in the y coordinate. There is a doubly infinite degeneracy—that is, all n and all k_y. Such states are all degenerate, and orthogonal for different n.

In the symmetric gauge $\hat{\boldsymbol{A}} = \frac{B}{2}(-y, x)$, the lowest Landau level with a perpendicular uniform magnetic field B reads,

$$\psi_{0,0}(x, y) = \eta \exp\left[-\tfrac{eB}{4\hbar c}x^2 - \tfrac{eB}{4\hbar c}y^2\right], \tag{15.14}$$

with $\eta = (\frac{2eB}{\pi\hbar c})^{1/4}$. This solution can clearly be moved anywhere in the plane and remain a ground-state wavefunction:

$$\psi_{x_0,y_0}(x, y) = \eta \exp\left[-\tfrac{eB}{4\hbar c}(x - x_0)^2 - \tfrac{eB}{4\hbar c}(y - y_0)^2\right], \tag{15.15}$$

constituting a double continuum degeneracy of the ground state for a single particle.

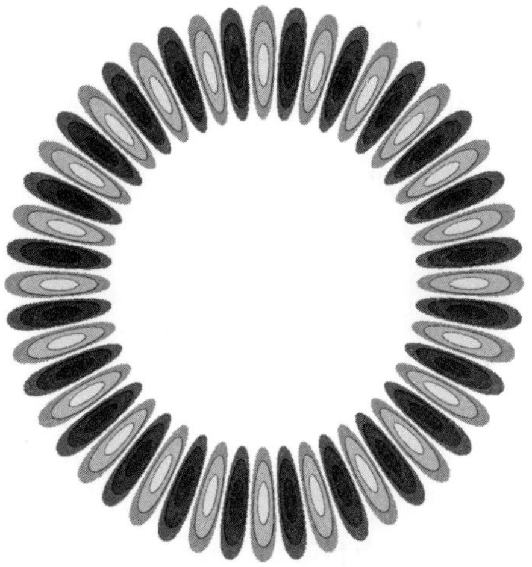

Figure 15.6. The wavefunction, equation 15.17, for $n = 24$. The nodal structure is due to canonical angular momentum, not mechanical angular momentum, so it does not represent rotation of the physical particle. The energy is that of the ground state.

It can be readily shown that for any function $f(z)$, $z = x + iy$,

$$\psi_{x_0,y_0}(x, y) = \beta\ f(z)\ \exp\left[-\tfrac{eB}{4\hbar c}(x - x_0)^2 - \tfrac{eB}{4\hbar c}(y - y_0)^2\right] \qquad (15.16)$$

is an eigenfunction, where β is a normalization factor. The traditional choice is $x_0 = y_0 = 0,\ f(z) = z^n$, so we have

$$\psi_n(x, y) = N_n\ z^n e^{-\frac{eB}{4\hbar c}(x^2+y^2)}, \qquad (15.17)$$

where $N_n = \left[n!\ \pi\ \left(\frac{2\hbar c}{eB}\right)^{n+1}\right]^{-1/2}$. One of these eigenfunctions, degenerate with the ground state energy, is shown in figure 15.6.

The very different looking solutions, equations 15.15 and 15.13 are merely distinct linear combinations (induced by the gauge) of the same reference set of degenerate states, these could be $\psi_{x_0,y_0}(x, y)$ for example. Many more possibilities exist, but a more interesting direction is two and many particle degenerate linear combinations that might be unchanged or nearly so when turning on electron repulsion.

15.3 And the Winner Is . . . .

In the author's view, the crowning achievement in the business of guessing robust (little change upon turning on electron repulsion) linear combinations of degenerate many body electronic states are Laughlin's wavefunctions [113] for infinitely many electrons in a plane, with an applied perpendicular magnetic field, used in understanding the fractional and integer quantum Hall effect. His wavefunctions are linear combinations of a staggering infinity of degenerate states for noninteracting electrons in a strong magnetic field in 2D, a spectacular example of the theme of this chapter.

Laughlin's wavefunction, equation 15.18, used for explaining the integer quantum Hall effect [113], became the key to rationalizing the fractional quantum Hall effect. We cannot delve into the fascinating lore surrounding the quantum Hall effect, integer or fractional, some of it depending on disorder in the 2D electron layer. It is noteworthy that a fully detailed understanding is still lacking. This is no surprise, since emergent phenomena in a many-body system can be arbitrarily complex. The Laughlin wavefunctions, the first and still the best of theoretical quantum Hall breakthroughs, remain phenomenological, although they have been shown to be very close to the true many-body ground state of the infinite system.

We imagine a perfect extended 2D flat potential, populated by electrons with a strong magnetic field B perpendicular to the electron plane. Laughlin's wavefunction for a filling factor of 1 (the maximum electron density of ground state Landau orbitals) is

$$\psi(z_1, z_2, z_3, \ldots, z_N, \ldots) = \lim_{N\to\infty} \eta_N \prod_{N\geqslant i>j\geqslant 1} \left(z_i - z_j\right) \prod_{k=1}^{N} \exp\left(- \mid z_k \mid^2\right), \quad (15.18)$$

with $z = \frac{1}{2l_B}(x + iy)$ and

$$\eta_N = \left[\frac{1}{2\pi\ \left(\sqrt{2}\right)^{2n}\ \sqrt{n!}\ \left(\sqrt{N-1}\right)^{n+1}}\right]^{\frac{N(N-1)}{2}}.$$

Laughlin's wavefunction is clearly Fermionic—that is, it is an antisymmetric filling of consecutive ground state Landau orbitals, concentric to the origin and exactly of the type seen in figure 15.6, since

$$\prod_{N\geqslant i>j\geqslant 1} \left(z_i - z_j\right) = \begin{vmatrix} 1 & z_1 & {z_1}^2 & {z_1}^3 & \cdots & {z_1}^N \\ 1 & z_2 & {z_2}^2 & {z_2}^3 & \cdots & {z_2}^N \\ 1 & z_3 & {z_3}^2 & {z_3}^3 & \cdots & {z_3}^N \\ \vdots & & & \ddots & \cdots & \vdots \\ 1 & z_N & {z_N}^2 & {z_N}^3 & \cdots & {z_N}^N. \end{vmatrix}. \quad (15.19)$$

This is the determinant of a so-called Vandermonde matrix consisting of antisymmetrized sums of products of successive powers of distinct z_j.

The direct product Vandermonde basis is seemingly "sloppy" about electron correlations, allowing electrons to approach each other along their circular tracks (see figure 15.6) with equanimity, apart (thanks to antisymmetry) from Fermi holes, no doubt a key factor. Could another anti-symmetrized superposition of degenerate states be created, perhaps using a generating state, that is even closer to the correct ground state? Indeed, the Laughlin wavefunction is certainly not the exact ground state; its energy is known to be off by about 4% in comparison with numerical calculations. This is either a big or a small error, depending on your perspective.

PART III

Applications to Spectroscopy

Chapter 16

Born-Oppenheimer Approximation and Its Implications

16.1 Inevitability of the Born-Oppenheimer Approximation

We are heading toward a time-dependent (but not necessarily semiclassical) Born-Oppenheimer formulation of spectroscopy. Spectra to be calculated, in equation 17.8 for example, need Born-Oppenheimer paraphernalia just to be evaluated.

The Born-Oppenheimer idea is one of those wonderful approximations that even in failure forms the basis for discussion and systematic corrections. Without the Born-Oppenheimer approximation as a foundation, there would be no molecular structure, no solid-state crystal structure, no molecular vibrations, no phonons, no electronic band structure, and so on. Why? Because it is the Born-Oppenheimer approximation that allows separation of electronic from nuclear motion. Without it, we appear to be lost in a soggy many-body "pea soup" or plasma of electrons and nuclei, where there is seemingly no structure at all, save the kind of structure one finds in a two-component liquid.

We would still need to invent the Born-Oppenheimer approximation even if we somehow had the exact many-body pea soup correlated wave functions given to us on a silver platter. First, the platter would have to be the size of Texas. The exact energies and particularly the eigenstates for the full interacting set of nuclei and electrons are an overwhelming amount of data, and we would need the simplification and intuition provided by the Born-Oppenheimer approximation, calling to mind molecules and solids with definite orientation and internal geometry. This is a far more powerful basis for understanding, even when it breaks down, than the exact wavefunctions would be.

We begin with a short discussion of basis set expansions in a simplified context of two variables, to introduce the slightly unusual expansions arising in the Born-Oppenheimer context. Suppose there is available a complete set of orthogonal functions $\phi_n(r)$, capable of describing any function $f(r)$ as $f(r) = \sum_n c_n\phi_n(r)$, and similarly another complete set $\chi_m(R)$ for R. In two variables r and R, any function $g(R, r)$ can be expanded (a basis set expansion) as

$$g(R, r) = \sum_{nj} c_{nj}\, \phi_n(r)\chi_j(R) \qquad (16.1)$$

in the complete "direct product" set (in the R, r space) $\varphi_{nj}(R, r) = \phi_n(r)\chi_j(R)$. If we happen to be solving for eigenstates of an operator then linear eigensystem equations for the coefficients c_{nj} would result.

These are completely standard points, but we will need two variations on this scheme. First, a modified expansion

$$g(R, r) = \sum_{nj} d_{nj}\ \phi_n(r)\chi_{nj}(R)$$

can be introduced. Here, we have assumed many *different* complete and orthogonal basis sets (one for each n) $\chi_{nj}(R)$, taking all j for each n. The coefficients d_{nj} can be found by projection in the usual way:

$$d_{nj} = \int dR \int dr\ \phi_n^*(r)\chi_{nj}^*(R)g(R, r).$$

The $\chi_{nj}(R)$ need not be orthogonal for different n, but $\phi_n(r)\chi_{nj}(R)$ is complete in the R, r space. This is not the usual "direct product basis", as in equation 16.1; in fact it is a correlated basis that takes on a different character for R, depending on the state ϕ_n that r finds itself in.

But we are not done yet. We suppose further that the basis $\phi_n(r)$ depends on R, but only parametrically in the following sense: $\phi_n(r)$ becomes $\phi_n(R; r)$, with the understanding that for each fixed R, the set $\phi_n(R; r)$ for all n is a complete orthogonal set of functions in the variable r. R is for this purpose only a parameter labeling a different complete set for each R, much as $\chi_{nj}(R)$ is a different complete set in R for each n. Then we write

$$g(R, r) = \sum_{nj} d_{nj}\ \phi_n(R; r)\chi_{nj}(R).$$

To see that this unusual expansion works and is a complete set over r and R, multiply from the left by $\phi_{n'}^*(R; r)\chi_{n',j'}^*(R)$ for any n' and integrate over r and R. Do the r integral first; this will project out a fixed $n = n'$ in the sum, erasing the parametric R dependence of $\phi_n(R; r)$, leaving $\sum_j d_{n',j} \int dR \chi_{n'j'}^*(R)\chi_{n'j}(R) = d_{n'j'}$. Thus, a projection by $\phi_{n'}^*(R; r)\chi_{n'j'}^*(R)$ will extract only $d_{n',m'}$, and the usual projection scheme for a complete orthogonal set works.

With this in mind, we address the full pea soup problem for molecules, solids, and in fact all ordinary matter. The Schrödinger equation for the kth many-body eigenstate, with the nuclei and electrons treated in a completely egalitarian fashion, is

$$H(\mathbf{R}, \mathbf{r})\Psi^k(\mathbf{R}, \mathbf{r}) = E_k\Psi^k(\mathbf{R}, \mathbf{r}), \tag{16.2}$$

with $H = T_{nuc}(\mathbf{R}) + T_{elec}(\mathbf{r}) + V(\mathbf{r}, \mathbf{R})$, the quantum Hamiltonian for the electrons $\mathbf{r}$ and nuclei $\mathbf{R}$, all interacting with each other by Coulomb forces, and $\Psi^k(\mathbf{R}, \mathbf{r})$ is an eigenstate of energy E_k. The interaction $V(\mathbf{r}, \mathbf{R})$ contains the all the potential energy—that is, electron attraction to the nuclei, nuclear-nuclear repulsion, and electron-electron repulsion. $T_{elec}(\mathbf{r})$ and $T_{nuc}(\mathbf{R})$, are responsible for the kinetic energies of the electrons and nuclei, respectively.

The Born-Oppenheimer Ansatz begins (without introducing any approximation as yet) by expanding the exact eigenstates $\Psi^k(\boldsymbol{R}, \boldsymbol{r})$ as

$$\Psi_k(\boldsymbol{R}, \boldsymbol{r}) = \sum_{nj} c_{nj}^k \, \varphi_n(\boldsymbol{R}; \boldsymbol{r})\chi_{nj}(\boldsymbol{R}). \tag{16.3}$$

Here, $\varphi_n(\boldsymbol{R}; \boldsymbol{r})$ are a complete set of electronic eigenstates for fixed nuclear configurations $\boldsymbol{R}$; they are fully defined without reference to the state index k and don't change if we examine another total eigenstate $\Psi_{k'}(\boldsymbol{R}, \boldsymbol{r})$ instead of $\Psi_k(\boldsymbol{R}, \boldsymbol{r})$. They satisfy

$$(T_{elec}(\boldsymbol{r}) + V(\boldsymbol{r}, \boldsymbol{R}))\, \varphi_n(\boldsymbol{R}; \boldsymbol{r}) = E_n(\boldsymbol{R})\varphi_n(\boldsymbol{R}; \boldsymbol{r}). \tag{16.4}$$

The index n specifies only the ordering of the electronic states starting with the lowest electronic state energy $n = 1$ for the given $\boldsymbol{R}$. If we follow say the sixth member of this list—that is, $n = 6$, as a function of $\boldsymbol{R}$, we map out a Born-Oppenheimer *potential energy surface* $E_6(\boldsymbol{R})$. We aren't using the states yet in any approximate way.

We choose the $\chi_{nj}(\boldsymbol{R})$, $j = 1, 2, \ldots$, in equation 16.3 to be a complete set of vibration-rotation eigenfunctions in $\boldsymbol{R}$ living on the nth potential energy surface, satisfying

$$\left(-\frac{\hbar^2}{2m}\nabla_{\boldsymbol{R}}^2 + E_n(\boldsymbol{R})\right)\chi_{nj}(\boldsymbol{R}) \;= E_{nj}\chi_{nj}(\boldsymbol{R}). \tag{16.5}$$

This smells like the Born-Oppenheimer idea, but still no approximation has been made because the electronic states $\varphi_n(\boldsymbol{R}; \boldsymbol{r})$ and the nuclear motion states $\chi_{nj}(\boldsymbol{R})$ are being used as complete basis sets in an unrestricted expansion, equation 16.3. We are of course relying on the theorem that all the eigenfunctions of a Hermitian operator (in the full space and not projected in any way onto a subspace) are a complete, orthogonal set of functions. We can contract equation 16.3 by summing over j to write (and defining $h_n^k(\boldsymbol{R})$):

$$\Psi_k(\boldsymbol{R}, \boldsymbol{r}) = \sum_{nj} c_{nj}^k \varphi_n(\boldsymbol{R}; \boldsymbol{r})\chi_{nj}(\boldsymbol{R}) \equiv \sum_n \varphi_n(\boldsymbol{R}; \boldsymbol{r})h_n^k(\boldsymbol{R}). \tag{16.6}$$

We now plug equation 16.6 into the full Schrödinger equation 16.2:

$$\begin{aligned}(T_{nuc}(\boldsymbol{R}) + T_{elec}(\boldsymbol{r}) + V(\boldsymbol{r}, \boldsymbol{R}))&\left(\sum_n \varphi_n(\boldsymbol{R}; \boldsymbol{r})h_n^k(\boldsymbol{R})\right)\\ &= \sum_n (T_{nuc}(\boldsymbol{R}) + E_n(\boldsymbol{R}))\left(\varphi_n(\boldsymbol{R}; \boldsymbol{r})h_n^k(\boldsymbol{R})\right),\\ &= E_k\left(\sum_n \varphi_n(\boldsymbol{R}; \boldsymbol{r})h_n^k(\boldsymbol{R})\right).\end{aligned} \tag{16.7}$$

Multiplying from the left by $\varphi_{n'}^*(\boldsymbol{R}; \boldsymbol{r})$ and integrating over electron coordinates, there results

$$\sum_n \int d\boldsymbol{r}\, \varphi_{n'}^*(\boldsymbol{R}; \boldsymbol{r})\frac{-\hbar^2}{2M}\nabla_{\boldsymbol{R}}^2\left[\varphi_n(\boldsymbol{R}; \boldsymbol{r})h_n^k(\boldsymbol{R})\right] + E_{n'}(\boldsymbol{R})\, h_{n'}^k(\boldsymbol{R}) = E_k\, h_{n'}^k(\boldsymbol{R}). \tag{16.8}$$

The $\nabla_{\boldsymbol{R}}^2$ derivative operator distributes itself across the products $\varphi_n(\boldsymbol{R};\boldsymbol{r})h_n^k(\boldsymbol{R})$ in an obvious notation as $\nabla_{\boldsymbol{R}}^2\left[\varphi_n(\boldsymbol{R};\boldsymbol{r})h_n^k(\boldsymbol{R})\right] = \varphi_n^{\boldsymbol{RR}}(\boldsymbol{R};\boldsymbol{r})h_n^k(\boldsymbol{R}) + 2\varphi_n^{\boldsymbol{R}}(\boldsymbol{R};\boldsymbol{r})\cdot h_n^{k,\boldsymbol{R}}(\boldsymbol{R}) + \varphi_n(\boldsymbol{R};\boldsymbol{r})h_n^{k,\boldsymbol{RR}}(\boldsymbol{R})$. The full Schrödinger equation is

$$\begin{pmatrix} H_1(\boldsymbol{R}) - E_k & N_{12}(\boldsymbol{R}) & N_{13}(\boldsymbol{R}) & \cdots \\ N_{21}(\boldsymbol{R}) & H_2(\boldsymbol{R}) - E_k & N_{23}(\boldsymbol{R}) & \cdots \\ \vdots & \vdots & \vdots & \cdots \end{pmatrix} \begin{pmatrix} h_1^k(\boldsymbol{R}) \\ h_2^k(\boldsymbol{R}) \\ h_3^k(\boldsymbol{R}) \\ \vdots \end{pmatrix} = \begin{pmatrix} 0 \\ 0 \\ 0 \\ \vdots \end{pmatrix}, \tag{16.9}$$

where $H_i(\boldsymbol{R}) = -\frac{\hbar^2}{2M}\nabla_{\boldsymbol{R}}^2 + E_i(\boldsymbol{R}) + N_{ii}(\boldsymbol{R})$, and

$$N_{lm}(\boldsymbol{R}) = \boldsymbol{C}_{lm}(\boldsymbol{R})\cdot\nabla_{\boldsymbol{R}} + D_{lm}(\boldsymbol{R}), \tag{16.10}$$

with

$$\boldsymbol{C}_{lm}(\boldsymbol{R}) = -\frac{\hbar^2}{M}\int \varphi_l^*(\boldsymbol{R};\boldsymbol{r})\nabla_{\boldsymbol{R}}\,\varphi_m(\boldsymbol{R};\boldsymbol{r})\,d\boldsymbol{r},$$

and

$$D_{lm}(\boldsymbol{R}) = -\frac{\hbar^2}{M}\int \varphi_l^*(\boldsymbol{R};\boldsymbol{r})\nabla_{\boldsymbol{R}}^2\,\varphi_m(\boldsymbol{R};\boldsymbol{r})\,d\boldsymbol{r}.$$

The diagonal correction $D_{ii}(\boldsymbol{R})$ is sometimes ignored or it can be incorporated into $E_i(\boldsymbol{R})$ to slightly adjust the Born-Oppenheimer potential energy surfaces. If the kth solution of these coupled equations is found with eigenvalue E_k, then equation 16.6 is the exact solution of the Schrödinger equation.

We finally arrive at the Born-Oppenheimer approximation if the terms $\boldsymbol{C}_{lm}(\boldsymbol{R})$ and $D_{lm}(\boldsymbol{R})$ involving derivatives of the electronic wavefunctions with respect to nuclear coordinates $\boldsymbol{R}$ are ignored (in the hope that the electronic states change slowly with $\boldsymbol{R}$, consistent with the old saw about "nimble" electrons and "sluggish" nuclei). Then $N_{lm}(\boldsymbol{R}) = 0$ and the different electronic states labeled by n become independent, and we have equation 16.5 $\left(-\frac{\hbar^2}{2m}\nabla_{\boldsymbol{R}}^2 + E_n(\boldsymbol{R})\right)\chi_{nj}(\boldsymbol{R}) = E_{nj}\chi_{nj}(\boldsymbol{R})$, for the nuclear wavefunctions, now elevated to the status of "exact" within the adiabatic Born-Oppenheimer approximation, earning the name adiabatic through the assumption that the electrons adiabatically adjust to the nuclear positions in a reversible way—for example, remaining in the 12th electronic state if that is where they started in spite of moving the nuclei or time evolution of the Born-Oppenheimer wavefunction.

The eigenstate index k becomes nj corresponding to the appearance of new "good" quantum numbers n and j belonging to the separation of the motion into electronic φ_n and vibrational χ_{nj} parts. The Born-Oppenheimer approximation to the eigenfunctions is

$$\Psi_{nj}^{B.O.}(\boldsymbol{R},\boldsymbol{r}) = \varphi_n(\boldsymbol{R};\boldsymbol{r})\chi_{nj}(\boldsymbol{R}), \tag{16.11}$$

with $\varphi_n(\boldsymbol{R};\boldsymbol{r})$ obeying equation 16.4 and $\chi_{nj}(\boldsymbol{R})$ obeying equation 16.5.

For clarity, we write down what has become of equation 16.9 now that the Born-Oppenheimer approximation is in place, especially since the subscripts should change to reflect the eigenstate status of the wavefunctions (within the Born-Oppenheimer approximation): Starting with equation 16.5 the eigenfunctions χ_{nj} can be written summarily as simply

$$\left(-\frac{\hbar^2}{2m}\nabla_{\boldsymbol{R}}^2 + E_n(\boldsymbol{R})\right)\chi_n(\boldsymbol{R}) = E_n\chi_n(\boldsymbol{R}). \tag{16.12}$$

This is just like writing $H\chi = E\chi$ instead of $H\chi_j = E_j\chi_j$, except that we already have the index n for the different electronic states with Hamiltonian H_n, so then $H_n\chi_n = E_n\chi_n$ is written to summarize $H_n\chi_{nj} = E_{nj}\chi_{nj}$:

$$\begin{pmatrix} H_1(\boldsymbol{R}) - E_1 & 0 & 0 & \dots \\ 0 & H_2(\boldsymbol{R}) - E_2 & 0 & \dots \\ \vdots & \vdots & \vdots & \dots \end{pmatrix} \begin{pmatrix} \chi_1(\boldsymbol{R}) \\ \chi_2(\boldsymbol{R}) \\ \chi_3(\boldsymbol{R}) \\ \vdots \end{pmatrix} = \begin{pmatrix} 0 \\ 0 \\ 0 \\ \vdots \end{pmatrix}. \tag{16.13}$$

ELECTRONIC TRANSITION MOMENTS

Now that we understand a bit about Born-Oppenheimer states, we should set the stage for transition between them by one of the most important agents: light. The electric dipole approximation is the low-order workhorse of light-matter interaction in molecules and solids, but we caution that many situations require going beyond it, including x-ray fields or whenever the wavelength is not necessarily small compared to the system. We restrict ourselves here to the electric dipole approximation, but use both equivalent "length" and "velocity" forms (see the following). The two forms are often not equivalent in practice: first, their relative ease of application varies with the situation. Second, the length and velocity (and there is a third: acceleration) forms are no longer mathematically equivalent when applied to approximate wavefunctions, such as Born-Oppenheimer states.

The dipole operator (length form) $\vec{\mu}$ for UV, visible, and infrared light-matter interaction of a system with both electrons $\boldsymbol{R}_g$ and nuclei $\boldsymbol{R}$ involves a summation over their charges and coordinates

$$\vec{\mu} = \vec{\mu}_{elec.}(\boldsymbol{r}) + \vec{\mu}_{nuc.}(\boldsymbol{R}); \tag{16.14}$$

with a given polarization of the light $\vec{E} = E_0\hat{e}$, we have

$$\vec{\mu}^{\hat{e}} = \sum_i q_i\, \hat{e}\cdot\boldsymbol{R}_g + \sum_k Q_k\, \hat{e}\cdot\boldsymbol{R}_k. \tag{16.15}$$

The Born-Oppenheimer approximation is key to making sense out of molecular and condensed matter spectroscopy. The wavefunctions consist as we just discussed of a nuclear wavefunction $\chi_{nj}(\boldsymbol{R})$, including both vibrational and rotational degrees of freedom of the molecule, and a wavefunction $\varphi_n(\boldsymbol{R};\boldsymbol{r})$ for the electrons $\boldsymbol{r}$ that depends adiabatically on the nuclear coordinates $\boldsymbol{R}$. Again, the Born-Oppenheimer wavefunction is $\Psi_{nj}^{B.O.}(\boldsymbol{R},\boldsymbol{r}) = \varphi_n(\boldsymbol{R};\boldsymbol{r})\chi_{nj}(\boldsymbol{R})$. Transition amplitudes between two Born-Oppenheimer states n, j and n', j', $t_{n,j\to n',j'}$, induced by light, can be expressed in first-order perturbation theory using the light-matter dipole interaction operator

$$t_{nj\to n',j'} = \langle\varphi_{n'}(\boldsymbol{R};\boldsymbol{r})\chi_{n'j}(\boldsymbol{R})|\,[\vec{\mu}_{elec.}(\boldsymbol{r}) + \vec{\mu}_{nuc.}(\boldsymbol{R})]\,|\varphi_n(\boldsymbol{R};\boldsymbol{r})\chi_{nj}(\boldsymbol{R})\rangle. \tag{16.16}$$

We define the electronic transition moment $\vec{\mu}_{n'n,elec.}(\boldsymbol{R})$ as (integrating only over the electrons, $\boldsymbol{r}$),

$$\vec{\mu}_{n'n,elec.}(\boldsymbol{R}) = \langle \varphi_{n'}(\boldsymbol{R};\boldsymbol{r})|\vec{\mu}_{elec.}(\boldsymbol{r})|\varphi_n(\boldsymbol{R};\boldsymbol{r})\rangle, \tag{16.17}$$

and the nuclear transition moment $\vec{\mu}_{n'n,nuc.}(\boldsymbol{R})$ analogously as

$$\vec{\mu}_{n'n,elec.}(\boldsymbol{R}) = \langle \varphi_{n'}(\boldsymbol{R};\boldsymbol{r})|\vec{\mu}_{nuc.}(\boldsymbol{R})|\varphi_n(\boldsymbol{R};\boldsymbol{r})\rangle. \tag{16.18}$$

Since Born-Oppenheimer states $\varphi_n(\boldsymbol{R};\boldsymbol{r})\chi_{nj}(\boldsymbol{R})$ are orthogonal for fixed $\boldsymbol{R}$ and different n, $\vec{\mu}_{nn',nuc.}(\boldsymbol{R})$ vanishes for $n \neq n'$.

If the electronic state is unchanged, $n = n'$, the expression becomes, after integrating over the electron coordinates,

$$t_{nj\to nj'} = \langle \chi_{nj'}(\boldsymbol{R})| \left[\vec{\mu}_{nn,elec.}(\boldsymbol{R}) + \vec{\mu}_{nn,nuc.}(\boldsymbol{R})\right] |\chi_{nj}(\boldsymbol{R})\rangle, \tag{16.19}$$

where $\vec{\mu}_{nn,elec.}(\boldsymbol{R}) + \vec{\mu}_{nn,nuc.}(\boldsymbol{R}) \equiv \vec{\mu}_{n,molecule}(\boldsymbol{R})$ is the electronic transition moment, a function of the nuclear coordinates $\boldsymbol{R}$. Unless symmetry dictates otherwise, $\boldsymbol{R}$ dependence is expected, since certainly the nuclear repulsions and electronic orbitals and their overlaps needed for the molecular dipole moment $\vec{\mu}_{molecule}(\boldsymbol{R})$ change as atomic distances and bond orientations, etc., change. $\vec{\mu}_{molecule}(\boldsymbol{R})$ controls microwave and infrared spectroscopy mainly, where vibrations and rotations are excited but the system remains on the same Born-Oppenheimer potential energy surface.

If $n' \neq n$,

$$t_{nj\to n'j'} = \langle \chi_{n'j'}(\boldsymbol{R})|\vec{\mu}_{n'n,elec.}(\boldsymbol{R})|\chi_{nj}(\boldsymbol{R})\rangle. \tag{16.20}$$

We will drop the "elec." part of the subscript in the future for Born-Oppenheimer state changing transitions, since it is understood to be present. The transition amplitude $t_{nj\to n',j'}$ is seen to be an integral over nuclear coordinates of a product of the transition moment $\mu_{n'n}(\boldsymbol{R})$ and two often very different vibration-rotation states. Most pairs of vibration-rotation states from Born-Oppenheimer surfaces of quite different shape have small overlap.

We will see that transition moment coordinate dependence and changes in ground versus excited Born-Oppenheimer potential surfaces can both cause vibrational excitation in molecules, or cause phonon excitation in solids. There are many cases where the potential surface shape changes little or not at all for different n, leaving transition moment coordinate dependence as the major or sole agent of vibrations or phonons within the Born-Oppenheimer approximation. A broad class of such cases arises when an electron is promoted to an excited orbital in a large conjugated molecule. Size or shape doesn't change at all in extended conjugated π electron systems such as graphene. In that case, as in molecular cases too, the now empty orbital the electron used to occupy is called a hole, and an electron-hole pair is said to have been produced. The promotion of a single very delocalized π electron to make an electron-hole pair can be of small or truly negligible consequence to the Born-Oppenheimer potentials, which are the result of the great majority of unexcited electrons. In other words, the ground and excited Born-Oppenheimer potential energy surface may be the same, or almost the same, leaving the transition moment coordinate dependence to do the work, so to speak, of causing vibrational excitation.

The velocity (or momentum) form of the transition moment is often convenient because momentum is a derivative in coordinate space, and changes of an extended function may be easier to handle than the function itself. The formal equivalence of the "length" and "velocity" forms of the transition moment arises because of the identity

$$\frac{i\hbar}{m}\boldsymbol{p} = [\boldsymbol{r}, \hat{H}_0], \tag{16.21}$$

where $\hat{H}_0$ is the system Hamiltonian not including the light-matter interaction. We have, ignoring errors in the Born-Oppenheimer approximation,

$$\begin{aligned} p_{n',n}(\boldsymbol{R}) &= \frac{m}{i\hbar}\int d\boldsymbol{r}\ \varphi^*_{n'}(\boldsymbol{R};\boldsymbol{r})[\boldsymbol{r}\hat{H}_0 - \hat{H}_0\boldsymbol{r}]\varphi_n(\boldsymbol{R};\boldsymbol{r}), \\ &= \frac{m}{i\hbar}(E_n(\boldsymbol{R}) - E_{n'}(\boldsymbol{R}))\, r_{n',n}(\boldsymbol{R}), \end{aligned} \tag{16.22}$$

with $p_{n',n}(\boldsymbol{R}) = \int d\boldsymbol{r}\ \varphi^*_{n'}(\boldsymbol{R};\boldsymbol{r})|\,\boldsymbol{p}|\varphi_n(\boldsymbol{R};\boldsymbol{r})$ and $r_{n',n}(\boldsymbol{R}) = \int d\boldsymbol{r}\ \varphi^*_{n'}(\boldsymbol{R};\boldsymbol{r})|\boldsymbol{r}|\varphi_n(\boldsymbol{R};\boldsymbol{r})$. If the potential energy surface is the same for both electronic states, n and n', apart from an energy shift, this becomes

$$p_{n',n}(\boldsymbol{R}) = \frac{m}{i}\ \omega_{nn'}\ r_{n',n}(\boldsymbol{R}), \tag{16.23}$$

where the energy shift is $(E_n(\boldsymbol{R}) - E_{n'}(\boldsymbol{R})) = \hbar\omega_{nn'}$.

It may be dangerous however to ignore the errors in the Born-Oppenheimer wavefunctions. The second line of equation 16.22 revealing the equivalence of the length and velocity forms of the transition moment is strictly true only when exact eigenstates of the Hamlitonian H_0 are used.

If exact eigenstates are not available (they almost never are), both the length and velocity forms may of course be in error, and there is no clear evidence for general superiority of one over the other. Nor is there agreement concerning the accuracy of either.

CORRECTIONS TO THE BORN-OPPENHEIMER APPROXIMATION

The Born-Oppenheimer approximation ignores the terms involving derivatives of the electronic states with nuclear position. We therefore have to be on the alert for rapid change in electronic character with changes in nuclear geometry. This happens quite commonly—for example, at avoided crossings and even more problematic places called conical intersections. Large non-Born-Oppenheimer coupling terms $N_{ij}(\boldsymbol{R})$ inhabit such places, just where the energy splitting is small and the coupling has its greatest effect. (See also section 11.5.) The situation is shown in figure 16.1, which depicts an avoided crossing of Born-Oppenheimer nuclear potential energy surfaces. The electronic make-up or character is indicated by the roman numerals I and II, and also by dashed or solid lines. Dashed is electronic character I, and solid is II. The so-called diabatic states are linear combinations of the Born-Oppenheimer adiabatic states and can be constructed so their energies cross. The diabatic states are of slowly changing electronic character (orbital structure) through the crossing. In contrast, the Born-Oppenheimer adiabatic levels avoid and their corresponding eigenstates rapidly switch from character I to II in the vicinity of the crossing. If the system is to remain, for example, on the upper adiabatic state as the nuclear coordinate moves through the

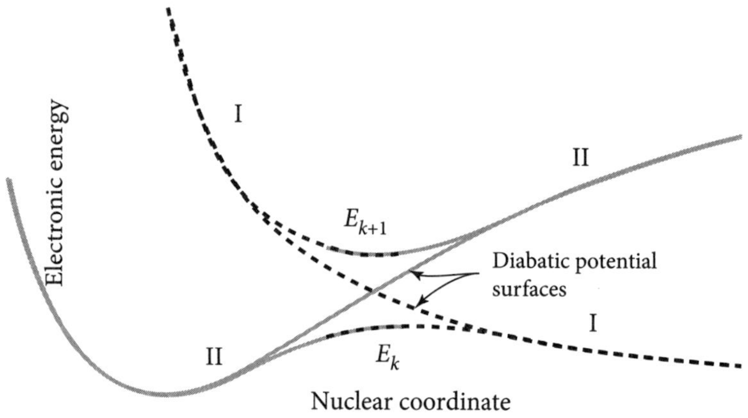

Figure 16.1. An avoided crossing of Born-Oppenheimer nuclear potential energy surfaces. The electronic makeup or character is indicated by the roman numerals I and II, and also with the dashes (I) and solid lines (II). The diabatic states cross and are of consistent character through the crossing; the adiabatic energy levels avoid each other and their corresponding eigenstates rapidly switch character in the vicinity of the crossing. If the system is to remain for example on the upper adiabatic state with energy E_{k+1} as the nuclear coordinate moves, the electrons are obliged to change character (orbital structure) abruptly, near the avoided crossing.

crossing, the electrons are obliged to change their configuration very quickly. Near such places, the notion that the nuclei move sluggishly compared to the electrons can break down, and corrections to the Born-Oppenheimer approximation can be large. The diabatic states are much nearer the truth, if the avoidance of the adiabatic potential energy surface is sudden and close.

In the diabatic representation near a close avoided adiabatic crossing, the electrons momentarily take on the role of the sluggish entities. This is not a complete role reversal; certainly we cannot "pin" the electrons only to let the nuclei move around them! As a consequence, there is a little freedom or perhaps really an ambiguity as to how to define a diabatic representation. In a case of just two states crossing, the diabatic representation is constructed by hand so to speak, as shown in figure 16.1, but defining a globally diabatic approximation is not really practicable.

The adiabatic states avoiding each other narrowly and pair-wise are (at the narrowest part of the gap) 50-50 linear combinations of the diabatic states that cross. If one is sure only two states are strongly interacting, it is easy to "undiagonalize" the adiabatic estates to reveal the two diabatic states $\Psi_I = 1/\sqrt{2}(\psi_{upper} \pm \psi_{lower})$ and $\Psi_{II} = 1/\sqrt{2}(\psi_{upper} \mp \psi_{lower})$. Defining the diabatic surfaces away from the crossing is more problematic, but can often be realized by ad hoc methods of keeping the electronic configuration from mixing, which is what results in adiabatic states. Even at the avoided crossing, potential surfaces seemingly staying clear of the avoided crossing fray can actually be responsible for much of the coupling between the two surfaces, complicating the analysis.

If a nuclear wavepacket on an adiabatic potential energy surface approaches a narrow avoided crossing, the electronic wavefunction will coherently split across both adiabatic states as the nuclei move through the avoided crossing, with some electron amplitude making the drastic change in character (and thus remaining on the adiabatic potential energy surface), and the rest of the electron amplitude keeping its prior character intact *jumping from the upper to the lower (or* vice versa*) adiabatic potential curves.* If the crossing is very narrowly avoided, most of the electron amplitude will jump from one adiabatic surface to another.

We consider the corrections to the Born-Oppenheimer approximation more carefully, using a 2×2 case coming from equation 16.9:

$$\begin{pmatrix} H_1(\boldsymbol{R}) - E_k & N_{12}(\boldsymbol{R}) \\ N_{21}(\boldsymbol{R}) & H_2(\boldsymbol{R}) - E_k \end{pmatrix} \begin{pmatrix} \chi_k^1(\boldsymbol{R}) \\ \chi_k^2(\boldsymbol{R}) \end{pmatrix} = \begin{pmatrix} 0 \\ 0 \end{pmatrix}. \tag{16.24}$$

We focus on the off-diagonal coupling term $\boldsymbol{C}_{12}(\boldsymbol{R}) = \frac{\hbar^2}{M} \int \varphi_1^*(\boldsymbol{R};\boldsymbol{r}) \nabla_{\boldsymbol{R}}\, \varphi_2(\boldsymbol{R};\boldsymbol{r})\, d\boldsymbol{r}$ in equation 16.10, the prefactor of the derivative term $\nabla_{\boldsymbol{R}}$ acting to couple $\chi_{k;1}$ with $\chi_{k;2}$. The revealing aspect of the integrand for constructing $\boldsymbol{C}_{12}(\boldsymbol{R})$ is the spike in $\nabla_{\boldsymbol{R}}\, \varphi_2(\boldsymbol{R};\boldsymbol{r})$ at a narrow approach of $E_1(\boldsymbol{R})$ to $E_2(\boldsymbol{R})$ (where the character of $\varphi_2(\boldsymbol{R};\boldsymbol{r})$ is changing quickly with $\boldsymbol{R}$). In fact, $\varphi_2(\boldsymbol{R};\boldsymbol{r})$ is rapidly acquiring the character of $\varphi_1(\boldsymbol{R};\boldsymbol{r})$ there, so inspection reveals that the integral $\boldsymbol{C}_{12}(\boldsymbol{R})$ will be large there. Thus, the coupling $\boldsymbol{C}_{12}(\boldsymbol{R})$ exhibits a peak just where $E_1(\boldsymbol{R})$ approaches $E_2(\boldsymbol{R})$ and it is easiest to couple $\chi_{k;1}$ with $\chi_{k;2}$. The mitigating factor is that the zone of strong coupling is narrow in space and, assuming a rate of traversal of the region, in time.

DIABATIC REPRESENTATIONS

Remember that an avoided adiabatic crossing entails a drastic change in the electronic wavefunction from well before to well after the avoided crossing. Therefore, a very close, narrow avoided crossing entails a *sudden* change of electronic configuration under a modest change of nuclear geometry. This suggests that the electrons may not be as nimble as they need to be to follow the nuclei adiabatically, and may instead tend to remain largely unchanged as the nuclei quickly pass through the avoided crossing. That is, the electronic state will nonadiabatically "jump" across the avoided crossing, making a diabatic instead of an adiabatic transition. There ought to be a way of handling this in the limit where this is very likely to occur, remaining cognizant of nasty intermediate cases where a 50-50 split in amplitude takes place at the crossing, half diabatic and half adiabatic.

While diabatic methods and approximations may not be so clearly defined as adiabatic ones, there are certainly rigorous treatments. A lot of the literature on this subject assumes a crossing as a function of a single nuclear coordinate, which is not general enough. Modern treatments exist, under the heading of "diabatization" [114, 115]. This is connected to the issue of conical intersections, section 16.2.

PHYSICAL EXAMPLE OF THE ADIABATIC VERSUS DIABATIC LIMITS

Figure 16.2 depicts a not uncommon situation involving a competition between adiabatic and diabatic electronic evolution [116, 117]. The molecule shown supports two π bonds susceptible to $\pi \rightarrow \pi^*$ transitions, one at each end. The ground state is unambiguously bonding on both sides and gives the potential energy contours shown as solid lines at the top left. If we *confine* the excitation to the left side we get the diabatic contour lines for π_L^*, revealing the potential energy surface with excitation forced to reside on the left, and similarly for the right.

This sounds artificial because the light will typically excite the left side and the right side with more or less equal amplitude, depending on its polarization. It is important

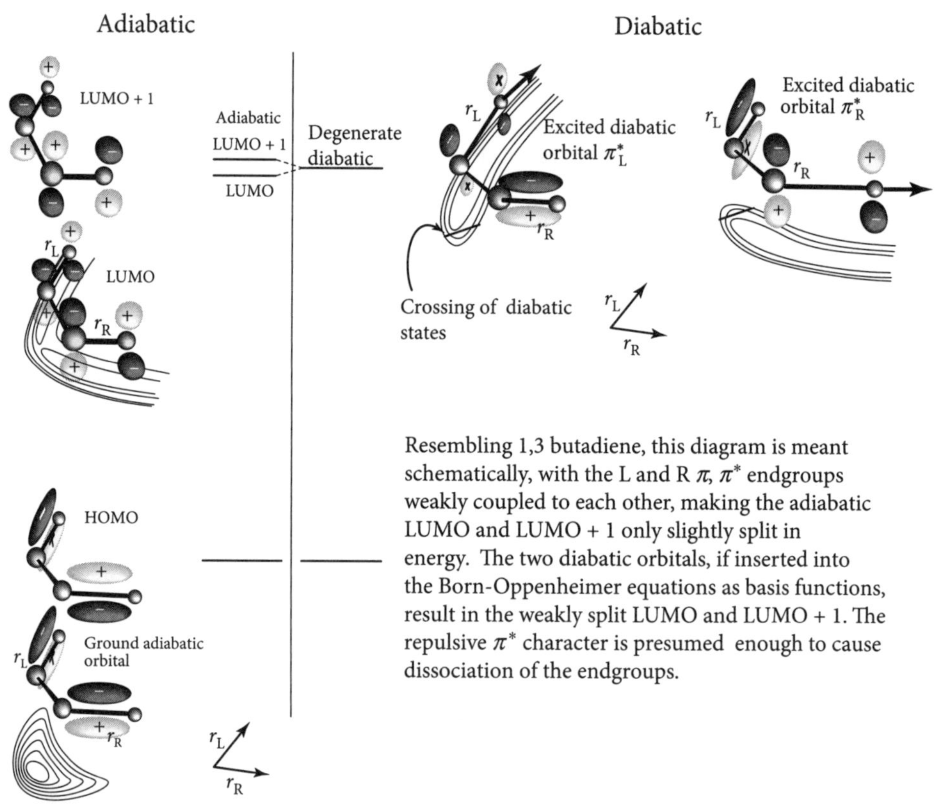

Figure 16.2. Suppose we have a molecule with the structure shown in the upper left, with π to π^* transitions possible at each end, *L* and *R*. The ground electronic state is simple, with bonding orbitals at both ends, and an attractive potential energy surface, holding extension of either bond in check. In the excited state, a photon can induce a π to π^* excitation at either end, but not both, which would require two photons. A π^* configuration at one end is degenerate with the same configuration at the other end, if the bond lengths r_L and r_R are equal. The lower (upper) Born-Oppenheimer adiabatic wavefunction *at equal bond lengths* consists of a symmetric (anti-symmetric) linear combination of the two bond-localized excitations. However, because of the relatively large physical separation of the two bonds, the energy splitting separating these two possibilities is small. In fact, the splitting is the minimum gap found at a narrow avoided crossing as a function of the anti-symmetric vibration in which the two bonds vibrate out of phase. The orbitals in the excited state adiabatic cases change rapidly depending on the relative magnitudes of r_L and r_R. The LUMO is occupied by one electron, the other remaining in a bonding ground-state orbital. The diabatic states have one electron in a π^* orbital on the *L*, and one in a π orbital on the *R*, or vice versa.

to think quantum mechanically: the photon is absorbed *either* at the right *or* the left, it cannot be absorbed at both places. That is, it cannot excite two electrons; computation of the two-electron transition moment vanishes within the usual approximations to the wavefunctions. The two distinct possibilities must necessarily be added coherently, much as a wavefunction for any single particle could be a coherent superposition of two spatially separate wavepackets.

The key to a deeper understanding of the situation is to exploit the fact that the Schrödinger equation is linear! This means we can work out the case of the excitation starting only on the left on Tuesday. On Wednesday, we can treat the case on the right (although here its just a reflection of the left case). Finally, we can add the Tuesday and Wednesday amplitudes on Thursday. The excitation from Tuesday, initially on the left, may or may not couple over to the right, and vice versa. We must solve exactly for what happens on Tuesday in any case, not restricting the solution. Thursday we must add the Tuesday and Wednesday amplitudes. They may or may not overlap strongly. If not, there is little interference between the two cases. Or they might overlap a great deal, possibly causing some processes present on Tuesday and Wednesday to disappear Thursday by destructive interference. Other processes might be enhanced by constructive interference.

In the present example, there is reason to believe there would be little interference, making the diabatic description appropriate: For excitation initially on the left, there is a tendency for that bond to lengthen, breaking the right-left degeneracy fast enough to allow the electronic energy to remain almost totally on the left. Competition is initially set up between the time τ_{LR} for energy to resonate from L to R, given by the adiabatic energy spitting ΔE at the avoided crossing where the two bonds attain equal length according to $\Delta E\ \tau_{LR} \sim \hbar$, and the time τ_U for the right bond to lengthen, detune the resonance, and shut down energy transfer to the left because the bonds have become unequal and nonresonant. If the left bond breaks or dumps energy fast to other modes not shown, resonance is never again (or not for a very long time) established, and the "left first" and "right first" histories are effectively decoupled and incoherent, even if they are formally coherent with each other.

The laser field is almost constant at any given moment across the molecule, and produces a symmetric linear combination of $\pi \to \pi^*$ excitation amplitude on the right (with the left side untouched), and $\pi \to \pi^*$ excitation amplitude on the left with the right side untouched. This gives a nuclear wavefunction residing on the LUMO *adiabatic* excited electronic state, shown at its most probable geometry in the upper right panel. Decomposing this in the Tuesday-Wednesday fashion, diabatic dynamics will prevail in both cases if the end bonds are weakly coupled. Adiabatic evolution would produce movement and extension along the symmetric stretch vibration coordinate.

Raman scattering is sensitive and informative in this situation [118, 119].

16.2 Aspects of the Born-Oppenheimer Approximation

Electron-phonon (or vibration) interactions are part of the Born-Oppenheimer approximation. We hesitate to use the diffuse term *vibronic interactions*, since it apparently means different things to different people. There are plenty of electron-nuclear interactions embedded in the Born-Oppenheimer approximation without having to violate it, because the electron configurations of course alter with nuclear position. The electrons are free to exert forces, even suddenly, within the Born-Oppenheimer approximation: when the electronic state is changed from n to n' by a photon for example, the nuclei can respond to the changed electronic states by accelerating and oscillating on a new potential energy surface $E_{n'}(\boldsymbol{R})$. This happens especially in molecules that are either small or have localized orbitals, or at impurities in crystalline material. In these cases, local electronic excitations can cause significant geometry changes in excited electronic states—including falling apart!

The vibrational state (phonons in the solid-state case) can be instantly changed (or phonons instantly created) upon photoabsorption or emission by the coordinate dependence of the electronic transition moment. We use the word *instant* for the changes caused by electronic transition moment $\vec{\mu}_{n'n}(\boldsymbol{R})$ because the nuclear wavefunction arrives on the new Born-Oppenheimer surface as $|\chi_{n'}(\boldsymbol{R})\rangle = \vec{\mu}_{n'n}(\boldsymbol{R})|\chi_{nj}(\boldsymbol{R})\rangle$. In terms of the vibrational states $|\chi_{n'j}(\boldsymbol{R})\rangle$ on the new surface n', this can be expanded as

$$|\chi_{n'}(\boldsymbol{R})\rangle = \vec{\mu}_{n'n}(\boldsymbol{R})|\chi_{nj}(\boldsymbol{R})\rangle = \sum_j c_j |\chi_{n'j}(\boldsymbol{R})\rangle.$$

This principle is clear in the first order time dependent formulation, and is a boundary condition before any motion occurs; see chapter 17. Even if the electronic Born-Oppenheimer potential surfaces n and n' are copies of each other, as is often the case in crystalline material, the vibrational or phonon content has been changed by $\vec{\mu}_{n'n}(\boldsymbol{R})$, instantly, upon photoabsorption.

Energy is subsequently exchanged between nuclei and electrons in the usual Born-Oppenheimer way, in the excited electronic state. Thus, as nuclei move downhill on a potential energy surface, they gain kinetic energy at the expense of the lowering of the electronic energy $E_{n'}(\boldsymbol{R})$, but the quantum number of the electronic state n' is not changing. If the nuclei return to the same configuration, the electronic state returns to its former self also (apart from possible geometric or Berry phases; see the following. There is no room for inelastic electron-phonon scattering; in the molecular case, this is called *internal conversion*) within the Born-Oppenheimer approximation. Both nonetheless happen, due to breakdown of the Born-Oppenheimer approximation, especially near surface crossings.

Correctly calculating quantum surface crossing dynamics is difficult. Wavepacket spawning methods [78, 120] have been invented that parse the amplitude well at difficult and crucial places. Alternatively, there is often a good argument to be made that the coherence between amplitude lying on different Born-Oppenheimer surfaces is not important with many degrees of freedom present; then essentially classical "surface hopping" methods can work extremely well [121].

NONCROSSING RULE, CONICAL INTERSECTIONS, AND BERRY PHASE

In figure 16.1, we plotted two Born-Oppenheimer energies as a function of a single coordinate x, that we suppose is a one-dimensional path in the multidimensional coordinate space $\boldsymbol{R}$. The Born-Oppenheimer potential energy surface is a function of all the nuclear coordinates (assuming the center of mass and overall rotation has been removed). The adiabatic potential surfaces are seen to avoid, not to cross. They avoid crossing because there is (almost) always some coupling between the two electronic configurations as they nominally reach degeneracy, causing them instead to split, as always with a 2×2 matrix with degenerate diagonal elements but nonzero off diagonal elements.

Can adiabatic Born-Oppenheimer surfaces ever cross? The noncrossing rule was given in 1929 by von Neumann and Wigner [122], after an earlier claim by Hund: "potential energy curves corresponding to electronic states of the same symmetry cannot cross." This claim is true only when geometry is changed by altering a single nuclear coordinate with others fixed arbitrarily. If even two coordinates are allowed to be varied, degeneracies may be found, called *conical intersections*, that take the

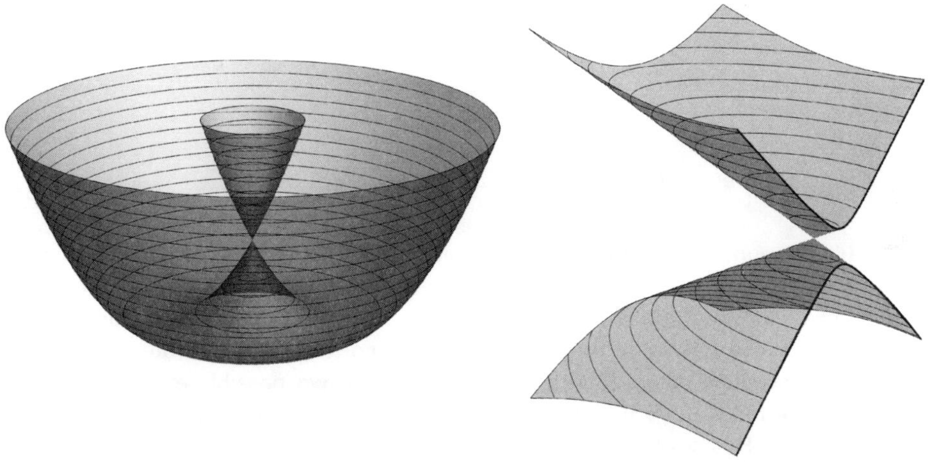

Figure 16.3. A conical intersection (left) and a slice through a conical intersection (right) displaying the two eigenvalues of a 2×2 matrix whose elements depend on the two parameters x and y in this plot (see text). A typical "slice," varying in only one parameter, is shown on the right as the cut-off, darkened edges. Those edges exhibit a generic avoided crossing. Picking the other parameter randomly, there is a 0% chance of passing exactly through the point where the cones touch—that is, an exact crossing as a function of one parameter. Clearly, the exact crossing can be found by varying two parameters.

form shown in figure 16.3: locally, two cones, one inverted, are touching at a point. Such regions are legion for allowing allowing diabatic transfer of amplitude from one adiabatic surface to another (adiabatic behavior would retain the amplitude on whichever cone it began on).

Conical intersections generate an important example of a wide class of problems having a *geometric phase*, or *Berry phase* [123]. Sir Michael Berry did not discover the first example of what is now called the Berry phase, but instead he was the first to recognize the ubiquity, mathematical framework, and universal features [123]. The problem is nicely illustrated (but not in its full generality) by the geometric phase (better known as the Berry phase) at a conical intersection. Among the most important antecedents is the work by Mead and Truhlar [124], which anticipated much of Berry's analysis. It also is associated with an extensive study by the same authors, and co-workers, of the performance of the adiabatic approximation under stress of electronic degeneracy.

The exposition in reference [123] is highly recommended, but one of the antecedents is particularly simple [125]. The model takes the form

$$\begin{pmatrix} A+kx-E & cy \\ cy & A-kx-E \end{pmatrix} \begin{pmatrix} c_1(x,y) \\ c_2(x,y) \end{pmatrix},$$

which is meant to represent the electronic Hamiltonian in a basis set of two orbitals as a function of two nuclear coordinates x and y. The energy eigenvalues are

$$E = A \pm \sqrt{k^2x^2 + c^2y^2},$$

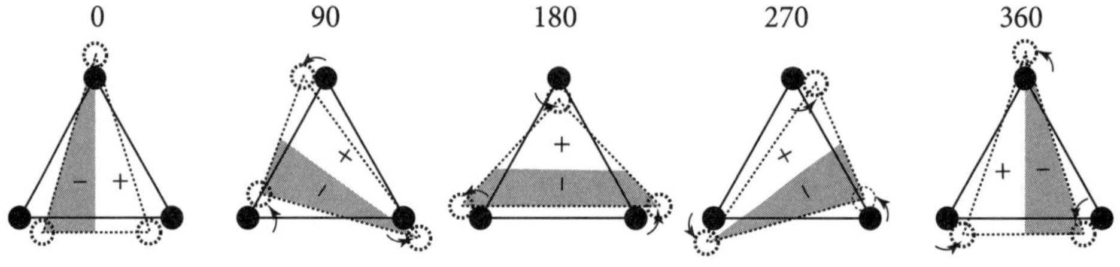

Figure 16.4. Accumulation of a Berry phase upon a circuit around the equilateral triangular geometry. In an excited electronic state near the E double degeneracy of H_3^+ the nuclei make a 360-degree circuit, moving adiabatically. Note the wavefunction changes sign upon a full circuit—a Berry phase effect. This corresponds to a round-trip on the upper cone (because this is the higher energy of the two possible excited states we could have started with on the left) in figure 16.3, left. The problem arises for a simple atomic basis for H_3^+, in which the two E-type degenerate molecular electronic states (at the equilateral triangular geometry) are the two basis states referred to earlier.

which is the equation of a double cone with it vertex at the origin. Defining

$$kx = R\cos\theta \text{ and } cy = R\sin\theta,$$

we find

$$\frac{c_1}{c_2} = \frac{-\sin\frac{1}{2}\theta}{1+\cos\theta} = -\tan\tfrac{1}{2}\theta,$$

and with it the inescapable conclusion that the wavefunction changes sign upon a circuit if of 2π (see figure 16.4).

It was H. C. Longuet-Higgins [126] who first addressed the question of multiple surface crossings properly: two coordinates must be varied to find crossings of two or higher dimensional potential surfaces. Figure 16.5 gives a spectrum of a two-dimensional nonlinear oscillator as a function of one of the frequency parameters. There are no crossings, but many close approaches.

It is worth going through a simple case. As two curves approach each other and look like they might cross, a 2×2 Hamiltonian matrix involving the two states becomes sufficient to examine the details of the crossing; the two states are usually interacting much more strongly with each other potentially than states much farther away in energy. The problem takes on the form

$$\begin{pmatrix} H_{11}(\lambda) - E & H_{12}(\lambda) \\ H_{21}(\lambda) & H_{22}(\lambda) - E \end{pmatrix} \begin{pmatrix} c_1 \\ c_2 \end{pmatrix} = \begin{pmatrix} 0 \\ 0 \end{pmatrix}.$$

In order for the two energy eigenvalues of this equation to coincide, two conditions must be met: $H_{11}(\lambda) = H_{22}(\lambda)$, and $H_{12}(\lambda) = H_{21}(\lambda) = 0$. Both conditions cannot be simultaneously reached, in general, by varying only one parameter. So as λ is varied, two eigenvalues may approach each other but will not cross, and instead pass through an "avoided crossing" (figure 16.1). This is the "noncrossing rule" for potential energy curves varying as a function of a single parameter. However, if we have two parameters to vary arbitrarily, say λ and μ, the condition $H_{11}(\lambda, \mu) = H_{22}(\lambda, \mu)$, and $H_{12}(\lambda, \mu) = H_{21}(\lambda, \mu) = 0$ could be arranged in general. Such a degeneracy point and its vicinity is called a *conical intersection*, since Taylor expansion of

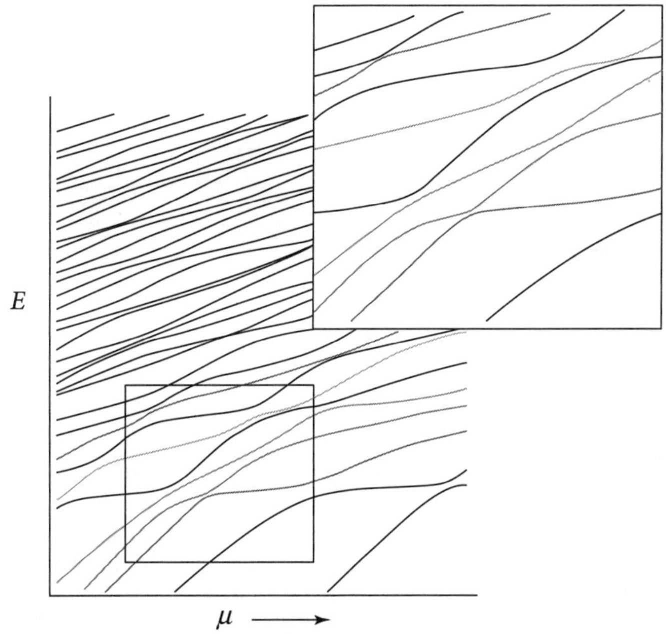

$$H = p_x^2/2 + p_y^2/2 + \mu^2 x^2/2 + \omega_y^2 y^2/2 + 0.2x^2 y$$

Figure 16.5. Part of the spectrum of the Hamiltonian is shown, as a function of the frequency parameter ω_x. Notice the multitude of avoided crossings; there are no true crossings, since only one parameter is being varied. The inset shows some detail inside the indicated box. If we replace 0.2 by λ and vary both λ and ω_x, conical intersections could be found, as in figure 16.3.

$H_{11}(\lambda, \mu)$, $H_{22}(\lambda, \mu)$, and $H_{12}(\lambda, \mu) = H_{21}(\lambda, \mu)$ around a degenerate (λ, μ) reveals the upper and lower potential energy surfaces to be conical in shape, with the tips of the cones touching at the degeneracy point, as in figure 16.3.

Longuet-Higgins also showed that more than two surfaces can cross at the same places in configuration space: for an n-fold crossing, $n(n+1)/2 - 1$ independent coordinates need to be varied to find such intersections. For example, threefold degeneracies (requiring variation of five coordinates) are possible in molecules containing four or more atoms (four atoms gives 12 nuclear coordinates; we subtract 3 for the center of mass and 3 more for orientation in space, leaving 6 coordinates). Three atoms leaves only three coordinates—still enough for two surfaces to cross everywhere on two-dimensional sheets, but there are no points with three energies degenerate. If there are many coordinates, crossing degeneracies happen not just at points but whole surfaces, albeit of lowered dimensionality. Entire K-2 dimensional surfaces support degeneracy of two electronic energies of the same symmetry in a K-dimensional coordinate space, for example.

FAST VERSUS SLOW

When we treat the electrons as nimble particles compared to nuclei, we are invoking a quantum version of the adiabatic approximation (see section 4.2 for a classical discussion). Electrons adjust adiabatically to the sluggish changes in nuclear position.

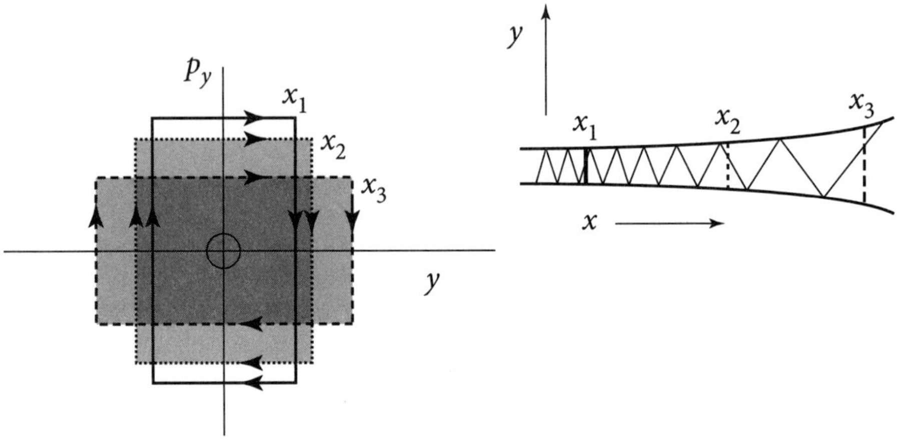

Figure 16.6. Classical adiabatic motion. As the bouncing trajectory makes its way to the right in x, the walls grow further apart in y. Yet the action in the p_y, y plane (area of the phase space boxes) remains nearly constant, even as the energy in y decreases, as evidenced by the decrease in momentum p_y during the free path between walls at increased x. This is a manifestation of the correspondence principle, since the classical action corresponds to a quantum number, as in $\oint p(y)\, dy = (n + \frac{1}{2})h$.

The adiabatic approximation works well classically too, as in figure 16.6 the particle steadily accelerates to the right, at the expense of the vertical component of the energy. We can say more than that, noting that the action of the classical y motion for fixed x remains nearly constant, even as the potential changes (here the distance between the vertical walls in y).

There are many other situations in quantum systems where some variables might be treated as fast compared to others. In problems involving vibrational motion only, some vibrational frequencies may be very slow compared to others, and a Born-Oppenheimer approximation may be derived just as earlier, greatly simplifying the solution, and providing intuition for it. In figure 16.7, a numerical simulation of a quantum wave traveling from left to right in a hard-walled waveguide widening slowly to the right is shown. The transverse motion plays the role of the fast variable, and the longitudinal motion the slow variable. An effective potential develops, sloping downhill to the right in the waveguide. The wave visibly accelerates to the right, with wavelength growing shorter in that direction.

The adiabatic approximation for this waveguide problem consists of treating x as the slow coordinate, since the potential in the y direction changes slowly with x and the y motion should be able to adjust to the x position without mixing different y states, found by fixing x. This is analogous to the electrons adjusting to the nuclear positions while remaining in states found at fixed nuclear positions. We write the wavefunction as

$$\Psi^k(x, y) = \psi_n(x; y)\chi_{nj}(x), \tag{16.25}$$

which as yet need not be approximate. The Born-Oppenheimer approximation arises if (1) $\psi_n(x; y)$ is defined by the y-dependent quantum solution holding x fixed; and

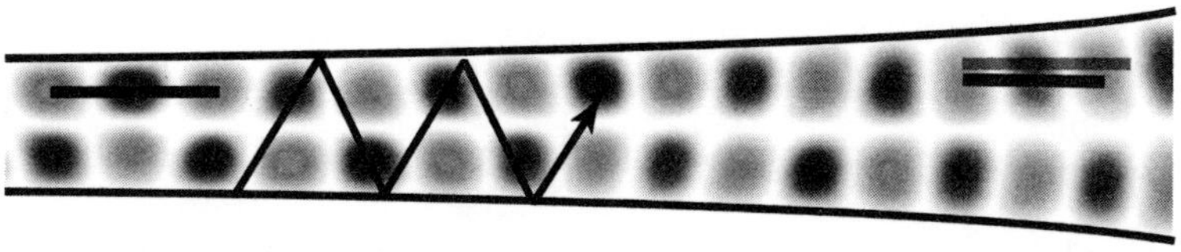

Figure 16.7. Adiabatic dynamics in a waveguide. A numerical simulation is shown of a quantum wave traveling from left to right in a waveguide that is widening slowly to the right (slowly compared to the wavelength). The wave begins at the left with one node in the transverse (vertical, *y*) direction, corresponding to the first excited state of a one-dimensional particle in a box in the *y* direction. Even though the speed longitudinally is comparable to the vertical speed, as seen from the wavelength in *x* compared to *y*, the particle bounces vertically many times in the classical sense (see bouncing path) before the width of the waveguide changes much, making the longitudinal motion the "slow" coordinate—that is, the width is changing slowly as a function of *x*. Note that the fast motion in *y* remains in its first excited state, as adiabatic theory would predict. This would not hold if the waveguide widened suddenly. The energy of a particle in the box goes as $a(x)^{-2}$, where $a(x)$ is the width of the box. The fast vertical motion does lose energy as $a(x)$ gets larger to the right; the horizontal motion picks up that energy: note that the longitudinal wavelength has become significantly shorter on the right, compared to the left. (Lines at both ends are fit to the local wavelength; the gray line at the right is the same length as the back line on the left.) Adiabatically, the longitudinal motion sees an effective downhill potential to the right, even though the waveguide has a perfectly flat bottom. The zero point energy in *y* is dropping to the right, and that lowered energy is being supplied to the longitudinal (*x*) motion.

(2) later ignoring x derivatives of $\psi_n(x; y)$. For the first, we have

$$-\frac{\hbar^2}{2}\frac{\partial^2}{\partial y^2}\psi_n(x; y) = E_n(x)\psi_n(x; y), \tag{16.26}$$

$$\psi_n(x; y) = \sqrt{\frac{2}{a(x)}}\sin\left(\frac{n\pi y}{a(x)}\right); \; E_n(x) = \frac{n^2\pi^2\hbar^2}{2a_n(x)^2}, \tag{16.27}$$

where $a_n(x)$ is the width of the waveguide at x. For the second, we write

$$\begin{aligned}-\frac{\hbar^2}{2}\left(\frac{\partial^2}{\partial y^2}+\frac{\partial^2}{\partial x^2}\right)\psi_n(x; y)\chi_{nj}(x) &= E_{nj}\psi_n(x; y)\chi_{nj}(x),\\ &\approx -\psi_n(x; y)\left(\frac{\hbar^2}{2}\frac{\partial^2\chi_{nj}(x)}{\partial x^2}+E_n(x)\chi_{nj}(x)\right),\end{aligned} \tag{16.28}$$

or dropping the common factor $\psi_n(x; y)$,

$$-\frac{\hbar^2}{2}\frac{\partial^2\chi_{nj}(x)}{\partial x^2}+E_n(x)\chi_{nj}(x) = E_{nj}\chi_{nj}(x). \tag{16.29}$$

This is a one-dimensional Schrödinger equation for a particle on the potential

$$E_n(x) = \frac{n^2\pi^2\hbar^2}{2a_n(x)^2},$$

which is downhill to the right if $a_n(x)$ is increasing to the right, as in figure 16.6. The prediction is that the state of y motion (in this case, $n = 2$) remains the same, but as its energy is lowered and transferred to the longitudinal motion, that is seen to be indeed developing a shorter wavelength in figure 16.6. The Born-Oppenheimer approximate solution is constructed as in equation 16.25, with the stated ways of finding $\psi_n(x; y)$ and $\chi_{nj}(x)$.

Chapter 17

Time-Dependent Formulations of Spectroscopy

A *spectrum* can mean many things, but very often we think of it as absorption or emission of radiation by a system, reported as a function of light frequency. The system can range from a bulk crystal, to mixtures in the gas, liquid, or solid states, to a pristine, isolated molecule in the gas phase.

17.1 Coherent Versus Single-Molecule Spectroscopy

When exciting an extended collection of molecules with coherent radiation, a collective coherence may develop with strict phase relationships across the sample. This coherence can strongly restrict the allowable frequency and momentum for absorption and emission of pump or probe radiation. The coherences may be very useful, especially for obtaining stronger signals, but they may also distract attention from the intrinsic internal dynamics of the individual species making up the bulk material.

For example, in coherent anti-Stokes Raman spectroscopy (CARS), the signal is enhanced by virtue of a coherent addition of amplitude from many individual molecules. The Stokes emission transition is pumped rather than spontaneous, populating an excited vibrational level and making the anti-Stokes emission upon excitation with another photon strong and coherent. The Stokes laser must be tuned again to see another transition; the whole spectrum is not spontaneously present, as it is in ordinary Raman scattering. The molecular levels are seen to exist, but their reasons for existing may become obscured.

A lucid discussion of CARS and other electronic spectroscopies using a wave packet formalism is given in David Tannor's book [1]. His discussion is essential reading; it puts much of the intuitive physics back into the game. Here, we will focus on individual system dynamics, neglecting sample coherences. The system can be large, or reside in a medium, but we do not consider coherences with neighboring identical systems.

For a many-body system, which might be a solid or even a modest sized molecule, eigenstates typically remain unresolved even in the best experimental spectra. The finest spectral feature in an experiment, as narrow as say $0.01\ \mathrm{cm}^{-1}$, might have hundreds or trillions of eigenstates contributing to it. Clearly, one does not want a theoretical description that requires knowing the individual eigenstates, even if they were available, which they are not.

Only finite times are needed to understand and compute emission and absorption spectra, depending on the time-energy uncertainty principle and the experimental resolution to be predicted or understood.

Even very sharp Raman spectra are often the result of remarkably short time processes, femtoseconds or less. We will explain why this is so in section 17.8. The time energy uncertainty principle does not apply to the excited state lifetime and Raman line widths!

The physics of these spectroscopies cries out for a short time description, returning to the frequency domain by a simple, one dimensional time-energy Fourier transform. The short time nature of the process already belies the need for eigenstates. On the contrary, many of the intuitive secrets of the dynamics and resulting features of the spectrum can go hidden if the eigenstates are used for the calculations (assuming one can find them).

Prof. Jeff Cina, through "wave packet interferometry," has achieved new levels of understanding of two- and three-photon pulsed spectroscopies (that are given less attention than they deserve in this book); a review of some of this work is given in reference [127].

17.2 First-Order Light-Matter Interaction

Here, we consider "one-photon" spectroscopy, which could involve absorption or emission. If a change of electronic state is involved, we use a two component wavefunction, giving the amplitude on each of two Born-Oppenheimer surfaces. The generalization to many surfaces (but still involving only a single photon) is formally quite straightforward.

Spontaneous emission occurs at a rate as if it were driven by an applied laser field, if the classical field strength is taken to be that of the quantum zero point energy of the field [128]. This is a small trick coming from quantum electrodynamics.

17.3 Transition Moment Operator

We work within the Born-Oppenheimer approximation for ground (or the initial) electronic state g and final electronic state f (or sometimes we use e if an excited electronic state is involved; see equation 16.11), with gj' and fj' labeling the vibrational state in the initial and final Born-Oppenheimer surface, respectively:

$$\begin{aligned}|f\rangle &= |\psi_{fj'}^{B.O.}(\boldsymbol{\xi}, \boldsymbol{r})\rangle = |\phi_f(\boldsymbol{\xi}; \boldsymbol{r})\rangle|\chi_{fj'}(\boldsymbol{\xi})\rangle,\\ |g\rangle &= |\psi_{gj}^{B.O.}(\boldsymbol{\xi}, \boldsymbol{r})\rangle = |\phi_g(\boldsymbol{\xi}; \boldsymbol{r})\rangle|\chi_{gj}(\boldsymbol{\xi})\rangle.\end{aligned} \tag{17.1}$$

We may write, for the transition moment between ground and excited electronic states

$$\langle f|\boldsymbol{D}^\rho|g\rangle = \langle \chi_{fj'}(\boldsymbol{\xi})|\langle \phi_f(\boldsymbol{\xi}; \boldsymbol{r})|\boldsymbol{D}^\rho(\boldsymbol{r})|\phi_g(\boldsymbol{\xi}; \boldsymbol{r})\rangle|\chi_{gj}(\boldsymbol{\xi})\rangle \tag{17.2}$$

as

$$\langle f|\boldsymbol{D}^\rho|g\rangle = \langle \chi_{fj'}(\boldsymbol{\xi})|\mu_{fg}^\rho(\boldsymbol{\xi})|\chi_{gj}(\boldsymbol{\xi})\rangle, \tag{17.3}$$

with

$$\mu_{fg}^\rho(\boldsymbol{\xi}) = \langle \phi_f(\boldsymbol{\xi}; \boldsymbol{r})|\hat{\boldsymbol{D}}^\rho(\boldsymbol{r})|\phi_g(\boldsymbol{\xi}; \boldsymbol{r})\rangle_{\boldsymbol{r}}. \tag{17.4}$$

The matrix elements of the dipole operator $\boldsymbol{D}$ between two Born-Oppenheimer electronic states is the *transition moment* $\mu^{\rho}_{fg}(\boldsymbol{\xi})$ connecting occupied electronic state g and empty electronic states f. $\mu^{\rho}_{fg}(\boldsymbol{\xi})$ is written for light polarization ρ; the subscript $\boldsymbol{r}$ indicates that only the electron coordinates are integrated. Note that $\mu^{\rho}_{fg}(\boldsymbol{\xi})$, is explicitly a function of nuclear coordinates $\boldsymbol{\xi}$ (if it does not vanish by symmetry, such as translational symmetry, e.g. momentum conservation).

ONE ELECTRONIC STATE

Slightly extending the notation of section 5.1 to apply to the Born-Oppenheimer problem, with radiation playing the role of the perturbation having a time dependence $v(t')$, we can write, starting in any ground vibrational state j,

$$\begin{aligned}\chi_{gj}(\boldsymbol{R},t) &= \chi^{(0)}_{gj}(\boldsymbol{R},t)+\chi^{(1)}_{gj}(\boldsymbol{R},t)+\cdots\\ &= \chi^{(0)}_{gj}(\boldsymbol{R},t)-\frac{i}{\hbar}\int_{-\infty}^{t} dt'\; G_0^{+}(t-t')\; V(\boldsymbol{R},t')\; G_0^{+}(t')\chi^{(0)}_{gj}(\boldsymbol{R},0),\end{aligned} \tag{17.5}$$

with

$$G_0^{+}(t')=e^{-iH_g^{B.O.}t'/\hbar}\quad\text{and}\quad V(\boldsymbol{R},t')=v(t')\mu_{gg}(\boldsymbol{R}), \tag{17.6}$$

where $\mu_{gg}(\boldsymbol{R})$ is the ground to ground state transition moment as described surrounding equation 16.20. Suppose the light intensity rises and falls with center frequency ω as in $v(t')=\exp[-\alpha(t'-t_0)^2-i\omega t']$, and we ask how much probability (according to first-order perturbation theory) has flowed into vibrational states j', including $j'=j$, starting with the system in the initial vibrational state j. The total first-order wavefunction is

$$\varphi^{(1)}_{gj}(\boldsymbol{R},t)=-\frac{i}{\hbar}\int_{-\infty}^{t} dt'\; e^{-iH_g^{B.O.}(t-t')/\hbar}\; e^{-\alpha(t'-t_0)^2-i\omega t'}\;\mu_{gg}(\boldsymbol{R})\chi^{(0)}_{gj}(\boldsymbol{R},t'), \tag{17.7}$$

and the total first-order probability can be calculated by taking the absolute value squared of $\varphi^{(1)}_{gj}(\boldsymbol{R},t)$ and integrating over $\boldsymbol{R}$. We start in the level j and compute the total probability $\Sigma_j(\omega)$ residing in first order wavefunction—that is, $\Sigma_j(\omega)\propto\int d\boldsymbol{R}\,|\phi^{(1)}_{gj}(\boldsymbol{R},\infty)|^2$. Passing to the limit of a very long (wide Gaussian) pulse and time t taken nonetheless after the pulse, a few manipulations give

$$\Sigma_j(\omega)\propto\int_{-\infty}^{\infty} dt\; e^{i\omega t}\langle\chi^{(0)}_{gj}|\mu_{gg}(\boldsymbol{R})\mu_{gg}(\boldsymbol{R},t)|\chi^{(0)}_{gj}\rangle, \tag{17.8}$$

where

$$\mu_{gg}(\boldsymbol{R},t)=e^{-iH_g^{B.O.}t/\hbar}\;\mu_{gg}(\boldsymbol{R})e^{iH_g^{B.O.}t/\hbar}. \tag{17.9}$$

We may write

$$\Sigma_j(\omega)\propto\int_{-\infty}^{\infty} dt\; e^{i\omega t}\langle\mu_{gg}\mu_{gg}(t)\rangle, \tag{17.10}$$

that is, the Fourier transform at frequency ω of the autocorrelation of the transition dipole μ_{gg}. This is a quantum autocorrelation, with

$$\mu_{gg}(t) = e^{-iH_g^{B.O.}t/\hbar} \cdot \mu_{gg} \cdot e^{iH_g^{B.O.}t/\hbar}. \tag{17.11}$$

There is an uninteresting "peak" in $\Sigma_j(\omega)$ at $\omega = 0$ corresponding to remaining in the initial state j. The biggest nontrivial peaks are usually at the frequency of the $0 \to 1$ transitions for vibrational modes.

If we expand $\mu_{gg}(\boldsymbol{R}) = \mu_{gg}(\boldsymbol{R_0}) + \nabla\mu_{gg}(\boldsymbol{R_0}) \cdot (\boldsymbol{R} - \boldsymbol{R_0}) + \cdots$, then $\mu_{gg}(\boldsymbol{R}, t) = \mu_{gg}(\boldsymbol{R_0}) + \nabla\mu_{gg}(\boldsymbol{R}_0) \cdot (\boldsymbol{R}(t) - \boldsymbol{R}_0) + \cdots$. The time-dependent operator $\boldsymbol{R}(t)$ governs the motion of nuclear or internuclear coordinates $\boldsymbol{R}$. Now we can write, for a diatomic, say where R is the internuclear distance, and keeping the nonzero frequency part,

$$\Sigma^e(\omega) \propto \int_{-\infty}^{\infty} dt\, e^{i\omega t} \langle \boldsymbol{R} \cdot \boldsymbol{R}(t) \rangle, \tag{17.12}$$

that is, the Fourier transform of the expectation value of the bond distance autocorrelation, or essentially the dipole moment autocorrelation.

The formula recalls a classical dipole autocorrelation, which indeed can be substituted as a semiclassical approximation (there are important improvements and subtleties; see for example references [129] and [130]).

There are a couple of interesting directions we can take this. The most important approach is to sum over thermally populated initial states (that is, not just one initial j), obtaining the observed spectrum at finite temperature. Another is extending the state space to include a bath, even of many degrees of freedom. If the latter is represented semiclassically, many degrees of freedom can indeed be included [131].

TWO ELECTRONIC STATES

An electromagnetic field, say $\vec{E}(t) = \vec{E}_0 \cos\omega t$, can induce absorption or stimulated emission between states that differ in energy by $\hbar\omega$. It is straightforward to apply our results on perturbation theory from section 5.1, adapted to a two-state basis—that is, an initial and final electronic state linked by radiation. We consider

$$\begin{pmatrix} \chi_{ej}(\boldsymbol{R}, t) \\ \chi_{gj}(\boldsymbol{R}, t) \end{pmatrix} = G_0^+(t) \begin{pmatrix} 0 \\ \chi_{gj}^{(0)}(\boldsymbol{R}, 0) \end{pmatrix} - \frac{i}{\hbar} \int_{-\infty}^{t} dt'\, G_0^+(t-t')\, V(t')\, G_0^+(t') \begin{pmatrix} 0 \\ \chi_{gj}^{(0)}(\boldsymbol{R}, 0) \end{pmatrix},$$

with

$$G_0^+(t') = \begin{pmatrix} e^{-iH_e^{B.O.}t'/\hbar} & 0 \\ 0 & e^{-iH_g^{B.O.}t'/\hbar} \end{pmatrix}, \tag{17.13}$$

and

$$V(t') = v(t') \begin{pmatrix} 0 & \mu_{eg}(\boldsymbol{R}) \\ \mu_{ge}(\boldsymbol{R}) & 0 \end{pmatrix}. \tag{17.14}$$

Define a time-propagated version of $\varphi(\boldsymbol{R}, 0) \equiv \mu_{eg}(\boldsymbol{R})\chi_g^{(0)}(\boldsymbol{R}, 0)$ arriving on the excited state as

$$\varphi(\boldsymbol{R}, t) \equiv e^{-iH_e^{B.O.}t/\hbar} \left(\mu_{eg}(\boldsymbol{R}) \cdot \chi_g^{(0)}(\boldsymbol{R}, 0)\right) = e^{-iH_e^{B.O.}t/\hbar}\varphi(\boldsymbol{R}, 0). \tag{17.15}$$

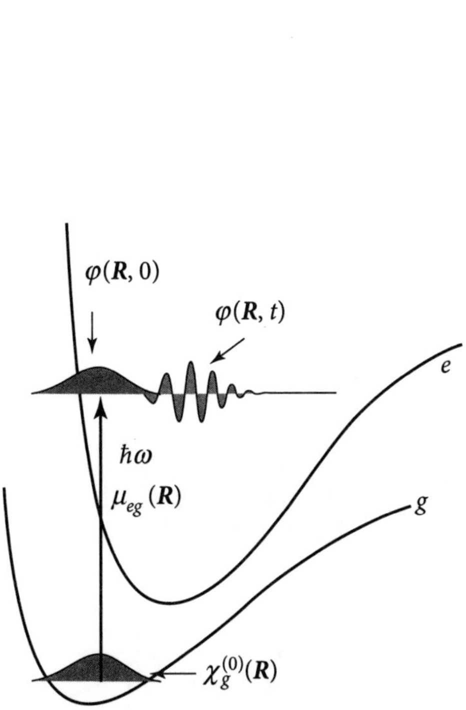

Figure 17.1. Early time dynamics of the wavepacket $|\phi(t)\rangle$ in the simple case of an electronic transition of a single bond. The state $\chi_g^{(0)}(\boldsymbol{R}, 0)$ is the ground-state vibrational wavefunction to be multiplied by the transition moment $\mu_{eg}(R)$. $\mu_{eg}(R)$ is usually too mild to shift the product $\mu_{eg}(R)\chi_g^{(0)}(\boldsymbol{R}, 0)$ very far from $\chi_g^{(0)}(\boldsymbol{R}, 0)$. Therefore, $|\varphi(\boldsymbol{R}, 0)\rangle$ finds itself perched on the side of the excited state Born-Oppenheimer potential surface just above the equilibrium geometry of the ground state. *In spite of the temptation to think of all this as a semiclassical wavepacket approximation of some sort, it is in fact exact Born-Oppenheimer theory plus first-order light-matter perturbation theory.* It becomes semiclassical only if semiclassical methods are used to propagate $\varphi(\boldsymbol{R}, t)$, which we do in chapter 18.

The complete first-order wavefunction on the excited Born-Oppenheimer surface is then

$$\begin{aligned}\phi(\boldsymbol{R}, \omega, t) &= -\frac{i}{\hbar}\int_{-\infty}^{t} dt' e^{-iH_e^{B.O.}(t-t')/\hbar} e^{i(\omega - E_0/\hbar)t'} \mu_{eg}(\boldsymbol{R})\chi_g^{(0)}(\boldsymbol{R}, 0),\\ &= \int_{-\infty}^{t} dt'\, e^{i(\omega - E_0/\hbar)t'} \varphi(\boldsymbol{R}, t - t'),\end{aligned} \tag{17.16}$$

with $v(t') = \exp[-i\omega t']$, a laser at frequency ω. The next step, which is very slightly delicate in an uninteresting way, is to take $t \to \infty$, place some smooth cutoffs on the laser field in the distant past and far future, and then compute the probability of being promoted to the excited state—that is, $\Sigma(\omega) \propto \int d\boldsymbol{R}\ |\phi(\boldsymbol{R}, \infty)|^2$. The photoabsorption cross section proportional to this probability becomes (see reference [132])

$$\Sigma(\omega) = \int_{-\infty}^{\infty} e^{i\omega t} \langle\varphi|\varphi(t)\rangle\, dt. \tag{17.17}$$

The ingredients for this formulation are shown in the simplest case of one dimension, in figure 17.1. There is no great difficulty in practice going to many dimensions, save perhaps knowing the Born-Oppenheimer potential.

It is crucial to recognize that the autocorrelation function is $\langle\varphi|\varphi(t)\rangle$ in its full complex glory is available by inverse Fourier transform of the absorption spectrum, certainly including experimental spectra. Looking ahead, this is shown for the case of benzophenone in figure 19.13.

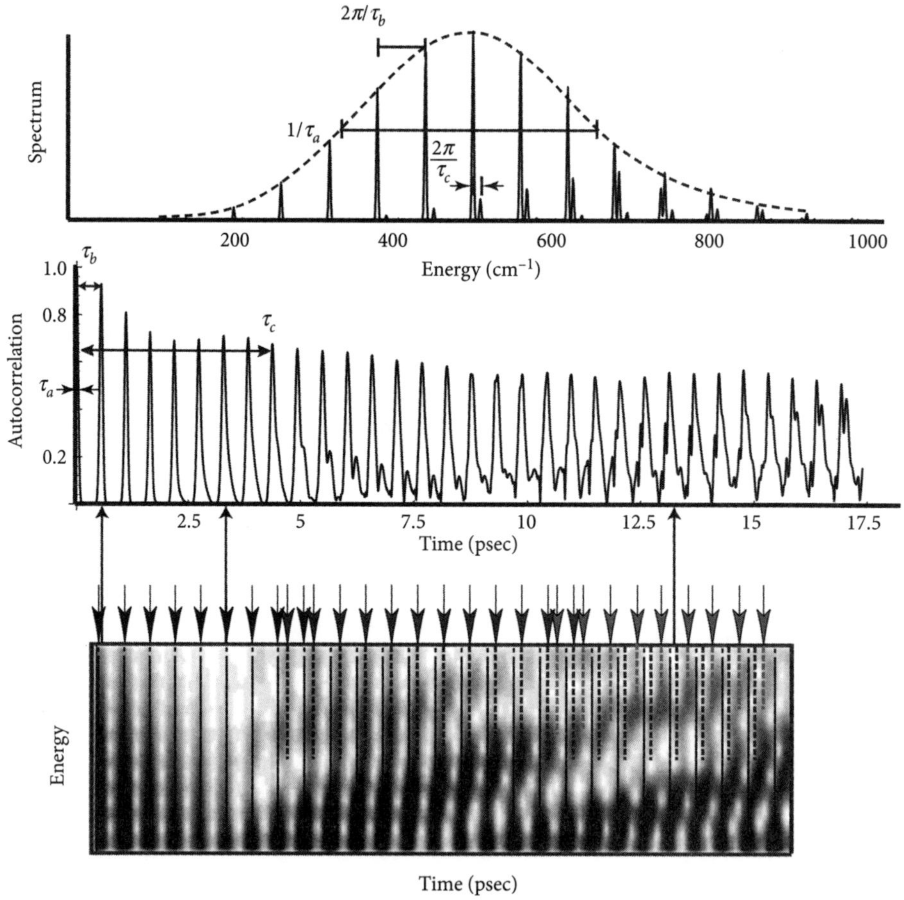

Figure 17.2. The energy spectrum, time autocorrelation function (magnitude), and time-energy "spectrogram" for the unstable periodic orbit corresponding to the spectroscopy of benzophenone. This figure illustrates the connections between the decays and recurrences in the time domain and spectral features in the frequency domain. Note the hierarchy: the shortest event in the time domain determines the broadest feature in the frequency domain; the next timescale determines the next frequency feature at higher resolution, and so on. A spectrogram is shown at the bottom and reveals that some of the later sets of recurrences are time delayed compared to the main period (black lines) and higher in energy. They tell a story of falling away from the periodic orbit and later returning to its vicinity—classically, this corresponds to homoclinic orbits. Doing that once gives the recurrence dark dashed set, and twice, the lighter dashed set. The time delay is a result of the *y* period being a little lower on average during the excursion away from the periodic orbit, causing a vibrational phase delay upon rejoining the vicinity of the periodic orbit. This is discussed more thoroughly in section 19.3.

Display of both the time pressure and the frequency spectrum are commonplace in acoustics. But one does not have to adhere to these two extremes; there is the intermediate time-frequency spectrogram, which can interpolate continuously between time-like and frequency-like limits, revealing secrets of the spectrum that are evident in neither extreme. For example, figure 17.2, bottom, is a spectrogram that exposes classes of amplitude that have fallen away from a period orbit, and returned, once, twice, and so on [133].

The reader is encouraged to examine figure 17.2, which shows all three types of spectral analysis (frequency spectrum, time autocorrelation, and spectrogram), for a spectrum very close to that of a part of the low-energy electronic spectrum of benzophenone, involving benzene ring twisting. See section 19.3 for much more discussion of this fascinating decoding of the vibrational coupling and order of events in this molecule.

Inserting a complete set of vibrational states $|\chi_{ej}\rangle$ on the excited potential surface e in equation 17.17, we have, starting in the state $|\chi_g^{(0)}\rangle$,

$$\Sigma(\omega) = c\int_{-\infty}^{\infty} e^{i\omega t}\langle\varphi|\varphi(t)\rangle\, dt = c\sum_j \int_{-\infty}^{\infty} e^{i\omega t}\langle\varphi|\chi_{ej}\rangle\langle\chi_{ej}|\varphi(t)\rangle,$$

$$= c\sum_m |\langle\chi_{ej}|\mu_{eg}(\boldsymbol{R})|\chi_g^{(0)}\rangle|^2\delta(\hbar\omega - E_{ej}). \tag{17.18}$$

The last form is the usual textbook Franck-Condon formula for the absorption spectrum, which has extreme disadvantages compared to the time-dependent expression equation 17.17 in all but the cases of very small, isolated molecules, where the $|E_{ej}\rangle$ may actually be found and the sharp energies known and seen in the spectrum. Put that small molecule in a gas and already one is at a loss to explain the line broadening with equation 17.18, but not so with equation 17.17, where the state $|\phi\rangle$ may be extended to include the gas as a bath, and its effect on the "system" molecule estimated. The gas will cause a decay of the autocorrelation function through dephasing—that is, depopulation plus pure dephasing (see equation 5.37). This decay causes a broadening of the spectrum and relieves us of needing to know the autocorrelation much beyond the decay time. The alternative of finding the system plus bath eigenstates is out of the question.

Returning to the Franck-Condon factors at the end of equation 17.18, we need to make several points. First, if the electronic Born-Oppenheimer potentials were the same in the ground and excited states, the states $|\chi_{gj'}\rangle$ and $|\chi_{ej}\rangle$ would be orthonormal. A strong resemblance between $H_g^{B.O.}$ and $H_e^{B.O.}$ is unlikely for a small molecule, where the excited electron has a major effect on a few bonds, but resemblance becomes increasingly true as conjugated molecules grow larger. Examples include the polyaromatic hydrocarbons (PAHs), which are well studied because there is a suspicion that they are responsible for light absorption and reflection in the interstellar medium. As the number of atoms over which a single electron orbital spreads increases, the effect of one excited electron on molecular geometry and force constants decreases. The ground and excited electronic states become identical in the large conjugated limit.

This leaves the transition moment $\mu_{eg}(\boldsymbol{R})$ to cause anything beyond the elastic $0 \to 0$ (or $j \to j$) vibrational transitions required by orthogonality. However, if $\mu_{eg}(\boldsymbol{R})$ is not a constant, then in general

$$\langle\chi_{ej}|\mu_{eg}(\boldsymbol{R})|\chi_{gj'}\rangle \neq 0,$$

even if $j \neq j'$. The transition moment often plays second fiddle to geometry change when the latter is significant.

The Franck-Condon factors are integrals that can be amenable to the stationary phase approximation; see section 18.1. Indeed, the advantages of the time-dependent formula over the usual time-independent Franck-Condon expressions in many dimensions are hard to overstate. Only the evolution of a wavepacket starting near the geometry of the ground state is needed, not the possibly tremendous numbers of overlaps of eigenstates as in equation 17.18, which may be far too numerous to calculate (or quite likely impossible to calculate even one of them). The vast majority of the eigenstate's coordinate dependence is likely to be spatially far removed from the locale of any dynamics important to the spectrum. It is senseless to do calculations in such regions, since the spectra do not depend on them, nor do the spectra give any information about them whatsoever. Even if we are trying to reproduce a high-resolution spectrum with sharp peaks, it is likely that the wavepacket explores only a tiny fraction of the available coordinate or phase space.

More often, the spectrum is not resolved down to the eigenstates, and the time-dependent formulation allows us to stop the time evolution whenever we have reached the desired frequency resolution. Moreover, the time-dependent expression teaches us that the spectrum is informative about the dynamics at all resolutions; the more we average the spectrum, the shorter the time we are probing.

Even if time-dependent formulations of spectroscopy often fairly cry out for semi-classical interpretation and approximations, the formulations themselves are not inherently semi-classical, and remain exactly equivalent to their energy counterparts. Sometimes confusion arises, because the time-dependent formulations are easily described in terms of wavepacket dynamics, and so on, but wavepackets are merely localized, moving quantum amplitude, and are not inherently semiclassical. Therefore, we need to distinguish what is an equivalent recasting of spectroscopy into the time domain on the one hand, and semiclassical approximations on the other. There is nothing semiclassical about equation 17.17, in spite of the urge to implement the wavepacket dynamics of $|\varphi(t)\rangle$ semiclassically. We will do this later to obtain qualitative properties of the spectrum. Numerically, one may get quantitative information by longer time quantum evaluation of the autocorrelation function. It is abundantly clear that the way to proceed is not to laboriously premap the entire potential energy surface by electronic structure calculations before commencing calculations, but rather to compute the potential on the fly, only where the wavepacket actually goes. It is not difficult to show that for 20 or 30 degrees of freedom, it can take days for the wavepacket to explore the entire phase space available, assuming even that it can.

17.4 Time Correlation Constraints on the Frequency Spectrum and Vice Versa

Figure 17.3 (and a different example in figure 19.6, later), reveals connections between features of the autocorrelation in the time domain and corresponding features in the Fourier transform frequency domain. In figure 17.3, a wavepacket finds itself displaced in both coordinates (dark oval, snapshot 1). The first event is the downhill departure of the wavepacket, causing a decay of the autocorrelation, almost to 0 if the excursion is large. This has already happened by snapshot 2. One is perfectly justified in examining the spectrum up to that point in time, integrating equation 17.17 up to the current time. If the time falls in one of the "quiet periods" at time T, the spectrum $\Sigma_T(\omega)$ is smooth and reflects a broadened version of higher resolution spectra to come later. The timescale for the first decay is revealed through the width of the spectrum

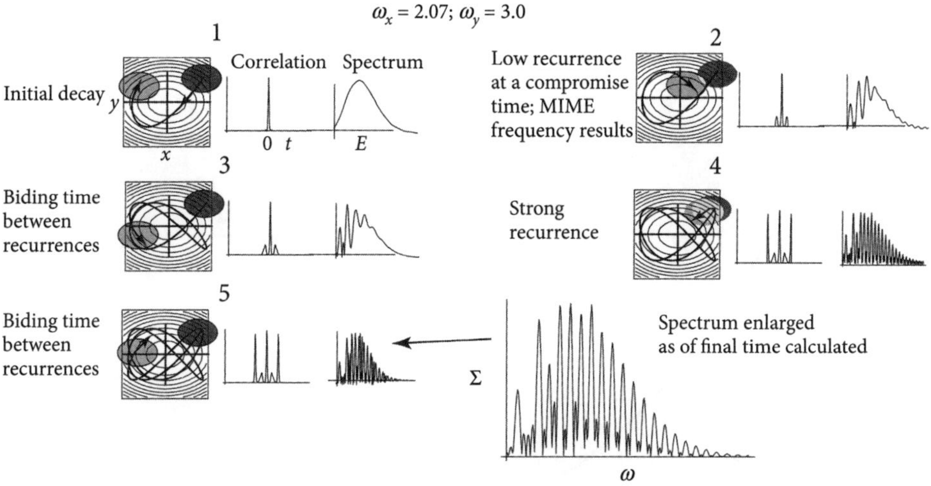

Figure 17.3. A coherent state wavepacket (a displaced ground state of the harmonic system, shown as dark gray) in a nondegenerate 2D harmonic oscillator evolves in time (light gray), exhibiting surprisingly instructive behavior. The overlap between the dark initial state $|\phi\rangle$ and the moving state $|\phi(t)\rangle$ decays rapidly at first (the decay is already over in snapshot 1) but develops recurrences later at various times and with various strengths. The spectrum at increasing resolution is the Fourier transform of the overlap of the fixed and moving wavefunctions over an increasing time interval. Notice that each stage of the development of the spectrum is a constraint on future stages, which can only add finer detail to what has already been "laid down." The first, relatively weak recurrence (snapshot 2) induces the big, low-energy progression of peaks. The peaks are not spaced at the frequency of either oscillator, but rather a compromise frequency reflecting the inverse of the time of the recurrence (the MIME effect; section 19.1). The spectrum develops in fairly sudden leaps, depending on when the recurrences happen. Finally, after the last frame shown here, the high-resolution peaks emerge at the allowed energy eigenvalues $E_{n,m} = (n + 1/2)\hbar\omega_x + (m + 1/2)\hbar\omega_y$, with intensities that are the Franck-Condon factors $|\langle n, m|\phi\rangle|^2$.

frequency, which can be extracted by inspection any time after the first decay. The mean energy of the initial (dark gray) state is revealed by its peak frequency together with any asymmetry. The spectrum obtained by Fourier transform up to the time of the snapshot is shown to the right of the snapshots.

An important point is that the decay is actually faster than the time for the dark and light wavepackets to separate completely in space. That is because the expectation value of the momentum of the light gray packet skyrockets quickly, well before the spatial overlap ceases. This is the major factor in the decay. This statement follows from several different perspectives. One is mere observation of the details of the evolving wavefunction; these are seen in the one dimensional example of figure 17.1, showing the wavepacket moving right developing oscillations as it departs, due to its growing momentum expectation. Those oscillations are developing well before this snapshot is taken, and cause the overlap to be small even as both wavepackets still share the same region in position space. This is shown in more detail in figure 17.4, where the time evolution of the real part of the wavefunction released on the given potential

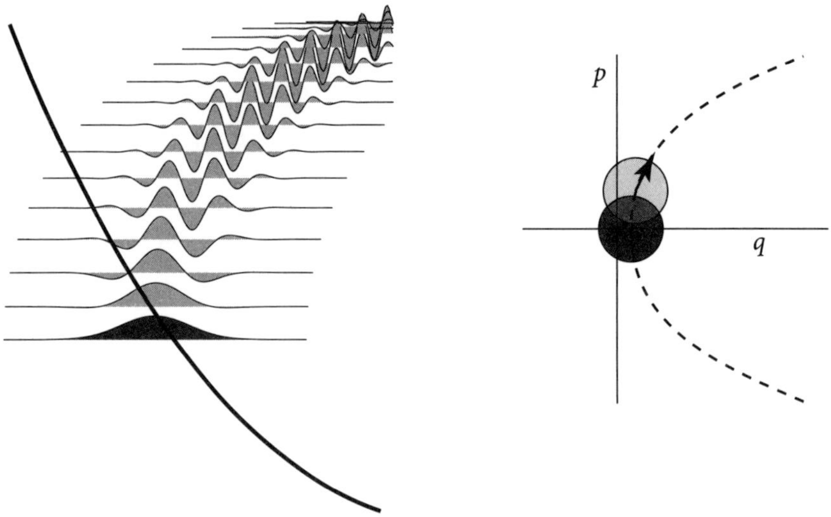

Figure 17.4. On the left, the real part of the wavefunction released on the given potential at $t = 0$ is shown in a sequence of equal time interval snapshots thereafter, plotting each snapshot as it appears in coordinate space, but on a horizontal axis raised by the average momentum of the evolving wavefunction. The phase space perspective of the same event is shown on the right, revealing that the decay is mostly due to momentum displacement initially. The reasons come from elementary short time classical dynamics.

is shown in a sequence of equal time interval snapshots, plotting each snapshot as it appears in coordinate space, raised vertically by the average momentum of the evolving wavefunction.

Another perspective is revealed in Wigner phase space. An initially stationary wavepacket on the side of a hill accelerates before it moves very far, just as in the classical equation of motion

$$\dot{p}(t) = -\frac{\partial V(q(t))}{\partial q}, \quad \dot{q}(t) = p(t)/m,$$

the short time solution for these initial conditions ($p(0) = 0$; $q(0) = q_0$) being

$$p(t) \approx -V'(q_0)t, \quad q(t) \approx q_0 - \frac{1}{2}V'(q_0)t^2.$$

It is seen in this semiclassical argument that the phase space movement goes linear in t, while the coordinate space displacement sluggishly starts off as t^2.

The correlation function-spectrum connections will be developed further as we proceed, by example. Looking ahead, an instructive example is given in figure 22.3. Other cases, such as in figure 17.2, figure 19.2, and figure 19.15 will be used to reveal aspects of the potential energy surface and dynamics.

PULSED SPECTROSCOPY

The author likes to make the remark that a CW laser is equivalent to a pulsed laser with a high repetition rate, so high in fact that the pulses overlap (see figure 17.5).

We have to distinguish two types of multiple pulse experiments. One or several pulses may coherently lead to the absorption or emission of a single photon, it is a

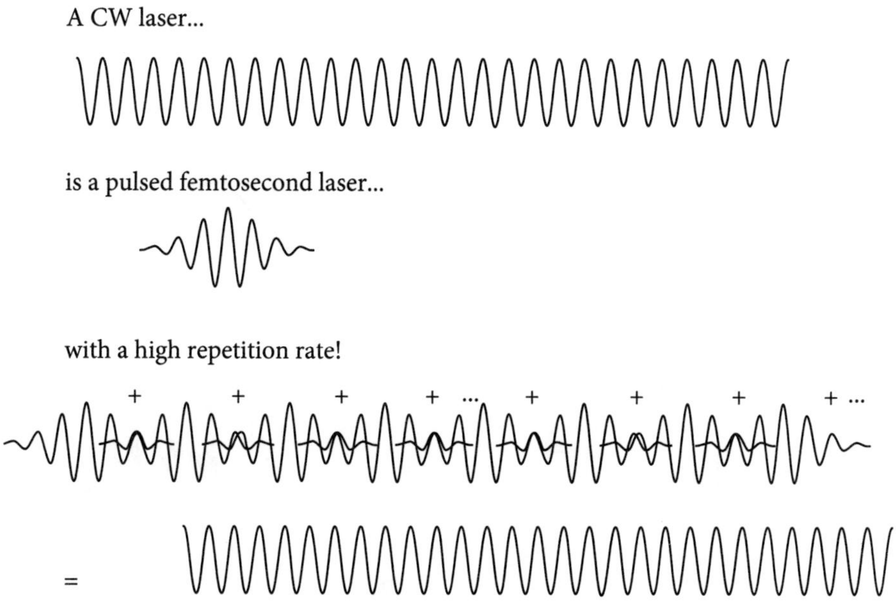
A CW laser...
is a pulsed femtosecond laser...
with a high repetition rate!
+ + + + ... + + + + ...
=

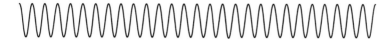

Figure 17.5. The relation between CW and pulsed sources.

major fallacy to suppose there has to be one photon (or none) per pulse. Later, we do consider two-photon spectroscopy—that is, Raman spectroscopy.

The excitation amplitudes created by two well separated pulses are shown in figure 17.6. In the perturbative, weak field regime, the total probability promoted to the excited state electronic is small in each pulse, so we have clearly exaggerated the excited state probability shown in figure 17.6 for clarity. Again, it is imperative that the reader understand that there is *at most* one photon absorbed after the two pulses have passed, in the weak field regime, where most spectroscopy is done. The amplitudes shown are coherent one-photon wavefunctions.

The same information is carried by both pulsed and continuous wave spectroscopy, supposing that the electromagnetic fields are not too strong. After scanning the spectrum over a large range of frequencies, CW lasers are often paradoxically much more informative than pulsed experiments about very short time dynamics, femtoseconds and shorter. This is poorly understood in many quarters because it is psychologically compelling to do things directly in the time domain. But very short pulses have large energy uncertainties, causing side effects like exciting the next few electronic states when only one was wanted. Lengthening the pulse to reduce this problem reduces its time resolution.

We have to distinguish two types of multiple pulse experiments. We first consider the case of one or several pulses that coherently lead to the absorption or emission of a single photon. Later, we consider two-photon spectroscopy, which corresponds to a pulsed version of the usual CW Raman spectroscopy.

It must be emphasized that true multiphoton two and three pulse experiments, with well-chosen delays and center frequencies, can reveal even more information than can any absorption, emission, or Raman spectrum. Although we do not do the multiphoton regime justice in this book (except for two-photon Raman scattering), we

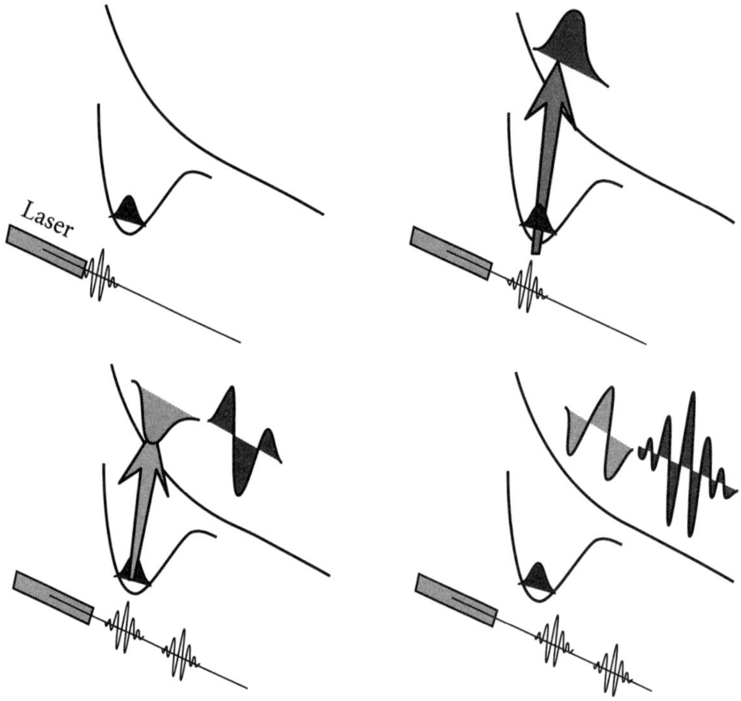

Figure 17.6. Short pulse excitation. Two laser pulses with the same center frequency pass in quick succession and each time promote a fraction of the ground state amplitude to the excited electronic potential energy surface. By the time a second pulse has done its work, the promoted amplitude coming from the first pulse has accelerated and moved away. The two pulses do not interfere here, because the excited state potential is steep. The single pulse case is easy to understand by omitting the second pulse. This is still one-photon spectroscopy: there is a small amplitude for absorption in each pulse.

also declare that if you don't understand the one-photon regime thoroughly, you are kidding yourself about understanding the multiphoton regime.

17.5 C.W. Spectroscopy from the Molecule's Point of View

We can Fourier transform an experimental spectrum to get the full, complex autocorrelation function:

$$\langle \varphi | \varphi(t) \rangle = \frac{1}{2\pi} \int_{-\infty}^{\infty} e^{-i\omega t} \Sigma(\omega) \, d\omega. \tag{17.19}$$

It is easy to go the other way, from autocorrelation function to spectrum. A peak in the autocorrelation function means part of the wavepacket is back at its birthplace at that time.

Suppose we examine closely what is happening as light arrives at the molecule, rather than focus on the autocorrelation function as in equation 17.17. Figure 17.7 depicts steady fixed-frequency laser radiation impinging on diatomic molecules. In the

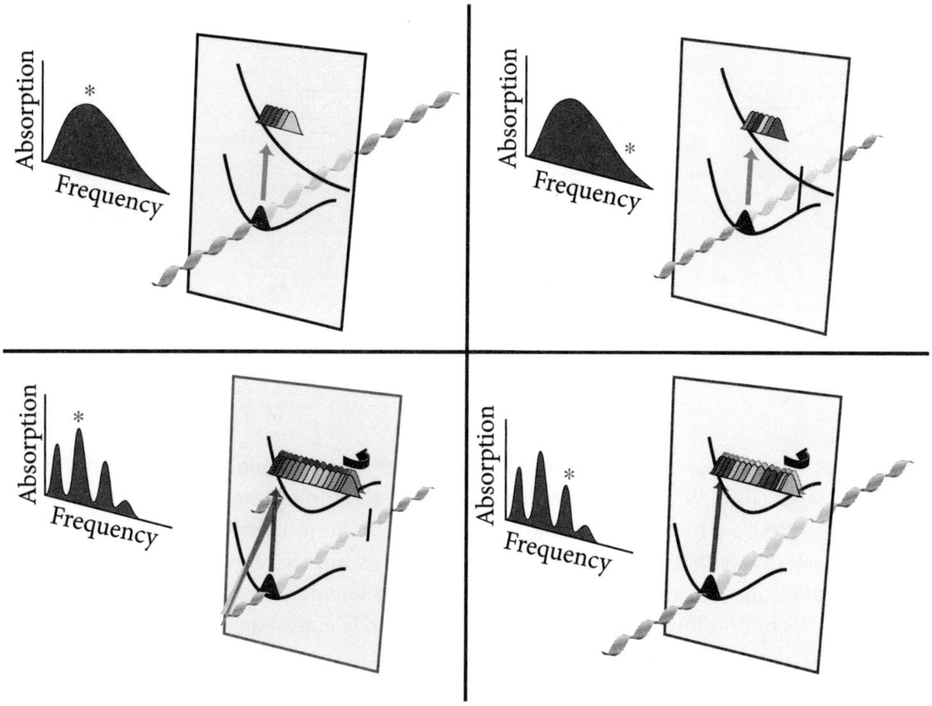

Figure 17.7. CW laser excitation. In the top two panels, photo absorption to a repulsive potential energy surface leads to dissociation. In the lower panels, the excited state potential is bound. In both cases, the effect of a laser frequency shift on the phase of successive wavepackets being promoted to the excite state is shown, represented by shade. The resulting new absorption strength is given under the star representing the laser frequency. A nonresonant case would correspond to frequency at a gap between absorption peaks. The cases shown are arriving in phase.

top two panels, the molecule has a repulsive excited state potential energy surface and dissociates. The exact absorption spectrum has no sharp features, and is a single broad peak with a center frequency, dispersion, asymmetry, and exponential fall-off in the tails. We can understand these things from the autocorrelation equation 17.17 and the dissociative behavior of $|\phi(t)\rangle$. The promoted wavepacket $|\phi(0)\rangle$ leaves directly, and thus no autocorrelation recurrences happen. The spectrum—that is, the Fourier transform of the autocorrelation—must be a pure unstructured continuum.

Figure 17.7 helps explain these features and their frequency dependence directly in terms of the sequence of events in the time domain while the light is arriving—that is, the sequential arrival of fresh, transition moment modified and phased (by the light) pieces of the ground state on the excited state surface, and their evolution on the excited state. The absorption strength is weak at high (upper-right panel) and low laser frequencies, because amplitude being continuously promoted from the ground state to the excited state arrives partially out of phase with what was promoted just prior, before the excited state wavepacket has moved away (including both position and momentum displacement). At the absorption strength maximum (upper-left panel), the fresh amplitude is in phase with amplitude just promoted. Constructive

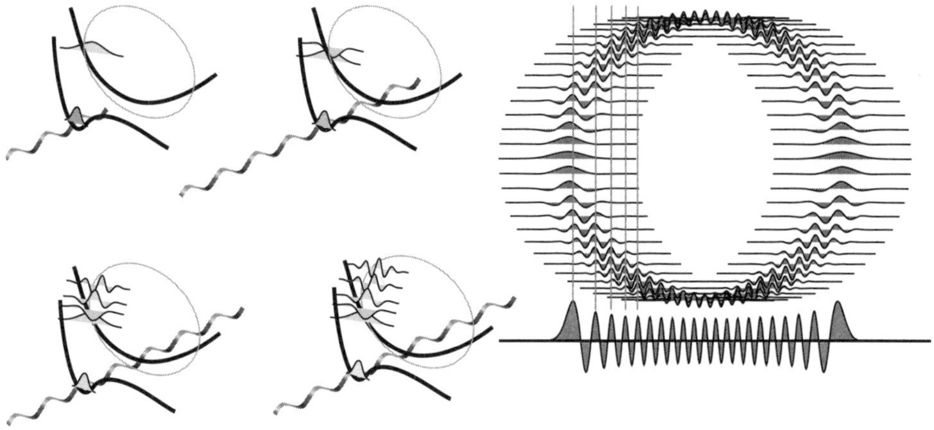

Figure 17.8. (Left) Four snapshots in the creation of the 60th eigenstate of a harmonic oscillator by exciting with a cw laser from a lower electronic state. The upper electronic state is taken to be a harmonic oscillator here, but any anharmonic potential would give analogous results with the laser frequency tuned to resonance with an excited vibrational state. The delicate way that the individual promoted wavepackets combine to make the eigenstate, including the phase imparted by the laser, is clarified. This may be viewed as very refined quantum control.

interference results, causing a larger population to develop on the excited potential energy surface, giving a larger cross section for absorption. There is only one short (usually under a femtosecond) chance for the freshly promoted pieces to interfere with older pieces.

In the lower panels, the excited state potential is taken to be overall attractive. Amplitude that had been promoted earlier makes a round trip and returns to the Franck-Condon region. As it does, new amplitude just being promoted adds in phase or out of phase with the returning amplitude, depending on the laser frequency. If it is in phase, constructive interference results in a build up of probability on the excited state; the frequency of the laser is thus tuned to an *absorption maximum*—that is, a peak in the absorption spectrum. If, on the other hand, the returning amplitude is out of phase with the recently promoted amplitude, the laser frequency corresponds to an absorption minimum. There is more than one way to be at a maximum or minimum, as shown in the lower right, by making extra full cycles in laser phase before the return. This of course is merely a matter of laser color, and corresponds to tuning to a different peak in the absorption spectrum. The returning wavepacket adds in phase, but earlier partial phase cancellation (as in the top panels) means the peak is not quite as high. Figure 17.8 contains four snapshots caused by the arrival of a long pulse of well-defined frequency, leading to the the creation of the 60th eigenstate of a harmonic oscillator by exciting from a lower electronic state. The completed eigenstate shown on the right, after the pulse has been on long enough, makes a more mathematical appearance in figure 17.8.

17.6 Spectroscopy at Finite Temperature

The time-dependent formulation has up to now assumed a single Born-Oppenheimer nuclear motion state as an initial condition. If the temperature is very low or special techniques have been used to isolate a particular initial state, this is sufficient. If temperature is higher (kT comparable to vibrational level spacings) it may be advantageous to treat the situation "by hand" so to speak, by separately running the single state time-dependent formulations and adding them up with the appropriate thermal weightings.

When that becomes cumbersome, the time-dependent formulation of spectroscopy extends to finite temperatures in a natural way [134]. It is not difficult to derive, from what we already know and some elementary thermodynamics, that the absorption spectrum is the Fourier transform of the correlation function

$$C_T(t) = \int d^{2n}\alpha \langle \Phi_{g,\alpha}(t - i\tau)|\Phi_{e,\alpha}(t)\rangle, \tag{17.20}$$

where

$$|\Phi_{e,\alpha}(t)\rangle = e^{-iH_e t/\hbar}|\alpha\rangle, \tag{17.21}$$

and

$$|\Phi_{g,\alpha}(t)\rangle = e^{-iH_g(t-i\tau)/\hbar}|\alpha\rangle. \tag{17.22}$$

The parameters α cover a $2n$-dimensional phase space. H_g is the initial Born-Oppenheimer potential surface, and H_e is the excited state Born-Oppenheimer potential. If more than one excited state is populated by the photoexcitation, the extension is straightforward. Other extensions apply to the possibility of more than one electronic state populated thermally.

There is one issue, thinly veiled: the state $|\Phi_{g,\alpha}(t)\rangle$ involves partial imaginary time $-i\tau = -i\hbar/kT$ propagation of the initial coherent state $|\alpha\rangle$. Generally speaking, this is rather benign. For calibration, $|\alpha(\beta)\rangle = \exp[-\beta H_g]|\alpha\rangle$ winds down toward the ground vibration-rotation state, the more so the larger β, as is easily seen by inserting a complete set of ground state nuclear motion eigenstates between $\exp[-\beta H_g]$ and $|\alpha\rangle$. Moreover, if $\langle\alpha|H_g|\alpha\rangle \gg kT$, the state $|\alpha(\beta)\rangle$ acquires a very small weight and may be eliminated from the integral.

17.7 Photoelectron Spectroscopy

Suppose we start with a molecule in its ground electronic state, and ionize it with a UV photon of known frequency. This will put the newly created ionic molecule in an excited electronic state, or a superposition of them. What vibrational-rotational-electronic eigenstates of the ion (that is, Born-Oppenheimer "vibronic" states) will be populated?

The total energy is quite fixed by the laser energy and the initial state of the molecule. The energy of the molecular ion state thus determines how much is left over for the photoelectron. The arithmetic is

$$E = E_{\mathrm{ng}} + h\nu = E_{\mathrm{ion}} + E_{\mathrm{photoelectron}}, \tag{17.23}$$

where E_{ng} is the energy of the nth state (the initial state) of the ground electronic Born-Oppenheimer potential energy surface.

A measurement is made of the distribution of electron energies (the photoelectron spectrum), revealing the quantum state of the ion, assuming the density of states is low enough to resolve them one from another. If the density of states is too high for that, one still has an "unresolved" photoelectron spectrum to explain, in the same sense that an unresolved absorption or emission spectrum carries information about the events following photoabsorption.

Imagine that the UV photon ejects the electron directly, creating the ion. What is the final electronic Born-Oppenheimer state? The key point is that there is formally a *different electronic state for each final Born-Oppenheimer vibronic state of the molecule*, since the ejected electron is part of the "electronic state," having a different energy (and also other properties—that is, angular distribution) for each final ionic state. Fortunately, the different Born-Oppenheimer potential energy surfaces do not depend on the far-away electron; it has no influence on the molecular ionic electronic energies regardless of its speed. The Born-Oppenheimer potential surfaces thus differ only by an energy shift given by the photoelectron energy, which in turn is $E_{\text{photoelectron}} = E_{1g} + h\nu - E_{\text{ion}}$. Figure 17.9 illustrates the situation.

The ground neutral Born-Oppenheimer potential energy surface is seen as the bottom solid curve, and we suppose starting in the lowest vibrational state, with energy E_{1g}. The ionic Born-Oppenheimer potential energy surface with no other electron present (or equivalently the photoelectron at infinity with no kinetic energy) is shown for reference as the dotted curve. The long vertical arrow shows the total energy available—that is, $E_{\text{total}} = E_{1g} + h\nu$. The solid curve depicts the ionic state with total energy apportioned so as to populate the third vibrational level of the ionic potential energy surface. The corresponding photoelectron energy is shown as the shorter arrow.

The photoelectron has taken on the role usually held by a variable photon energy, even though the photon is of fixed frequency. The best analogy is with photoemission, where we have no control over the frequency of the emitted photons (just as we have no control over the photoelectron energy), but collectively the photoemission photons reveal a spectrum of the final state (say the emission is to the ground electronic state).

In the example shown in figure 17.9, the nth line has a strength that is given by a Franck-Condon factor, but we prefer writing this again as

$$\Sigma_{1g \to ne(\text{ion})} = \int_{-\infty}^{\infty} dt\, e^{iE_{ne}t/\hbar} \langle \phi_1 | \phi_1(t) \rangle, \tag{17.24}$$

where $E_{ne} = E_{\text{total}} - E_{\text{photoelectron}}$ with energy measured on the (dotted) reference Born-Oppenheimer potential energy surface f, but all excited surfaces are the same except for an energy shift. Dynamics $|\phi_1(t)\rangle$ takes place on surface e—that is,

$$|\phi_1\rangle = \mu_{eg}(\mathbf{R}) |\chi_{1g}\rangle,$$

$$|\phi_1(t)\rangle = e^{-iH_{e,\text{ion}}t/\hbar} \mu_{eg}(\mathbf{R}) |\chi_{1g}\rangle.$$

It is now clear that no matter what the size of the molecule, or whether the final states are resolved or not, the wavepacket autocorrelation $\langle \phi_1 | \phi_1(t) \rangle$ controls the photoelectron spectrum, through equation 17.24. We have assumed here that only one ionic Born-Oppenheimer potential energy surface is created, but if more are involved, we must sum the spectra from each, with a weight given by the fraction owing to that state's population. Very often the spectra are not overlapping due to very different

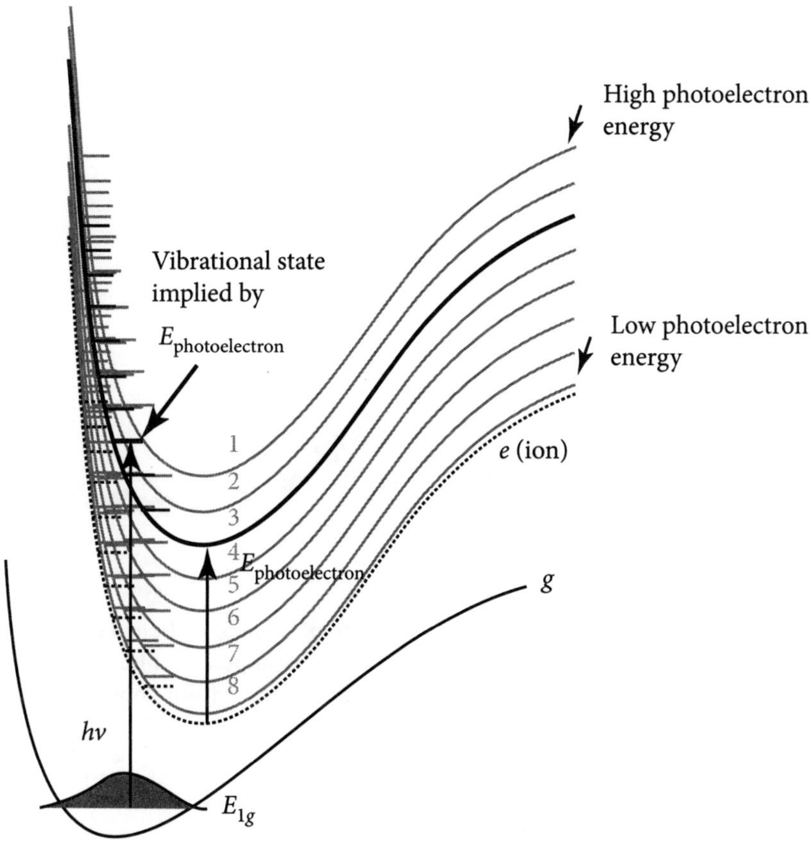

Figure 17.9. Schematic of photoelectron spectroscopy for a simple diatomic molecule and a modest-UV photon energy. The energy of the photoelectron may range greatly, and whatever it doesn't possess remains with the molecule. If the experiment were done in a synchrotron, the photon energy could be much higher. Low photoelectron energy at a given laser frequency corresponds to larger vibrational energy remaining in the molecule.

electronic ionic state energies, depending on whether a core or valence electron has been ejected. If a core electron is involved, a big chunk of energy is used just getting the electron out, shifting any vibrational structure to much lower photoelectron energy. If such a core electron is ejected, spectra are further broadened by the time evolution of many participating electronic states as the electrons relax, perhaps ejecting more electrons (Auger processes), or even subsequent dissociation of the molecule.

In fact, photoelectron spectroscopy is often used to map out low-resolution spectra showing poorly vibrationally resolved electronic bands, thus revealing the different ionic electronic state energies. Although usually ignored, even these spectra have information about the slopes of potential energy surfaces in the "vertical," or Franck-Condon region.

The beautiful synchrotron radiation photoelectron spectrum of methane is shown in figure 17.10. A carbon 1*s* orbital electron has been ejected, changing the C-H potential equilibrium and force constants in the ionic state and causing a progression

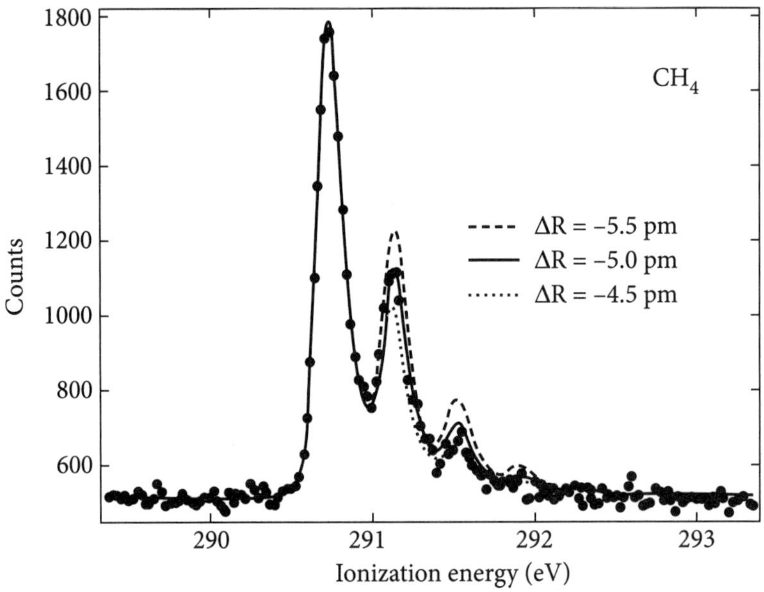

Figure 17.10. Carbon 1*s* photoelectron spectrum of methane. Calculated vibrational structure is shown with three different values of the change in CH bond length due to ionization, compared with the experimental spectrum points, from Thomas et al. [135].

in the ionic C-H stretch mode, spacing about 0.4 ev as compared to a photoelectron energy of 291 ev.

17.8 Second-Order Light-Matter Interaction—Raman Spectroscopy

Two-photon Raman scattering expressions are due originally to Kramers, Heisenberg, and Dirac (KHD) in 1924–1927. It is now straightforward to derive the KHD expressions from the Born-Oppenheimer approximation and second-order time-dependent perturbation theory for light-matter interaction. (One-photon light absorption and emission obey the same statement, with "second-order" replaced with "first-order.") The idea is to start in some vibrational level of the ground electronic state, apply the light-matter perturbation, and collect all amplitude (according to the second-order perturbation theory), arriving back on the initial electronic state after having journeyed, either virtually or resonantly, to one or more excited electronic states. In doing so, the amplitude will end up occupying one or more of the allowed vibration-rotation eigenstates on that initial electronic surface. These can be the same state that the system started in, leading to what is called *Rayleigh scattering*, or a higher level, leading to Stokes Raman scattering, or a lower vibration-rotation level, leading to anti-Stokes Raman scattering. The light emitted in the return to the initial electronic state must be redder (less energetic) than the light sent in for a Stokes Raman scattering event, and bluer (more energetic) than the light sent in for an anti-Stokes Raman scattering event. The frequency shift of the light can reveal exactly what transition took place overall on the initial electric state.

The probability of a Stokes or anti-Stokes event is found by squaring and integrating the corresponding second-order perturbation theory amplitude. Raman scattered light must have been emitted if one finds a finite population in some vibrational level higher or lower than the initial level on the ground electronic state.

It should be noted that the Born-Oppenheimer approximation is not required within KHD second-order light-matter perturbation theory. More accurate many-body ground and excited states could be used in the expressions but these are rarely available. Certainly, there is an opportunity to include corrections to the Born-Oppenheimer level of approximation if needed.

Raman scattering is almost always given in its energy domain form, even though it would be impossible to actually implement in all but the smallest systems. We have already seen the difficulties in one-photon spectroscopy, where the sum-over-states, Franck-Condon form can easily involve say 10^{15} states for a modest system, almost all of which cannot be calculated in the first place. But even a 10-atom molecule typically does not have a spectrum resolved down to eigenstates anywhere above the lowest energies, since the density of states grows dramatically with energy. This means much if not all of the spectrum as seen experimentally can be computed from short time dynamics within a time-dependent formulation, possibly implemented semiclassically.

Raman light scattering often depends only on very short time dynamics, perhaps just femtoseconds, especially for large molecules. Even so, the Raman lines seen in emission can be extremely sharp—they do not reflect the short lifetime in the excited state. We will explain why as we proceed.

Raman scattering discussions in textbooks almost always focus on the special case of off-resonant Raman scattering, where the incident laser frequency is too low to cause a lasting transition to an excited electronic state. Often this is done without mentioning the restriction! Being "off-resonance" of course is possible only if there is a gap between the starting (usually the lowest) electronic state and the next one up, especially at the equilibrium geometry. This is true in gapped semiconductors, and most molecules, but not for gap-less solids like a sheet of graphene. The rule given in such standard discussions is: "a Raman band in mode R_g can be activated only in the presence of nonzero derivative of the system polarizability $\alpha(\boldsymbol{R})$ with respect to the normal coordinate of the mode R_k: $\partial\alpha(\boldsymbol{R})/\partial R_k \neq 0$." This rule holds only for off-resonant excitation. We will have more to say about off-resonance Raman scattering shortly.

The KHD formula for the total Raman cross section can be used on or off resonance, although multiple excited electronic states will need inclusion off resonance. The polarizability derivative rule just mentioned can be derived in the off-resonance limit from the KHD formula [136]. The value of $\partial\alpha(\boldsymbol{R})/\partial R_g$ depends critically on the coordinate dependence of the transition moments $\partial\mu^{\rho}_{gf}(\boldsymbol{R})/\partial R_g$ where ρ specifies polarization of the dipole operator.

The KHD Raman cross section $\Sigma(\omega_I)$ for incident frequency ω_I and polarization ρ, scattered frequency ω_s and polarization σ, between initial Born-Oppenheimer state $|i\rangle$ and final Born-Oppenheimer state $|f\rangle$ via intermediate Born-Oppenheimer states $|n\rangle$ reads

$$\Sigma^{\rho,\sigma}_{i\to f} = \frac{8\pi e^4\omega_s^3\omega_I}{9c^4}\left|\alpha^{\rho,\sigma}_{i,f}\right|^2;$$

$$\alpha^{\rho,\sigma}_{i,f} = \frac{1}{\hbar}\sum_n\left[\frac{\langle f|\boldsymbol{D}^{\dagger,\sigma}|n\rangle\langle n|\boldsymbol{D}^{\rho}|i\rangle}{E_i - E_n + \hbar\omega_I - i\Gamma_n} + \frac{\langle f|\boldsymbol{D}^{\dagger,\sigma}|n\rangle\langle n|\boldsymbol{D}^{\rho}|i\rangle}{E_i + E_n + \hbar\omega_I + i\Gamma_n}\right]. \quad (17.25)$$

Whether $\hbar\omega_I$ is resonant or off resonant, the KHD expressions are the same, except that the denominators do not nearly vanish off resonance. This expression need not imply the Born-Oppenheimer approximation version of $|i\rangle$, $|f\rangle$, and intermediate states $|n\rangle$. Instead, these could be more accurate many-body states, or diabatic states if appropriate, and so on. Fleshing out this expression within the Born-Oppenheimer approximation, with atom coordinates $\boldsymbol{\xi}$ and electron coordinates $\boldsymbol{r}$, we write

$$\begin{aligned}
|f\rangle &= |\psi_{f,j}^{B.O.}(\xi, \boldsymbol{r})\rangle = |\phi_f(\xi; \boldsymbol{r})\rangle|\chi_{fj}(\xi)\rangle, \\
|g\rangle &= |\psi_{g,j}^{B.O.}(\xi, \boldsymbol{r})\rangle = |\phi_g(\xi; \boldsymbol{r})\rangle|\chi_{gj}(\xi)\rangle, \quad \text{and} \\
|n\rangle &= |\psi_{n,j}^{B.O.}(\xi, \boldsymbol{r})\rangle = |\phi_n(\xi; \boldsymbol{r})\rangle|\chi_{nj}(\xi)\rangle.
\end{aligned} \tag{17.26}$$

$|\phi_g(\xi; \boldsymbol{r})\rangle$ is the Born-Oppenheimer to the initial electronic ground state; $|\phi_n(\xi; \boldsymbol{r})\rangle$ is an excited electronic state. $|\phi_n(\xi; \boldsymbol{r})\rangle|\chi_{nj}(\xi)\rangle$ is a complete Born-Oppenheimer intermediate state, including the vibrational wavefunction $|\chi_{nj}(\xi)\rangle$ labeled by the electonic state n and the phonon or vibrational state j. The intermediate states may range freely over complete sets of functions, although some or most matrix elements may vanish by virtue of symmetry.

17.9 Raman Amplitude

In terms of the transition moments, the Raman amplitude reads (using only the resonant term of the two)

$$\begin{aligned}
\alpha_{fk,gj}^{\rho,\sigma}(\omega_I) &= \frac{1}{\hbar}\sum_{n,j'}\left[\frac{\langle\chi_{fk}(\xi)|\mu_{fn}^{\sigma}(\xi)^{\dagger}|\chi_{nj'}(\xi)\rangle\langle\chi_{nj'}(\xi)|\mu_{gn}^{\rho}(\xi)|\chi_{gj}(\xi)\rangle}{E_{gj} - E_{nj'} + \hbar\omega_I - i\Gamma_n}\right], \\
&\equiv \frac{1}{\hbar}\sum_{n,j'}\left[\frac{\langle\varphi_f(\xi)|\chi_{nj'}(\xi)\rangle\langle\chi_{nj'}(\xi)|\varphi_g(\xi)\rangle}{E_{gj} - E_{nj'} + \hbar\omega_I - i\Gamma_n}\right].
\end{aligned} \tag{17.27}$$

The extreme difficulties in implementing this formula in all but simple cases are manifest: One has to find all the excited state Born-Oppenheimer electron-nuclear eigenfunctions, form matrix elements and sum over every one. Yet what emerges, assuming this is possible to find and sum say 10^{15} or 10^{21} terms (to pick some arbitrary but typical numbers), is a just single value for each final vibration-rotation state $\langle\chi_{fk}(\xi)|$ and initial vibration-rotation state $|\chi_{gj}(\xi)\rangle$. The irretrievable compression of all this information into so few results suggests that there is a way to shortcut the whole program. There is a way: by casting everything in the time domain.

In the off-resonant case, the excited electronic states are reached only transiently, with a transient lifetime given by the time-energy uncertainty principle. The relevant energy ΔE is the deficit between the laser energy and what it would need to be to reach resonance. Typically ΔE could be 1 to a few volts and Δt range from sub-femtosecond to a few femtoseconds.

17.10 Raman Amplitudes from the Time Domain

The equivalent time domain expression, derivable either by directly employing the second-order time-dependent perturbation theory expression, or by Fourier transform

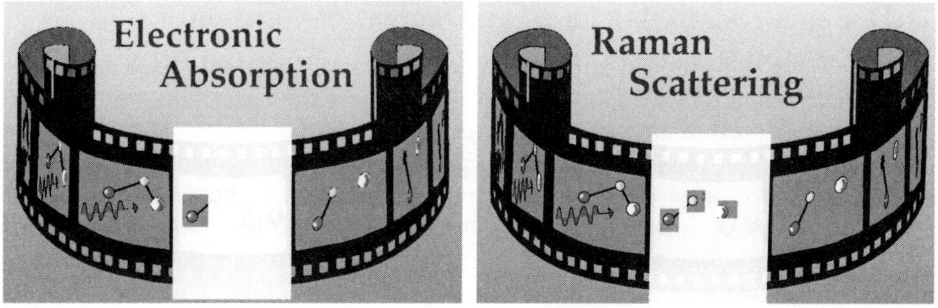

Figure 17.11. Although we would love to see every nuance of dynamics following absorption or emission of a photon taking place in polyatomic or solid state spectroscopy, absorption and emission spectroscopy (left) gives us only one window to observe the dynamics: the wavefunction $|\phi\rangle$, which is initial nuclear wavefunction $|\psi\rangle$ before the photon is involved, modified by the electronic transition moment—that is, $|\phi(\boldsymbol{R})\rangle = \mu(\boldsymbol{R})|\psi(\boldsymbol{R})\rangle$. It evolves as $|\phi(\boldsymbol{R}, t)\rangle$ on the new electronic state. The single "window" we refer to is the projection, $\langle\phi|\phi(t)\rangle$. This is one complex function of time, and is handed to us on a small silver platter, as the Fourier transform of the measured spectrum. Still, much can be inferred by looking through a single keyhole over a span of time. Raman spectroscopy (section 17.8) gives us many keyholes, or windows, one for each final state that we follow in the Raman displacement spectra as a function of incident laser frequency.

of equation 17.27, is (see references [38, 137])

$$\alpha^{\rho,\sigma}_{fk,gj}(\omega_I) = \sum_e \int_0^\infty dt\, e^{iE_I t/\hbar} \langle \varphi_{fk}| e^{-i(H_e^{B.O.} - i\Gamma_e)t/\hbar} |\varphi_{gk}\rangle = \sum_e \int_0^\infty dt\, e^{iE_I t/\hbar} \langle \varphi_{fj}|\varphi_{gk}(t)\rangle, \tag{17.28}$$

where $E_I = E_{gj} + \hbar\omega_I$.

Simplifying the notation, we write $|\varphi_{fk}\rangle \equiv |\varphi_f\rangle$. For a single, dominant excited state e, we write

$$\alpha^{\rho,\sigma}_{\boldsymbol{m}_f,\boldsymbol{m}_g}(\omega_I) = \int_0^\infty dt\, e^{iE_I t/\hbar} \langle \varphi_f| e^{-i(H_e^{B.O.} - i\Gamma_e)t/\hbar} |\varphi_g\rangle = \int_0^\infty dt\, e^{iE_I t/\hbar} \langle \varphi_f|\varphi_g(t)\rangle. \tag{17.29}$$

This remarkable simplification has many intuitive and computational advantages. Note the key changes compared to the absorption/emission formula. The expression involves cross-correlation functions corresponding to an initial vibration-rotation state and ending in a different, final vibration-rotation state. Only a half Fourier transform is involved. Computationally, we have already mentioned that normally only very short times are needed (equation 17.17), just a small fraction of coordinate space is accessed, and the required information can be gathered "on-the-fly" according to where the wavefunction wants to go. Many degrees of freedom can be treated if semiclassical methods are used to propagate the wavefunction. Raman scattering affords many more windows on the dynamics than the single window available in absorption of emission spectroscopy (figure 17.11, right).

17.11 Raman Scattering for a Few Degrees of Freedom

If the system being probed by Raman scattering is small, say two or a few atoms, and the molecule remains intact upon photoabsorption, then the absorption spectrum may reveal resolved spectra indicating many recurrences of the autocorrelation function over long times. For resonant photon energies, the Raman fundamental and overtone spectra are expected to be strong functions of frequency, just as the absorption spectrum is. Indeed, when the laser it tuned to a particular resolved absorption line, the corresponding vibration-rotation eigenstate is strongly populated. Clearly the Raman spectrum then becomes essentially a Franck-Condon emission spectrum from that state. The "game" has changed, and even the name: this is called resonance fluorescence, not Raman scattering. If the laser is tuned to somewhere between sharp absorption lines, the game is on again, although long time cross-correlation functions are important. The concept of the Raman wavefunction helps here (next section). For small systems on or near resonance, the time-independent KHD sum over states formula may involve few enough calculable terms to be implementable.

Two situations open even small molecules to very useful insight from the time domain perspective. Suppose the laser is tuned below resonance, below any significant absorption strength, whether or not there is a resolved resonance absorption spectrum. Again the Raman wavefunction (next section) lends insight, and we defer this off-resonance Raman discussion a moment.

If molecular dissociation happens directly, the autocorrelation and cross-correlation functions peak and decay quickly, never to revive. This makes time domain calculations much simpler. The alternative energy domain calculations could be cumbersome, even for a small molecule, due to the continuum scattering states required.

17.12 Raman Wavefunction

The *Raman wavefunction* is the half Fourier transform of the entire time evolving wavefunction $|\varphi(t)\rangle$ [53, 117]. We can write equation 17.29 as

$$\alpha^{\rho,\sigma}_{m_f,m_g}(\omega_I) = \int_0^\infty dt\; e^{iE_I t/\hbar}\langle\varphi_f|\varphi_g t)\rangle = \langle\varphi_f|\int_0^\infty dt\; e^{iE_I t/\hbar}|\varphi_g(t)\rangle. \tag{17.30}$$

This suggests we collect a complete Raman wavefunction

$$|\mathcal{R}_g(E_I)\rangle = \int_0^\infty dt\; e^{iE_I t/\hbar}|\varphi_g(t)\rangle \tag{17.31}$$

and project it onto various final states $\langle\varphi_f|$ at our leisure:

$$\alpha^{\rho,\sigma}_{m_f,m_g}(\omega_I) = \langle\varphi_f|\mathcal{R}_g(E_I)\rangle. \tag{17.32}$$

The Raman wavefunction is seen to be the half Fourier transform of the time evolving wavefunction $|\varphi(t)\rangle$. The Raman wavefunction is also the solution to the time-independent inhomogeneous Schrödinger equation

$$(H_e^{B.O.} - E_I)|\mathcal{R}_g(E_I)\rangle = i\hbar|\varphi_g(0)\rangle \tag{17.33}$$

that can form the basis of a numerical approach to avoid the sum over a continuum of states, because it can be solved at each E_I. It is perhaps not as intuitive as watching

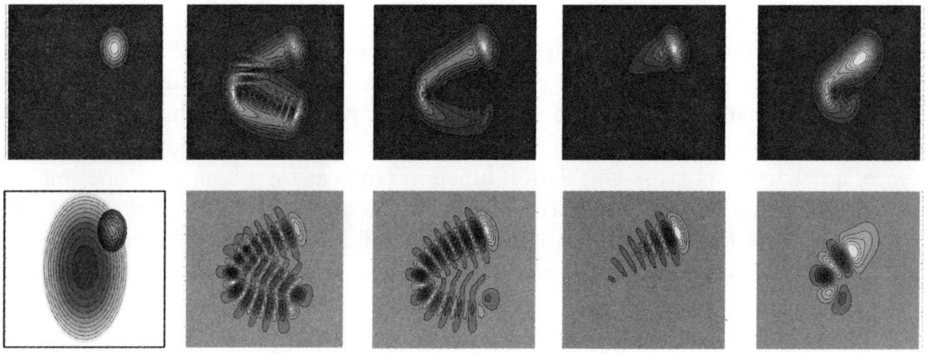

Figure 17.12. (Top row) $|\mathcal{R}_g(E_I)|^2$. (Bottom row) Re$[\mathcal{R}_g(E_I)]$. Raman wavefunction for the initial wavefunction and excited state potential is shown at the left. The next three columns show the absolute value and the real part of the Raman wavefunction near resonance with increasing Gaussian self-damping. The far right displays a well-below-resonance driving frequency with moderate damping as in the third column. Note how the Raman wavefunction is contracting to its origin.

the evolution of $|\varphi_i(t)\rangle$ to see how $|\mathcal{R}_g(E_I)\rangle$ builds up, which may often be done quite accurately by semiclassical methods [53, 117]. However, one determination of $|\varphi_i(t)\rangle$ serves all purposes: every Raman line is obtained at every energy E_I by simple Fourier addition of $|\varphi_i(t)\rangle$ over time, and every Raman intensity is obtained by the projection of the result onto different final states. The Raman wavefunction acts just like an eigenstate of the upper electronic potential surface for the purpose of calculating the Franck-Condon emission (see figure 17.12).

The time-dependent picture together with the Raman wavefunction allows a raft of questions about Raman scattering to be answered and understood intuitively. For example:

- Suppose we photodissociate a diatomic molecule on resonance with an electronic absorption—what do we expect of the Raman emission? The Raman wavefunction will be an Airy function–like wavefunction on the repulsive excited state surface, and will generally have overlap with every excited state of the diatomic, giving a long progression of overtones, perhaps attenuated by the fall-off of the transition moment, which multiplies the Franck-Condon emission of the Raman wavefunction.
- Suppose the same photodissociation took place in liquid helium. The Raman wavefunction may be taken to include the coordinates of all the helium atoms. Suppose we could begin in one of the eigenstates of diatomic + liquid helium. The collision of the diatomic with the helium in the excited state dissociation is expected to move some helium atoms out of the way, causing a damping-factor like decay of the Raman cross-correlation functions. This will cut off the long series of overtones by truncation of the Raman wavefunction in space. Averaging over many thermally populated eigenstates of diatomic + liquid helium, not much will change about this statement. Thus embedding in liquid helium will shorten the overtone progression somewhat. This may be viewed as decoherence due to the helium bath.

- Now think of the same diatomic alone in the gas phase, but we tune below resonance, below even the dissociation threshold. The time will again be effectively shortened, in this case by the off-resonant half Fourier transform of the cross-correlation functions. Some insight into how this happens can be had by realizing that below the energy of any excited states, the half-Fourier transform of any late-onset cross-correlation function (such as for a high overtone) may as well be replaced by a full Fourier transform, because the cross-correlation is zero at $t = 0$ and for a while after. However, the full Fourier transform extracts exact energy components, and below threshold there aren't any. So the high overtone intensity must get much smaller below resonance. By extension, all overtones are affected, the lower ones to a lesser extent.

A very beautiful and thorough study of Raman scattering by the water molecule using Raman wavefunctions was presented by Zhang and Imre [138].

17.13 Complete Raman Emission Spectrum

Even though it is not an eigenstate of the excited Born-Oppenheimer potential, the Raman wavefunction $|\mathcal{R}_g(E_I)\rangle$ acts like it is for the purpose of calculating Raman emission to final states, as equation 17.32 shows. Just is in Franck-Condon photoabsorption to specific excited vibrational states, we may obtain the whole spectrum at once (here the emission spectrum from a Raman wavefunction) by Fourier transform of an autocorrelation function. The Raman intensity scattered at frequency ω_S can be derived as (with $\omega_S = \omega_I + (E_i - E_f)/\hbar$ and apart from overall constants)

$$
\begin{aligned}
I(\omega_s) &= 2\pi\omega_I\omega_S^3 \sum_f \delta(\omega_I - \omega_s + (E_i - E_f)/\hbar)|\langle\varphi_f|\mathcal{R}_g(E_I)\rangle|^2, \\
&= \omega_I\omega_S^3 \int_{-\infty}^{\infty} dt\, e^{i(\omega_S-\omega_I-E_i/\hbar)t}\langle\mathcal{R}_g(E_I)|\mu_{fg}^{\sigma}\, e^{-iH_g^{B.O.}t/\hbar}\, \mu_{fg}^{\sigma}|\mathcal{R}_g(E_I)\rangle, \\
&= \omega_I\omega_S^3 \int_{-\infty}^{\infty} dt\, e^{i\tilde{\omega}t}\langle R_g(E_I)|\, e^{-iH_g^{B.O.}t/\hbar}\, |R_g(E_I)\rangle, \\
&= \omega_I\omega_S^3 \int_{-\infty}^{\infty} dt\, e^{i\tilde{\omega}t}\langle R_g(E_I)|R_g(E_I, t)\rangle.
\end{aligned}
\tag{17.34}
$$

Suppressing the initial state energy E_I, we have

$$
I(\omega_s) = \omega_I\omega_S^3 \int_{-\infty}^{\infty} e^{i\tilde{\omega}t}\langle R_g|R_g(t)\rangle\, dt, \tag{17.35}
$$

where $|R_g(t)\rangle = e^{-iH_g^{B.O.}t/\hbar}|R_g\rangle$ is the transition moment modified Raman wavefunction propagated on the ground-state potential energy Born-Oppenheimer surface. This expression gives the Raman spectrum at all frequencies starting from the excited state Raman wavefunction and its dynamics on the ground potential energy surface.

Intuitively, equation 17.35 seems quite reasonable. We imagine that, by applying the incident frequency, we create a wavefunction $|\mathcal{R}_g(E_I)\rangle$ on the excited surface that is not an eigenfunction there. This wavefunction can then emit a photon and return to the ground surface, just as if it were an eigenfunction. We expect a full Fourier transform of the autocorrelation $\langle R_g | R_g(t)\rangle$ to give the emission spectrum, which here is a Raman spectrum.

It is easily shown that $|\mathcal{R}_g(E_I)\rangle$ obeys the inhomogeneous (driven) time-independent Schrödinger equation

$$(H - \tilde{\omega}_I - i\Gamma)\mathcal{R}_g(E_I)\rangle = -i|\varphi_g\rangle. \quad (17.36)$$

This form is useful in time-independent basis set calculations, but is not nearly so easy to implement in many degrees of freedom.

17.14 Raman Scattering for Many Degrees of Freedom

The same advantages accrue to equation 17.35 in computing Raman emission as benefited the time-dependent expression for absorption spectroscopy, equation 17.17—namely, that when the Raman spectrum devolves into a myriad of partially resolved or merged lines, there is no need to build up the spectrum by projections onto perhaps billions of eigenstates in a large molecule. Rather, one computes the autocorrelation function equation 17.35 once, and computes the Raman spectrum at any new ω_I by a single Fourier transform of this stored quantity.

We give an example of modest dimension. Suppose a molecule with 15 vibrational degrees of freedom experiences a total displacement of 0.1 Å in the excited state. This displacement could be in one mode or divided among several modes. Assume an average vibrational frequency of 1000 cm^{-1}. Suppose a given coordinate has been "sampled" at a point on the potential surface if $|\varphi_i(t)\rangle$ has visited there with an amplitude of at least 1% of its maximum. Define "available configuration" as points on the excited Born-Oppenheimer surface such that the potential energy lies below the total average energy of $|\varphi_i(t)\rangle$ on the excited surface. These are both conservative metrics. All of these configurations are sampled by the totality of eigenstates contributing to the energy form of the absorption and Raman scattering. Then, in 10 vibrational periods (typically much less time is needed to determine the Raman scattering) the wavefunction $|\varphi_i(t)\rangle$ samples less than 1 part in 10^5 of the available coordinate space. Often one needs to know only the Franck-Condon regions of both potentials and a little bit in the direction of steepest descent on the excited state potential.

The picture evoked by equation 17.29 is embodied in figure 17.13. The propagated wavefunction is shown as dropping back down at time t, and there it is displaced and can overlap many different excited vibration-rotation wavefunctions on the ground electronic state. The case of two coordinates shown in the figure is meant to suggest many more, where the same principles apply.

It is very frequently true that the observed Raman spectrum to all the lowest fundamentals and overtones is determined before any recurrences happen in the autocorrelation function, or only weak recurrences may have happened. Sure-fire evidence for this is in the absorption spectrum, if that carries little or no structure beyond the broad envelope representing the initial autocorrelation decay. We can use the short time evolution of a multivariate Gaussian and its Fourier transform

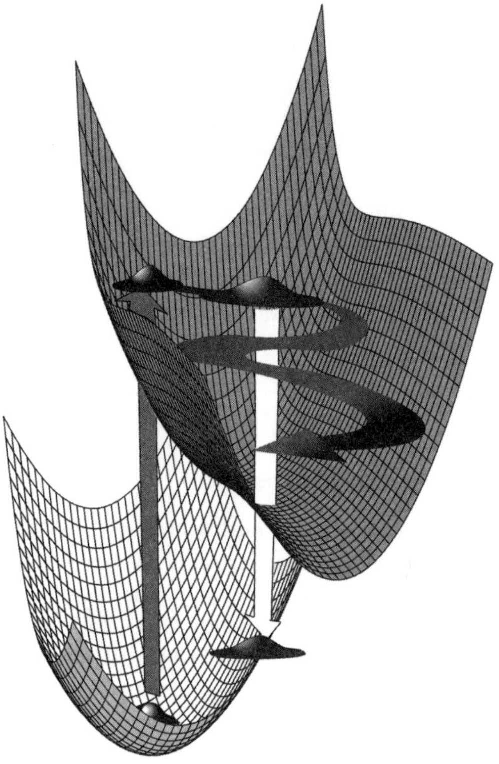

Figure 17.13. In Raman scattering, the time-dependent formulation promotes a copy (transition moment modified) of the ground-state vibration-rotation wavefunction to one (or more) excited electronic states at time $t = 0$. The propagation of this wavefunction on the excited electronic state can be projected onto various ground electronic state vibration-rotation wavefunctions; the half Fourier transform of these time-dependent projections gives their Raman amplitude at the incident frequency.

to say a great deal about the information in the Raman spectrum. The simplest clear-cut example of this is direct photodissociation, including polyatomic molecules. The Raman fundamentals and first overtones (transitions to two or a few quanta of excitation in a given mode) are determined in femtoseconds.

It is a good time to explain why the Raman lines remain sharp in spite of the short time events and short time autocorrelations in the excited states. This is already implied of course by the meaning of the expressions for $\alpha^{\rho,\sigma}_{m,m_f}(\omega_I)$: it is the amplitude to arrive at a given final vibration-rotation state on the ground potential energy surface. If that is a long-lived state not itself embedded in a quasi-continuum of nearby states also accessible by Raman scattering, then the Raman emission must be a very narrow band, since the only energies in the problem are the initial and final electronic state energy, which cancels out, the laser energy, and the energy of the final vibration-rotation state. There is no leeway for lifetime broadened lines!

Chapter 18

Semiclassical Spectroscopy

18.1 Semiclassical Franck-Condon Factors

CLASSICAL PROBABILITY DENSITIES

Both one-photon and radiationless transitions between two Born-Oppenheimer potential energy surfaces require Franck-Condon overlaps between nuclear wavefunctions involved in the transition.

A recurrent theme in this book connects generating functions including classical action functions with classical probability densities and overlaps in phase space (see discussion around figure 3.2 and figure 3.4, especially section 3.2), and making the connection with quantum mechanical probability amplitudes (figures 6.5, 7.3, 7.5, and surrounding discussion). The reader is encouraged to review these images and discussions.

Recalling these principles in the present context of Franck-Condon factors will prove especially useful, in part because they are so closely related to understanding related problems such as the "curve crossing" problem of Born-Oppenheimer theory, and topics introduced below such as radiationless transitions, Landau-Zener curve crossing, and nonadiabatic surface hopping. These problems are closely related because they all involve Born-Oppenheimer potential surface crossings. Franck-Condon transitions between vertically separated potentials can also involve crossing surfaces, once we draw the lower potential energy surface up by $\hbar\omega_{laser}$ and include the transition moment, as in figure 18.1 (the same can be done for electronic transition by light emission). (When the surfaces do not cross in any of these situations—that is, Franck-Condon or radiationless transitions, and so on, tunneling processes will prevail, as will be discussed shortly.)

Figure 18.1 decomposes a Franck-Condon factor (see equation 17.18) in a revealing way. The lower, bound (attractive) potential energy surface has been raised up by the laser energy $\hbar\omega$ to allow the vibrational state (modified by transition moment coordinate dependence) on that surface to coincide with a Franck-Condon partner on the repulsive, upper surface. The Franck-Condon integral is just the product of the two wavefunctions integrated over space. They are also shown by phase space diagrams, where it is seen that their respective phase space paths cross at the q position of the crossing of the two potential energy surfaces involved. Importantly, the two wavefunctions, having the same or similar momentum in that vicinity, oscillate in phase in a way that accounts for most of the integral. Away from that region, the two waves are effectively cancelling any further contributions, just as the phase space paths have also diverged from each other. Once again the diamond shaped areas are overlaps

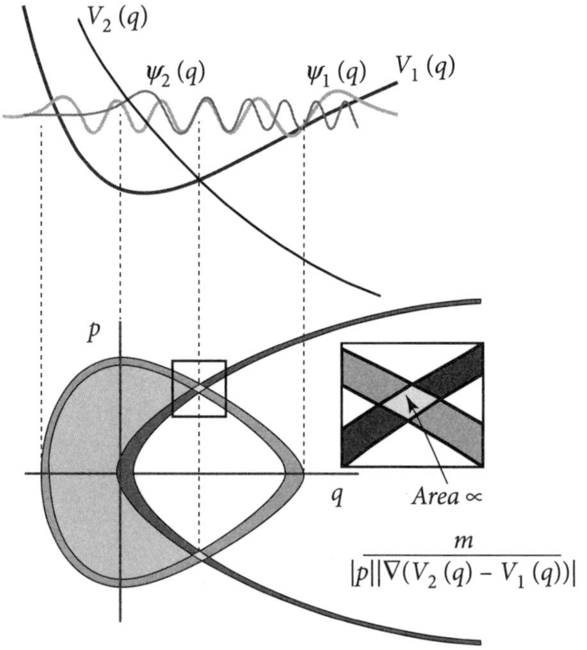

Figure 18.1. Anatomy of a Franck-Condon factor. Electronic spectroscopy involving two different potential energy surfaces gives rise, as we shall see, to overlap integrals between vibrational wavefunctions on each surface. Ignoring the transition moment, which for such very different potential surfaces is often of secondary importance, the Franck-Condon factor is the spatial overlap of $\psi_1(q)$ and $\psi_2(q)$. The phase space path, including "thickness" to correctly describe the probability of each of the two wavefunctions shown is also given. In that picture, the overlap involves intersections of the paths, at a place in q where the wavefunctions oscillate with very similar wavelength. Two coincidences of position and momentum for the two states are seen in the phase space picture. The semiclassical Franck-Condon formula for the overlap is given in the text. Semiclassically, the phase difference between them is given by the shaded area an action difference between the two curves divided by $\hbar$.

of classical probability densities: $P(E_1, E_2) = \mathrm{Tr}[\delta(E_1 - H_1(p, q))\delta(E_2 - H_2(p, q))]$. This is the classical probability measure for points to be found to belong to both of the states—that is, their overlap, one of energy E_1 in the Hamiltonian H_1 and the other of energy E_2 in the Hamiltonian H_2. More specifically,

$$\begin{aligned} P(E_1, E_2) &= \int dpdq\; \delta(E_1 - \frac{p^2}{2m} - V_1(q))\; \delta(E_2 - \frac{p^2}{2m} - V_2(q)), \\ &= \int dq\; \frac{dp^2}{2|p|}\; \delta(E_1 - \frac{p^2}{2m} - V_1(q))\; \delta(E_2 - \frac{p^2}{2m} - V_2(q)), \\ &= \int dq\; \frac{m}{|p|} \delta(E_1 - E_2 - V_1(q) + V_2(q)), \\ &= \sum_k \frac{m}{|p_k||\nabla(V_2(q_k) - V_1(q_k))|}, \end{aligned} \tag{18.1}$$

where the sum is over all intersection points of the two energy distributions, p_k is the momentum at the intersection, and $V_2(q_k) - V_1(q_k)$ is the potential difference at the intersection point. A cautionary note: the interference between different contributions k has not been included, though it could be put in by hand so to speak, and the phase difference is approximately the shaded area between the two curves, divided by $\hbar$.

FRANCK-CONDON FACTORS BY STATIONARY PHASE

Now we examine such Franck-Condon overlaps by a more traditional semiclassical approach. We assume the transition moment operator is independent of nuclear position (the so-called Condon approximation). Figure 18.1 and figure 18.4, later (which reveals a case where the phase space trajectories do not cross), show the potential energy surfaces involved raised or lowered by the photon energy. For example, in the case of absorption, the lower potential energy surface has been "dressed"—that is, raised by $\hbar\omega$, the photon energy. This has the advantage of automatically making clear the relationship to radiationless transitions, where no photon and no shift of the potential surfaces is involved—that is, the two surfaces fall where they are shown, under the Born-Oppenheimer approximation, without being dressed. This is certainly possible, and at higher energy the probability of such near approaches and crossings of Born-Oppenheimer surfaces increases. Suppose the wavefunction starts on a bounded potential surface, but is subject to transfer decay to the repulsive surface, which in this case results in breakup or dissociation of the system. Since there are very significant differences between the two potential surfaces, we are presently ignoring the coordinate dependence of the transition moment, which ordinarily takes a backseat in such situations.

Very interesting questions are posed by these two pictures and generalizations of them to be considered soon. It is clear from the figures that the overlap integrals may or may not possess real stationary phase points if the wavefunctions are represented semiclassically. When there is a stationary phase point, what is the formula for the overlap? Can one write down convenient expressions when the stationary phase point is a saddle point in the complex plane? What is the connection between these two spectroscopy and/or radiationless transition problems and the so-called surface crossing problem, or Landau-Zener transitions?

The derivation reads

$$\int dq \; \psi_2^*(q)\psi_1(q),$$

written with

$$\psi_j(q) = \sum_k \left| \frac{\partial^2 S_k^j(q, E)}{\partial q \partial E} \right|^{1/2} e^{i\frac{S_k^j(q,E)}{\hbar}} \sim \sum_k \frac{1}{|p_j^k(q)|^{1/2}} e^{-\frac{i}{\hbar}\int^q p_j^k(q')\,dq'}, \tag{18.2}$$

where j refers to the Hamiltonian $H_j = H_1$ or H_2 and k refers to the kth stationary phase contribution to the wavefunction. The Franck-Condon overlap integral I is then performed by stationary phase:

$$I = \int dq \frac{1}{|p_1(q)p_2(q)|^{1/2}} e^{-i\int^q (p_2(q')-p_1(q'))\,dq'/\hbar}. \tag{18.3}$$

A stationary phase point is given by $p_2(q) = p_1(q)$, and the integral evaluates as

$$\left|\int dq\ \psi_2^*(q)\psi_1(q)\right|^2 \sim \sum_k \frac{1}{p^2\ |\partial p_2/\partial q - \partial p_1/\partial q|} = \sum_k \frac{1}{|p||\nabla(V_1(q) - V_2(q))|}. \tag{18.4}$$

in agreement (apart from constants we are ignoring) with equation 18.5, discussed next.

CONNECTION TO THE LANDAU-ZENER TRANSITION

The Landau-Zener transition probability P between two *adiabatic* potential curves [139] is particularly elegant:

$$P = e^{-2\pi\gamma},\quad \gamma = \frac{\epsilon^2}{\hbar\frac{d}{dt}|V_2(q_t) - V_1(q_t)|} = \frac{m\epsilon^2}{\hbar|p||\nabla(V_2(q) - V_1(q))|}, \tag{18.5}$$

where the first form is Zener's and the second is Landau's. 2ϵ is the narrowest splitting between the adiabatic potentials.

The connection of this problem to the classic Landau-Zener "surface hopping" transition problem can now be made. The probability of a diabatic transition (such as staying on the same manifold with no abrupt slope changes through the crossing in figure 18.1) is, according to Landau-Zener theory,

$$\begin{aligned} P &= e^{-2\pi\Gamma}, \\ \Gamma &= \frac{a^2/\hbar}{\left|\frac{\partial}{\partial t}(V_2 - V_1)\right|} \propto \frac{a^2/\hbar}{\left|p(q_s)\frac{\partial}{\partial q}(\Delta V)\right|}. \end{aligned} \tag{18.6}$$

One thing is still missing: interference effects due to the two branches of the solution. We have so far given only the magnitude of one of them. If a particle were to approach from infinity at negative momentum on potential $V_2(q)$ and turn around, heading out with positive momentum, it would have two chances to cross and become trapped in the bound state potential $V_1(q)$. These two amplitudes can interfere, according to the actions $e^{-i(S_1^{(1)}(q,E)/\hbar - S_1^{(2)}(q,E)/\hbar) + i\frac{\pi}{2}(\nu_1-\nu_2)}$ and $e^{-i(S_2^{(1)}(q,E)/\hbar - S_2^{(2)}(q,E)/\hbar) + i\frac{\pi}{2}(\nu_1-\nu_2)}$. In analogy with the issue of relative phase of the two contributions to the Airy function in the coordinate representation (see figure 6.4, earlier, and surrounding discussion), the relative phase of the two contributing stationary phase points is given by the area enclosed by the two classical manifolds, divided by $\hbar$ (shaded region in figure 18.1).

Comparing equation 18.6 to equation 18.1, we see striking similarities. Equation 18.1 does not weigh the coupling strength between the two states involved, rather only the overlap in phase space. If we assigned a probability of hopping from one track to another on a given pass, the formulas would correspond, except we must discuss the exponential form of the Landau-Zener result. P is defined as the probability to make a nonadiabatic hop between adiabatic potential surfaces. The Franck-Condon factor on the other hand refers to a diabatic process. Suppose the adiabatic avoided crossing is narrow—that is, 2ϵ is small. The Landau-Zener result predicts a near unit probability jump between adiabatic surfaces, as befits a narrow avoided crossing (involving a breakdown of the Born-Oppenheimer approximation). We may write the probability

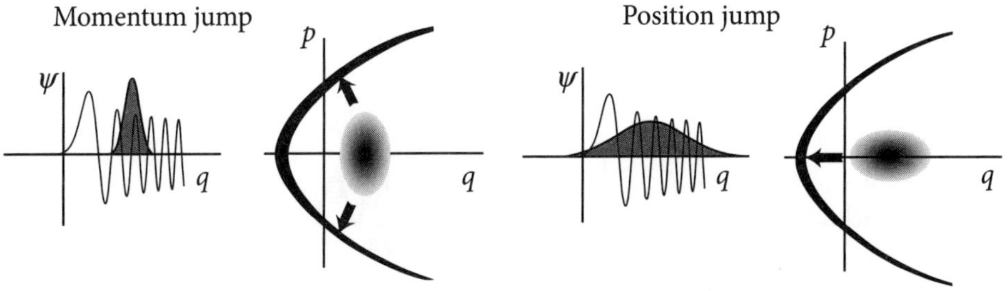

Figure 18.2. By varying the uncertainty in position versus momentum, the Gaussian nonclassical overlap with the wavefunction living on the parabolic manifold shown finds most of its propensity to tunnel in momentum on the left, and in position on the right. It is interesting that the tunneling and momentum (mostly) could be enhanced or suppressed by interference effects coming from the two contributions.

of *not* jumping then as

$$1 - P \sim \frac{m\epsilon^2}{\hbar |p| |\nabla(V_1(q) - V_2(q))|}; \tag{18.7}$$

this just corresponds to the *diabatic* probability of making the transition—that is, proportional to the phase space overlap. Thus, Landau-Zener theory is connected to a direct semiclassical—that is, stationary phase—evaluation of the Franck-Condon factor (see equation 18.1).

18.2 Classically Forbidden Franck-Condon Factors

If doing a Franck-Condon overlap by stationary phase succeeds in finding stationary phase points, we can safely say the overlap is "classically allowed," with phase portraits like the one in figure 18.1.

If there are no (real) stationary phase points, the classical portraits of the wavefunction involved fail to overlap. The Franck-Condon overlap will be small, but not vanishing. A classically forbidden, tunneling process involving the shortest path between the states will dominate. The propensity of a wavefunction to reach out in position, or alternately momentum, to find the "largest of the small probability pathways" can be seen in the cases shown in figure 18.2.

CLASSICALLY FORBIDDEN PHOTOABSORPTION AND EMISSION

The lasers can often be tuned to reach classically allowed or forbidden regimes depending on frequency—that is, as seen in figures 18.1, 18.3, and 18.4. In figure 18.3, a ground vibrational state resides on an attractive Born-Oppenheimer potential energy surface. It is promoted by photoabsorption to a repulsive, dissociative potential energy surface. The cross section for light absorption is shown on the left, as a function of laser frequency. The lower Born-Oppenheimer potential energy surface can be dressed, or raised, by the various laser frequencies. It is seen that the overlap of the ground vibrational state with the various degenerate (including the laser energy) dissociative states goes from classically forbidden at low frequency, to classically allowed, to forbidden again at high frequency. This is a continuum spectrum characteristic of a dissociative potential. There is no reason not to extend this tunneling concept to discrete spectra, where in the fading wings of the spectrum we may view the transition

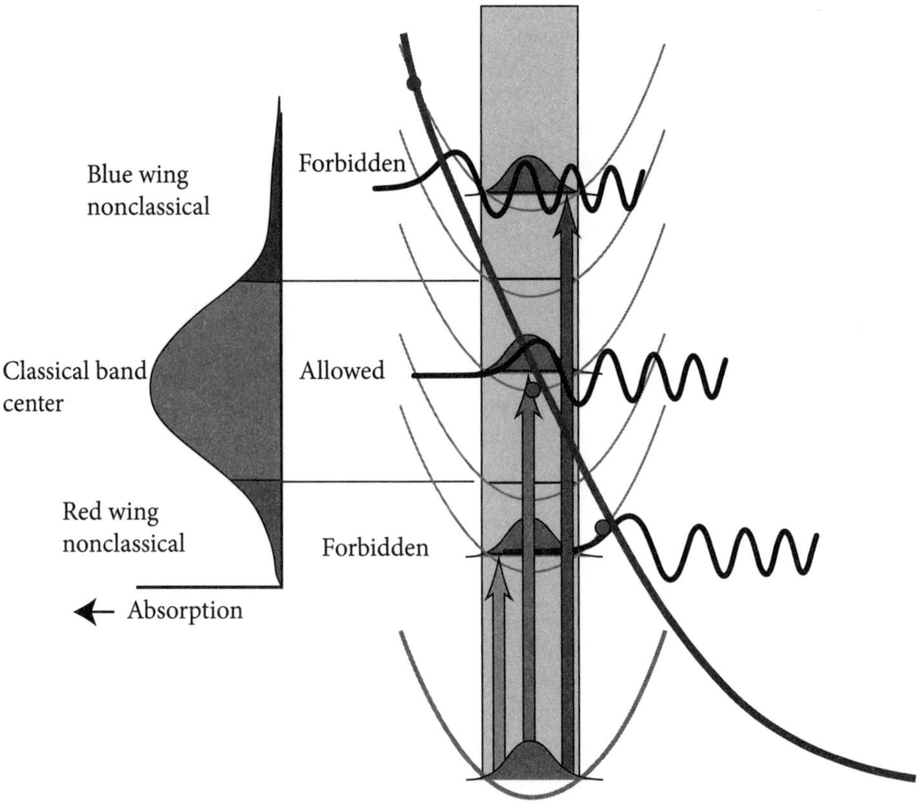

Figure 18.3. A ground vibrational state residing on an attractive Born-Oppenheimer potential energy surface is promoted by photoabsorption to a repulsive, dissociative potential energy surface. The cross section for light absorption is shown vertically on the left, as a function of laser frequency. The lower Born-Oppenheimer potential energy surface is shown dressed, or raised, by various laser frequencies. It is seen that the overlap of the ground vibrational state with the various degenerate (including the laser energy) dissociative states goes from classically forbidden at low frequency, to classically allowed, and again to classically forbidden at high laser frequency. This same principle—that is transitions being classically allowed or forbidden—applies even if the upper surface is also bound, with the spectrum composed of sharp peaks. The same statements hold in many spatial dimensions, except what is at stake is much more interesting.

as nonclassical—that is, tunneling. The same principles also hold in many spatial dimensions, except that the possibilities are much more interesting.

We reach the inescapable conclusion that the absorption spectrum divides into three regions: a central, "classically allowed" region of largest spectral intensity, and two classically forbidden "wings" one to high and one to low frequency, both of lower intensity.

In the nonclassical region the states involved in the Franck-Condon factors may differ in unexpected ways. Unusual jumps in phase space may be required in these nonclassical wings. We illustrate this point next.

Consider figure 18.4, continuing the subject of "classically forbidden" Franck-Condon factors. There is no real position where both momenta are the same. The Franck-Condon factor will be small(er), and the stationary phase points will become

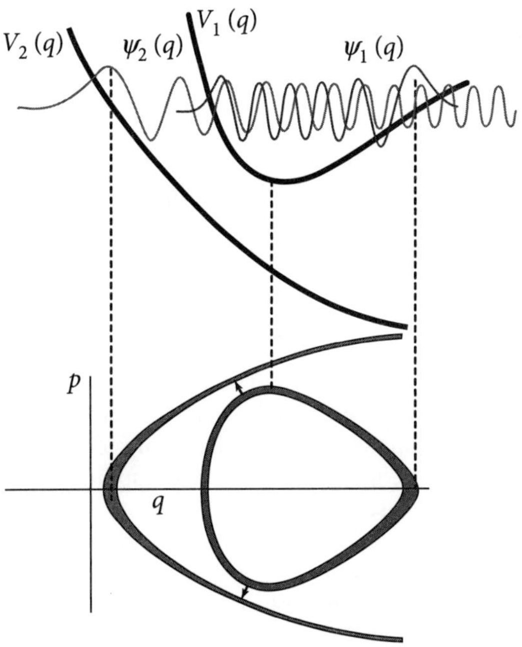

Figure 18.4. There are no (real) stationary phase points in the integral overlap of these two vibrational eigenstates. The phase space picture in the lower portion, showing the semiclassical nature of the two states, demonstrates that the tracks do not intersect, nor do the wavefunctions above ever oscillate in synchrony. A strong suspicion arises that there are well-defined "jumping off" places to tunnel, as shown by arrows. Whereas this may be not terribly important in one dimension, the choices available in two or more dimensions correspond to widely varying initial conditions that are critical to the subsequent dynamics. See references [140–142].

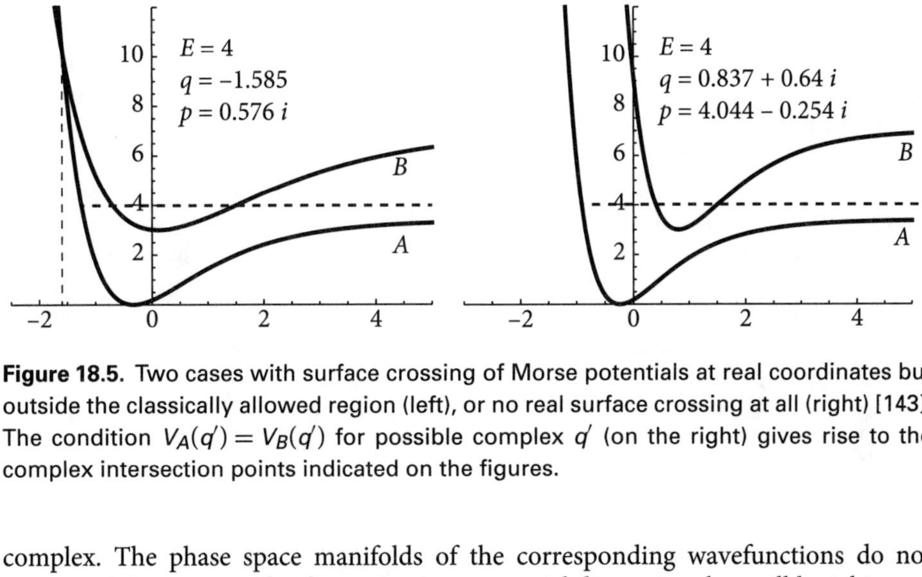

Figure 18.5. Two cases with surface crossing of Morse potentials at real coordinates but outside the classically allowed region (left), or no real surface crossing at all (right) [143]. The condition $V_A(q') = V_B(q')$ for possible complex q' (on the right) gives rise to the complex intersection points indicated on the figures.

complex. The phase space manifolds of the corresponding wavefunctions do not intersect. We continue the discussion in one spatial dimension, but will be taking up the interesting issue of many dimensions in this section.

Referring to figure 18.5, writing the semiclassical wavefunctions and performing their overlap by stationary phase, we look for positions q' where, expressing the actions $S(q', E)$ as integrals over momentum, the stationary phase condition is

$$\frac{d}{dq}\left(-\int^{q} p_A(q')\,dq' + \int^{q} p_B(q')\,dq'\right) = 0 \tag{18.8}$$

on surfaces A and B where $p_A(q') = \pm\sqrt{2m(E - V_A(q'))}$, with solution

$$p_A(q') - p_B(q') = 0. \tag{18.9}$$

For the cases at hand, there is no stationary real value of q'. Nonetheless, (1) equation 18.9 will have a solution if the surfaces cross—that is, $V_A(q') = V_B(q')$ for real q' *outside* the classically allowed region. Or (2) a pair of complex q' may be found if the potential surfaces do not cross. In case (1), an imaginary momentum results at the intersection. The real crossing in q signals the position at which a jump occurs between the two wavefunctions, or expressed differently, a region where most of the (very small) integral of the product of the two wavefunctions arises. If the surfaces do not cross for real q, complex q crossings can be found. The corresponding real parts of p and q represent the momentum jump and position jumps, respectively, (tunneling events) that connect the two wavefunctions (jumps are required because their manifolds do not cross in phase space). The imaginary parts are responsible for the attenuation leading to the small stationary phase integral.

In figure 18.5, two cases with surface crossing of Morse potentials either outside the classically allowed region (left) or no real surface crossing at all (right) are considered. On the left, having a real surface crossing, the solution to $V_A(q') = V_B(q')$ at $E = 4$ gives $q' = -1.585$, implying $p' = 0.576i$ on either surface. Tunneling of both wavefunctions to this imaginary momentum region produces a very small semiclassical estimate of the overlap of the two wavefunctions, one on surface A and one on B, at the same energy $E = 4$.

At the right, the Morse potentials never cross for any real q', but do cross at $q' = 0.837 + 0.64i$, $p' = 4.04466 - 0.254077i$. Both semiclassical wavefunctions must be carried to these coordinates, with a concomitant very small semiclassical estimate. The real part of the position and momentum may be taken as the "hand off" point in phase space connecting the two wavefunctions, and the imaginary parts of the position and momentum are responsible for tunneling attenuation of the resulting overlap estimate to reach those imaginary values. In the case on the right, for example, the scenario for flow from surface B to A is for an eigenstate trapped temporarily on potential B to suddenly appear on potential A in the region near $q = 0.84$, arriving with considerable momentum, $p \approx 4$ at that moment. Although this affects only the vibrational phase on surface A in one dimension, it is easy to see that in several dimensions, various competing position or momentum jumps can have drastically different dynamical outcomes, depending on the "launch point" in phase space.

NONCLASSICAL JUMPS IN TWO AND MORE DIMENSIONS

The question of classically allowed versus classically forbidden transitions is a nontraditional and underrated topic in Franck-Condon theory. When the incident laser frequency puts the Franck-Condon system in a nonclassical realm, or for a nonradiative transition with no good "classical" places to hop to a new surface, the system must find ways to jump, or tunnel, in phase space to often surprising new places. These jumps, depending on the nature of the potential surfaces, the energies, and the states involved, may involve combinations of position and momentum tunneling [140–142, 144].

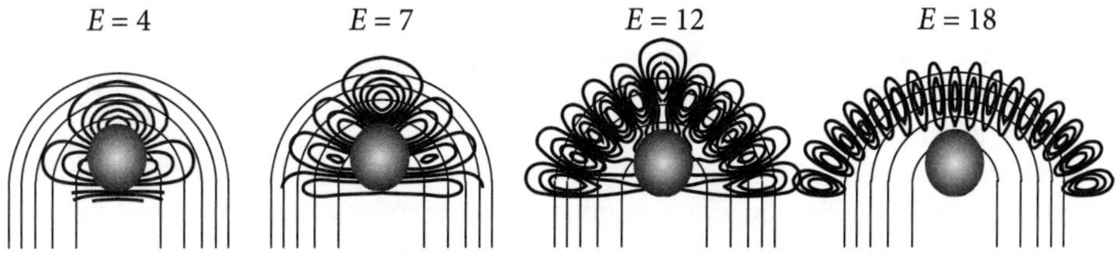

Figure 18.6. The wavefunction contours show the eigenstate of maximum overlap with the Gaussian state, with energy of the degenerate states increasing to the right. The eigenstate reveals that a jump in position (in the vertical or y direction) and in momentum (in the horizontal or x direction) takes place in order to reach the other surface. The disposition of the eigenstate in phase space can be read off the contour locations and their nodal structure. The tendency to position or momentum jump depends on the ground state wavefunction and excited state potential. Every possible combination of position and momentum jump is feasible depending on their properties.

We can make intuitive diagrams best in two spatial dimensions, but what happens is clear enough, making implications for many dimensions evident. We begin with a two-dimensional ground vibrational wavefunction—that is, a two-dimensional Gaussian—even though much more general situations could be considered. This is a wavefunction on the excited potential surface. Or, it could be a wavefunction placed or trapped by relaxation to the lowest energy region of an excited triplet surface, very unlikely to relax further by radiative decay, but also having no clear classically allowed relaxation pathway available. How shall we determine the propensity to jump in various directions in phase space onto new potential energy surfaces in order to relax further? Here, to illustrate what can happen, we use a trick, and construct a degenerate set of states to which to jump or tunnel. The states span many different parts of phase space but do so at the same energy, and so allow the initial state to show us where it most wants to jump. By making a linear combination of the degenerate states using coefficients determined by their projection onto the initial state, we reveal the nature of the state to which the initial state wants to jump. The badly positioned initial state, which must tunnel, thus acts as a generator state in the sense of section 15.1. As always, the linear combination of degenerate states will make a new eigenstate (all linear combinations are still eigenstates) that most resembles the generating state, except here it is the nearly invisible and subtle wings of the generating state that are revealed. This resemblance is impossible for the eye to see, but the linear combination uncovers its hidden propensities.

Using the two-dimensional scattering potential whose contours are shown in figure 18.6, degenerate linear combinations of the total energy shown are made with projections on to the initial Gaussian:

$$|\psi\rangle = \sum_n |n\rangle\langle n|\phi\rangle. \tag{18.10}$$

In this equation, $|\phi\rangle$ is the Gaussian generating state, the $|n\rangle$ are a set of degenerate eigenfunctions of the excited state potential, labeled by n (in this case, representing the partition of energy on the potential into transverse vibration longitudinal translation), and $|\psi\rangle$ is the resultant function that reveals the tendency that the generative state has projected onto the degenerate eigenstates. While this may seem rather artificial—that is, depending on the excited state to have a set of degenerate levels (that the scattering potential shown does have)—the propensities are so dramatic that clearly they reveal tendencies to jump in momentum, or hop in position, even differently in different degrees of freedom, which would prevail on more general potential surfaces. Specifically, figure 18.6 clearly displays a propensity for the system to make a jump in position in the vertical, or y direction, and a sudden jump in momentum in the horizontal, or x direction, as the transition goes to the blue of resonance.

Chapter 19

Semiclassical Dynamics and One-Photon Spectroscopy

The central idea of semiclassical spectroscopy is to cast the spectrum as the Fourier transform of an appropriate correlation function (see equation 17.18), and then calculate that correlation function semiclassically, meaning ultimately that classical trajectories determine the outcome through the time-dependent wavefunction they control. The reward is often deep insight into the spectral features in terms of visualizable classical motion of the system, exactly as we will do for benzophenone (see section 19.3). The Fourier transform of the time correlation data into the energy domain need not be done semiclassically, since it is just a one-dimensional integral no matter how many degrees of freedom there are. It is simple to do numerically, and constitutes a uniformization.

We start with the expression for the absorption rate in terms of the time-dependent dynamics (repeated here from the discussion leading to equation 17.17):

$$\Sigma(\omega) = \int_{-\infty}^{\infty} e^{i\omega t} \langle \varphi | \varphi(t) \rangle \, dt. \tag{19.1}$$

This is not a semiclassical expression; it is identical to the Born-Oppenheimer approximation and first order light-matter perturbation theory. Given that $|\varphi(t)\rangle$ is often a localized wavepacket launched on the side of a hill (again, this is not a semiclassical statement or approximation) semiclassical approximations become tempting.

Even without implementation by semiclassical techniques, the time-dependent formulation of electronic absorption spectroscopy avoids having to know any eigenstates or go through $H\psi = E\psi$ at all, a tremendous advantage for systems of more than a few degrees of freedom.

One of the best arenas for semiclassical time-dependent approaches is electronic spectroscopy, especially involving large molecules, including those embedded in liquids or solids and crystalline solids. This includes electronic absorption, emission, and Raman scattering. Suppose we find the lowest and next lowest Born-Oppenheimer electronic potential energy surfaces (see chapter 16) as a function of nuclear coordinates. Nuclear eigenfunctions (vibrational states, or phonons) reside on these Born-Oppenheimer surfaces. If we bring light into the picture, treated with first-order light-matter interaction perturbation theory (see section 17.2), transitions can be induced from one potential surface to another. Light-matter perturbation theory is extremely accurate for low to moderate light intensities. In most textbooks, such transitions are treated in the energy domain as a problem connecting vibrational eigenstates living on one potential energy surface to those on another, with propensities to make transitions governed by the so-called Franck-Condon principle, equation 17.18. The Franck-Condon principle is fine, but finding the vibrational eigenstates well up in the spectrum is, well, so yesterday.

Originating as relatively informal comments by Franck and Condon, the Franck-Condon principle expressed the idea that when a photon is absorbed, the sluggish nuclei don't move while the nimble electrons quickly change state. This is not actually a new principle; it is exactly equivalent to the Born-Oppenheimer approximation (that already has sluggish nuclei built in, see section 16.2) and first-order light-matter perturbation theory (see section 17.2).

19.1 Pseudo or MIME Modes in Spectra

Low-resolution spectra may have very regular, perhaps evenly spaced peaks that cannot be trusted to reflect anything but a transient effective vibrational or rotational frequency. Here, we give a clear example of how this can happen—in fact, it is commonplace. The effect was discovered in connection with an inorganic molecule with 14 vibrational degrees of freedom displaced in an excited electronic state relative to the ground state, WCO_5Py, or tungsten penta-carbonyl pyridine [145, 146]. The acronym MIME stands for *missing mode effect*, reflecting a vibrational progression seen for a mode that doesn't exist.

Suppose a Franck-Condon absorption takes place between the ground vibrational level of the ground electronic state and a displaced and distorted (compared to the ground Born-Oppenheimer potential energy surface) excited electronic state. Suppose the molecule is fairly large and there are many modes k that are displaced by δq_k in equilibrium from the ground state. There is likely a new normal mode basis in the excited state, rotated from the ground state normal modes (Duschinsky effect). In that case the displacements δq_k are figured from the equilibrium geometry of the ground state wherever it lies on the excited state potential energy surface, and the modes k refer to the excited state modes. (If instead we are considering emission from the ground vibrational mode of the excited electronic state, as often happens after photoexcitation, the roles are simply reversed.)

After the vibrational wavepacket falls away from its birthplace on the new potential surface, it will some time τ_R later have a first, partial autocorrelation or overlap recurrence $\langle \phi(0)|\phi(t)\rangle$. The broad feature already "burned in" by the initial drop-off of the autocorrelation will develop substructure as the Fourier transform time includes the recurrence. Nearly equally spaced peaks with a width on the order of or wider than their spacing appear, at a frequency interval of $\omega = 2\pi/\tau_R$. (The peaks are resonances but do not add to total intensity; they gather their strength by robbing the broad spectrum next to themselves. There is a sum rule—in fact, smoothing a sharp spectrum corresponds to shortening the time of the autocorrelation Fourier transform. Smoothing it enough will cut it off before the recurrence ever happens.) For a large molecule, this first recurrence is often the last, or close to it: as the wavefuction winds its way in many dimensions, it becomes more unlikely to have many displaced modes (that are perhaps starting to set even more modes in motion in $|\varphi(t)\rangle$) return

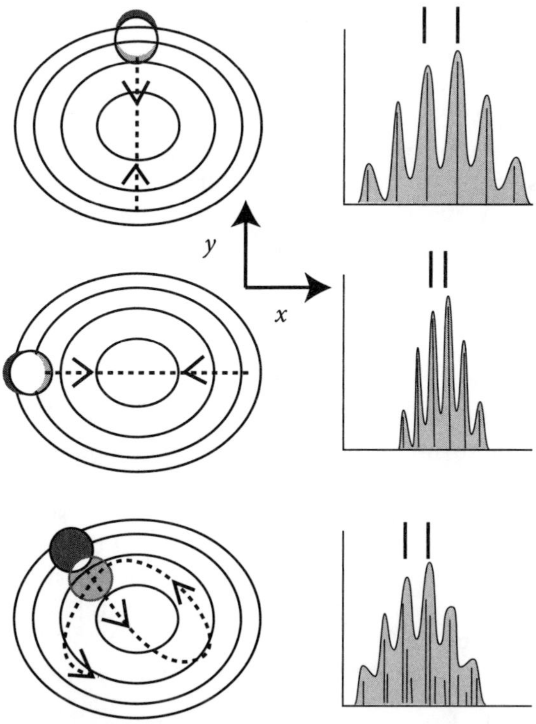

Figure 19.1. For the same quadratic two-dimensional potential energy surface, and the same stationary wavepacket (except for its mean initial position), the effect of the position of the wavepacket on the resulting spectrum at low and high resolution is shown. At the bottom, a case involving a MIME (missing mode effect) is shown, with a weak recurrence at a time intermediate between the two harmonic periods, and a resulting medium resolution progression spaced at a frequency in between the harmonic frequencies. The MIME spectrum at medium resolution might be all that is seen, due to other effects such as inhomogeneous broadening, or the presence of many other modes that are also displaced but much lower in frequency. For a longer time sequence of a similar situation, displaying a MIME effect, see figure 17.3.

even approximately all at once to the Franck-Condon birthplace where $|\varphi(0)\rangle$ "lives," enough for the two to have a significant if partial overlap as time increases.

Two dimensions is already enough for this scenario to happen, as seen in figure 19.1. Note that the higher frequency y mode has a faster decay of the initial overlap due to its steeper potential than does the lower frequency x mode at the same displacement. Thus, a broader low-resolution spectral band applies to the high-frequency mode (for which the autocorrelation decayed more quickly). The single-mode fine structure in the first two cases (displacement along the high- and low-frequency modes only) is reflected in the medium-resolution structure. Since the two modes have different frequencies, the spacing of the peaks is different in the two cases, presenting no mystery. The bottom case at medium resolution however is at first puzzling, since the perfectly uniform spacing is that of neither the x or y oscillators alone. All the high-resolution peaks under the broader features of course correspond to exact combination mode energies of the two separable oscillators, but the medium-resolution peaks are equally spaced at neither frequency and hint of the presence a nonexistent mode.

The spectrum as always is the Fourier transform of the autocorrelation—namely, the overlap of the initial vibrational wavepacket promoted from the ground electronic state with the time evolution of that state on the new, excited state Born-Oppenheimer potential energy surface. In fact, it is not difficult to show that effective MIME frequency, given the frequencies and displacements that are intrinsically present, is approximately

$$\bar{\omega} = \frac{\sum_j \omega_j^2 \delta q_j^2}{\sum_j \omega_j \delta q_j^2 N_j}, \tag{19.2}$$

where $N_j = [\omega_j/\bar{\omega}]$ is the nearest integer to the indicated ratio. The formula is derived as a Taylor series approximation to the time of the dominant early peak in the autocorrelation function, its inverse being $\bar{\omega}/2\pi$.[1]

A qualitatively different example of the MIME effect is shown in figure 19.2. A wavepacket $|\phi\rangle = |\chi(x)\rangle|\chi(y)\rangle$ appears symmetrically at the vertical Franck-Condon position on an excited state potential energy surface, consisting of two identical and orthogonal (in x and y) Morse oscillators. The exact, high resolution spectrum, the Fourier transform of $\langle\phi(0)|\phi(t)\rangle = \langle\phi(x,0)|\phi(x,t)\rangle\langle\phi_y(0)|\phi_y(t)\rangle$, i.e., $\Sigma(\omega) = \sum_{n,m} f_n f_m\, \delta(\hbar\omega - E_n - E_m)$, must be the combination spectrum of the two separate oscillators, both of which are excited in the symmetric displaced wavepacket (see figure 19.2; the high-resolution spectrum is shown by dense vertical spectral lines). The Franck-Condon factors are $f_n = |\langle\psi_n|\chi(x,0)\rangle|^2$, where $|\chi(x,0)\rangle$ is the ground state of the Morse oscillator in the x direction, and similarly for y. It is intuitively clear that the wavepacket will return as soon as it completes one vibration along the 45-degree symmetric coordinate $(x+y)/\sqrt{2}$—that is, both separate oscillators oscillating in phase since they were identically excited. Nonetheless, distortion of the evolving wavefunction in the separable coordinates x and y appears partly as spreading along the perpendicular coordinate $(x-y)/\sqrt{2}$; this will diminish the overlap $\langle\phi(0)|\phi(t)\rangle$ at the first and more so at subsequent recurrence times, but far from completely. The coordinate $(x+y)/\sqrt{2}$, with $(x-y)/\sqrt{2} = 0$ is interesting: it is a Morse oscillator along that slice, a combination of both of the separable Morse coordinates. It is different than either of the separable Morse oscillators present, with twice their dissociation energy since along that coordinate both separable oscillators are pulled apart. Motion along this periodic coordinate is unstable but not exponentially so.

The two-dimensional wavepacket behaves as if it is moving periodically in the effective $(x+y)/\sqrt{2}$ oscillator, but decaying along the perpendicular $(x-y)/\sqrt{2}$ coordinate. Thus, figure 19.2 shows, at the lower right, a beautiful progression of frequencies that belong to a periodic but unstable nonexistent mode, the symmetric slice along the potential energy surface. However, the short time autocorrelation function reflects this symmetric motion, and the low resolution spectrum in figure 19.2 has peaks consistent with this effective 1D oscillator. The stick spectrum shows how

[1] We cannot resist mentioning that the MIME formula has done double duty now, in connection with the autocorrelation of sounds and the perception of pitch of musical and other tones. See the chapter on pitch perception in the author's book, *Why You Hear What You Hear*, Princeton University Press, 2012.

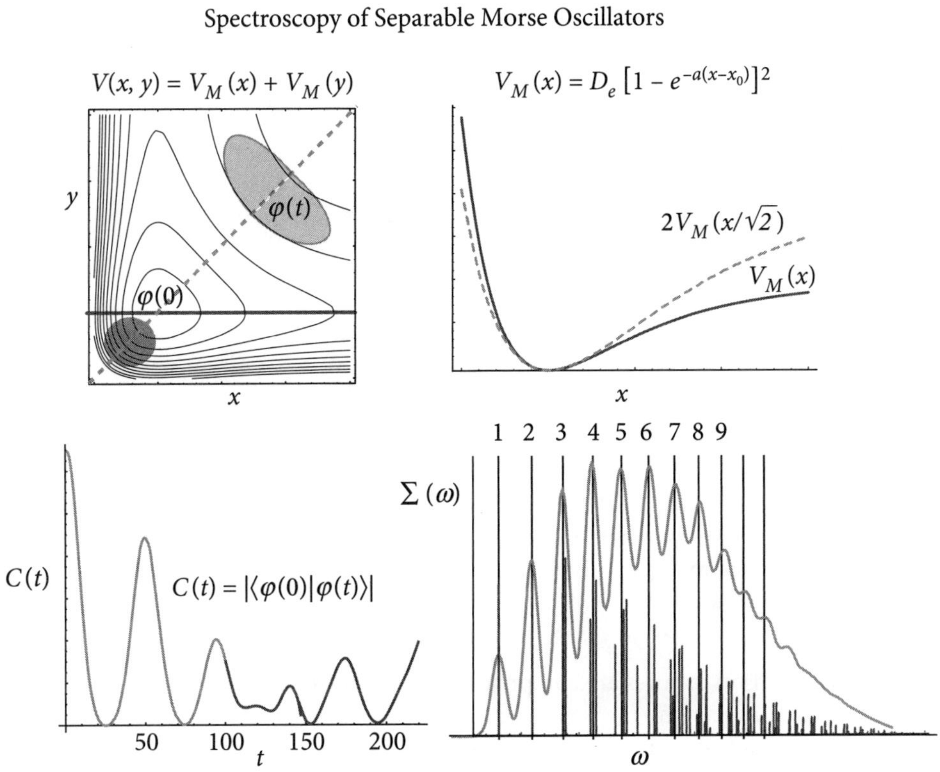

Figure 19.2. Franck-Condon spectrum of the symmetric, separable 2D Morse oscillator $V(x, y) = V_M(x) + V_M(y)$, given the initial wavepacket $\phi(x, y, 0)$. The Franck-Condon factors are projections of the eigenstates $\psi_{n,m}(x,y) = \chi_n(x)\chi_m(y)$ onto the initial wavepacket $|\phi(x, y, 0)\rangle$: $f^{\phi}_{n,m} = |\langle\psi_{n,m}|\phi(0)\rangle|^2$ [147]. The resolved spectrum is $\Sigma(\omega) = \sum_{n,m} f_n f_m \, \delta(\hbar\omega - E_n - E_m)$, seen as the dense forest of lines at the lower right. The upper-left panel reveals the contour map of $V(x, y)$ with a dashed slice of the potential along the 45-degree line $x = y$, and a another slice along the x Morse potential at the minimum y value of the y oscillator as a solid line. The analytic form of the dashed potential is given, and seen to be Morse in form with the same force constant and twice the dissociation energy (since moving out at 45 degrees breaks both Morse oscillator bonds). The initial wavepacket $|\phi(0)\rangle$ is shown as a circle, and an outline of that wavepacket $|\phi(t)\rangle$ propagated for short time is seen some distance away, having moved along the $x = y$ line but also spread perpendicular to it. The wavepacket returns to its birthplace in distorted form at multiples of the oscillation period along the diagonal slice of the potential. At the lower left, we see the absolute value of the autocorrelation function, which following its initial decay after $t = 0$ experiences recurrences at periods of motion along the effective 45-degree slice. The recurrences are responsible for the vibrational progression of the effective Morse oscillator $2V_M(r/\sqrt{2})$ labeled by broader peaks 1 through 9. The lines show the exact eigenvalues of $2V_M(r/\sqrt{2})$. The high-resolution lines lie as they must at the separable 2D Morse energies, but the low-resolution progression instead features its peaks at the energies of the effective Morse oscillator $2V_M(r/\sqrt{2})$, with the coordinate r along the symmetric stretch dashed line. The symmetric mode is a marginally unstable periodic orbit.

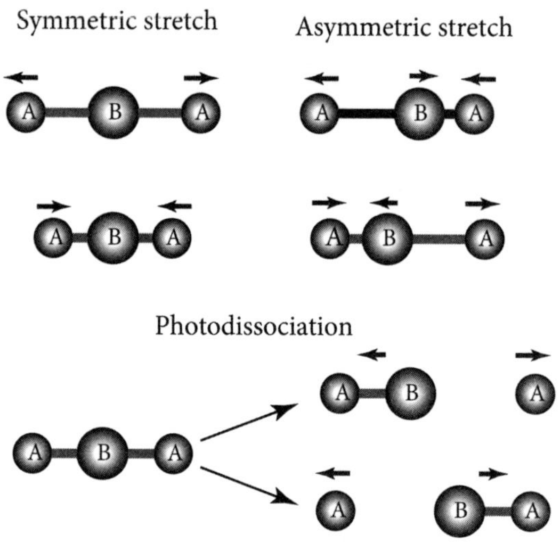

Figure 19.3. (Top) The symmetric and asymmetric stretch normal modes of a linear symmetric triatomic molecule. (Bottom) Two equivalent channels in the photodissociation of the molecule.

these low-resolution features are comprised of the intensities and positions of the exact high-resolution spectrum.

The low-resolution spectrum agrees with the 1D oscillators at low energy, but deviates at higher energy consistent with its lower onset of anharmonicity and higher dissociation energy.

19.2 Photodissociation of a Molecule

Consider the photodissociation of a linear symmetric triatomic molecule. The situation is illustrated in figures 19.3 and 19.4. The ground electronic state potential energy surface has a simple, nearly harmonic bowl (near the bottom) giving rise to a ground state wavefunction with the shape of $\phi(0)$. When such a molecule absorbs UV light, the new electronic state might allow a bond to break from the Franck-Condon region where $\phi(0)$ is born. In the case of a symmetric triatomic molecule with two equivalent bonds, there is typically insufficient energy for both bonds to break. But which bond breaks? The evolving wavefunction $\phi(t)$ remains symmetric. It develops *amplitude* for breaking the lefthand bond but not the righthand one, and equal amplitude for breaking the righthand bond but not the lefthand one. This is not the same as both bonds breaking of course.

The minimum in the potential along the symmetric stretch is extended to longer bond lengths in the excited state, with the result that the wavepacket $|\varphi\rangle$ finds itself displaced on the upper surface and heading downhill. Classically, the motion along the symmetric stretch is a periodic orbit, albeit an unstable one: a small deviation will lead away from the region. Both bonds stretch in unison along this orbit, but there is insufficient energy to dissociate directly, which would correspond to both bonds breaking at the same time.

For simplicity we take the potential to be separable in the vicinity of the saddle region, $V(x, y) \approx V_x(x) + V_y(y)$. This is a good approximation, since (1) $|\varphi(0)\rangle$ is

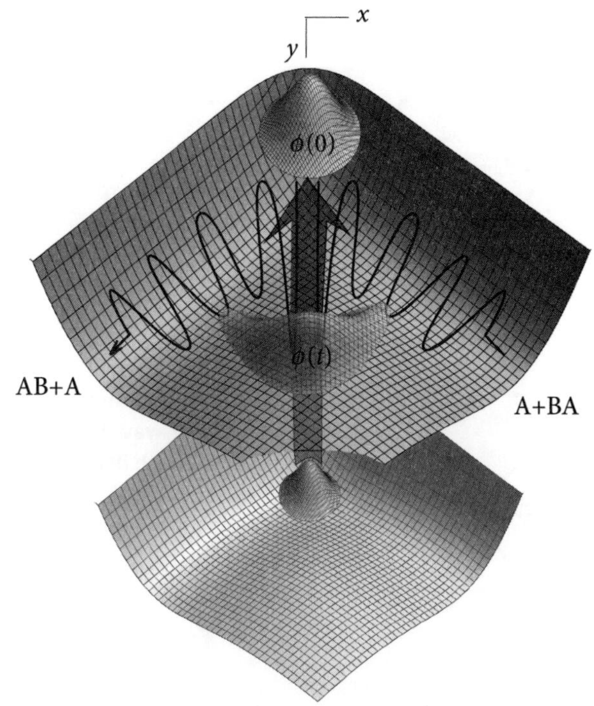

Figure 19.4. The potential energy surfaces and the initial wavepacket dynamics for the linear symmetric triatomic photo-induced dissociation.

localized in the saddle region; (2) the spectrum is given by the overlap $\langle\varphi(0)|\varphi(t)\rangle$; and (3) the escaping portion of $|\varphi(t)\rangle$ (falling away toward dissociation) does not return to the region of $|\varphi(0)\rangle$. Because x and y are normal modes on the lower surface, the initial state $\varphi(x, y)$ factors as $\varphi(x, y, 0) \approx \varphi_x(x)\varphi_y(y)$ Together with the separability of the upper potential energy surface this means that $\varphi(x, y, t) \approx \varphi_x(x, t)\varphi_y(y, t)$, and $\langle\varphi(0)|\varphi(t)\rangle \approx \langle\varphi_x(0)|\varphi_x(t)\rangle\ \langle\varphi_y(0)|\varphi_y(t)\rangle$ where $\varphi_x(x, t)$ ($\varphi_y(x, t)$) propagates on $V_x(x)$ ($V_y(y)$). Since the correlation function is a product of the x and y correlations, the spectrum will be a convolution of the individual spectra.

We consider the y motion first—that is, the symmetric stretch coming toward the viewer in figure 19.4. The potential wavefunctions $\varphi_y(y, t)$ and phase space pictures are shown in figure 19.5. There will be a decay in $\langle\varphi_y(0)|\varphi_y(t)\rangle$ owing to the motion of $|\varphi_y(t)\rangle$. The phase space picture shows the initial motion and overlap decay of $|\langle\varphi(0)|\varphi(t)\rangle|$ is nearly complete even while the two wavepackets $|\varphi_y(0)\rangle$ and $|\varphi_y(t)\rangle$ are still overlapping; the decay takes place because of momentum displacement, manifested as oscillation developing in the displacing wavepacket, even before it changes much in average position. This is of course just what happens to an accelerated particle starting at rest.

The early time development $|\varphi_y(t)\rangle$ follows within the thawed Gaussian approximation (see section 11.1) from the dynamics of the guiding trajectory,

$$y_t = y_0 - \frac{1}{2m}\frac{\partial V_y}{\partial y}t^2, \qquad p_{yt} = -\frac{\partial V_s}{\partial y}t, \tag{19.3}$$

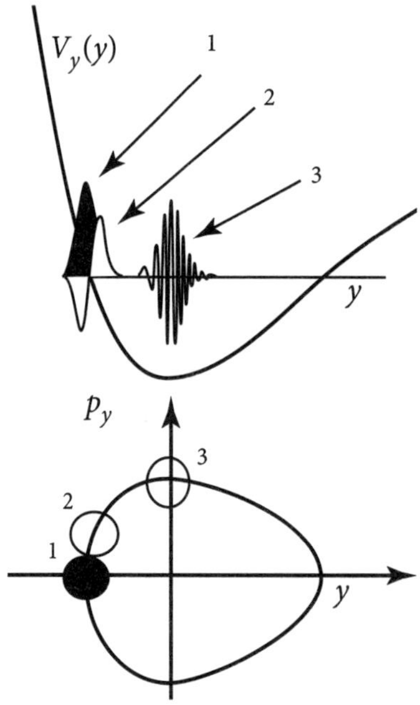

Figure 19.5. Early time dynamics of the symmetric stretch coordinate wavefunction, and its autocorrelation (overlap) with itself at $t = t_1 = 0$. Note that the initial wavepacket (black) and the propagated wavepacket at time t_2 have almost no overlap in the sense of $\langle \phi_1 | \phi_2 \rangle$, yet the two wavepackets still overlap spatially. Coordinate space, top, and phase space, bottom.

where y is the symmetric stretch coordinate. The momentum goes as t for short time, the position as t^2, so the momentum controls the short time autocorrelation function, at least if the potential is steep. Neglecting the spreading of the wavepacket in either direction (this is also a t^2 effect), we have

$$\langle \varphi(0) | \varphi(t) \rangle \approx \exp\left(-t^2/2\delta^2 - i E_0 t/\hbar\right), \tag{19.4}$$

where

$$\delta^2 = \frac{2m\hbar\omega}{(\partial V_y/\partial y)^2}, \tag{19.5}$$

and ω is the frequency of the normal mode along the symmetric stretch direction, which determines the width of the Gaussian in that direction, as $\exp[-m\omega\, y^2/2\hbar]$. The Fourier transform of equation 19.4 gives the low-resolution envelope as

$$\Sigma_T(E) = \exp[-\delta^2 (E - E_0)^2/2\hbar^2]. \tag{19.6}$$

We have linked the earliest time dependence of the autocorrelation function and the broadest feature of the spectrum, with key parameters like the ground state symmetric stretch frequency ω and the slope of the excited state potential $\partial V_s/\partial s$ at the ground state equilibrium geometry. The steeper the excited state slope, the broader the spectrum, in proportion to the slope, and inversely as the square root of the ground state frequency. Whatever happens at later times can only add detail to the spectrum by local "digging" and adjacent "piling up" of spectral density, without affecting any of the low resolution character of the spectrum.

If the wavepacket moves away and the initial decay is complete, the low resolution envelope is "burned in." A recurrence at the classical period of s-motion will take place next, as the wavepacket $\varphi(t)$ returns to the vicinity of $\varphi(0)$. It is a straightforward

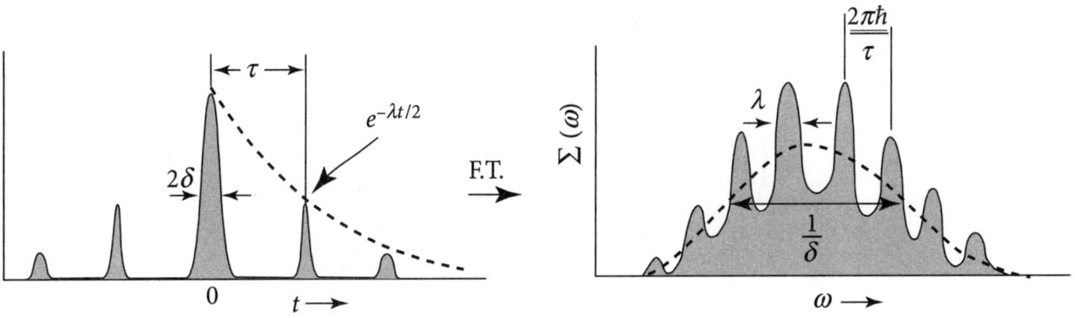

Figure 19.6. Symmetric photodissociation (see figure 19.4) is a paradigm of time-frequency feature connections. All the major characteristics of the time autocorrelation function, from the shortest to the longest times given here, are recorded in the frequency domain, from the broadest to the narrowest, respectively. The absorption spectrum will look something like this, with no higher resolution features hidden, because of the early and permanent decay of the autocorrelation function.

exercise to show that for a 1D harmonic oscillator the overlap $\langle\varphi(0)|\varphi(t)\rangle$ is given by

$$\langle\varphi(0)|\varphi(t)\rangle = \exp\{-\Delta^2/2[1-\exp(-i\omega t)] - i\omega t/2\}, \tag{19.7}$$

where $\Delta = \sqrt{m\omega/\hbar}\, s_0$, and s_0 is the initial displacement. Since $\partial V_s/\partial s = m\omega^2 s_0$, this checks with equation 19.5. The magnitude of $\langle\varphi|\varphi(t)\rangle$ and the spectrum $\Sigma_T(E)$ are shown in figure 19.6.

We turn our attention to the asymmetric stretch direction—that is, the u motion and its spectrum. In the u coordinate, the dynamics is approximated by an inverted oscillator. The initial wavepacket $|\varphi_u\rangle$ spreads rapidly (see equations 10.61 through 10.63), and develops strong position-momentum correlation. The coordinate and phase space pictures are shown earlier in figure 10.11. If the inverted oscillator is of the form $V(u) = -(1/2)m\lambda^2u^2$, and the initial wavepacket is $\varphi(u,0) = (m\lambda_0/\pi\hbar)^{1/4}\exp[-m\lambda_0\, u^2/2\hbar]$, then the overlap is

$$\langle\varphi_u|\varphi_u(t)\rangle = \left[\cosh(\lambda t) + i\left(\frac{\lambda_0}{\lambda} - \frac{\lambda}{\lambda_0}\right)\sinh(\lambda t)\right]^{-1/2}. \tag{19.8}$$

If $\lambda = \lambda_0$, this simplifies to

$$\langle\varphi_u|\varphi_u(t)\rangle = [\cosh(\lambda t)]^{-1/2} \approx e^{-\lambda t/2}. \tag{19.9}$$

Thus, the correlation for the inverted oscillator gives exponential decay with the exponent given by the parameter λ, which is also the classical Lyapunov exponent. The Fourier transform of this approximately exponential decay is a near Lorentzian, with width λ.

The full spectrum, a convolution of the x and y spectra, is a series of Lorentzians spaced by the energy levels of the y oscillator. Alternatively, we can multiply the correlation functions together and Fourier transform the complete $\langle\varphi|\varphi(t)\rangle$, with the same result. This is a spectrum of an unstable periodic orbit, having many of the properties of the interesting case of periodic orbits embedded in chaotic dynamics.

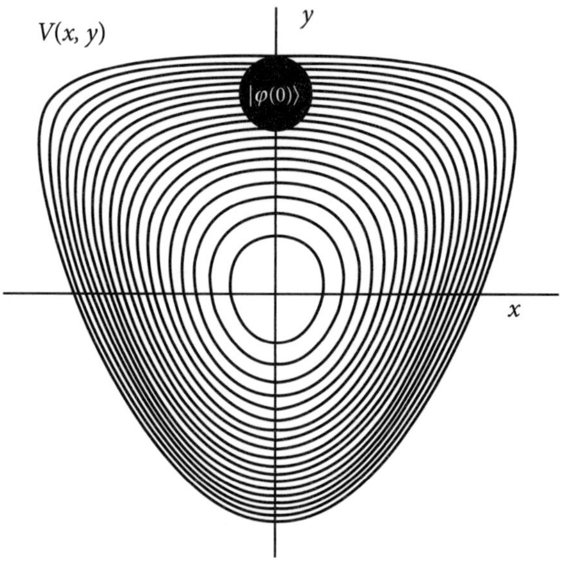

Figure 19.7. Contours of the potential of equation 19.10, and an initial nonstationary wavepacket.

It also has experimental realization in photodissociation of symmetric triatomic molecules, such as H_2O.

19.3 Benzophenone: Transition from Stable to Unstable Motion

Consider the problem of a wavepacket launched on periodic orbit on the two-dimensional potential [148, 149]:

$$V(x, y) = \tfrac{1}{2}\omega_x^2 x^2 + \tfrac{1}{2}\omega_y^2 y^2 + \gamma x^2 y. \tag{19.10}$$

We take $\omega_x = 1.1$, $\omega_y = 1.0$, and $\gamma = -0.11$. As artificial as this might seem, it accidentally generated a complex spectrum almost identical to that of the molecule benzophenone. Equipotential contours of the potential are shown on figure 19.7. The black circle gives the location and size of a nonstationary state generating the spectrum. The full Hamiltonian is $H = p_x^2/2 + p_y^2/2 + V(x, y)$. By symmetry, a periodic orbit rides along the y-axis, with $x = 0$, $p_x = 0$. We can excite this motion by the initial conditions $x_0 = 0$, $p_{x_0} = 0$, $p_{y_0} = 0$, $y_0 = y_0$. The subsequent classical motion is $y(t) = y_0 \cos(\omega_y t)$, $x(t) = 0$.

Is the periodic orbit stable to small deviations, $x = \epsilon \neq 0$, or $p_x = \delta \neq 0$? That is, will such deviations remain small, or will they grow, perhaps enormously? The answer depends on the effect of the $\gamma x^2 y$ coupling term, which in turn depends on how large γ is and how large y gets, and thus on the maximum displacement y_0. If $y(t) \sim y_0 \cos(\omega_y t)$, the term $\gamma x^2 y(t)$ is seen to induce a time-dependent force constant to the x harmonic oscillator, one that could resonantly couple x and y motion, depending on the resonance mismatch of the pumping of the x motion by the y oscillation, and whether the $x - y$ coupling strength controlled by γ and y_0 can overcome a given mismatch.

It is convenient to organize the 4×4 stability equation (see section 7.49) as follows:

$$\frac{d}{dt}\begin{pmatrix}\delta p_x\\ \delta x\\ \delta p_y\\ \delta y\end{pmatrix} = \begin{pmatrix}0 & -V_{xx} & 0 & -V_{xy}\\ 1 & 0 & 0 & 0\\ 0 & -V_{yx} & 0 & -V_{yy}\\ 0 & 0 & 1 & 0\end{pmatrix}\begin{pmatrix}\delta p_x\\ \delta x\\ \delta p_y\\ \delta y\end{pmatrix}.$$

Along the y periodic orbit, this reduces to

$$\frac{d}{dt}\begin{pmatrix}\delta p_x\\ \delta x\\ \delta p_y\\ \delta y\end{pmatrix} = \begin{pmatrix}0 & -V_{xx} & 0 & 0\\ 1 & 0 & 0 & 0\\ 0 & 0 & 0 & -V_{yy}\\ 0 & 0 & 1 & 0\end{pmatrix}\begin{pmatrix}\delta p_x\\ \delta x\\ \delta p_y\\ \delta y\end{pmatrix}, \tag{19.11}$$

taking a 2×2 block form. The stability matrix $\boldsymbol{M}$ also has the same form, and we may separately consider the x and y stability problems along the y periodic orbit. There, $V_{yy} = \omega_y^2$ at all times, so the motion along the periodic orbit is a stable harmonic oscillator. Attention turns to the x motion, where,

$$V_{xx}(t) = \omega_x^2 + 2\gamma y(t) = \omega_x^2 + 2\gamma y_0 \cos(\omega_y t). \tag{19.12}$$

The effective force constant $V_{xx}(t)/2$ of the x motion is time dependent, modulated sinusoidally with the frequency ω_y and with an amplitude that is determined by the coupling constant γ and by the amplitude of the y motion, y_0. If that modulation is sufficiently resonant with the x oscillation, instability of the y orbit will ensue and energy flows from y motion into x.

By making the substitutions

$$\tau = \omega_y t/2, \qquad \sqrt{a} = 2\omega_x/\omega_y, \qquad q = -4\gamma y_0/\omega_y^2,$$

the equation for $\delta \dot{p}_x = \delta \ddot{x}$ becomes

$$\delta \dot{p}_x + [a - 2q\cos(2\tau)]\delta x = 0. \tag{19.13}$$

This is the standard form for the Mathieu equation, which describes this situation and also the Schrödinger equation for Bloch wave quantum eigenstates in a sinusoidally period potential, with $\delta x \to \psi$, $\tau \to x$, and $2a \to E$, and many other physical problems. The unstable regions we soon discuss correspond to band gaps of the spectrum in that potential. In the scaled time, the stability equations have the form

$$\frac{d}{d\tau}\begin{pmatrix}\delta x\\ \delta p_x\end{pmatrix} = \begin{pmatrix}0 & 1\\ -a + 2q\cos(2\tau) & 0\end{pmatrix}\begin{pmatrix}\delta x\\ \delta p_x\end{pmatrix}. \tag{19.14}$$

The stability matrix obeys (see equation 3.75)

$$\frac{\mathrm{d}}{\mathrm{d}\tau}\boldsymbol{M}(\tau) = \boldsymbol{K}(\tau)\boldsymbol{M}(\tau). \tag{19.15}$$

It is important to know the stability matrix $\boldsymbol{M}(\pi)$, at the period of the orbit, here $\tau = \pi$, in scaled time. Its *eigenvalues* can be read from figure 19.8 as a function of a and q, or more precisely, as a function of $\sqrt{a}$ and $2q/a = 2\gamma y_0/\omega_x^2$.

The light regions correspond to unstable motion (hyperbolic with reflection)—that is, $\mathrm{Tr}[\boldsymbol{M}(\pi)] = \mu_+ + \mu_- < -2$. The level curves are labeled with the values $|\exp[\lambda\tau_y]| = |\exp[i\nu\pi]|$. Three zones of unstable motion are shown, corresponding to 1:2, 1:1, and 3:2 ($\omega_x : \omega_y$) resonances. They begin with a narrow base, where with small γ the resonance condition—for example, $2\omega_x = \omega_y$ needs to be close to cause instability.

The shaded regions correspond to stable motion, $\mathrm{Tr}[\boldsymbol{M}(\pi)] = \mu_+ + \mu_- < 2$. (Note: Since $\det[\boldsymbol{M}] = 1$, then $\mu_- = \mu_+^{-1}$, so that $\mu_+ + \mu_-$ is real and equal to $2\cos$ [real angle].) In the stable regions, the dashed lines give the value of ν, where

$$\mu_+ = e^{i\nu\pi} = e^{i\omega_x^{\text{eff}}\tau_y + 2i\pi n}.$$

In the first exponent here, we have used the standard form of the stability eigenvalue, recognizing that the scaled period is π. In the second form, we have defined τ_y = *unscaled* period of y motion—that is, $\tau_y = 2\pi/\omega_y$. In the stable region, the elliptical island surrounding the periodic orbit corresponds to some effective (stable) harmonic oscillator; this oscillator has a frequency

$$\omega_x^{\text{eff}}\tau_y + 2\pi n = \nu\pi,$$

where the integer n can be chosen to put ω_x^{eff} into the vicinity of ω_x. Choosing $n = -1$, we have

$$\omega_x^{\text{eff}} = (1 + \frac{\nu}{2})\omega_y. \tag{3.18}$$

To check this, we set the coupling $\gamma = 0$, and we read off $\nu = 0.2$ at $\sqrt{a} = 2(1.1)/1 = 2.2$, $q = 0$. This gives $\omega_x^{\text{eff}} \to 1.1$ as $\gamma \to 0$ (or $y_0 \to 0$), which is correct.

Figure 19.9 summarizes a great deal of classical, quantal and experimental information. At the top we see four eigenstates plotted as $\psi^2(x, y)$, corresponding to the peaks indicated with arrows in the spectrum. They are chosen as the states with the largest spectral intensity in their respective regimes. Trajectories (plotted as dot densities) and Poincaré surfaces of section (x, p_x) at the energies ($E = 4.95,\ 6.95,\ 7.95,\ 11.95$) are given. Adding the zero point energy of 1.05, these correspond to the quantum states near $E = 6, 8, 9, 13$. (We run the trajectories with the zero point energy subtracted off.) On the coordinate space plots, a single trajectory (plotted as a dot density at discrete times) is shown; each had a small δx_0 displacement. (The initial condition, showing x_0 and y_0, is shown as a larger dot in each case. Completing the initial conditions are p_{x0} and $p_{y0} = 0$.) The surface of section plots (third row) for three trajectories each are shown in the vicinity of the periodic orbit. These transition from stable, elliptic patterns at low energy to unstable, hyperbolic patterns at higher energy. On the stability diagram (see figure 19.8), this transition occurs at a value of $-2q/a$ of 0.67 ($\sqrt{a} = 2.2$).

The theoretical absorption spectrum of the displaced Gaussian is given below the classical trajectory data. The eigenstates $|E_n\rangle$ were determined in a variational basis set calculation and are nearly exact. The absorption spectrum consists of the intensities p_n^{φ} against E_n.The peaks have been broadened artificially; their area is proportional to the Franck-Condon factors $f_n^{\varphi} = |\langle E_n|\varphi(0)\rangle|^2$. The bottom row reveals the experimental

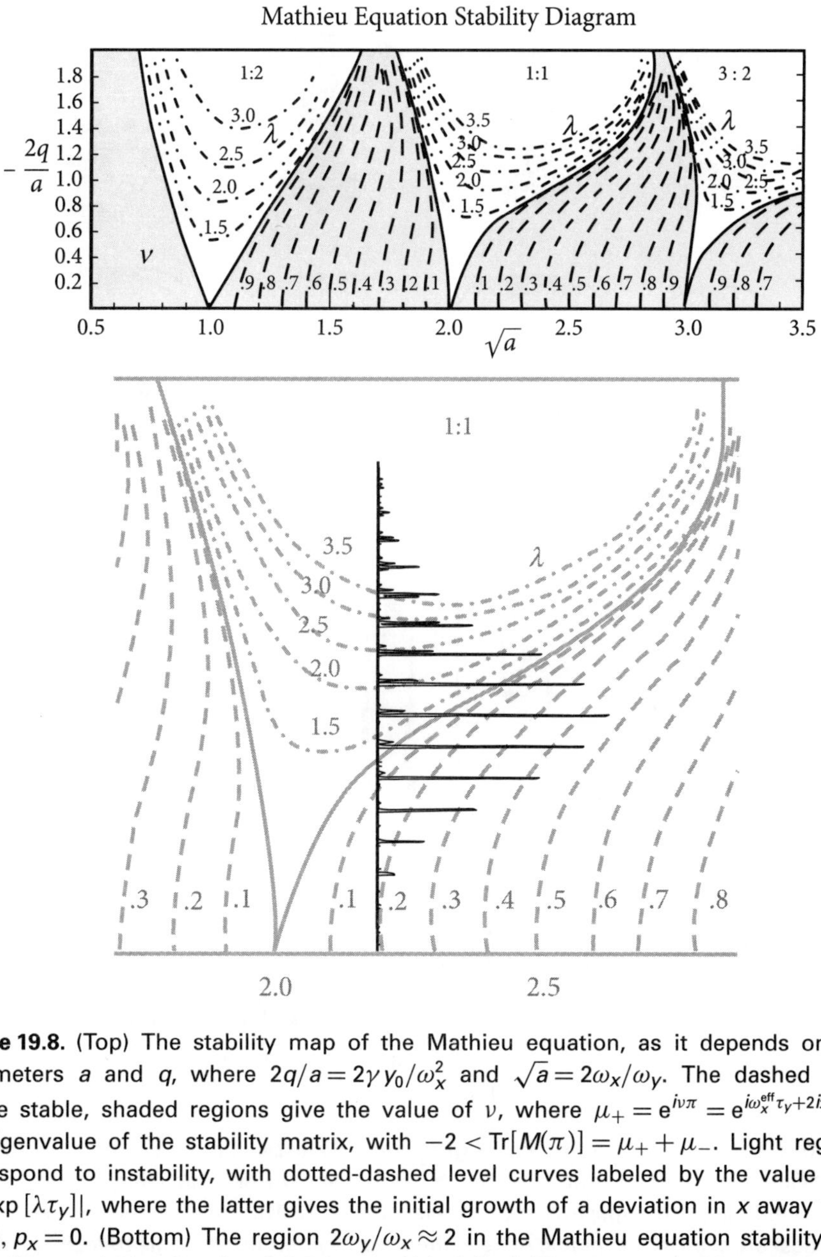

Figure 19.8. (Top) The stability map of the Mathieu equation, as it depends on the parameters a and q, where $2q/a = 2\gamma y_0/\omega_x^2$ and $\sqrt{a} = 2\omega_x/\omega_y$. The dashed lines in the stable, shaded regions give the value of ν, where $\mu_+ = e^{i\nu\pi} = e^{i\omega_x^{\text{eff}}\tau_y + 2i\pi n}$ is an eigenvalue of the stability matrix, with $-2 < \mathrm{Tr}[M(\pi)] = \mu_+ + \mu_-$. Light regions correspond to instability, with dotted-dashed level curves labeled by the value of λ in $|\exp[\lambda\tau_y]|$, where the latter gives the initial growth of a deviation in x away from $x = 0$, $p_x = 0$. (Bottom) The region $2\omega_y/\omega_x \approx 2$ in the Mathieu equation stability diagram. The model has $2\omega_y/\omega_x = 2.2$ (vertical black line). The spectrum of the wavepacket placed on the periodic orbit spans energies from very stable (gray zone) at low energy to very unstable (whiter zone) at higher energy. The character of the spectrum reflects the transition to unstable, with the developing sidebands revealing a coupling to x motion. The sideband splitting reveals the time for the return to the symmetric twisting after a falling into asymmetric twisting.

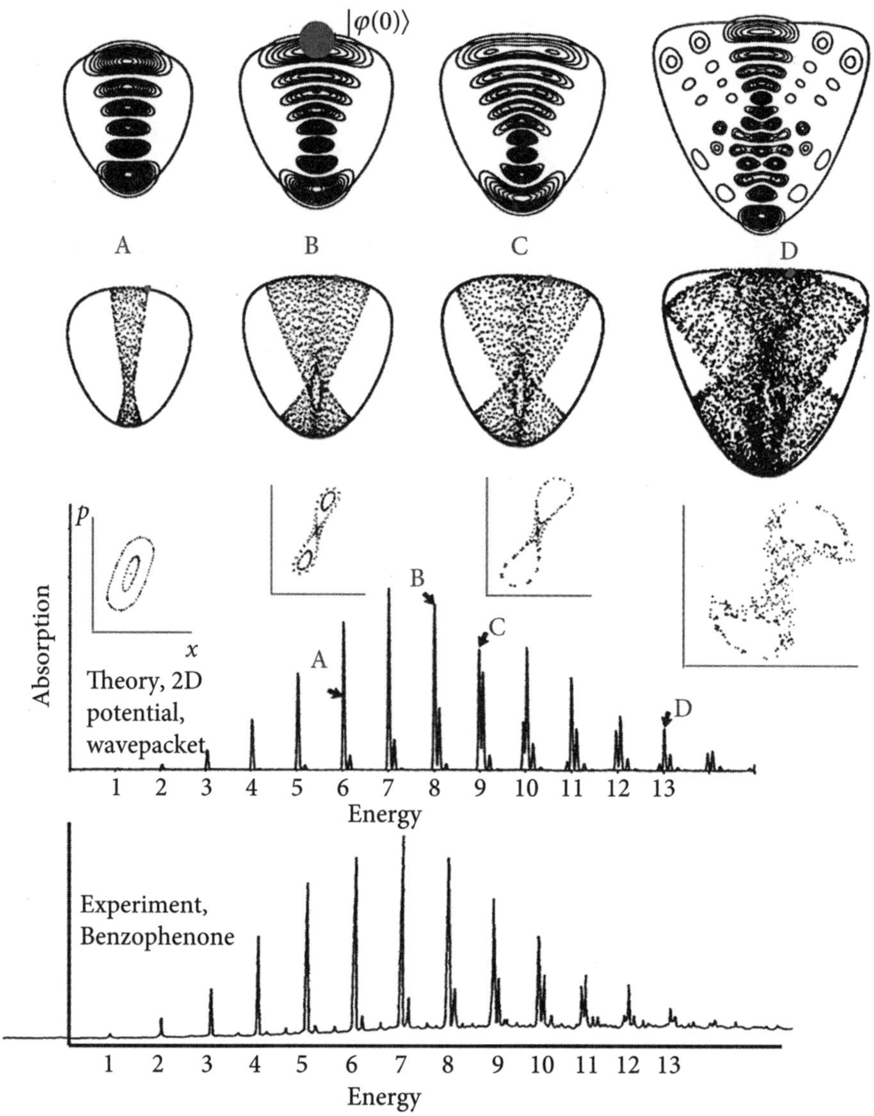

Figure 19.9. This figure summarizes a great deal of classical, quantal, and experimental information. At the top four eigenfunctions are shown as probability contour maps corresponding to the spectral lines labeled (A, B, C, D). Trajectories (plotted as dot densities) are seen in the middle rows at those same energies. A single trajectory (plotted as a dot density) is shown; each had a small δx_0 displacement. (The initial condition, showing x_0 and y_0, is shown as a larger dot in each case. Completing the initial conditions are p_{x_0} and $p_{y_0} = 0$.) Poincaré surfaces of section (x, p_x) at four energies ($E = 4.95, 6.95, 7.95, 11.95$) are shown just below. The plots show stable, elliptic patterns turning to unstable, hyperbolic motion at higher energy. Second from the bottom we see a Franck-Condon spectrum for the initial wavefunction $|\varphi(0)\rangle$, serendipitously calculated before it was known that the nearly identical experimental spectrum for benzophenone, bottom, had been taken [150].

spectrum of benzophenone [150], which we shall explain [149]. The state $\varphi(x, y)$ was taken to be

$$\varphi(x, y) = \left(\frac{\omega_x \omega_y}{\pi^2}\right)^{1/4} e^{-\omega_x^2 x^2/2 - \omega_y^2 (y-4.0)^2/2},$$

which is the ground state of the harmonic part of the potential, displaced in y by 4.0.

The state $\varphi(x, y)$ corresponds to a localized Gaussian wavepacket on the potential surface with the initial conditions $x_0 = 0$, $p_{x_0} = 0$, $p_{y_0} = 0$, $y_0 = 4.0$. We know that the dynamics of the wavepacket $\varphi(x, y, t)$ will determine the spectrum via Fourier transform of $\langle\varphi|\varphi(t)\rangle$. The linearized dynamics of the Gaussian wavepacket time evolution corresponds to a single trajectory with classical energy 8.0, and its stability matrix. However, the full spectrum spans a rather large energy range, from $E = 1$ to $E = 15$ or more. The single periodic trajectory and thawed Gaussian dynamics at $E = 8.0$ cannot reveal all the changes that we will see take place in this energy range. A range of improved semiclassical methods, all the way to the VVMG propagator, which would give a very accurate, full spectrum, remains at our disposal (see sections 11.1 and 11.2).

The single trajectory (and the Gaussian wavepacket it controls) does get qualitative, short time information correctly, resulting in an accurate position and shape of the smoothed envelope of the spectrum. It also gets the positions and total intensities of the smoothed clumps spaced by $\hbar\omega_y = 1.0$. The trajectory with displacement $y_0 = 4.0$ is unstable, so the Gaussian wavepacket spreads indefinitely, causing $\langle\varphi|\varphi(t)\rangle$ to decay permanently. Thus, the structure at higher resolution that is present in the spectrum is missed by the single trajectory guiding the thawed Gaussian at $E = 8.0$.

By $E = 13$, a significant chaotic zone has formed near the primary unstable periodic orbit. Its Lyapunov exponent has grown, from 0 at the transition to unstable motion ($E = 6.8$), to $\pi^{-1}\ln(2.4) = 0.28$ at $E = 13$, $2q/a \approx 1.0$, $\sqrt{a} = 2.2$. The stable periodic orbits in the "eyes" still exist at $E = 13$. Starting near the original periodic orbit, significant chaotic energy exchange takes place between x and y motion, but the system is not ergodic.

The expected quantum energy of the instability, $E = 7.9$, is indeed the locale of a qualitative change in the spectrum. New side peaks emerge that would be absent if $\gamma = 0$. The spacing of the peaks above the instability can only be due to a *return* of amplitude to the region of $\varphi(x, y)$ after the initial escape. The figure-eight surfaces of section and the double-well analogy make it clear why this happens. *Nonlinear* classical dynamics succeeds here in explaining a spectral feature where a *linearized* dynamics fails.

We can still use classical mechanics and indeed the classical stability analysis to understand a good deal more of the spectrum. We adjust the energy of the y axis periodic orbit to the region we are considering: around quantum energy E we examine the behavior of the periodic trajectory with energy $E - E_0$, where E_0 is the quantum zero point energy.

If $\gamma = 0$, the high-resolution spectrum would be a single progression of equally spaced (by $\hbar\omega_y$), isolated lines with a Poisson distribution of intensities, characteristic of a one-dimensional harmonic oscillator. The y oscillator initial state is simply the ground state displaced by $y_0 = 4.0$. For $\gamma = 0$, the eigenstates of the Hamiltonian are product states of a harmonic oscillator $|n_x\rangle$ in x and a harmonic oscillator $|n_y\rangle$ in y. Only (n_x, n_y), where n_x is even, can have nonvanishing intensity. Odd states are forbidden because $\varphi(x, y)$ is even in x. The x parity is maintained for $\gamma \neq 0$, so again the odd states have no intensity.

Already at $E = 4$ in the calculation shown in figure 19.9, which had $\gamma = -0.11$, there are signs of the effects of the $x - y$ coupling in the form of a side peak (more visible near $E = 5,\ 6$). Assigning quantum numbers to the peaks, the large peak is $(0, n_y)$, and the side peak is $(2, n_y - 2)$. The uncoupled, unpertubed $\gamma = 0$ position of this peak would be 0.2 above the $(0, n_y)$—that is,

$$(2 + \tfrac{1}{2})\omega_x + (n_y - \tfrac{3}{2})\omega_y - \tfrac{1}{2}\omega_x - (n_y + \tfrac{1}{2})\omega_y = 2\omega_x - 2\omega_y = 0.2. \tag{19.16}$$

The actual spacing is *less* than this—somewhat surprising if you think only in terms of 2×2 matrices and the universal repulsion of the unperturbed diagonal energies by an off-diagonal perturbation.

LEVEL ATTRACTION

We now discuss these interesting preresonant *level attractions* [149]. From the most qualitative point of view, if there is a 1:1 resonance, the energy doesn't change if we substitute two quanta of one mode for two of the other. Thus, as we approach 1:1 resonance, which is the mechanism of instability of the Mathieu equation in this region, the spacings are expected to decrease. The y-mode frequency ω_y is just the frequency of the periodic orbit, which in this case does not change with energy in the y mode. The effective x frequency, $\omega_x^{\text{eff}} \cdot \tau_y + 2\pi n = \nu\pi$, depends on ν. Notice that as E increases for fixed γ, ω_x and ω_y, we travel vertically up the graph in figure 19.8 for fixed $\sqrt{a} = 2.2$, passing through smaller and smaller values of ν. The parameter ν decreases slowly with energy at first, then more rapidly as the unstable region is approached, until at y_0^*, the critical value at which the periodic orbit becomes unstable, ν is 0. This means $\omega_x^{eff} = \omega_y$. That is, the frequency of the periodic orbit is the same as for the x motion perpendicular to it: the two motions are in exact resonance, so naturally the motion becomes unstable! The side peak corresponding to the exchange of two quanta is now spaced by

$$2\omega_x^{eff} - 2\omega_y = \nu\omega_y = \nu(E) < 0.2.$$

Figure 19.10 shows the numerical spacings (dots) and the spacing predicted by finding ν as a function of the energy of the trajectory. There is *attraction* of the levels as energy is increased toward the critical energy of $E = 7.9$ where the trajectory goes unstable. The theoretical curve is predicted from the Mathieu equation behavior of ν, and the dots are the found spacings. The deviation is attributable to the finite value of $\hbar$: as $\hbar \to 0$, the found and the theoretical values would agree completely.

UNSTABLE REGIME

At $E = 7.9$, the orbit has become unstable (hyperbolic). Energy exchange is taking place resonantly between the x and y modes, which for the molecule means between the symmetric (along the y axis) and asymmetric twisting. The local mode twisting, where one benzene ring remains almost stationary, cannot be accessed from the symmetric mode twist initial condition, as the path leading from the unstable separatrix in figure 19.11, left, shows. The Mathieu equation stability matrix analysis correctly predicts that the y motion, symmetric mode periodic orbit at this energy is unstable.

As figure 19.11 shows, two stable islands have been born above resonance; these are the "eyes" of the figure-eight pattern seen in the Poincaré surface of section in figure 19.9. The stable eyes are not predicted by a linearized analysis based on the y-periodic orbit. The stable periodic orbits in the center of the eyes are shown in

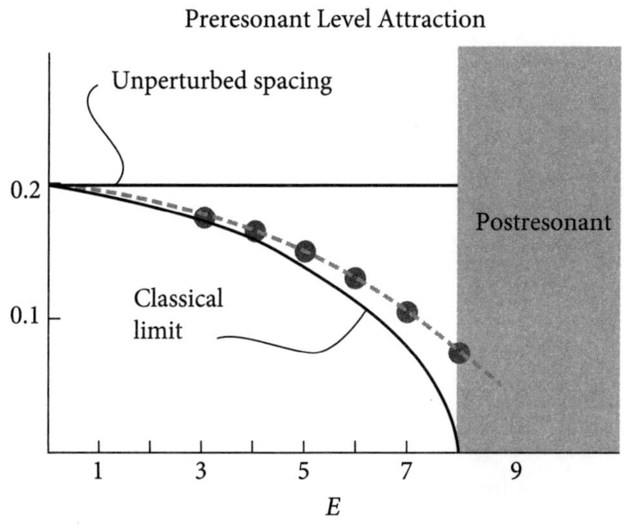

Figure 19.10. As the system approaches 1:1 resonance at energy above 7.9, the energy splitting of a swap of two quanta in x for two quanta in y approaches 0. However, residual interactions (essentially a quantum avoided crossing) keep the splitting finite, as evidenced by the numerical splitting (dots) versus the splittiing predicted by classical mechanics (solid line).

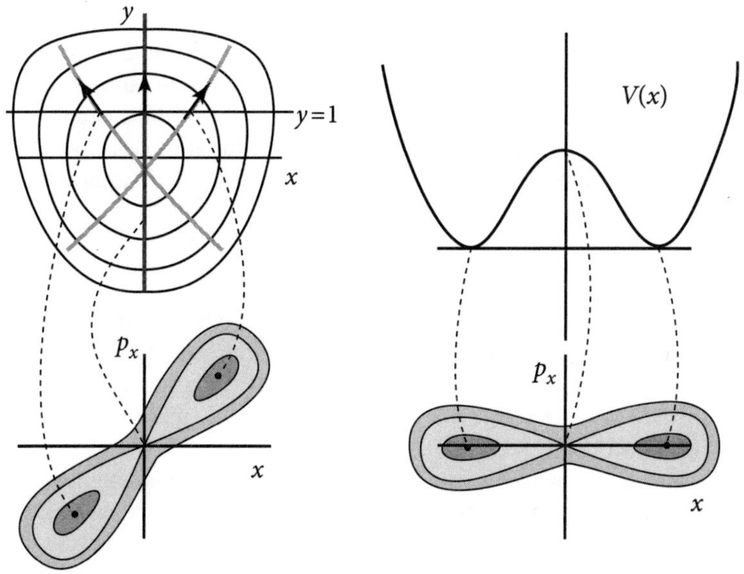

Figure 19.11. (Left) Stable and unstable periodic orbits are shown on the two-dimensional potential and below on a constant $y = 1$ surface of section. At this energy the periodic vertical $x = 0$ oscillation in y has become unstable, giving birth to two stable periodic trajectories whose x, p_x coordinates are the two dots shown for $y = 1$ and positive y motion. (Right) Coordinate and phase space depictions of the one dimensional double well potential analogy of the situation on the left.

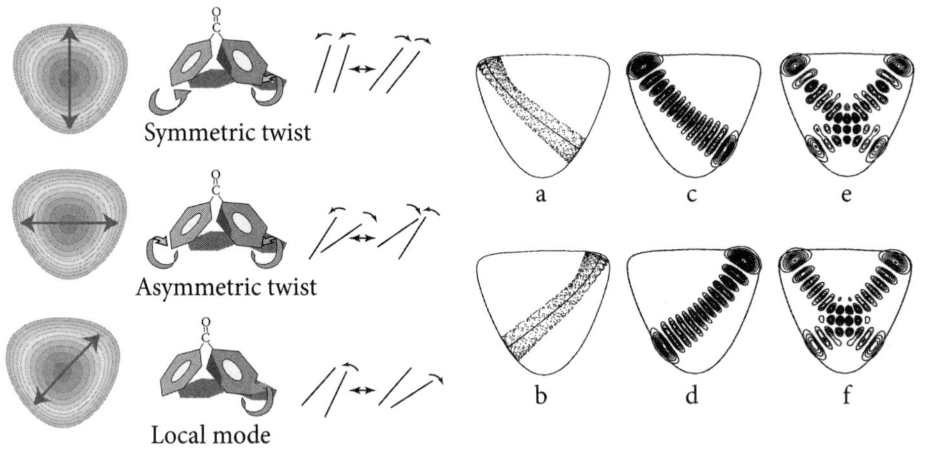

Figure 19.12. (Far left): Classical paths for symmetric, asymmetric, and local mode motion on the benzophenone benzene ring twist potential energy surface. (Next column) Visualization of the three types of benzene ring twist motion. The local mode shown at the bottom corresponds to a single ring twisting with the other stationary. This motion becomes a stable periodic orbit above about 500 cm^{-1} of ring twisting. (Next column) Edge-on schematic views of the benzene rings for clarity. The slice of the potential along the asymmetric twist is itself (somewhat confusingly) symmetric about 0 displacement, and the potential along the symmetric twist has no particular symmetry. (Right) (a) and (b) are classical local mode trajectories on and near the stable local mode periodic orbit that forms after the central periodic orbit goes unstable. (e) and (f) are a pair of quantum eigenstates, at an energy above the instability onset at ~7.9. (e) is even and (f) is odd about x; the two are split by a very small tunneling energy. If (e) and (f) are added, the nonstationary state (c) results, and if subtracted, (d) results. If the system is initiated in (c), for example, dynamical tunneling will take place to (d) and back to (c) sinusoidally, with a time scale determined by the energy splitting of the even and odd pair of states, (e) and (f).

configuration space in the left panel of figure 19.11. They are sometimes known as "local modes," being linear combinations of the former normal modes along the x and y directions at lower energy. Between the eyes sits the original periodic obit, now unstable, with the requisite stable and unstable hyperbolic manifolds leading from it (see figures 10.11 and 3.25).

There is a direct analogy with a double well potential. This analogy suggests the presence of tunneling pairs of states, narrowly split in energy. This is clearly manifested in panels (a) through (f) in figure 19.12, where two eigenstates—namely, (e) that is even about x and (f) that is odd—are split by a very small energy. If the state (e) and (f) are added together, the northwest-southeast trending nonstationary state is seen in (c), and if subtracted, the mirror image state is seen in (d). They both look very much like the stable classical trajectories launched in that region, (a) and (b). The situation is exactly analogous to a double well potential, except here there is no potential barrier separating the stable classical regions from each other, but rather a dynamical one [93, 96]. If the system is initiated in the nonstationary state (c), for example, resonant tunneling will take place to (d) and back to (c) sinusoidally, with a time scale determined by the energy splitting of the even and odd pair of states, (e) and (f). This is a clear example of dynamical tunneling; see sections 12.3, 13.1, 13.2, and 13.3.

A revealing analysis of the computational data is given in figure 17.2, starting with the spectrum shown at the top, for competeness. Next, the absolute value of the Fourier transform of the spectrum—that is, the absolute value of the autocorrelation function, is shown, with three clearly discernible and qualitatively distinct timescales revealed. The first, called τ_a, is the initial decay of the autocorrelations giving the broadest aspect of the spectrum. Next comes τ_b, the recurrence of the y-periodic orbit, be it stable or unstable. Finally, there is the time τ_c for amplitude to return to the vicinity of the periodic orbit, having fallen away from the periodic orbit above its instability threshold. These are the signatures of homoclinic recurrences (see figure 3.28, for example) showing up after 5 picoseconds.

It is well known in many fields (such as acoustics for example) that a compromise picture somewhere between the pure time or pure frequency limits can be very informative. The bottom panel in figure 17.2 is such a "spectrogram," revealing the energy character of various recurrences. Here, the first and second generation of the homoclinic recurrences (falling away and returning once, and then twice, from the periodic orbit) are seen. A remarkable aspect of these returns shows up—namely, that the act of falling away lowers the average y period, so that the trajectories that re-join the central periodic orbit after such an excursion do so with a time or phase lag in the y motion. This lag is then doubled if two excursions have been made.

Autocorrelation function extracted from an experimental absorption spectrum is seen in figure 19.13. Very similar features are seen, and with the same interpretation, as in the theoretical autocorrelation spectrum, figure 17.2.

19.4 Serendipity—the Toy Spectrum Is Real

In 1986, Holtzclaw and Pratt [150] obtained an ultraviolet spectrum of the molecule benzophenone. The benzophenone absorption band is used as a UV block in some suntan lotions. Holtzclaw and Pratt had taken a high-resolution spectrum using fluorescence excitation of a supersonic jet expansion for the first part of the ultraviolet electronic absorption; figure 19.9, bottom, is the very clean spectrum they produced. The agreement with the toy model of reference [149] is pure coincidence, since the model was done without knowledge of the experiment.

If the benzene rings were not so close to each other in the molecule, the two modes would have the same frequency. But the fact that they repel each other makes for a zero-order frequency ratio of about 1.1:1. The first possible interaction term in the polynomial expression of an anharmonic potential of this symmetry is $\gamma x^2 y$. Again, through sheer coincidence, benzophenone had (in reduced units) $\gamma \sim -0.11$. Finally, y_0 (also in reduced units) was about 4.0! The result was an experimental spectrum almost identical to the model spectrum in its major features. The experimental "grass" at higher energy is due to small displacements and/or interaction with other modes like the low frequency bending angle between the C-benzene bonds (there are 66 modes, but even at the highest energy of this spectrum, most cannot be excited, due to energy considerations). The ring twist frequency is $\sim 57\,\text{cm}^{-1}$, some 50 times slower than a C–H vibration frequency.

How does the symmetric ring twist get excited in photoabsorption? The molecule is not planar in either the ground electronic or the excited electronic states (the benzene rings are twisted even at rest), but the excited state is actually 14 degrees more nearly

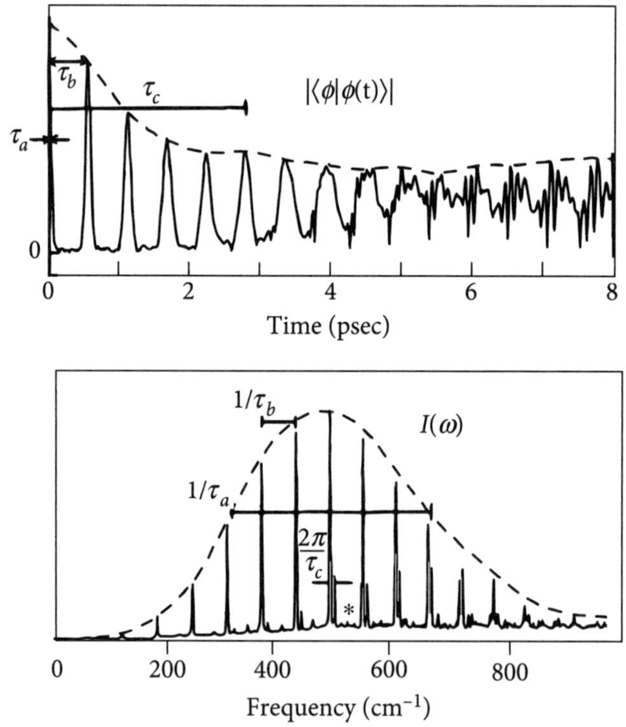

Figure 19.13. (Top) Absolute magnitude of the experimental autocorrelation function $\langle\phi|\phi(t)\rangle$ computed via the Fourier transform of the spectrum in the lower panel. The dashed line represents an "envelope" for the peaks, while $\tau_{a,b,c}$ identify three time scales relevant to different features in the spectrum. (Bottom) Experimental benzophenone spectrum with features correlated to time scales in the autocorrelation function. The asterisk at 560 cm^{-1} indicates the onset of the classical 1:1 resonance between the torsion modes.

co-planar than in the ground state. This means the wavepacket for the ground state finds itself 14 degrees displaced on the excited state, or in reduced units almost $y_0 = 4.0$. The asymmetric twist has no mean displacement initially, by symmetry. The local-mode tunneling doublet states corresponding to the single well depicted in figure 19.12 must exist, but even the totally symmetric member of the doublet is not seen directly in the experimental spectrum since the local mode motion is remote from the symmetric stretch Franck-Condon birthplace. The local modes correspond to one benzene ring twisting while the other is at rest, or nearly so.

Due to the symmetry of the benzene rings, if we twist both rings in the symmetric mode (both twisting in the same direction) over an energy barrier somewhat higher than the highest energy seen in the spectrum, we arrive at a symmetrically equivalent "double" of the first well. There are eight such wells, all equivalent, found in the range $(0, 2\pi)$ of both twist angles (see figure 19.14). It would be interesting to investigate this more energetic dynamics, but other modes would no doubt get in the act at higher energies, greatly complicating the analysis.

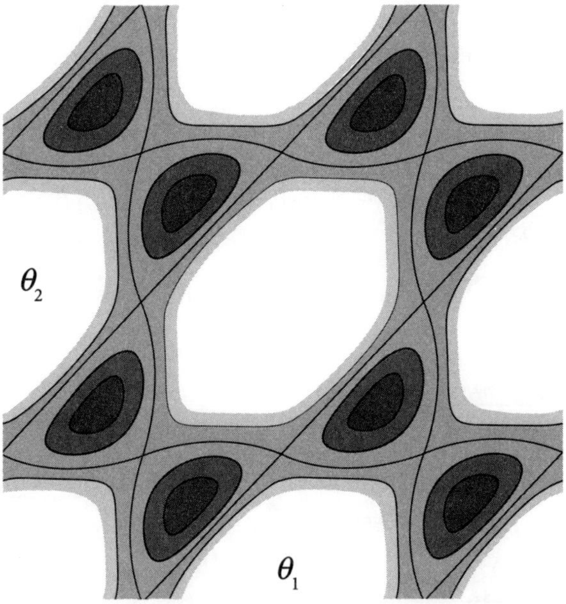

Figure 19.14. A model potential for the eight equivalent potential wells for twisting of the two benzene rings in benzophenone, each one twisting though 2π. An ultra high resolution spectrum would show many very closely spaced tunneling multiplets, requiring a symmetry analysis to fully understand. We have been working in the broken symmetry basis of just one of these wells, but these possess the dynamical tunneling local mode doublets; see figures 19.11 and 19.12.

19.5 Recapitulation

The investigation of the model potential and its spectral features has allowed us to understand this piece of molecular nonlinear dynamics extremely well. We have pinned down the vibrational evolution of a molecule in great detail following electronic photoabsorption, revealing events ranging from 10 fs to 10 ps— that is, three orders of magnitude in time. The hierarchy is clearly revealed in figure 19.15.

Turning now to figure 19.16, we see that the shortest event in time after photoabsorption in the 20,000 cm^{-1} region seen in A is the acceleration and lengthening of the CO bond and the resultant decay (in about 10 femtoseconds) of the autocorrelation function. The small lumps on the 8000 cm^{-1} CO $n \to \pi^*$ peak seen in A and B at the 1700 cm^{-1} CO stretching frequencies at about 30 fs, representing the earliest recurrence in the autocorrelation function. In C, we see two new timescales. The benzene rings have a different, smaller equilibrium twist angle in the excited state, and they start twisting in that direction immediately, but the twisting force together with heavy mass of the rings conspire to make this a slower motion. The wavepacket overlap in the twisting mode decays in about 100 fs as is responsible for broadening the C stretch bands. At 550 fs the symmetric twist has returned, at its period, giving the progression of bands and their spacing seen in C.

But the show is not over. Above the energy of the instability of the periodic pure symmetric twist mode, the asymmetric twist gets into the act. The broadening of the bands seen in C, and width of the clusters seen in D, come from the unstable motion

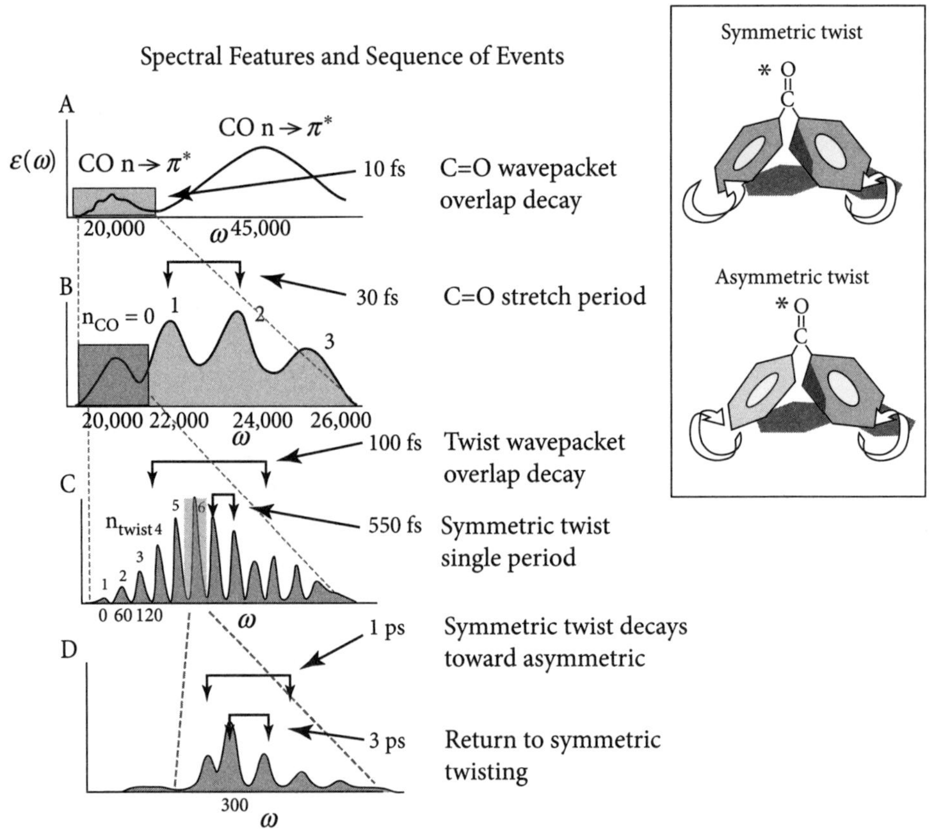

Figure 19.15. Time sequence and corresponding spectral features generated by vibrational coupling and relaxation in benzophenone. Even though this is a traditional study involving an absorption spectrum, not using femtosecond pulses, or certainly 2D spectroscopy, it remains, to our knowledge, the most detailed understanding of polyatomic vibrational evolution for the longest time span to date [149].

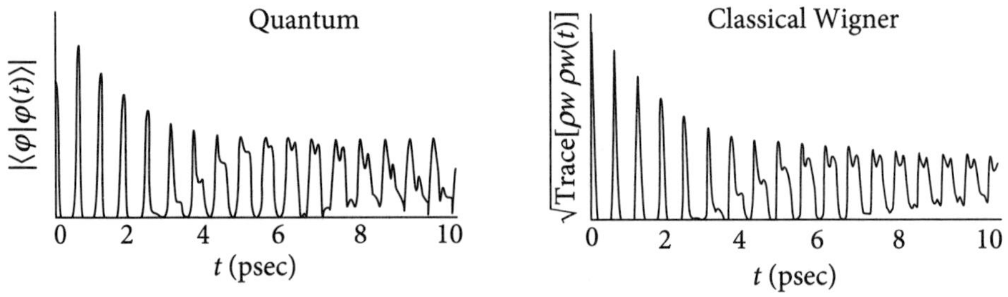

Figure 19.16. The absolute value of the quantum autocorrelation function for the benzophenone model on the left, and the square root of the "classical Wigner" autocorrelation, obtained as explained in the text.

leading from the symmetric twist to the asymmetric one—that is, the timescale for the spreading of the wavepacket in the asymmetric mode direction, or about 1 ps judging from the band width. There is still more: the timescale for the energy to return to the symmetric twisting, having flowed into the asymmetric twist, is about 3 ps.

There is another way of confirming that so much of the quantum mechanical autocorrelation function and its Fourier transform, the spectrum, is classical in nature. We form the Wigner transform of the initial state, $|\varphi\rangle\langle\varphi| \xrightarrow{\text{Wigner}} \rho_W(\boldsymbol{p}, \boldsymbol{q}, 0)$, then represent it by a dense set of trajectories with initial density following $\rho_W(\boldsymbol{p}, \boldsymbol{q}, 0)$, and propagate the trajectories classically to give $\rho_W(\boldsymbol{p}, \boldsymbol{q}, t)$. At each time step, the overlap is taken between the delta function trajectories and the smooth initial density $\rho_W(\boldsymbol{p}, \boldsymbol{q}, 0)$. The square root is taken to give a classical "amplitude" and this is compared in figure 19.16 to the quantum absolute value of the amplitude [151–153].

Chapter 20

Semiclassical Dynamics and Two-Photon Spectroscopy: Raman Scattering

The time-dependent formulation of Raman scattering [38] is independent of any semi-classical implementation, but we dive into semiclassical ideas here. The starting points in the literature are an *Accounts of Chemical Research* article [41], and an application to a 15 degree of freedom harmonic system with large mode displacements (very difficult to do with the energy version of Kramers-Heisenberg-Dirac [KHD]) [39], and David Tannor's wonderful book [1].

It is often a very good starting point, surprisingly close to the truth, to assume the initial vibrational state is a ground vibrational multivariate Gaussian (equation 10.34) of the ground electronic state. It may be expressed in terms of normal modes. We reproduce equation 10.34 here for convenience, with some initial $\boldsymbol{p}_0 = 0$ and $\boldsymbol{q}_0$. The potential energy surface of the ground state is reflected in the matrix $\boldsymbol{A}_0$, which is diagonal in the normal mode basis of the ground electronic state, and assuming a smooth excited state potential energy surface, we may take the thawed Gaussian approximation to the short time evolution—that is,

$$\psi_t(\boldsymbol{q}) \equiv \psi_t(\boldsymbol{q}_t, \boldsymbol{p}_t; \boldsymbol{q}) = \exp\left[\frac{i}{\hbar}\{(\boldsymbol{q} - \boldsymbol{q}_t) \cdot \boldsymbol{A}_t \cdot (\boldsymbol{q} - \boldsymbol{q}_t) + \boldsymbol{p}_t \cdot (\boldsymbol{q} - \boldsymbol{q}_t) + s_t\}\right]. \quad (20.1)$$

Importantly, we certainly do not need to assume that the excited Born-Oppenheimer potential energy surface is a harmonic oscillator. We need only first and perhaps second derivatives of the excited state potential at the ground-state geometry. In fact, we have not assumed the ground state is a multivariate harmonic oscillator either, only that the bottom of the ground Born-Oppenheimer potential energy surface is well approximated by its quadratic expansion. This is often the case.

Consider the Raman intensity to a particular fundamental of the kth normal mode on the ground-state Born-Oppenheimer surface, starting in the ground vibrational state. We ignore any coordinate dependence of the transition moment, which often plays a back seat in molecules with heathy reorganization energies—that is, relative equilibrium geometry changes in the ground and excited states. Suppressing polarization dependence, we write, in a transparent notation,

$$\alpha_{0k}(\omega_I) = \eta \int_0^\infty dt\, e^{iE_I t/\hbar} \langle \varphi_k | \varphi_0(t) \rangle, \tag{20.2}$$

where η normalizes the initial states and $|\varphi_k\rangle \sim q_k|\varphi_0\rangle$. Whether raw intuition or semiclassical wavepacket propagation is brought to bear, this formula holds the key to all fundamental, overtone, and combination mode intensity. Asking why the amplitude $\langle \varphi_k | \varphi_0(t) \rangle$ should or should not grow and what its half Fourier transform should look like, using "local knowledge" around the Franck-Condon region, is extremely useful and informative. There is no justification for being concerned with entire global eigenstates on far flung parts of a potential energy surface.

We summarize the important results:

- Self-damping effects
  - For more than one degree of freedom, moderate to fast decay (self-damping) of the cross-correlation takes place, due to the presence of possibly many gradients in the excited state potential along ground state normal modes, and possibly many force constants that have changed, including Duschinsky rotations (wherein new "normal mode" directions pertain, as given by the second derivatives in the excited state at the ground state geometry).
  - The cross-correlation amplitude for exciting one quantum of any given mode goes as t, except for contributions from linear dependence of the transition moment along that coordinate, which go as t^0.
  - The cross-correlation amplitude for exciting a two quantum overtone of any given mode goes as t^2, except for (1) changes in second derivatives, which give a t^1 component, and (2) quadratic contributions from dependence of the transition moment along that coordinate, which go as t^0.
  - The higher the power of t for a given process, the more attenuation of that process that results from strong self-damping (because the damping sets in before the amplitude grows, the more so the higher the power of t). A strong suppression of overtones and even fundamentals results from strong self-damping. Transitions going at t^0 coming from coordinate dependence of the transition moment suffer the least self-damping suppression.
  - A key point is the suppression of overtone intensity in larger molecules with many displaced modes. The reason is again the self-damping early shut down of the cross-correlation function, which has a suppressive effect on overtone amplitudes that appear as higher powers of t and are therefore late. They are susceptible to quenching in the time domain by the overall many mode decay. The common result is that many fundamentals will appear, but weak or nonexistent overtone and combination intensity.
- Off-resonance effects
  - Another effect limiting the strength of Raman amplitudes is the time-energy uncertainty effect of probing ΔE below (or rarely, above) resonance, $\Delta E\, \delta t \sim \hbar$.
  - Again, the higher the power of t of the developing amplitude, the more attenuation is suffered by being off resonance, due now to another source of time cutoff—that is, time-energy uncertainty. The t^0 behavior, coming from

coordinate dependence of the transition moment, suffers the least off resonance suppression.

- Far off resonance, only the coordinate dependence of the transition moment matters, in accord with the Placzek polarizability expression [136]. As the incident frequency falls below resonance, it is inevitable that the higher excited electronic states, also somewhat more off resonance but great in number, start to dominate. The Raman intensities become controlled by the derivatives of the Placzek polarizability, which is a sum over contributions from all excited electronic states.

If we compare the entire many-coordinate autocorrelation with the analogous one-dimensional autocorrelation in mode k alone, the striking observation emerges that the other $\mathrm{N}-1$ modes apart from k act as a possibly strong exponential and Gaussian damping factor on the kth mode cross correlation, reducing the intensity for the $0 \to 1$ kth mode Raman fundamental. The self-damping has nothing to do with any small Γ damping taken in our expressions. More empirical approaches have often been forced to assume large Γ in the absence of treatment of all the degrees of freedom explicitly. The danger in this is that the actual many-body damping can have both t and t^2 dependence. The relative amount of each depends on the steepness of the potential and changes in its second derivatives in going to the excited state at the Franck-Condon position.

Another important conclusion follows from strong self-damping, even with the incident radiation on resonance: an effectively larger attenuation is imposed on Raman amplitudes that take longer to develop in time. Specifically, since coordinate dependence of the transition moment develops in time as t^0, fundamentals as t^1, and first overtones as t^2, and so on it is seen that overtones and even fundamentals not caused by coordinate dependence of the transition moment become weaker in a many-coordinate context. Put another way, overtones tend to suffer low intensity relative to fundamentals in a polyatomic context [137]. True, the attenuation is weaker if only one ground-state normal mode coordinate lies along the steepest descent in the excited state, but some attenuation persists unless all the second derivatives of the two potentials, ground and excited, are also the same.

Supposing Gaussian wavepackets propagate on excited state potentials that are expandable quadratically in the Franck-Condon region (the potentials do not need to be globally harmonic), no matter how many coordinates there are, we can derive analytic expressions for the intensities and relative intensities of various Raman lines in terms of the Franck-Condon region of the Born-Oppenheimer potential surfaces involved [137]. The short time dynamics of $\boldsymbol{A}_t$ is (see equation 10.35 and following)

$$\begin{aligned} \boldsymbol{A}_t &= \frac{1}{2}\boldsymbol{P}_{Zt} \cdot \boldsymbol{Z}_t^{-1}, \\ \boldsymbol{P}_{Zt} &\approx 2\boldsymbol{A}_0 - \boldsymbol{V}''(0)t, \\ \boldsymbol{Z}_t &\approx 1 + \tfrac{1}{m}\boldsymbol{A}_0 t. \end{aligned} \tag{20.3}$$

We have also, approximately,

$$\begin{aligned} \boldsymbol{q}_t &\approx \boldsymbol{q}_0 - \tfrac{1}{2}\nabla_q V\, t^2, \\ \boldsymbol{p}_t &\approx -\nabla_q V\, t, \end{aligned} \tag{20.4}$$

and

$$s_t = s_0 + i\,\hbar\frac{1}{2}\mathrm{Tr}\left[\log\left(Z_t\right)\right] + S_t,$$
$$S_t = \int_0^t L_t\,dt \approx -V(0)\,t. \tag{20.5}$$

From these expressions, analytic approximations may be derived for the intensities and ratios of intensities of fundamentals and overtones [137].

Since the duration of the initial overlap $\langle\varphi_k|\varphi_0(t)\rangle$ is typically a few femtoseconds, we are led immediately to the following conclusions, which will be valid whenever at least one of the preceding reasons applies: (1) Since the excited state amplitude is born in the Franck-Condon geometry, the information about the upper potential surface needed to calculate the Raman intensities is localized near the Franck-Condon region, and the classical paths leading from there according to the potential. This is a tiny fraction of the coordinate space for many degrees of freedom systems. The opportunity for "on-the-fly" methods is clear. (2) Electronic structure calculations in a very limited coordinate domain (FC region) are sufficient to determine the Raman intensities. (Any approximations made to the upper surface should be biased toward achieving accuracy in the Franck-Condon region and not, for example, in the region of the potential minimum of the excited state.) (3) It is a waste of effort to generate wavefunctions for the upper surface that span regions not sampled by the early motion of the wavepacket.

PART IV

Chaos and Quantum Mechanics

Chapter 21

Aspects of Quantum Chaology

The mystery of how classical chaos relates to quantum mechanics was subject to many attempts to define *quantum chaos*. For example, the great mathematician and physicist John von Neumann stated that all quantum systems are ergodic, save for those with degenerate eigenstates. This would include, for example, a two-dimensional harmonic oscillator with incommensurate frequencies, a very integrable, non-ergodic classical system. By this he meant that any initial quantum state goes everywhere available to it in the sense that all terms in the expansion of the initial state over eigenstates will dephase (randomize their relative phase) over time, unless they have the same energy and no dephasing occurs. If one wants to be absolutely mathematically rigorous, this conclusion may be necessary. But it is also fairly unhelpful. The von Neumann definition does have analogues in classical ergodic theory, which we discuss later.

Some people claimed there could be no quantum chaos, implying the subject was dead on arrival, since as everyone knew the Schrödinger equation is linear and cannot support nonlinear evolution, the essence of chaos. They were quite right of course in a strict sense, but it is important to explore the subject of how classical chaos relates to the behavior of quantum systems, and to meet the challenges of treating classically chaotic systems semiclassically. There need be no definition of *quantum chaos.*

Almost all classical trajectories with nearby initial conditions separate exponentially from each other in phase space if the dynamics is chaotic. But two different quantum states, ψ_1 and ψ_2, do not separate at all in the sense of their overlap, since

$$\langle\psi_1|\psi_2\rangle = \langle\psi_1|e^{iHt/\hbar}e^{-iHt/\hbar}|\psi_2\rangle = \langle\psi_1(t)|\psi_2(t)\rangle. \tag{21.1}$$

Classical systems have the same "problem" if we consider distributions and their overlaps: for classical densities $\rho_1(0)$, $\rho_2(0)$ and their time evolutions $\rho_1(t)$ and $\rho_2(t)$, we have

$$\int \rho_1(0)\rho_2(0)d\Gamma = \int \rho_1(t)\rho_2(t)d\Gamma \tag{21.2}$$

for all t, where $\int d\Gamma$ implies integration over all classical phase space. This is the correct analog of equation 21.1; classical distributions are also the correct analog to quantum wavefunctions.

It is also important to note that in the limit $\hbar \to 0$, any phase space localized wavepacket will follow a classical trajectory for as long as you care to look. Clearly, quantum mechanics has to "know" something about classical chaos.

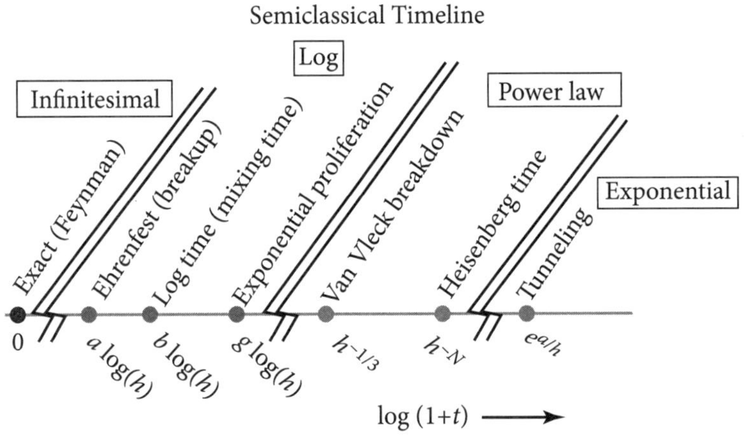

Figure 21.1. Timeline of quantum and semiclassical phenomena for a typical classically chaotic system, in the limit $\hbar \to 0$. Breakup refers to wavepacket spreading and dissolution. Mixing refers to classical spreading over all of the phase spaces from a 'ball' of trajectories of volume h^N. Breakdown refers to failure of semiclassical VVMG. The Heisenberg time is the time required to resolve eigenstates one from the next.

Sir Michael Berry cleared the air [154, 155] when he stated: "There is no quantum chaos, in the sense of exponential sensitivity to initial conditions, but there are several novel quantum phenomena that reflect the presence of classical chaos. The study of these phenomena is quantum chaology." Quantum chaology might seem an esoteric pursuit, of interest to a few semiclassical enthusiasts. This is far from the case. Quantum chaology is very rich territory, yielding new physical and mathematical insights into many quantum systems that would otherwise have escaped attention [156]. Figure 21.1 can be helpful to connect the size of Planck's constant with the order of events in time.

21.1 Semiclassical Timeline

We have already encountered a little quantum chaology in section 8.3 and chapter 22 continues with more aspects of quantum chaology.

Purely classical issues that inform a sophisticated understanding of this subject are found in chapter 3, section 3.8, and chapter 4, section 4.5. Perhaps the most important background for this chapter is understanding the semiclassical Van Vleck-Morette-Gutzwiller (VVMG) propagator and wavepacket dynamics derived from it. We need it right now, to set the stage for understanding quantum time evolution (and later its Fourier transform into the energy domain) in classically chaotic systems.

Suppose we have a classically chaotic Hamiltonian system (figure 21.2), which we will take to imply energy (and perhaps angular momentum) conserving dynamics that is everywhere unstable in the sense that almost all infinitesimally nearby trajectories separate exponentially from each other in time. We pause to define the measure of this separation, the Lyapunov exponent.

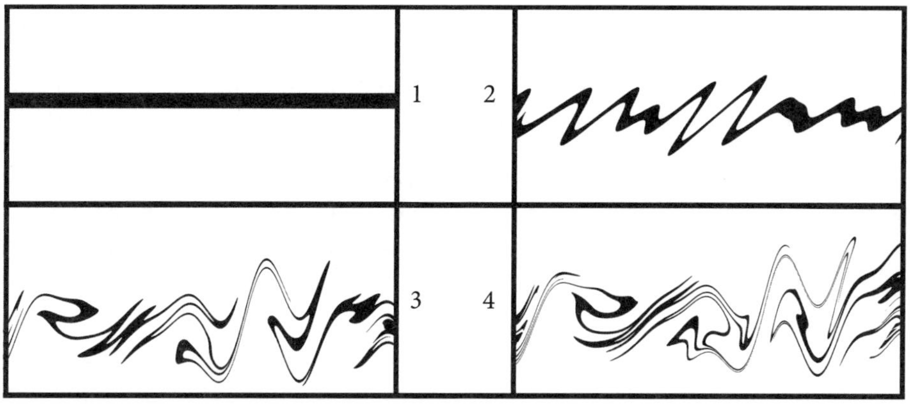

Figure 21.2. Four snapshots of the evolution of an area-preserving chaotic map of a strip in the phase plane.

LYAPUNOV EXPONENT (REVISITED)

Define a small (ultimately infinitesimal) separation in phase space between two trajectories as $\delta \boldsymbol{X}(t)$. Exponential separation in phase space means

$$|\delta \boldsymbol{X}(t)| \approx e^{\lambda t} |\delta \boldsymbol{X}(0)|, \tag{21.3}$$

where λ is the Lyapunov exponent, properly defined as involving a very long time t, and with $\lim_{t \to 0} |\delta \boldsymbol{X}(t)| \to 0$. More precisely, the *maximal* Lyapunov exponent is defined as (we repeat equation 3.87 here for convenience)

$$\lambda = \lim_{t \to \infty} \lim_{\delta X_0 \to 0} \frac{1}{t} \ln \left[\frac{|\delta \boldsymbol{X}_t|}{|\delta \boldsymbol{X}_0|} \right]. \tag{21.4}$$

An equivalent definition of the maximal Lyapunov exponent in terms of the stability matrix is

$$\lambda = \lim_{t \to \infty} \lim_{\delta X_0 \to 0} \frac{1}{t} \ln \| \boldsymbol{M}_t(\boldsymbol{X}_0) \cdot \delta \boldsymbol{X}_0 \|. \tag{21.5}$$

Even though there are $2N$ Lyapunov exponents for N degrees of freedom, the fastest growth will dominate at large times.

In section 3.8 we learned that the Lyapunov exponents come in pairs $\mu_k = e^{\lambda_k t}$ and $\mu_k^{-1} = e^{-\lambda_k t}$. As an example, we have seen that an unstable periodic orbit in two spatial dimensions has an unstable axis or manifold with nearby trajectories separating as $e^{\lambda t}$, and a stable axis or manifold, with trajectories attracting as $e^{-\lambda t}$ (see figures 3.25, 3.27, 22.3, and section 3.13). This exponential separation is true of arbitrary close pairs of trajectories in a chaotic system—that is, they don't have to lie on periodic orbits.

21.2 Anticipating the Stages of Evolution in a Chaotic System

INFINITESIMAL TIME

Suppose we begin with a minimum uncertainty wavepacket centered on a classical phase point in a chaotic system. The correct classical analog is the corresponding

Gaussian Wigner distribution. There are three time evolutions we need to follow: (1) the exact quantum wavefunction; (2) the Van Vleck level semiclassical one (here the most direct is the generalized Gaussian wavepacket dynamics [GGWPD], section 11.2); and (3) the purely classical evolution of the initial Gaussian phase space density. We refer to the timeline shown in figure 21.1.

The first time segment is the infinitesimal, where for $t = 0$ to $t = \epsilon$ all three agree essentially exactly, even in a rather ragged potential. The Van Vleck methods are essentially exact for short times, before any underlying classical manifolds can curl back on themselves and cause singularities. Locally linear approximation (thawed Gaussian dynamics) is accurate in this regime.

EHRENFEST TIME

Ehrenfest's theorem is often somewhat misunderstood. There is no theorem that says that average positions and average momenta computed from a quantum wavefunction obey Newton's laws, except for globally linear systems such as the harmonic oscillator, linear ramp, and constant potentials. The theorem is approximate for the general potential, since, if $H = \mathbf{p}^2/2m + V(\mathbf{q})$,

$$\frac{d}{dt}\mathbf{p} = \frac{1}{i\hbar}[\mathbf{p}, H] = -\nabla \cdot V(\mathbf{q}), \text{ then}$$

$$\frac{d}{dt}\langle \mathbf{p} \rangle = -\langle \nabla \cdot V(\mathbf{q}) \rangle. \tag{21.6}$$

If the system starts out in a well-collected wavepacket, perhaps initially we can write

$$\langle \nabla \cdot V(\mathbf{q}) \rangle \approx \nabla \cdot V(\langle \mathbf{q} \rangle). \tag{21.7}$$

Thereafter, up to the Ehrenfest time, it may be true that both $\langle \mathbf{q} \rangle$ and $\langle \mathbf{p} \rangle$ nearly obey classical equations of motion. When the wavefunction has broken up or spread too far, $\langle \nabla \cdot V(\mathbf{q}) \rangle \approx \nabla \cdot V(\langle \mathbf{q} \rangle)$ will no longer hold, and the Ehrenfest regime has come to a close.

As the Ehrenfest regime ends, the wavepacket has spread beyond linearizable regimes. The thawed Gaussian approximation (TGA) starts to fail. The time this takes to happen depends on the degree of local anharmonicity, the initial wavepacket, and $\hbar$. The uncertainty in position and momentum of the initial wavepacket each shrink as $\sqrt{\hbar}$; however, this is a weak defense against exponential spreading. With subscript $\mathcal{E}$ denoting Ehrenfest limits (that is, when things start to break down in the Ehrenfest sense), $\Delta q_{\mathcal{E}}$ increases as $\Delta q_{\mathcal{E}} = \sqrt{\hbar}\, e^{\lambda t_{\mathcal{E}}}$—that is, $\lambda t_{\mathcal{E}} \sim \log \hbar$. The time $t_{\mathcal{E}}$ is called the Ehrenfest time, and is typically associated with wavepacket breakup. It used to be thought that $t_{\mathcal{E}}$ would herald the demise of semiclassical approximations, but this is very far from the truth. $t_{\mathcal{E}}$ heralds the breakdown of the Ehrenfest approximation,

$$\begin{aligned} \frac{d}{dt}\langle p \rangle &\sim -\langle V'(q) \rangle, \\ \frac{d}{dt}\langle q \rangle &\sim \frac{\langle p \rangle}{m}, \end{aligned} \tag{21.8}$$

where the expectation values are taken over the spreading wavepacket.

LOG TIME

In the next regime, times that go as $\log(1/h)$, any detailed semblance of the classical distribution and the quantum wavefunction (or its Wigner transform) is lost. The classical distribution becomes stretched and convoluted, with many branches contributing to each position, as has started to happen in the third and fourth panels of figure 21.2. The several or many contributions to each position interfere with each other and cause deviations of the quantum and purely classical distribution. Semiclassical contributions in the sense of terms in the Van Vleck sum over classical paths may become very numerous compared to the number of Planck cells in the problem, or put another way, each Planck cell may have thousands of contributions. Even so, the accuracy of the Van Vleck or GGWP propagator is holding, assuming $\hbar$ is small compared to the hairpin areas seen in figure 21.2. This happened even in the integrable Morse oscillator at long times (section 8.1), where quantum revivals were successfully predicted semiclassically. Using wavepacket initial conditions rather than classical manifolds like position or momentum eigenstates can provide additional smoothing over classical caustics.

Classically, phase space is soon accessed everywhere: The mixing time for the exponentially unstable dynamics, such that most cells of size h in phase space are accessed by a typical initial state, also goes as $\log(1/h)$.

EXPONENTIAL PROLIFERATION TIME

As the number of branches contributing to each position proliferates exponentially, in very short order a computationally tractable 100,000 terms becomes an intractable 10^{10} terms. This is the *exponential proliferation* regime, also taking a time of order $\log(1/h)$ to reach. These are not Feynman paths in spite of their colossal number, but rather a huge number of classical ways to getting from the initial distribution to some final positions. If they could all be found and added up *the Van Vleck or GGWP is still doing very well even in the proliferation regime for small enough $\hbar$.* Diffractive errors do build up at the hairpin ends, but these are manageable [58] until a much longer time. The problem in the proliferation regime is practical, not fundamental.

DIFFRACTION DEGRADES THE VAN VLECK PROPAGATOR

Eventually, the diffractive errors start to mount, causing a problem for the accuracy of semiclassical approximations. In reference [58], it was shown this happens in a time on the order of $\hbar^{-1/3}$.

HEISENBERG TIME

For a system with N degrees of freedom, the Heisenberg time of order $t_H \sim \hbar^{-N}$ is the next milestone as time increases. t_H is the time required to resolve all the eigenstates (save those involved in tunneling—see the next regime) from each other by Fourier analysis: since the density of states rises as $\hbar^N$, the time-energy Heisenberg uncertainty principle time required to distinguish them is of order $\hbar^{-N}$. The Van Vleck propagator has likely already broken down, implying the eigenstates of a many-body system cannot be determined accurately semiclassically.

TUNNELING TIME

If tunneling is important, energy levels can be split by much less than the mean spacing, in fact by $\delta E \sim \exp[-a/\hbar]$, requiring a time of order $\exp[a/\hbar]$ to resolve, where a is a classical action. Finally, after this time, all the essentially new dynamics that the quantum system is capable of has finally taken place.

21.3 Understanding Quantum Ergodicity

Classical ergodicity is the notion that all eligible phase space cells thought to be eligible are in fact visited, and uniformly so. Ergodic or not, wherever classical flow decides to go, it does so uniformly. This is easy to show: average a classical Hamiltonian trajectory over an infinite time, and just for good measure apply an infinitesimal smoothing or "coarse graining." This does not take the distribution anywhere new except within $\epsilon \to 0$ of where it has been. The resulting distribution must be uniform on whatever subdimensional structure it decided to explore. For example, suppose the system is ergodic, so the distribution should be $\rho_\infty(\boldsymbol{p}, \boldsymbol{q}) = \delta(E - H(\boldsymbol{p}, \boldsymbol{q}))$. This distribution is time invariant:

$$\frac{d\rho_\infty}{dt} = \{H, \rho_\infty\} = \sum_i \left[\frac{\partial H}{\partial q_i} \frac{\partial \delta(E - H(\boldsymbol{p}, \boldsymbol{q}))}{\partial p_i} - \frac{\partial H}{\partial p_i} \frac{\partial \delta(E - H(\boldsymbol{p}, \boldsymbol{q}))}{\partial q_i} \right] = 0. \tag{21.9}$$

Now suppose the distribution ρ_∞ is weighted nonuniformly by some function $\alpha(\boldsymbol{p}, \boldsymbol{q})$, while still confined to the energy hypersurface:

$$\rho_\alpha(\boldsymbol{p}, \boldsymbol{q}) = \alpha(\boldsymbol{p}, \boldsymbol{q})\delta(E - H(\boldsymbol{p}, \boldsymbol{q})). \tag{21.10}$$

It is simple to show that

$$\frac{d\rho_\alpha}{dt} = \{H, \rho_\alpha\} = \sum_i \left[\frac{\partial H}{\partial q_i} \frac{\partial \alpha(\boldsymbol{p}, \boldsymbol{q})}{\partial p_i} - \frac{\partial H}{\partial p_i} \frac{\partial \alpha(\boldsymbol{p}, \boldsymbol{q})}{\partial q_i} \right], \tag{21.11}$$

so unless $\alpha(\boldsymbol{p}, \boldsymbol{q})$ is itself time-invariant, ρ_∞ cannot be so weighted. The only time-invariant functions $\alpha(\boldsymbol{p}, \boldsymbol{q})$ are themselves a function only of the Hamiltonian, and therefore do not favor one region over another on the energy hypersurface. Similar arguments apply to motion on reduced manifolds that reflect other conserved quantities such as angular momentum or good actions.

ERGODICITY *REDUCTIO AD ABSURDUM*

Suppose we have a wildly gyrating, extremely nonlinear mechanical system floating freely with no external torques. We might be tempted to think that the system ergodically accesses its entire energy hypersurface over time. But the system must necessarily fall far short of this: angular momentum is also conserved, and this restricts the motion to a surface down by three dimensions from the entire energy surface. Of course, the immediate fix, which could have been considered to begin with, is to redefine ergodicity as uniform population of this lower dimensional surface in phase space. This new ergodicity may indeed be fulfilled by the system.

Or, there may be yet another constant of the motion, besides the energy and the angular momentum. Now the system is no longer ergodic according to the old definition. The system may now be checked with respect to ergodicity on the new surface in phase space reduced by one more dimension. It may or may not be found to be ergodic on this surface. If it is not, there are more constants of the motion to be discovered.

In this way the discovery of new constants and redefinition of ergodicity continues. For example, if the system is completely integrable, it will normally (but still not always!) be ergodic on the invariant tori characterizing the system.The exceptions happen when two or more classical frequencies find themselves locked in exact rational ratios. Then, the covering of tori is not uniform, and in fact the dynamics covers lower dimensional "spiral" like objects. The classical motion is then finally ergodic on the spiral surfaces!

By this logic, it is correct to say that every classical dynamical system is completely ergodic, with respect to all the constraints governing its motion. This is not the game we want to play, and is ultimately uninformative as a blanket maxim. In the quantum world, von Neumann had an analogous answer to the question or ergodicity—namely, that all quantum nondegenerate systems are ergodic. This is analogous to the classical maxim just expressed and is equally uninformative. The fix is straightforward: specify known prior constraints (the level of prior knowledge) and measure ergodicity with respect to them.

QUANTUM SYSTEMS

In order to learn more about a quantum system's evolution, we may have to adopt small amounts of imprecision and measures that are essentially semiclassical in nature. For example, we can measure flow and ergodicity by starting in small corners of phase space, as defined by a Gaussian wavepacket or other nonstationary state.We follow the dynamics to see if it appears in every corner of phase space. But we have to take care to be sure those corners are not "ineligible" on some grounds (including angular momentum conservation again) that we should have incorporated. The issue of prior constraints thus also rears its head quantum mechanically, but we will see not as sharply as it does in classical mechanics.

If energy is sharp quantum mechanically, nothing moves, even though energy can be sharp classically with much still in the balance. We must consider nonstationary states with an energy spread as measures of interesting quantum flow. Classical analogs to this exist and are well defined, involving distributions spanning a range of energy shells.

A nonstationary quantum state has an energy expectation value and dispersion discoverable from short time dynamics. This is also true of any final state we would like to use as a measure of phase space flow. Nonstationary states with nonoverlapping energy envelopes cannot develop flow from one to another over time, and they are thus uninteresting as a pair, either classically or quantally. If there is only partial energy overlap of the states, measures need to be developed to properly bias the measure, making the outcome informative.

CLASSICAL LIMIT

Before we begin, we can insist that whatever measures we develop would agree with classical mechanics as $\hbar \to 0$. There lurks an important subtlety. Classical mechanics has an infinite amount of time to explore the phase space available to it. In other words, the dynamics might be ergodic, but with a very small Lyapunov exponent, or even perhaps an algebraic rate of exploration of phase space.

On the other hand, no matter how small the value of $\hbar$, a double well oscillator will tunnel, in a very nonclassical way, as time goes to ∞. Tunneling is nonclassical, of course, and by taking $t \to \infty$ first we can take $\hbar \to 0$ without recovering classical mechanics.

This problem is easily fixed: take $\hbar \to 0$ before the limit $t \to \infty$. But experiments and calculations are trapped in the opposite sequence of limits. Time can always be taken to be very long, but one cannot take $\hbar \to 0$ in nature. In some favorable cases it may be possible to cut off the dynamics early enough that long time tunneling processes (including dynamical tunneling and diffraction) have not happened, but late enough that almost all other classical-like phase space flow has taken place. In such instances, the system may well reproduce its classical analog. There is one final caveat, however, if we are to expect complete classical-quantum agreement: the initial state and the final state will have to be averaged over small regimes sufficiently large to wash out multiple path interference effects.

ERGODICITY OF INDIVIDUAL EIGENSTATES

Suppose a classical system is ergodic. In the $\hbar \to 0$ limit, will its eigenstates be somehow uniformly spread over phase space? This question violates the order of limits we just stated: time must be taken to infinity before $\hbar \to 0$ to work with eigenstates; in the other order of limits one has stacks of eigenstates superposed. The stacks would have to cover the energy hypersurface evenly, if the classical system does, because the wavepacket remained narrow and followed an ergodic trajectory its entire life.

That does not stop us from being curious about individual eigenstates as $\hbar \to 0$. Theoretical work by Schnirelman, Colin de Verdiere, and Zelditch [157–159] answered this question, but interesting holes remain in their results. The holes are not a defect in their logic; it is probably not possible to do better. Their theorem allows rigorous statements about ergodicity of spatial domains anywhere within eigenstates, shrinking to infinitesimal as $\hbar \to 0$. Yet the domains do not shrink as fast as $\hbar$; rather something like $\sqrt{\hbar}$. Such a shrinkage rate means their result pertains to averages over zones infinitely many wavelengths across as $\hbar \to 0$.

In other words, the Schnirelman, Colin de Verdiere, and Zelditch theorem leaves open the possibility of eigenfunctions becoming Swiss cheese with serious holes as $\hbar \to 0$, leaving large but not macroscopic gaps (or build-ups) on the energy hypersurface. The theorem separately leaves open the possibility of a subset of macroscopically nonergodic eigenstates, which however must approach zero measure (zero percent of all eigenstates) as $\hbar \to 0$.

In one study, that of a nearly rectangular box with one side tilted at an irrational angle, the eigenstates badly failed to be uniform on the energy hypersurface. Classically, this system exhibits ergodicity, on a very slow timescale [160]. Most eigenfunctions of modest energies (and therefore, effectively still fairly large $\hbar$) shown in figure 21.3 have not yet begun to show their ultimate Schnirelman, Colin de Verdiere, and Zelditch behavior. Almost every one is favoring a small range of momenta, reflecting the very slow classical momentum transport.

In another study—namely, the stadium billiard—the highly non-ergodic "bouncing ball" eigenstates trapped in vertical bouncing motion between the straight walls can be proven to persist even at infinite energy ($\hbar \to 0$) [161]. The proof strongly suggests they become an ever smaller percentage of the total number of states in that limit, agreeing with Schnirelman, Colin de Verdiere, and Zelditch (see figure 21.4).

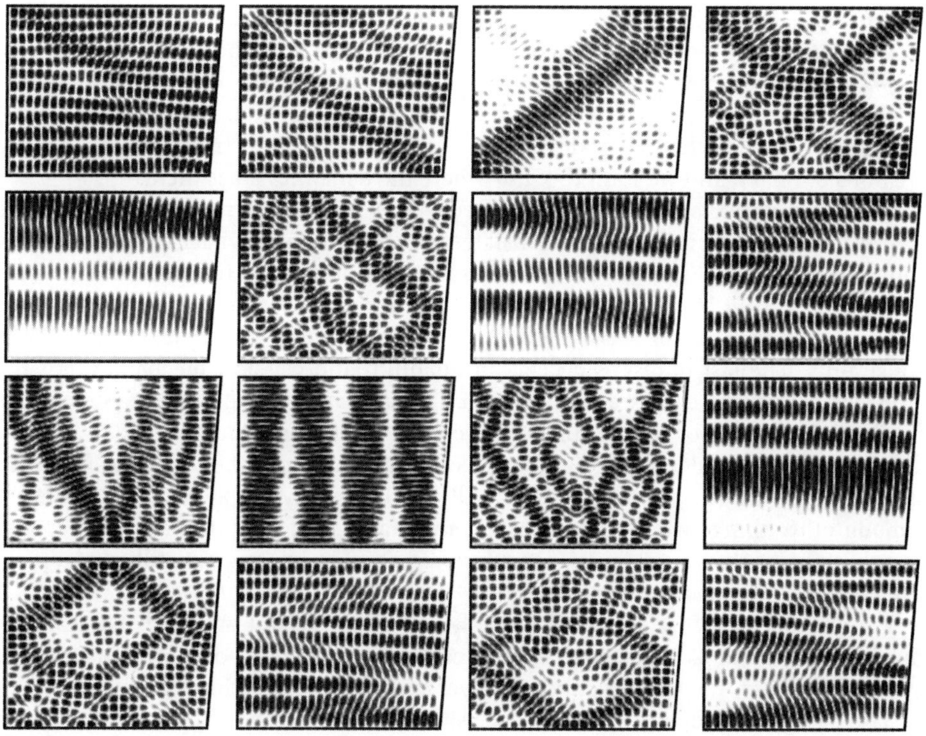

Figure 21.3. Successive eigenstates in a nearly rectangular box with one side tilted at an irrational angle. Classically, this system exhibits ergodicity, on a very slow timescale. They have not reached their ultimate Schnirelman, Colin de Verdiere, and Zelditch behavior, which similar states at much higher energy will presumably do.

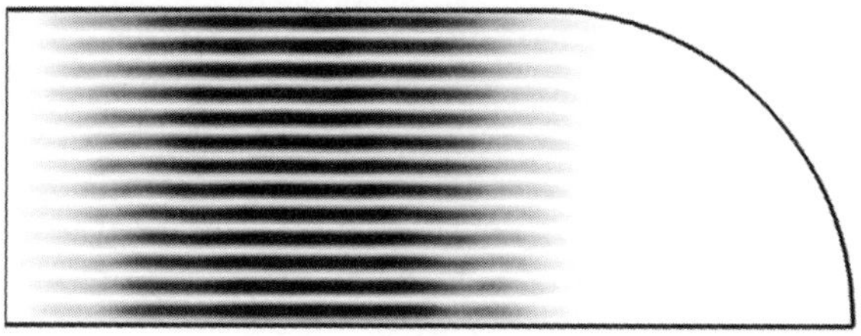

Figure 21.4. Absolute square of a so-called vertical bouncing ball eigenstate in the classically chaotic quarter stadium billiard, of the type that persists (with an increasing number of nodal lines) up to infinite energy, becoming a smaller and smaller fraction of the total number of states, until finally this unusual type of state, and even ones like it with a finite number of vertical nodes, indicating some progress right and left as well as up and down, become 0% of the eigenstates at infinite energy (or $\hbar \to 0$).

21.4 Measures of Ergodicity in Quantum (and Classical) Mechanics

MEASURES OF ERGODICITY IN CLASSICAL MECHANICS

We consider transport between test states (or density matrices) ρ_a, ρ_b that are both manifestly localized somewhere in an eligible part of the phase space, so that as far as we know ρ_b is accessible to the dynamics starting at ρ_a (or equivalently, vice versa). "As far as we know" is a key phrase, and our goal is to check to see if the flow really happens. By *phase space*, we mean as defined by Wigner transforms, and by *localized*, we mean of course within the confines of the uncertainty principle. States may be interesting that lie well outside the uncertainty principle bounds. Such distributions $\rho_a, \rho_b, \ldots$ are nonstationary, and we want to know where the dynamics takes them.

There are important differences between pure state density matrices, like those arising from a quantum nonstationary state, and mixed densities, such as an incoherent sum over such states. The latter is much less susceptible to local interference effects.

It is easiest (but by no means necessary) to consider Gaussian wavepackets as test states. We form a set of them $\rho_a, \rho_b, \ldots$ covering phase space subject to known constraints. The last phrase is crucial, in quantum and classical mechanics. For the latter we would not expect a trajectory in a conservative Hamiltonian system to be found in the vicinity of another phase space point of different energy.

We are forced to deal with distributions that span a range of energies. A Gaussian spans such a range quantum mechanically or classically, as a Wigner transform. Among other things we need to consider test states with only partially overlapping distributions, such as two Gaussians with different mean energy, but still overlapping. We raised these issues in section 21.3.

For example, suppose energy is the only known constraint on the dynamics; how much are $\rho_a(t)$ and ρ_b expected to overlap each other, averaging over long times, if the system is ergodic, and their energy envelopes do or do not coincide exactly?

The energy envelope of the initial density matrix is computable from its Wigner density ρ_a^W as

$$S_a(E) = \int d\boldsymbol{p}d\boldsymbol{q}\; \rho_a^W(\boldsymbol{p}, \boldsymbol{q})\delta(E - H(\boldsymbol{p}, \boldsymbol{q}))]. \tag{21.12}$$

For a minimum uncertainty Gaussian, $\rho_a = |a\rangle\langle a|$ this will often be very nearly equal to the Fourier transform of autocorrelation $\langle a|a(t)\rangle$ up to the time of the first decay, assuming it decays steadily to near 0 after a time T commensurate with the energy width of $S_a(E)$:

$$S_a(E) = \frac{1}{2\pi}\int_{-T}^{T} e^{iEt/\hbar}\langle a|a(t)\rangle\, dt, \tag{21.13}$$

where the time T is after the decay to 0 but before any recurrences.

Also,

$$\int_{-\infty}^{\infty} S_a(E)\, dE = \mathrm{Tr}[\rho_a^W] = 1.$$

If the system is classically ergodic, what will the phase space distribution starting from $\rho_a^W(\boldsymbol{p}, \boldsymbol{q})$ become after a long time? Clearly, it should be a weighted integral over microcanonical populations $\delta(E - H(\boldsymbol{p}, \boldsymbol{q}))$, and certainly the weight must involve the energy envelope $S_a(E)$ that measures the energy content of $\rho_a^W(\boldsymbol{p}, \boldsymbol{q})$. The total density of states

$$D(E) = \int d\boldsymbol{p}d\boldsymbol{q}\; \delta(E - H(\boldsymbol{p}, \boldsymbol{q}))] \tag{21.14}$$

must also come in, so that if the dynamics is ergodic,

$$\rho_a^e(\boldsymbol{p}, \boldsymbol{q}) = \lim_{T\to\infty} \frac{1}{T} \int_0^T \rho_a(t)dt = \int dE\; \frac{S_a(E)\, \delta(E - H(\boldsymbol{p}, \boldsymbol{q}))}{D(E)} \tag{21.15}$$

becomes the long time phase space distribution (with superscript *e* for ergodic) arising out of the classical initial distribution $\rho_a^W(\boldsymbol{p}, \boldsymbol{q})$, Note that $\rho_a^e(\boldsymbol{p}, \boldsymbol{q})$ is unit normalized after tracing over all phase space, thanks to the denominator $D(E)$. To see another reason why $D(E)$ needs to be present, suppose $D(E)$ jumps up at some energy because a large "arena" opens up in coordinate space above that energy. If ρ_a does not start in that arena, and is localized, it knows nothing about it. $S_a(E)$ is oblivious to it, and needs to be divided by $D(E)$ so as not to overweight the sum over all the high density of states that arises as the arena opens up.

Now to answer the question posed earlier, "How much are $\rho_a(t)$ and ρ_b expected to overlap each other, averaging over long times, if the system is ergodic?" we look for the overlap of $\rho_a^e(\boldsymbol{p}, \boldsymbol{q})$ and $\rho_b(\boldsymbol{p}, \boldsymbol{q})$ (with a superscript $c.e.$ for "classical ergodic"):

$$\begin{aligned} P^{c.e.}(a|b) &= \mathrm{Tr}[\rho_b(\boldsymbol{p}, \boldsymbol{q})\rho_a^e(\boldsymbol{p}, \boldsymbol{q})], \\ &= \int \frac{S_a(E)\,\mathrm{Tr}[\rho_b(\boldsymbol{p}, \boldsymbol{q})\delta(E - H(\boldsymbol{p}, \boldsymbol{q})]}{D(E)}\, dE, \\ &= \int \frac{S_a(E)S_b(E)}{D(E)}\, dE. \end{aligned} \tag{21.16}$$

This is the proper weighting of the energy envelopes $S_a(E)$ and $S_b(E)$ for ρ_a and ρ_b to give an estimate of their averaged long time overlap under classical ergodic dynamics.

MEASURES OF ERGODICITY IN QUANTUM MECHANICS

We now find all the quantum analogs to the components above, with the intent of comparing the quantum $P(a|b) = \sum_n p_n^a p_n^b$ with an ergodic estimate, armed not only with the initial energy envelope but perhaps also more refined envelopes, showing substructure gleaned from intermediate time dynamics.

The quantum mechanical analog to $S_a(E)$, the classical energy envelope function, is

$$S_T^a(E) = \frac{1}{2\pi} \int_{-\infty}^{\infty} e^{iEt/\hbar} \hat{\Omega}_T(t) \mathrm{Tr}\left[e^{-iHt/\hbar}\rho_a\right]\, dt, \tag{21.17}$$

where a variable energy window ΔE is implied by the cutoff time T in $\hat{\Omega}_T(t)$, figure 21.5—that is, $T\;\Delta E \sim \hbar$. $S_T^a(E)$, given in equation 21.17, shows that it carries information about short time dynamics of ρ_a, up to time $t = T$. If $S_T^a(E)$ is examined as a function of increasing T, the spectrum will appear with increasing resolution; each $S_T^a(E)$ is consistent with, and implies, all earlier time, lower resolution spectra, and

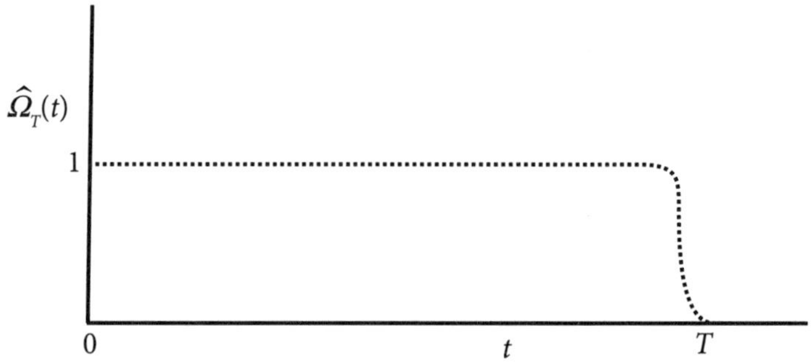

Figure 21.5. The time cutoff function $\hat{\Omega}_T(t)$. It is symmetric about $t = 0$.

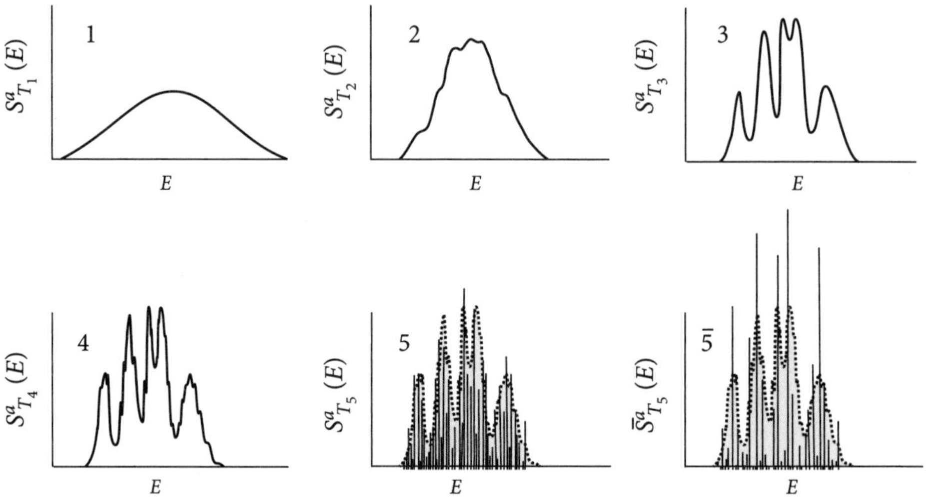

Figure 21.6. Time evolution of the spectral envelope $S^a_T(E)$ as defined in equation 21.17 for ever longer times for an initial, generic nonstationary density ρ_a. All longer time spectra smooth to their shorter time ancestors. In the last two frames, bottom right, two different high-resolution scenarios $S^a_{T_5}(E)$ and $\bar{S}^a_{T_5}(E)$ are consistent with the lower resolution envelopes; the last shows a nonergodic, nonrandom set of p^a_n's. Tick marks show the energies of all the eigenstates below the horizontal axis.

each $S^a_T(E)$ is consistent with, but does not imply, any later time, higher resolution spectra; see figure 21.6.

The total quantum density of states, determined according to the same energy window, is

$$D_T(E) = \sum_{n'} \Omega_T(E - E_{n'}), \tag{21.18}$$

where

$$\Omega_T(E') = \frac{1}{2\pi} \int_{-\infty}^{\infty} \hat{\Omega}_T(t) e^{-iE't/\hbar}. \tag{21.19}$$

The ergodic estimate of a spectral intensity p_n^a, call it $\bar{p}_n^a$, would be

$$\bar{p}_n^a = \frac{S_T^a(E_n)}{D_T(E_n)}, \tag{21.20}$$

and the "quantum ergodic" estimate of $P(a|b)$, the analog of the classical $P^{c.e.}(a|b)$, equation 21.16, is

$$\begin{aligned} P^e(a|b) &= \sum_n \bar{p}_n^a \bar{p}_n^b = \sum_n \frac{S_T^a(E_n) S_T^b(E_n)}{D_T(E_n) D_T(E_n)} \\ &\sim \int \frac{S_T^a(E) S_T^b(E)}{D_T(E_n)} \, dE, \end{aligned} \tag{21.21}$$

the direct analog of equation 21.16. An important difference is that the quantum version applies to various energy windows ΔE corresponding to different extents of prior knowledge.

21.5 Time Development of Dynamical Constraints

As a quantum system evolves in time, we naturally learn more and more about it. Keeping in mind that the definition of ergodicity is up to the user, we could arrive at von Neuman's definition that all quantum systems are ergodic save those that have degeneracies. In a word, as we learned more about the system, we would find it (in a kind of tautology) doing all that it could possibly do, thus earning the designation as ergodic. This is clearly not very useful.

Ignoring all prior short time knowledge about a system could be equally uninformative. For example, suppose we start in a nonstationary state with a mean energy of 5 and an energy dispersion of 0.2. It would be "unfair" to check if the evolution populated two other nonstationary states, one with a mean energy of 3 and a dispersion of 2, and the other with a mean energy of 6 and an energy dispersion of 0.2. The average energy and dispersion of an initial nonstationary state is revealed by time evolution. The key point is that the average energy and energy dispersion can be gleaned (if $\hbar$ is small enough) before any critical or interesting dynamics occurs, such as hitting a wall in a billiard. The "uninteresting" and trivial dynamics would be the flat potential traversed before the wall is encountered. More generally, the average energy and energy dispersion is a function only of the initial state chosen, and local properties of the Hamiltonian near the initial state's location. This can be taken into account without creating any tautological problems.

Sometimes more subtle choices have to be made. A Gaussian wavepacket is a fairly innocuous choice with very little information revealed about the system in the subsequent very short time dynamics. It may be appropriate for some questions arising in a polyatomic molecule, representing displacement of a would-be normal mode. Given the anharmonic interactions, the mode excite states will not be stationary states. As we observe the subsequent decay, the mean energy and energy dispersion of the state are revealed. The timescale for the initial decay is already interesting, and we may discover that the spectrum consists of narrow populous "bright" energy domains separated by wide "deserts" between the different overtones of the normal mode. It will be impossible for vibrational normal mode A to communicate with B if B has a nonoverlapping energy "envelope." Whether this is prima facie evidence of nonergodicity is up to the user. There is no right or wrong here, unless there is a failure to understand these issues. The interest might focus on what is happening within the bands.

There may be thousands of eigenstates within one narrow, bright molecular spectral band. Are they all populated statistically, or are only 0.5% of them activated? Taking a band as a prior constraint—that is, taking into consideration that the narrow bright band exists—there still remain very interesting questions about the high-resolution spectrum inside that band.

Consider the time-averaged probability $P(a|b)$ of being in the state $|b\rangle$, with density matrix ρ_b, if initially the system is in the density ρ_a. This is

$$
\begin{aligned}
P(a|b) &= \lim_{t\to\infty} \frac{1}{t} \int_0^t |\langle b|e^{-iHt/\hbar}|a\rangle|^2 dt = \lim_{t\to\infty} \frac{1}{t} \int_0^t |\langle b|a(t)\rangle|^2 dt, \\
&= \lim_{t\to\infty} \frac{1}{t} \int_0^t \mathrm{Tr}[\rho_b \rho_a(t)] dt, \\
&= \lim_{t\to\infty} \frac{1}{t} \sum_{n,n'=1}^{\infty} \int_0^t \langle b|n\rangle\langle n|a\rangle\langle a|n'\rangle\langle n'|b\rangle e^{-i(E_n - E_{n'})t/\hbar}, \\
&= \sum_{n=1}^{\infty} |\langle b|n\rangle|^2 |\langle n|b\rangle|^2 = \sum_{n=1}^{\infty} p_n^a\, p_n^b,
\end{aligned} \tag{21.22}
$$

assuming no degeneracies among the E_n. This formula is worth highlighting in a box:

$$
\boxed{P(a|b) \equiv \lim_{t\to\infty} \frac{1}{t} \int_0^t \mathrm{Tr}[\rho_b \rho_a(t)] dt = \sum_{n=1}^{\infty} p_n^a\, p_n^b.} \tag{21.23}
$$

This has the elegant interpretation that the time-averaged probability $P(a|b)$ of being found in the state $|b\rangle$, or a density matrix ρ_b, if initially the system is in the density ρ_a or the state $|a\rangle$ is the inner product of their respective spectra or eigenstate probability distributions. One probability distribution is $p_n^a = |\langle n|a\rangle|^2$ (or $p_n^a = \langle n|\rho_a|n\rangle$ in the case of a density matrix ρ_a) of being in the eigenstate n if in the nonstationary state a and similarly for p_n^b. (These read equally of course "the probability of being in the nonstationary state a if in the eigenstate n," and so on).

To give significance to values of $P(a|b)$, we must develop prior constraints and thus frame the questions of ergodicity that remain interesting.

Summarizing, short and medium time dynamics constrain the set of p_n^a's, increasingly as time increases. One can examine the ergodicity from some fixed time onward by looking to see if "random" spectra fill the envelope $S_T^a(E)$ determined up to that time. For example, in figure 21.6, panels 5 and $\bar{5}$, the envelope determined up to time T_5 is resolved into sharp peaks, revealing the p_n^a's. In panel 5, the spectrum is consistent with ergodic dynamics for times beyond T_4 (we will discuss what a "random" spectrum is in a moment). In panel $\bar{5}$, for a different Hamiltonian and long time dynamics that

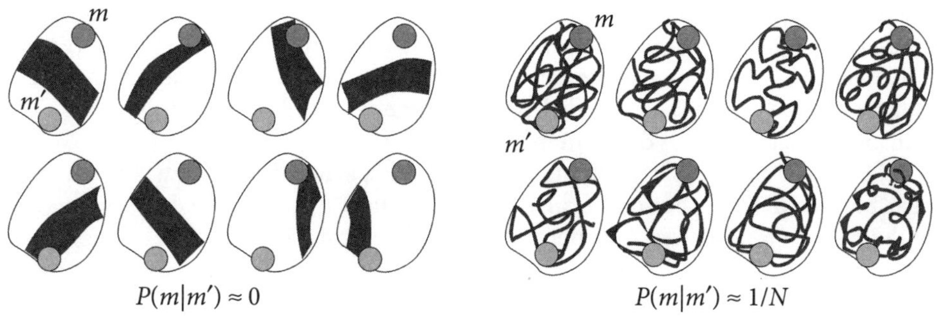

Figure 21.7. The localized states $|m\rangle$ and $|m'\rangle$ are indicated schematically as circles, and a sampling of the spatial disposition of typical eigenstates is shown. On the left, the eigenstates belong to an integrable system, and the localized states overlap m, or m', or neither, but not both. On the right, the states belong to an "ergodic" system, and each eigenstate overlaps both $|m\rangle$ and $|m'\rangle$.

agrees up to panel 4 with the previous case, the limited participation of levels (only a few large p_n^a). The $\bar{S}_{T_5}^a$ now reveals that the dynamics failed to be ergodic after time T_4.

21.6 Random Spectra and Quantum Ergodicity

The qualitative reason for associating a sparse filling of levels (a few large p_n^a and the rest small, as for most initial ρ_a in figure 21.6, panel $\bar{5}$) with nonergodicity is obvious from figure 21.7. We can be much more quantitative about what we expect from a "random" or ergodic spectrum. There are several ways to proceed, all more or less equivalent and all based on the expected quantum response to classical chaos or ergodicity.

One approach is to appeal to random matrix theory, asserting that the Hamiltonian (apart from whatever structure gave the short time dynamics correlation functions and structure in S_T^a) shares many of the properties of random matrices, which were introduced into physics by Eugene Wigner as a model for the nuclei of heavy atoms [162]. Wigner's idea was that in ensembles of nuclei, Hamiltonians varying from one to the next, the spectral statistics based on neutron collision spectra of many nuclei might resemble those of a random matrix ensemble. At least three factors made this a lasting contribution: (1) there is no known Hamiltonian much less any numerical solution giving a quantitative spectrum of any nucleus, making Wigner's model seem to be desperation, were it not for the next two points; (2) the nucleons in the nucleus appear to be a rather chaotic system if there is any merit to a classical analog; and (3) statistical properties of nuclear energy levels, their spacing distributions, and spectral fluctuations might be more revealing than knowledge of specific eigenvalues and eigenfunctions, especially if nuclear data and properties of random matrices seem to match.

A random real Hermitian matrix has Gaussian randomly chosen matrix elements H_{ij}, with $H_{ij} = H_{ji}$. The variance of the elements of H is controlled by the probability distribution for members of an ensemble of such random matrices, given by [163]:

$$P(H) = C\, e^{\mathrm{Tr}[H^2]/4\sigma^2}. \tag{21.24}$$

The two most important ensembles in physical systems are the Gaussian orthogonal ensemble, or GOE, for real symmetric Hamiltonians, corresponding to time reversible dynamics, and the Gaussian unitary ensemble, or GUE, for complex random but Hermitian matrices, as are needed for non-time-reversible magnetic systems. The statistics of eigenvalues is different in the two cases.

We review Wigner's arguments for the level spacing distribution and strength distribution (the distribution of their energy spacings $E_{n+1} - E_n$ and the fluctuation of the p_n^a) of a random real symmetric matrix in the next section.

AMPLITUDES AND SPECTRA

The eigenvectors of a random matrix are Gaussian random amplitudes a_{jn} for the nth eigenvector (except for the constraint that the a_{jn} be normalized, and that a_{jn} is orthogonal to $a_{jn'}$) for $n \neq n'$. With no constraints on H other than equation 21.24, the spectral intensities $p_n^a = |a_n|^2$ must be distributed as random, central limit processes—that is, squares of real Gaussian random amplitudes (in the Gaussian orthogonal ensemble, GOE—that is, real, symmetric Hamiltonians). This generates a χ^2 distribution of one degree of freedom—that is,

$$P_{GOE}(p_n) = \frac{e^{-p_n/2}}{\sqrt{\pi p_n}}. \tag{21.25}$$

This is the expected global distribution for a set of p_n^a from a random GOE Hamiltonian with no constraints. For the Gaussian unitary ensemble, GUE, the result changes, becoming

$$P_{GUE}(p_n) = \frac{e^{-p_n/2}}{2p_n}. \tag{21.26}$$

ENERGY-LEVEL REPULSION—THE WIGNER SURMISE

Random matrix theory dates to the early twentieth century in mathematics, and was connected midcentury by Eugene Wigner to nuclear spectroscopy. Results were derived for statistical ensembles of random matrices, rather than any given matrix with random entries.

Random matrices have certain universal properties such as level spacing distributions of their eigenvalues and statistical properties of the wavefunctions. Individual eigenvalues and resonances play second fiddle to the statistical properties of the spacings between observed scattering resonance eigenvalues, and the distribution of resonance widths.

Wigner asked the question: what are the statistics of the nearest neighbor spacings between the eigenvalues of a random matrix? Given a specific energy eigenvalue, what is the distribution of energy gaps to its nearest neighbors? Might that look like the nuclear physics data? It did, as seen later in figure 21.10. The nearest neighbor distribution need not come from a single nucleus, assuming they share similar properties and the data are scaled for the mean density of states.

If the eigenvalue spectrum had been random, with eigenvalues falling anywhere with impunity consistent with the average density of states, the level spacing distribution would be Poissonian. Energy levels could fall extremely close to each other, since the position of any of them would be uncorrelated to the others. A famous example of a Poisson distribution early in its history was to examine the number of soldiers in the Prussian army killed accidentally by horse kicks. In 1898, Ladislaus Bortkiewicz showed the distribution was Poisson. One might imagine that the distribution would not be Poisson, if soldiers were more careful for a while after a comrade died from a horse kick. This might be true in a single regiment, but with many widely distributed regiments, and the time between events fortunately long, the 122 deaths from horse kicks over 20 years were found to be uncorrelated—that is, Poisson.

For a Poisson process, given an event at $t = 0$, the probability of having no following event up to time t is $p(t) = \exp[-\lambda t]$—that is, the probability of no new event decays exponentially with time. If it was otherwise, say with a lower probability per unit time just after an event, the probability rising later, the events would manifestly be correlated. A histogram of the nearest neighbor distance of random and uncorrelated events will be exponential, peaking at zero spacing.

In an interacting quantum system, however, we expect some kind of energy-level repulsion. Suppose two diagonal energies of a random Hamiltonian matrix are very close. Any off-diagonal coupling will push them apart, as happens in a symmetric 2×2 matrix with any nonzero off-diagonal coupling. Wigner derived an approximation to the distribution of level spacings of the eigenvalues of a real, symmetric infinite dimensional random matrix by working out the properties of 2×2 random matrix ensembles. While this seems a little optimistic, it turns out to be extremely close to the exact results for large matrices, derived by much more difficult methods.

Consider a 2×2 matriix specified by

$$H = \begin{pmatrix} H_{11} & H_{12} \\ H_{21} & H_{22} \end{pmatrix}, \tag{21.27}$$

with eigenvalues

$$\lambda = \frac{H_{11} + H_{22}}{2} \pm \tfrac{1}{2}\sqrt{(H_{11} - H_{22})^2 + 4H_{12}H_{21}}. \tag{21.28}$$

The level splitting is $S = \sqrt{(H_{11} - H_{22})^2 + 4H_{12}H_{21}}$. Wigner introduced a matrix ensemble with the elements of the matrix random, except that the matrix was Hermitian, $H_{12} = H_{21}$. We need a probability distribution where the matrix elements are Gaussian random—for example,

$$P(H_{11}, H_{22}, H_{21} = H_{12}) = e^{-\gamma H_{11}^2 - \gamma H_{22}^2 - \gamma H_{12}^2}. \tag{21.29}$$

But there is a problem with this: the defined probability distribution is not invariant to orthogonal transformations on H; however, it must be invariant, since no basis should be special for an ensemble of random matrices. Consider a 2×2 general real orthogonal transformation,

$$T = \begin{pmatrix} \cos\theta & \sin\theta \\ -\sin\theta & \cos\theta \end{pmatrix}.$$

The form

$$H_{11}^2 + H_{22}^2 + H_{12}^2$$

is not invariant, but $\mathrm{Tr}[H^2] = H_{11}^2 + H_{22}^2 + 2H_{12}^2$ *is* manifestly invariant, since it is the trace of the matrix, invariant under unitary transformation. Therefore, we take

$$P(H_{11}, H_{22}, H_{21} = H_{12}) = Ce^{-\mathrm{Tr}[H^2]} = Ce^{-\gamma H_{11}^2 - \gamma H_{22}^2 - 2\gamma H_{12}^2}. \tag{21.30}$$

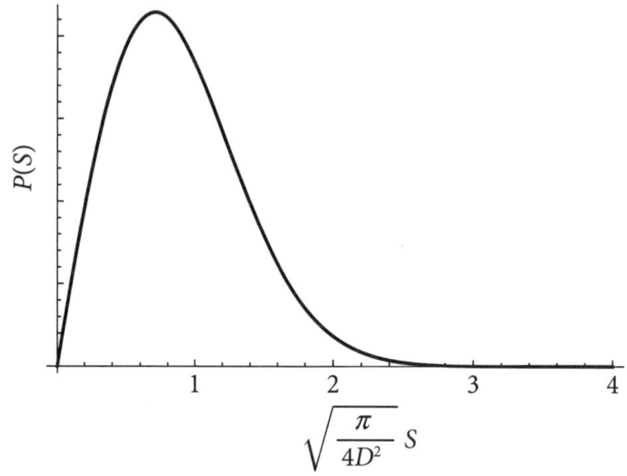

Figure 21.8. The Wigner surmise for the distribution of nearest neighbor level spacings in a random matrix ensemble, showing the level repulsion as a low probability, starting at 0, of levels being close together.

To determine C, we want

$$C\int e^{-\mathrm{Tr}[H^2]} dH_{11} dH_{22} dH_{12} = C\sqrt{\frac{\pi}{\gamma}}\sqrt{\frac{\pi}{\gamma}}\sqrt{\frac{\pi}{2\gamma}} = 1;\; C = \frac{\gamma^{3/2}\sqrt{2}}{\pi^{3/2}}.$$

We need to determine the distribution of splittings $S = \sqrt{(H_{11}-H_{22})^2 + (2H_{12})^2}$. Given the probability density (equation 21.30), the distribution $g_1 = (H_{11}-H_{22})$ goes as $\int \delta(g_1 - (H_{11}-H_{22}))e^{-\gamma H_{11}^2 - \gamma H_{22}^2 - 2\gamma H_{12}^2}\, dH_{11}\, dH_{22}\, dH_{12} \sim \exp[-\gamma g_1^2/2]$. Accounting for the 2 multiplying H_{12}—that is, $g_2 = 2H_{12}$, the required distribution $P(S)$ becomes the square root of the sum of two squared Gaussian random variables:

$$\begin{aligned} P(S) &= \frac{\gamma}{2\pi}\int dg_1 dg_2\, \delta\left(S - \sqrt{g_1^2 + g_2^2}\right)\, e^{-\frac{\gamma}{2}g_1^2 - \frac{\gamma}{2}g_2^2} \\ &= \frac{\gamma}{2\pi}\int_0^{2\pi} d\theta \int_0^{\infty} u\, du\, \delta(S-u)\, e^{-\frac{\gamma}{2}u^2}, \\ &= \gamma\, S\, e^{-\frac{\gamma}{2}S^2}. \end{aligned} \tag{21.31}$$

The mean spacing $D \equiv \langle S\rangle$ is

$$\begin{aligned} D &= \int_0^{\infty} S\, P(S)\, dS = \gamma\int_0^{\infty} S^2 e^{-\frac{\gamma}{2}S^2}, \\ &= \sqrt{\frac{\pi}{2\gamma}};\ \text{thus } \gamma = \frac{\pi}{2D^2}, \end{aligned} \tag{21.32}$$

so that (see figure 21.8)

$$P(S) = \left(\frac{\pi}{2D^2}\right)\, S\, e^{-\frac{\pi}{4D^2}S^2}. \tag{21.33}$$

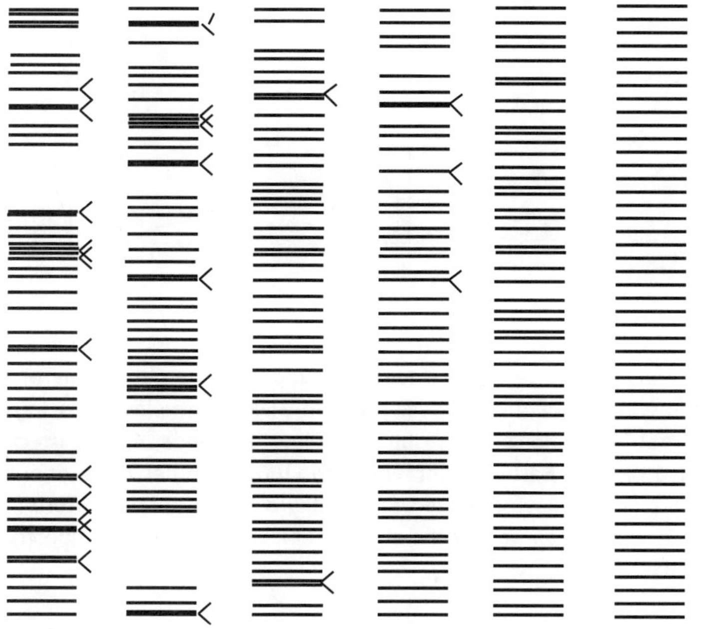

Figure 21.9. Typical spectra showing consecutive energy levels in six very different cases, both numerical and experimental. Redrawn from reference [164]. The distinctions among the cases are mostly clear. "Sinai" refers to numerical quantum eigenvalues found for a Sinai billiard (a square billiard with a circular hard wall in the center, the first system proven to be classically chaotic), and the zeros of the Reimann Zeta function ζ (s), which plays a role as paradigm and center of speculation for quantum chaos. Of the six cases, the spectrum of the Sinai billiard and of resonances in neutron scattering from Erbium ^{166}Er are thought to resemble each other due to underlying chaos. The marks denote energy levels too close to distinguish visually at this resolution.

Figure 21.9 illustrates found or calculated energy-level distribution in widely distributed circumstances. The nuclear spectrum of resonances for neutrons impinging on Er, the energy levels of a quantum Sinai billiard (the first system to be proven classical ergodic), and other cases are explained in the caption.

The notion arose in the 1980s that eigenfunctions and spectra of a single Hamiltonian, especially those connected with classical chaos, might bear resemblances to those of random matrices. Although the idea was certainly already motivating much research, the paper by Bohigas, Giannoni, and Schmit crystallized the conjecture [165] that eigenvalue statistics of classically chaotic systems resemble those of random matrices; see figure 21.10.

The intuitive concept behind the energy-level repulsion evident in the GOE case is the interaction of pairs of levels that otherwise would have fallen much closer to the same energy; such interaction always leads to repulsion.

The observation of level repulsion is not proof that the underling system is classically chaotic. Other statistics are important, such as the Dyson and Mehta Δ_3

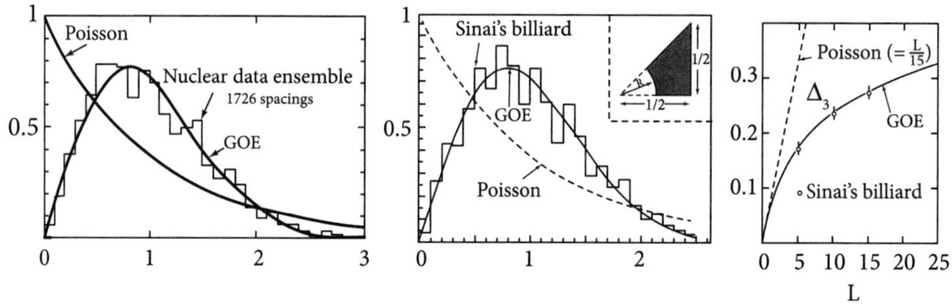

Figure 21.10. (Left) Nearest neighbor level spacing distribution of the "nuclear data ensemble" involving the resonance level spacing over 30 sequences of 27 different nuclei, plotted with the Wigner surmise (labeled GOE) for the level spacing distribution of a random matrix.The Poisson distribution of level spacings that are uncorrelated and random is also plotted. Partially redrawn from [166]. (Middle) 740 numerical levels of the classically chaotic desymmetrized Sinai billiard (inset, shaded), from [165]. (Right) Poisson and GOE Δ_3 measure (L is the number of levels in a rescaled spectrum) with the Sinai billiard data shown, from [165].

statistic,

$$\Delta_3 = \{\substack{\min \\ A,\, B}\} \frac{1}{\Delta E} \int_E^{E+\Delta_E} [N(E) - AE - B]^2 \, dE, \tag{21.34}$$

which depends on the "staircase function" $N(E)$, as it measures the deviation of the actual eigenlevels in an interval ΔE, given by the staircase function $N(E)$, against a rising straight line. $N(E)$ is a staircase jumping up by one unit at each new eigenvalue. If the levels are fairly evenly spaced, the staircase will deviate less from a straight line. The deviation Δ_3 grows as the interval E enlarges, but in very different ways for a Poisson and a GOE spectrum.

DYNAMICAL CORRECTIONS TO RANDOM MATRIX THEORY

The short time autocorrelation function may reveal a structured or peaked local density of states envelope $\Sigma(E)$. The peaked density implies that the dynamics is nonergodic. Ergodicity produces an unstructured, smoothed version of a uniform χ^2 distribution. The definition of ergodic flow can be adjusted to account for prior knowledge of the existence of the peaked envelope, so that the p_n^a are a χ^2 distribution of one degree of freedom with an average *local* strength $\Sigma(E)$. It is clear that such corrections for prior constraints can involve even finer detail gleaned from dynamics at longer times. The last two panels in figure 21.6 show one case meeting the ergodicity requirement including the constraint envelope up to time T_5. The lower right-hand case shows a spectrum having the same T_5 envelope but failing to be ergodic with respect to it.

Accounting for the spectral envelope $\Sigma(E)$ is a key concept that enriches the notion of quantum ergodicity, and yet retains a modified, weighted form of random matrix theory at its foundation.

QUANTUM MEMORY

Individual probabilities p_n^a for different n will fluctuate one to the next—for example, as a χ^2 distribution if the system is ergodic—while maintaining the correct average smoothed spectrum, or "local density of states." It is very interesting that formula 21.23 predicts a rather large memory effect, purely due to quantum interference, for the probability that the system has returned to the initial state $|a\rangle$, averaged over long times:

$$\boxed{P(a|a) = \sum_{n=1}^{\infty} (p_n^a)^2.} \tag{21.35}$$

This is a purely coherent measure of average return to the initial state, involving all degrees of freedom. If some degrees of freedom return often but others do not, then the whole does not return often. One can choose to label this as decoherence, formalizing it by tracing over the degrees of freedom that have not returned. If the p_n^a fluctuate as a χ^2 distribution of one degree of freedom, $P(a|a)$ is expected to be three times larger than the typical $P(a|b)$. The reason is that every (by chance) large p_n^a gets squared in the sum, whereas $P(a|b)$ has no such systematic enhancement for a typical $|b\rangle$. (However $|b\rangle = |a(t)\rangle$ is not typical!) If the amplitudes had been complex, the p_n^a are χ^2 of two degrees of freedom (one picks a squared number twice, making very small results much less likely). The predicted $P(a|a)$ is now only twice as big as the average $P(a|b)$. These are large deviations from classical ergodicity (with a complete memory loss about the initial state of the system)—or perhaps not, depending on your point of view. The state $|a\rangle$ is one of the N states but it is not the only one exhibiting a significant anomaly: any state $|a(t)\rangle$ has the same anomaly, although if t is very large, the state $|a(t)\rangle$ may itself be distributed widely in phase space and be a poor test state for ergodicity.

$P(a|a)$ has aptly been called a dilution factor [167, 168] and correctly interpreted as the inverse of the number of states accessed in the course of the dynamics, to be compared with the total number of states available. There are two caveats: first, there is the quantum memory factor of 3 enhancement of $P(a|a)$ versus the typical $P(a|b)$ even in a quantum ergodic system, so using

$$P(a|a) \sim \frac{1}{\mathcal{N}_\infty}, \tag{21.36}$$

where $\mathcal{N}_\infty$ is the number of states, underestimates that number by a factor of 3. Second, some implicit choice has been made of the prior constraints, by using the experimental band envelopes such as the one seen below for propynol. The narrowness of those bands is usually taken for granted, but some thought should be given to why they are so narrow and what kinds of motion have been excluded from the flow because the band envelopes are so restricted.

Stewart and MacDonald [168] found that the number of states present in high-resolution spectra increases almost linearly with the density of states, for intramolecular vibrational relaxation from fundamental C-H stretches in different molecules. This means that the degree of ergodicity remains about the same through a series of similar situations in different molecules.

Perhaps the molecules were nearly ergodic given the spectrum, the state density, and the prior constraint of the narrow absorption bands, but it would have required accurate knowledge of the full density of states to demonstrate this. Pate and

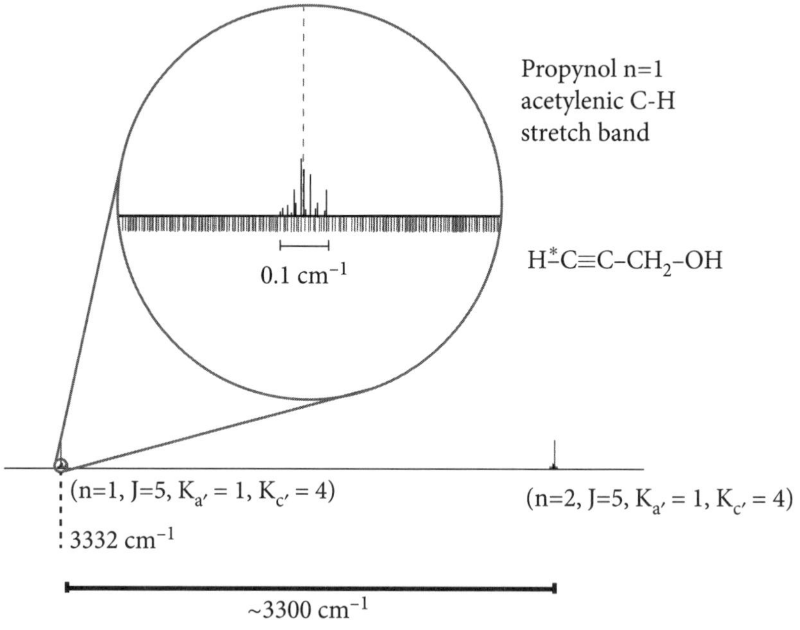

Figure 21.11. Interpretation of the propynol spectroscopy of Pate et al. [169] with inclusion of some of their data. By double resonance and astonishing high-resolution spectroscopy, the authors were able to display the first C-H stretch band at fixed rotational quantum numbers $J = 5$, $K_a = 1$, $K_c = 4$. The whole band is incredibly narrow at about 0.1 cm^{-1}, revealing a 400 ps decay time of the C-H stretch or 6300 C-H vibrational periods. The total available density of states for the spectrum shown here is about 200 cm^{-1}.

co-workers [169], to be discussed next, were able to accurately measure the high-resolution spectrum and estimate the density of states, showing the system was essentially ergodic within the constraint of narrow spectral bands. This kind of behavior was successfully modeled and rationalized by Bigwood, Gruebele, Leitner, and Wolynes [167] and Gruebele and Wolynes [170] and related papers.

It would be good to understand the nature of the coupling causing the ergodic-like behavior. Dynamical tunneling between states with otherwise good quantum numbers might be responsible. The tunneling mechanism is consistent with the work just mentioned, and does give a more specific reason for the couplings. The discussion of Pate's propynol results, next, makes clear why this is the suspicion.

21.7 Propynol

A relevant example illustrates the electronic absorption spectroscopy of medium-size molecules and moderate energies. A wonderful case, worked out in the group of Brooks Pate at the University of Virginia, involves fundamental C-H acetylenic stretch vibration in propynol (propargyl alcohol, H−C≡C−CH$_2$−OH) [169]. Figure 21.11 shows the results. The spectrum, totally cleaned up of thermal and rotational contamination by double resonance, consists of incredibly narrow bands. Nonetheless, these are resolved into a couple dozen or more molecular eigenstates. The density of populated states checks with estimates of the total density of states of the proper symmetry.

If the initial quantum of C-H stretch vibration had been a stationary state, only a single line (tall dashed line drawn on actual data) would have been seen for each vibration, for fixed rotational quantum numbers. The spectrum shows instead that such pure C-H character decays due to "anharmonic" (see the following) interactions on a 400 ps timescale, fractionating into about 16 states, *which is very close to all the possible states available in the approximately 0.1 cm^{-1} bandwidth.* The total density of available states is about 200 cm^{-1}.

To give this some perspective, it takes some 6300 C-H vibrations to dump half the energy out of the initially excited C-H stretch. This leads to the great deserts lacking any spectral intensity surrounding the narrow 0.1 cm^{-1} envelope harboring all the spectral density, representing a huge failure of ergodicity if one did not know about the tight constraint of the slow initial escape. If one accepts the envelope as a prior constraint, then the system is very close to ergodically accessing all the possible states within it!

It not necessary to know what actually caused the C-H stretch to couple to every available state under its narrow envelope and finally decay in order to assess ergodicity. However it is strongly suggested the decay is caused by dynamical tunneling—that is, *the flow would not have occurred classically on the same potential.* The 6300 vibration periods are fairly convincing evidence that classical evolution on the same Born-Oppenheimer potential energy surface would simply not have decayed at all. The quantum decay would be found to be classically dynamically forbidden [93, 96, 106, 171, 172]. Motion in regions not connected by classical time evolution can quantize to nearly degenerate energies and then by "diffractive" or classically forbidden tails of wavefunctions can communicate and eventually tunnel into one another. See sections 13.1, 13.2, and 13.3. Taking 400 ps to half empty the C-H stretch is a serious failure of classically driven IVR (intramolecular vibrational redistribution). Benzophenone's slowly twisting benzene rings will have passed energy from symmetric twisting to asymmetric twisting and back again about 200 times in that 400 ps (see section 19.3). A time of 400 ps sounds like the time required to tunnel, not to pass energy around in a small molecule classically.

It is interesting to consider the conclusions in reference [169] regarding the eventual fate of the C-H stretch quantum of energy:

> Comparison of the calculated and measured values [of the state density] suggests that all energetically accessible states participate in the intramolecular dynamics. It is interesting to note that 87% of the bath states in this energy region involve excitation of the -OH rotor. Although the energy redistribution rate may be controlled by interactions localized in the acetylenic region of the molecule, the energy eventually gets redistributed into vibrational states with high torsional excitation.

IMPURE DENSITY MATRICES

We have been neglecting density matrices $\rho_a, \rho_b, \ldots$ that are impure ($\mathrm{Tr}[\rho_a]^2 < 1$). For example, ρ_a could be an incoherent sum over n nearby but almost nonoverlapping pure Gaussian density matrices $\rho_a = \sum_\alpha |\alpha\rangle\langle\alpha|$. This collection can become an infinitesimal zone in phase space as $\hbar \to 0$, and similarly for ρ_b, and so on. For large n, $P(a|a)$ loses its coherent interference and memory properties. If all density matrices are constructed similarly, with ρ_a and ρ_b having similar energy envelopes, then every $P(a|b)$ including $P(a|a)$ is nearly the same in a quantum ergodic system, likely reflecting the underlying classical ergodicity, now stripped of interference effects.

ERGODICITY OF MAPS

Systems living in finite-dimensional Hilbert spaces simplify some of the complications of unbounded systems and help make some statements about ergodicity more precise. We take a quantum system with N orthogonal basis states in the Hilbert space, $|m\rangle$, $(m, 1, N)$. An $N \times N$ unitary matrix $\hat{U}$ defines the dynamics; after k time steps or iterations of a vector $|m\rangle$, we have

$$|m(k)\rangle = \hat{U}^k|m\rangle. \tag{21.37}$$

The autocorrelation of the initial state $|m\rangle$ is

$$\langle m|m(k)\rangle = \langle m|\hat{U}^k|m\rangle, \tag{21.38}$$

with cross-correlations

$$\langle m'|m(k)\rangle = \langle m'|\hat{U}^k|m\rangle. \tag{21.39}$$

Presuming eigenstates of $\hat{U}$ are

$$\hat{U}|j\rangle = e^{iE_j}|j\rangle, \tag{21.40}$$

it is simple to show that the time average of the flow (long time average over iterations) from $|m\rangle$ to $|m'\rangle$ under the evolution of the unitary operator $\hat{U}^k$ is (assuming no degeneracies)

$$P(m|m') \equiv \lim_{L\to\infty} \frac{1}{L} \sum_{k=1}^{L} |\langle m'|m(k)\rangle|^2 = \sum_{j=1}^{N} p_j^m \cdot p_j^{m'}, \tag{21.41}$$

where $p_j^m = |\langle j|m\rangle|^2$ and $\sum_{j=1}^N p_j^m = 1$; $\sum_{m=1}^N p_j^m = 1$—that is, p_j^m is the overlap probability of the mth basis state with the jth eigenstate. The p_j^m form the "spectrum" of the mth basis state, as in

$$\sigma_m(\epsilon) = \sum_j p_j^m \delta(\epsilon - E_j). \tag{21.42}$$

Again as with unbounded systems, the time-averaged flow from nonstationary state $|m\rangle$ into another, $|m'\rangle$, is given by the inner product of their spectra, equation 21.41. Figure 21.12 demonstrates several important characteristics of this "ergodicity calculus," highlighting issues arising in quantum ergodicity. For reference, in the upper-left panel, the positions of all N eigenvalues are shown as short black tick marks. Spectra can have population at these places, but depending on the underlying dynamics and phase space exploration, very different populations can occur.

Twelve spectra are shown in figure 21.12 reflecting different underlying dynamics induced by the unitary operator $\hat{U}$. Half of the spectra are shown inverted for comparison and clarity. In each case the inverted spectra are generated by the same unitary transformation as the spectrum immediately above them; the difference is in the initial nonstationary state m. The top row of pairs shows typical quantum manifestations of classically integrable dynamics. Close inspection of the energy

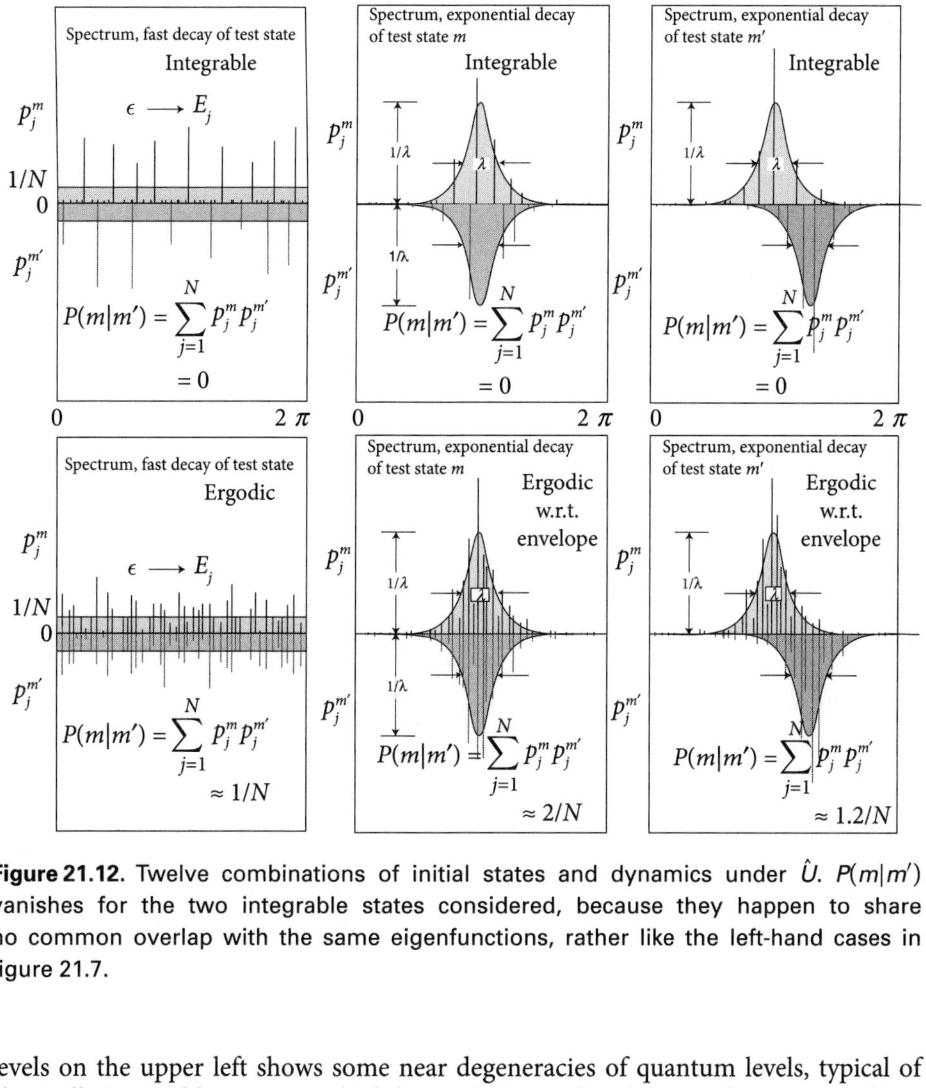

Figure 21.12. Twelve combinations of initial states and dynamics under $\hat{U}$. $P(m|m')$ vanishes for the two integrable states considered, because they happen to share no common overlap with the same eigenfunctions, rather like the left-hand cases in figure 21.7.

levels on the upper left shows some near degeneracies of quantum levels, typical of classically integrable systems, which have a Poisson distribution of nearest neighbor level spacings. The bottom three pairs reveal a much stronger mixing tendency, and level repulsion as well, as seen by the rarity of near degeneracies. In all twelve cases the number N of eigenstates and eigenvalues is the same. A uniform energy envelope characterizes the case on the left, and a Lorentzian envelope is prominent for the middle and right cases. Ideally, ergodic motion (that is, unattainable due to statistical fluctuations) would imply $p_j^m = 1/N$ for all m and j, and $P(m|m) = \sum_j (1/N)^2 = 1/N$. For real valued eigenfunctions, and given fluctuations that are consistent with random matrix theory, the result for the most ergodic possible motion is $\langle P(m|m)\rangle = 3/N$, and $\langle P(m|m')\rangle = 1/N$. The latter, cross-correlation result is possible because the statistical fluctuations p_j^m and $p_j^{m'}$ are uncorrelated. Because p_j^m is obviously correlated with itself, large fluctuations are squared, leading to the result $P(m|m) = 3/N$.

Both cases on the left have spectra that span some (in the integrable case at the top) or nearly all (for the ergodic case below) of the states without any constraint envelope at work in the full energy range from lowest (0) to highest (2π). In these cases $|m\rangle$

decays to a typical long time value of the autocorrelation in just one time step, which we can show as follows:

$$\langle m|m(t=1)\rangle = \langle m|e^{-i\hat{U}}|m\rangle = \sum_{k=1}^{N} p_k^m \, e^{-iE_k},$$
$$|\langle m|m(t=1)\rangle|^2 = |\langle m|U|m\rangle|^2 = \sum_{k,k'=1}^{N} p_k^m p_{k'}^m \, e^{-i(E_k-E_{k'})}. \tag{21.43}$$

This is a set of positive probabilities added with uncorrelated phases spanning $(0, 2\pi)$. Qualitatively, the same distribution of probabilities with random phases modulo 2π also occurs later:

$$\langle m|m(t=n)\rangle = \langle m|e^{-in\hat{U}}|m\rangle = \sum_{k=1}^{N} p_k^m \, e^{-inE_k/\hbar}. \tag{21.44}$$

That is, the systems typically decay to their long time behavior in one time step. The integrable system decays to $|\langle m|m(t=n)\rangle|^2 \sim 2/K$, and the random ergodic system would decay to typically $|\langle m|m(t=n)\rangle|^2 \sim 3/N$ in one step. $3/N$ is also the long time average for GOE random matrices. The factor of 3 arises from squaring the p_n^a's when forming $P(a|a)$, enhancing the fluctuations (if all p_n^a were $1/N$—that is, no fluctuations, then $P(a|a) = 1/N$.) Clearly, recurrences happen sooner and more often for the integrable system to maintain the higher average self-overlap $P(m|m)$.

N iterations are required to reach the Heisenberg time, $T_H \sim N$. After that time, essentially all the exploration than can occur will have occurred, and neighboring eigenstates are dephased and resolvable, one from the other, by Fourier analysis. How much exploration will have occurred in the two cases? $|\langle m|m(t)\rangle|^2$ hovers at $3/N$ in the ergodic case, and $2/K$ in the integrable one. Thus, their averages will be $3/N$ and $2/K$, respectively—that is, the ergodic dynamics will sample all N states and the integrable dynamics only $K < N$ states (because the cross-correlation $|\langle m'|m(t)\rangle|^2$ hovers at $1/K$ and $1/N$, respectively).

We now turn our attention to the cases that do not decay immediately, with spectra as seen in the middle and right-hand panels in figure 21.12. The early time evolution goes as

$$|\langle m|m(t=n)\rangle|^2 \sim e^{-\lambda n}. \tag{21.45}$$

The time it takes to decay causes a Lorentzian band shape to develop in the spectrum, with a width determined by the decay lifetime. From the perspective of an egalitarian spread of the spectrum over the whole range $(0, 2\pi)$, the Lorentzian lump is a failure of ergodicity. Yet, if we incorporate this information and adopt an updated prior constraint envelope $S_\phi(E)$, accounting for the Lorentzian band (perhaps assigning it to some trivial short time dynamics), then the new question becomes: is the system ergodic with respect to the revised envelope? And, now that there are envelopes, what about the flow between two states with somewhat different envelopes $S_a(E)$ and $S_b(E)$? Clearly, if there is not much energy overlap of two envelopes, there cannot be much flow between the corresponding states. This is not a failure of ergodicity if we have incorporated the envelopes as prior constraints. Under the original "colorless" constraints, the lack of overlap of states with very different envelopes would register

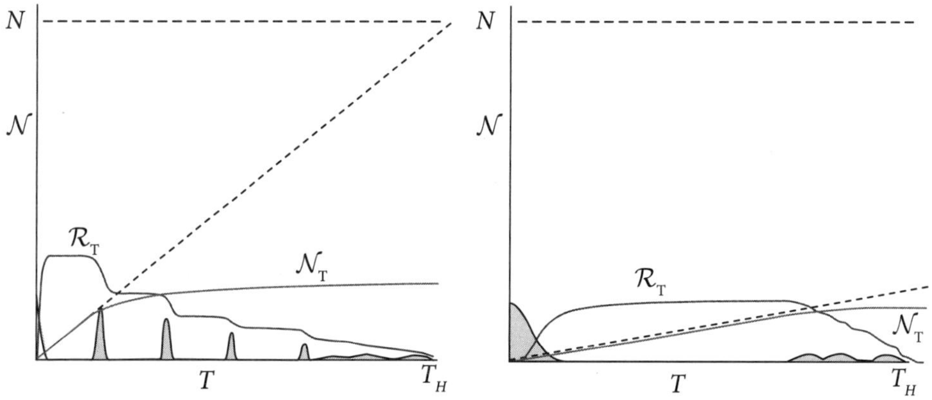

Figure 21.13. The rate of exploration of phase space and how it is affected by the initial decay rate and subsequent recurrences is depicted here, in two cases, supposing the same number of states N are theoretically available in both cases. The ideal unconstrained ergodic limit of accessing all N available states (available according to unconstrained random matrix estimates) is not reached in either case. On the left, early and several recurrences slow down $\mathcal{R}_T$ and thus reduce $\mathcal{N}$. On the right, recurrences do not happen until near the Heisenberg time T_H; however, the "0th" recurrence—that is, a slow initial decay—has set the rate of exploration low, making it impossible to approach N states explored by the time T_H. The slow decay implies a narrowed spectral envelope.

as a nonergodicity. If we incorporate the envelopes into the prior constraints, we can compensate for partial overlaps and say what ergodic flow would be between them.

21.8 Rate of Exploration of Phase Space

We have seen that early recurrences in the autocorrelation function $\langle a|a(t)\rangle$ cause the spectrum to develop envelope structures constraining the p_n^a. These constraints cause a failure of complete ergodicity compared to expectations based without knowledge of the envelope. These failures are real, in that not as much phase space will be explored due to the early recurrences. As stated earlier, we can incorporate these constraints, recalibrating our estimate of ergodicity to see if the system might be ergodic with them in place.

That an unexpected recurrence or an unexpectedly slow decay can cause a failure of ergodicity (see figure 21.13) is readily understood with two principles. First, there is a finite time to explore. Time is essentially limited to the Heisenberg time, when eigenstates have all been dephased with respect to each other, even with their neighbors with nearby eigenvalues—that is, the eigenstates could be resolved one from another by Fourier analysis. Suppose there are N states total, as in our map example. There are also, by unitary transformation, N localized phase space cells, one of which might be the initial state $|a\rangle$. It is readily seen that the Heisenberg time is N iterations of the map. But this means that the system must explore a *new* phase space cell each iteration, until N iterations have been reached. Second, a strong early recurrence is devastating: the recurrence means $|a\rangle$ has been at least partially repopulated, and that repopulated fraction is destined to explore in future steps what $|a\rangle$ has already explored—that is, old territory. Since the evolving system explores at a fixed total rate, a recurrence and the implied exploration of old territory will slow down the rate of exploration of new phase space. The same can be said for failure to completely leave the initial cell in one step—that is, failure to decorrelate after one step, a factor already mentioned earlier. In fact, slow decay of the initial state is essentially a recurrence already at the first step.

For a system with a large number of states, something else might rightly define a unit of time after which the system should have decayed if it is to be ergodic in the original sense without knowledge of the recurrence or slow decay. For the C-H vibration in propynol, section 21.7, the natural timescale is one vibrational period of the C-H stretch. If the initial state decorrelated in that time, the vibrational states 3300 cm^{-1} apart would be broadened into obscurity. Instead, the C-H hangs on for 6600 vibrations, leaving 0.01 cm^{-1} envelopes separated by deserted gaps of 3300 cm^{-1}.

QUANTIFYING THE RATE OF EXPLORATION AND ITS EFFECTS

At what rate is phase space explored by a moving density $\rho(t)$, given a nonstationary initial density $\rho(0)$? Nonstationary distributions $\rho(t)$ will be shown below to sweep out phase space at a rate $\mathcal{R}$ that can be determined from the autocorrelation function $P(t) = h^N \, Tr[\rho(0)\rho(t)]$. It is demonstrated that, after the initial decay of $P(t)$, subsequent recurrences (increases) of $P(t)$ cause permanent slowdowns in the rate $\mathcal{R}$. Such slowdowns cause localization of quantum eigenstates, whereas although the same slowdown of exploration of new phase space happens classically, no nonergodic implications follow, since the classical system has an infinitely long time available to it.

The concept of the rate of exploration of phase space, and the allied concepts of the number of phase space cells accessed in the infinite time limit and the fraction $\mathcal{F}$ of available phase space explored, are useful extensions to the usual participation ratio ideas. The latter focus on the "participation" of individual eigenstates in a set of localized states, calculated as $\mathrm{IPR}_a = 1/\sum_m (p_m^a)^2$, whereas the former emphasizes the eigenstate content of particular localized states. The two points of view are intimately related but by no means identical.

Consider a normalized phase space distribution ρ, that is,

$$\mathrm{Tr}[\rho] \equiv \int \rho(\boldsymbol{p}, \boldsymbol{q}) d\boldsymbol{p}\, d\boldsymbol{q} = 1. \tag{21.46}$$

The density ρ must have the dimensions of $(\mathrm{action})^{-N}$ for N degrees of freedom. Therefore, $\mathrm{Tr}[\rho^2]$ has the dimension of $(\mathrm{action})^{-N}$, or in other words an inverse of a volume in phase space.

The density $\rho(\boldsymbol{p}, \boldsymbol{q})$ could be a Wigner transform of a pure state

$$\rho(\boldsymbol{p}, \boldsymbol{q}) = \left(\frac{1}{\pi\hbar}\right)^N \int_{-\infty}^{\infty} e^{2i\,\boldsymbol{p}\cdot\boldsymbol{s}/\hbar} \varphi^*(\boldsymbol{q}+\boldsymbol{s})\varphi(\boldsymbol{q}-\boldsymbol{s})\, d\boldsymbol{s}). \tag{21.47}$$

More generally, ρ may derive from a quantum density ρ_Q:

$$\rho(\boldsymbol{p}, \boldsymbol{q}) = \left(\frac{1}{\pi\hbar}\right)^N \int_{-\infty}^{\infty} e^{2i\,\boldsymbol{p}\cdot\boldsymbol{s}/\hbar} \rho_Q(\boldsymbol{q}-\boldsymbol{s}, \boldsymbol{q}+\boldsymbol{s})\, d\boldsymbol{s}. \tag{21.48}$$

We can define the volume that a density ρ with $Tr(\rho) = 1$ occupies in phase space as follows:

$$\text{Volume} = \frac{1}{\text{Tr}[\rho^2]}. \tag{21.49}$$

For a pure state density,

$$\int \rho^2(\boldsymbol{p}, \boldsymbol{q})\, d\boldsymbol{p}d\boldsymbol{q} = h^{-N} \equiv \text{Tr}[\rho^2], \tag{21.50}$$

and the volume occupied is one cell of volume h^{-N}, as it should be. If $\rho = \rho_A/2 + \rho_B/2$ with ρ_A and ρ_B both normalized to 1 but not overlapping, then ρ is also normalized, but

$$\text{Tr}[\rho^2] = \text{Tr}[\rho_A^2] + \text{Tr}[\rho_B^2] = 1/2h^N, \tag{21.51}$$

that is, two phase space cells are occupied.

Suppose now that $\rho(t)$ is a nonstationary distribution evolving under the influence of the Hamiltonian H. How rapidly does $\rho(t)$ sweep out new regions of phase space that it has not visited before? To answer this question, we need to assess where $\rho(t)$ has visited. A natural indicator is the average of where it has been—that is,

$$\rho^{av}(\boldsymbol{p}, \boldsymbol{q}, T) = \frac{1}{T}\int_0^T \rho(\boldsymbol{p}, \boldsymbol{q}, t)dt. \tag{21.52}$$

This average density will define the measure of flow $\mathcal{N}_T$:

$$h^N \text{Tr}[(\rho^{av})^2] \equiv \frac{1}{\mathcal{N}_T}. \tag{21.53}$$

This defines $\mathcal{N}_T$, the number of phase space cells accessed. The density $\rho^{av}(\boldsymbol{p}, \boldsymbol{q}, T)$ clearly is nonvanishing only where $\rho(\boldsymbol{p}, \boldsymbol{q}, t)$ has visited, but both distributions "feather out" rather than cut off abruptly in phase space. That is to say, there is no sharp distinction between a region that has been visited by $\rho(\boldsymbol{p}, \boldsymbol{q}, t)$ and one that has not, but the number of phase space cells visited still has meaning.

Combining the preceding equations and using the fact that $\text{Tr}[\rho(t)\rho(t')] = \text{Tr}[\rho(0)\rho(t'-t)]$, we have

$$\frac{1}{\mathcal{N}_T} = h^N \text{Tr}[(\rho^{av})^2] = \frac{2}{T}\int_0^T (1 - \frac{\tau}{T}) P(\tau)\, d\tau, \tag{21.54}$$

where the survival probability $P(\tau)$ is $P(\tau) = h^N \text{Tr}[\rho(0)\rho(\tau)]$).

21.9 The Rate of Exploration $\mathcal{R}$

QUALITATIVE FEATURES

Typically, there are three regimes in the history of $\mathcal{N}_T$, for a bounded system with an initially localized density ρ: (1) the initial transient, including a rapid rise of $\mathcal{R} = d\mathcal{N}_T/dt$ followed possibly by some decline, and leading to (2) a plateau in

$\mathcal{R}$—that is, a linear increase in $\mathcal{N}_T$, the number of phase space cells accessed—and (3) the onset of the decline of the rate $\mathcal{R}$, which may occur in steps caused by recurrences in $P(t)$. The rate may temporarily become negative (see the following), but it will approach zero and remain there asymptotically (refer to figure 21.13, earlier).

These three regimes are simple to understand if one imagines the nature of an initially localized distribution ρ and its subsequent classical dynamics. The quantal case will differ from the classical, especially in regime 3, but the distinct regimes will be present in both. In figure 21.13, we see two cases of initially localized distributions that start to explore phase space, thus $P(t)$ declines and $\mathcal{N}_T$ increases. In the case on the left, the initial region is visited by part of the moving distribution in a decaying-periodic way, which is a common behavior. On the right, the system revisits the initial region only after a long sojourn, and when it does start to arrive back it is quite disorganized, leading to the $P(T)$ shown.

The rate $\mathcal{R}$ is easily shown, by differentiation of equation 21.54, to be

$$\mathcal{R}(T) = \frac{d\mathcal{N}_T}{dT} = \frac{\frac{1}{2}\int_0^T (1-\frac{2\tau}{T})P(\tau)d\tau}{\left(\int_0^T (1-\frac{\tau}{T})P(\tau)\,d\tau\right)^2}. \tag{21.55}$$

The rate $\mathcal{R}$ is not guaranteed to be strictly positive. Recurrences or "lapping" of previously explored regions of phase space leads to an uneven distribution ρ^{av}, which temporarily causes $\mathcal{N}_T$ to diminish.

From equation 21.55 it is seen that the steady state rate (for large enough T and before any recurrences) approaches

$$\mathcal{R} = \frac{1}{2\int_0^T P(\tau)\,d\tau} = \frac{1}{\int_{-T}^T P(\tau)\,d\tau}. \tag{21.56}$$

RATE $\mathcal{R}$ AND THE ENERGY DISTRIBUTION OF DENSITY ρ

The initial decay of $P(t)$ sets the initial rate of exploration of phase space. We see this trend from equation 21.56, supposing no recurrence up to time T beyond the initial decay. For $P(t) = \exp[-\lambda t]$, we have $\mathcal{R} = \lambda/2$. For $P(t) = \exp[-t^2\sigma^2]$, we have $\mathcal{R} = \sigma/\sqrt{\pi}]$.

We can establish a link between $\int_{-T}^{T} P(t)\,dt$ and the envelope function $S_T(E)$, which carries the information about the energy distribution. The key ingredient is the equation

$$\begin{aligned}\langle\varphi|\varphi(t)\rangle &= \int_{-\infty}^{\infty} e^{-iE't/\hbar}S(E')\,dE', \\ &= \int_{-\infty}^{\infty} e^{-iE't/\hbar}S_T(E')\,dE',\end{aligned} \tag{21.57}$$

provided $[-T \leq t \leq T]$. This follows directly from the definitions and the properties of Fourier transforms. From this and the equation

$$\int_{-\infty}^{\infty} \Omega_T(E - E')\Omega_T(E' - E'')\, dE' = \Omega_T(E - E'') \tag{21.58}$$

follows

$$\int_{-T}^{T} P(t)\, dt = 2\pi\hbar \int_{-\infty}^{\infty} S_T(E)^2\, dE, \tag{21.59}$$

and thus the rate as defined in equation 21.56 is

$$\mathcal{R} = \frac{1}{2\pi\hbar \int_{-\infty}^{\infty} S_T(E)^2\, dE}. \tag{21.60}$$

Chapter 22

Quantum Scars of Classical Periodic Orbits

The trace formula revealing the energy eigenvalues is very egalitarian: it seems every periodic orbit contributes to every eigenvalue. Although orbits clearly do not contribute exactly equally, since phase interference and the stability of orbits both play a role, nothing jumps out as an obvious role for a particular periodic orbit for any given eigenvalue. Although the eigenfunctions (as opposed to the eigenvalues) were under intense scrutiny, no one had surmised from the trace formula that any one periodic orbit cold play an outsized role compared to the infinity of other periodic orbits. Amplifying this point, the eigenstates are obviously not the result of a trace over all positions, yet it was the trace that reduced the eigenvalues to the contribution of periodic orbits only. This makes it even more unlikely that a particular periodic orbit would have an outsized contribution to a particular eigenstate. Each classical periodic orbit in a fully chaotic system is of measure 0 among all typical orbits. To add to the unlikelihood of a given periodic orbit having a starring role, we note there is an uncountable infinity of them, and they all appear in the trace formula at every energy. To seemingly further demote any role of the periodic orbits in forming the eigenstates, all together they occupy 0% of phase space, much like the rationals occupy 0% of the real number line.

Numerical studies at Los Alamos National Laboratory, which had the best graphics capability in the world at the time, proved these expectations wrong. They revealed surprisingly large footprints of individual periodic orbits affecting some eigenstates of the Bunimovich stadium billiard problem, a proven classically chaotic system. The discovery was published with a proof of their existence and a theory for their strength [37]. They look a bit like scars on the wavefunctions, thus the name. Semiclassical wavepackets were key in proving the existence of scars, and are the simplest and most quantitative route to understanding them.

The time average of a typical (that is, nonperiodic) classical orbit is uniform on the energy hypersurface—it fills out the volume $\delta(E - H)$. Periodic orbits are of measure 0 among all the orbits, but scarred states are of finite measure, for finite $\hbar$. Thus, there is no analog of scars in classical mechanics. Some analogous structures appear even in random superpositions of plane waves (see section 23.1), so we must proceed carefully so as not to confuse a statistical fluctuation with scarring. First, we need a definition of the scar phenomenon [37, 173, 174].

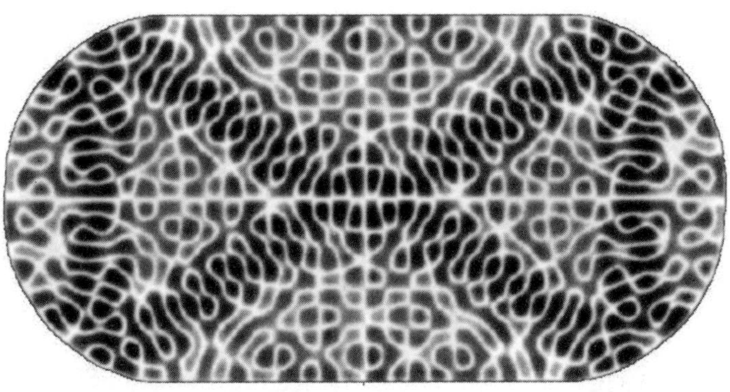

Figure 22.1. A quantum eigenstate $|\psi|^2$ of the stadium billiard is shown, scarred by the "bowtie" shaped classical periodic orbit. Larger values of $|\psi|^2$ are shaded darker.

> ***Definition***: A quantum eigenstate of a classically chaotic system is *scarred* by a periodic orbit if its density on the classical invariant manifolds near and all along that periodic orbit is systematically enhanced above the classical, statistically expected density along that orbit.

The statistical comparison is with a fixed energy random wave on the potential, in the sense of the Berry conjecture, stated in section 23.1. In figure 22.1, we see a scarred state along the unstable "bowtie" periodic orbit of the stadium-shaped billiard. The stadium is a well-known example of "hard" classical chaos (no quasiperiodic zones or stable trajectories anywhere). Typical trajectories are wildly chaotic in this system (see figure 3.17). Higher energy examples are seen in figure 22.2.

We will see that scars are associated also with the stable and unstable manifolds of the associated orbits, not just the central orbit; this makes it more plausible that individual orbits could have such an effect—it's not just them, it's their vicintiy.

Reference [37] showed that the strength S, measuring the enhancement of the squared amplitude or projection of scarred eigenstates onto a wavepacket located on the orbit, is a function only of the Lyapunov exponent for one period of the periodic orbit, λ, specifically $S \propto \omega/\lambda$ for small λ, where ω is the ω is the classical angular frequency of the orbit. We give the argument shortly. The scar strength is independent of $\hbar$, yet it is often said that scars vanish as $\hbar \to 0$. This is not quite true, although scars will become more localized as $\sqrt{\hbar}$ around their parent periodic orbits and in that sense they occupy less "real estate" as $\hbar \to 0$. The strength S, once again, does not diminish as $\hbar \to 0$. Also, long orbits have larger λ and smaller ω and thus a smaller S and tend not to scar the eigenstates so visibly.

When a wavepacket is launched inside the stadium billiard (a classically chaotic system), for example, the first of a hierarchy of spectral envelope shapes is almost immediately determined, as the overlap $\langle\phi(0)|\phi(t)\rangle$ decays due to the free flight of the wavepacket, before it hits any walls. Consider

$$S_T(E) \equiv \frac{1}{2\pi\hbar}\int_{-T}^{T} e^{iEt'/\hbar}\langle\phi(0)|\phi(t')\rangle\, dt', \tag{22.1}$$

that is, the Fourier transform of the autocorrelation function of the wavepacket up to time T (see section 5.3 and chapter 17). If that time is long enough for the wavepacket

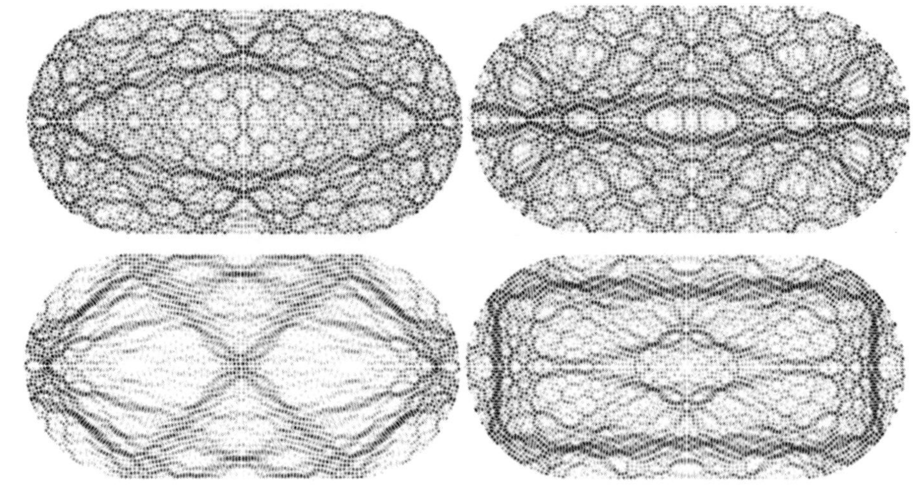

Figure 22.2. Four higher energy (than in figure 22.1) examples of eigenstates of the stadium, three of which are scarred. The fourth, at the upper left, is not scarred in a typical way, since the eye-shaped band of higher intensity is not linear, and therefore not representative of an unstable orbit in the stadium. Many-bounce classical trajectories however have segments that together make a very similar pattern, skirting the football shaped region in the center, the curved shape resulting from a bundle of many straight ray paths traveling tangent to the region. This kind of pattern in the wavefunctions and its detailed connection to classical trajectories is not very well understood.

to have left its original location, but too short for it to have returned—that is, $t < \tau$, the first and broadest feature of the spectrum will have been "burned in." No subsequent dynamics can cause the spectrum to depart from this averaged shape and location, even as the spectrum sharpens with increasing T. This early time is devoid of interesting dynamics, and the decay of the autocorrelation leads an envelope of constraint on the spectrum (figure 22.3, lower-right dashed). The envelope contains the information about the average energy of the wavepacket and its dispersion, something determined by the choice of wavepacket rather than anything interesting about the dynamics.

The spectrum's finer features built on the short time envelope of support reveal aspects of longer time dynamics. After something very close to one classical period τ, a wavepacket $|\phi(t)\rangle$ launched on the periodic orbit returns to its initial phase space launch point (the same vicinity with the same momentum), albeit spread by an amount given by the Lyapunov exponent of the orbit, and in directions controlled by the stable and unstable manifolds associated with the orbits. The associated recurrences or blips in the autocorrelation $\langle\phi(0)|\phi(t)\rangle$ at time $\tau, 2\tau, \ldots$ impose more constraints on the spectrum of the wavepacket, which at highest resolution is $\Sigma(E) = \sum_n |\langle\phi(0)|E_n\rangle|^2 \delta(E - E_n) = \sum_n p_n^\phi \; \delta(E - E_n)$, where the $|E_n\rangle$ are the eigenstates. The consequences of the recurrences and spreading governed by the Lyapunov exponent are reflected in the smooth "fingers" developed in the dashed envelope of the spectrum, see figure 22.3, where a section of the high-resolution peaks—that is, the p_n^ϕ's and E_n's—is also shown.

The magnitude of the recurrence at time τ is a function of the stability matrix and the initial form of the wavepacket $|\phi(0)\rangle$. We learned in equation 19.9 that

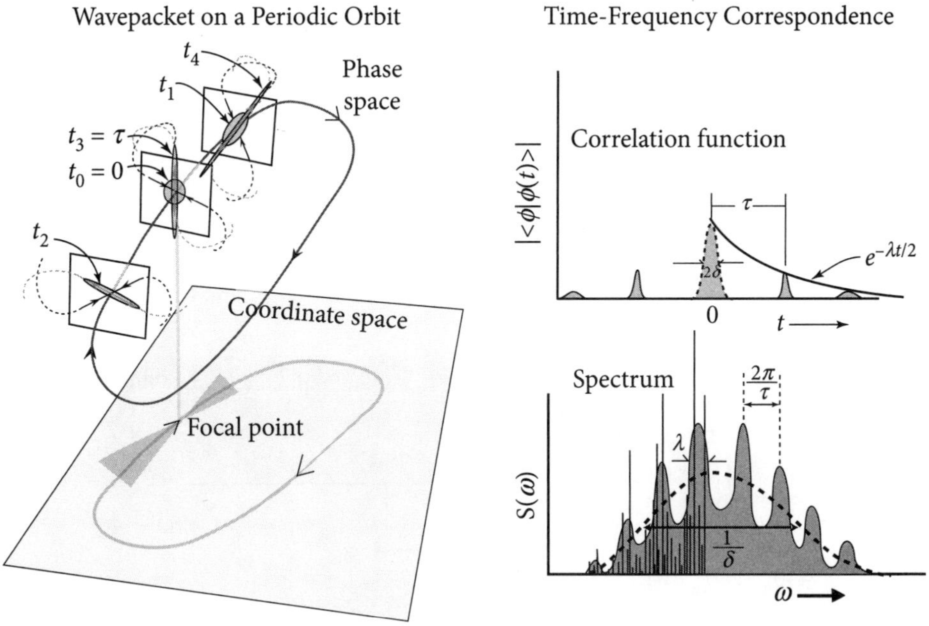

Figure 22.3. (Left) The stable and unstable manifolds attached to an unstable periodic orbit as they affect the dynamics of a wavepacket in phase space, initially the wider oval at $t_0 = 0$. Its early dynamics is shown as the slightly stretched and rotated oval. When it returns at one period $t_3 = \tau$ to its "birthplace," it is significantly stretched along the unstable manifold, and compressed along the stable one. See also figures 22.4 and 22.5, later. (Right) Correlation function and spectrum of the initial wave packet.

$\langle \varphi_u | \varphi_u(t) \rangle = [\cosh(\lambda t)]^{-1/2} \approx e^{-\lambda t/2}$, where λ is the classical Lyapunov exponent. The nth recurrence at time $n\tau$ decays as $\exp[-n\tau\lambda/2]$. These recurrences cause oscillations of width λ in the spectral envelope, as seen in figure 22.3, right. The subbands are spaced (locally) by $\hbar\omega$, where ω is the classical frequency of the periodic orbit. The width of the sub-bands is the Lyapunov exponent λ. The bands stand out only if $\omega/\lambda > 1$, and since ω/λ is a purely classical quantity, the spectral oscillations survive the classical limit $\hbar \to 0$ (the linearized wavepacket analysis holds rigorously in this limit for any finite time). The existence of the local density of states oscillations implies that some or all of the states under the peaks must have a larger projection onto the initial wavepacket than would have been anticipated incorporating the initial broad envelope constraint.

If the orbit is stable, the localization will be even stronger, but this is by definition not scarring, and the relation $\omega/\lambda > 1$ is ill defined and moot anyway. In the stable case, classical localization exists. There is not classical localization around the periodic orbit in the unstable case, and that is why quantum localization there is interesting.

The scar enhancement is predicted all along the periodic orbit, since the eigenvalues of the stability matrix M_τ are independent of the starting point on the orbit. We cannot say precisely which particular states (the 275th?) are scarred and by how much, rather that *some* states are scarred by the periodic orbit within certain energy zones and by *at least* a factor of ω/λ, the local density of states enhancement. "Egalitarian" sharing of this scar intensity is not expected in general. The smooth fingers of spectral density,

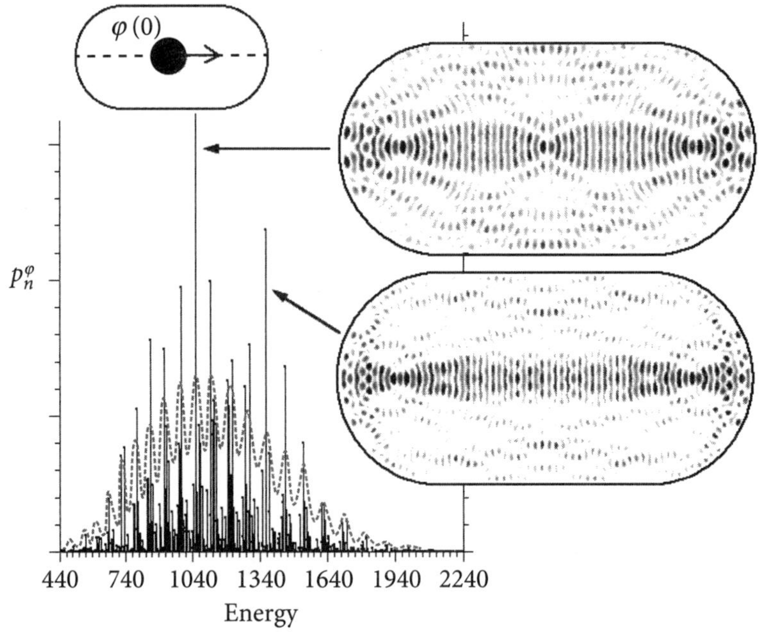

Figure 22.4. A wavepacket $|\phi(0)\rangle$ (dark circle with arrow showing momentum in inset) is placed on the horizontal bounce periodic orbit (dashed line in inset). Its spectrum $p_n^{\phi} = |\langle E_n|\phi(0)\rangle|^2$ (dark vertical lines) reflects the broad envelope (average energy and energy dispersion) that are a matter of the initial conditions given the wavepacket, not the shape of the billiard, since the first autocorrelation decay happens before any walls are reached. The narrower fingers seen in the dashed line are due to the first recurrence in $\langle\phi(0)|\phi(t)\rangle$ at the period τ after twice bouncing off the end caps. Scarred states are revealed by especially large overlaps—that is, p_n^{ϕ}, with this Gaussian. These fingers are necessarily a refined constraint envelope on the high resolution spectrum. Two large individual peaks p_n^{ϕ} have been selected to show the eigenstate $|n\rangle$ associated with them. Indeed, as expected, they are strongly scarred by the symmetric horizontal periodic orbit. Note (1) the self-focal points, as first noted by Eugene Bogomolny, and (2) the higher energy state indeed has more nodes along the orbit but the same location of the self-focal points (see figure 22.3). The cause of the self-focal points is seen and explained in figure 22.5. These results are from a numerical calculation of the eigenstates, energies, and overlaps.

in the absence of longer time information, are expected to become the new envelopes of support about which more random fluctuations occur, leaving some states much more heavily scarred than others within each dashed spectral density oscillation. Some states within the dashed peaks need not be scarred at all, so long as the totality of their strengths supports the existence of the oscillation [174].

What is perhaps just as intriguing as scarring is a necessary consequence—namely, *antiscarring*. Over a large enough energy range ΔE, the collective, average probability density of the eigenstates within that range must be uniform in phase space. This is easy to prove, as follows: we measure uniformity in phase space by projecting all the eigenstates onto sufficiently spatially narrow Gaussians $|\phi_{\Gamma_0}(0)\rangle$, where $\Gamma_0 = (x_0, y_0, p_{x_0}, p_{y_0})$, distributed over phase space. The expectation value of the energy of each is E_0, and all share the same ΔE, chosen large enough so that no

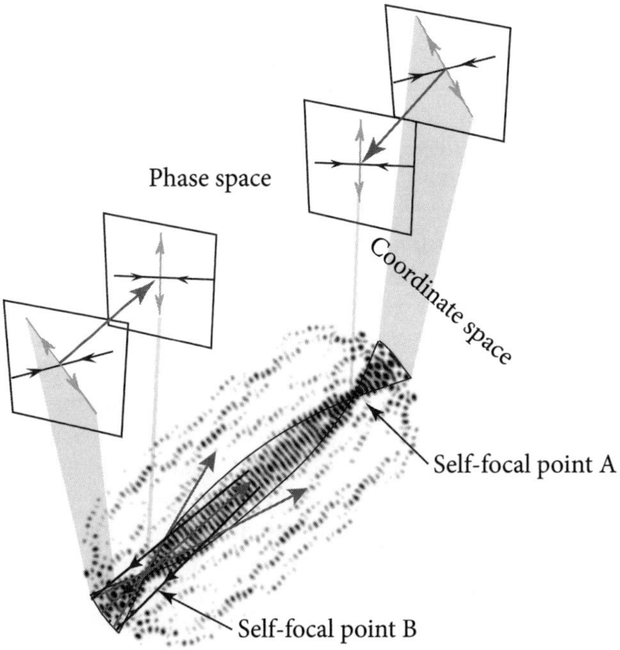

Figure 22.5. Two self-focal zones A and B are evident in this eigenstate of the stadium billiard, scarred by the unstable periodic orbit bisecting the billiard on its long axis. A spread of trajectories coming from focal zone A, after reflecting off the end near focal zone B, pass through the self-focal zone B. Continuing, these rays, after a bounce off the other end near A, pass through self-focal point A going back toward B, and the process repeats, with a thinning density of surviving rays that started near the periodic orbit. Orbits that homoclinically rejoin the vicinity of the period orbit again contribute to self focusing, assuming they are in phase with the original set. They will be in phase at the energy of the scarred eigenfunction. The classical manifolds rotate and stretch under free particle motion, and are abruptly reoriented upon collision with the end caps. The unstable manifold, along which the wavefunction is stretching, stands vertically at A and B, casting a sharp shadow in coordinate space—that is, the self-focal points A and B.

test wavepacket can return to overlap itself when propagated in under time Δt, where $\Delta E\,\Delta t \sim \hbar$. This works even near the edge of a billiard, for example, because after a bounce the wavepacket has changed momentum, giving almost no overlap even as it passes back over itself in position space within time Δt. Then summing over a set of adjacent eignestates within $\Delta E/2$ of E_0,

$$\langle \phi_{\Gamma_0}(0)\rangle \equiv \sum_{n,\,E_n\in\Delta E} |\langle \phi_{\Gamma_0}(0)|E_n\rangle|^2 = \int\limits_{E_0-\Delta E/2}^{E_0+\Delta E/2} dE \int e^{iEt/\hbar}\langle \phi_{\Gamma_0}(0)|\phi_{\Gamma_0}(t)\rangle\, dt,$$

$$= \int\limits_{-\infty}^{\infty} C_T(t) e^{iE_0 t/\hbar}\langle \phi_{\Gamma_0}(0)|\phi_{\Gamma_0}(t)\rangle\, dt, \tag{22.2}$$

where $C_T(t)$ is the transform-limited time cutoff corresponding to the energy band ΔE. The expectation $\langle \phi_{\Gamma_0}(0) \rangle$ is the same for every Γ_0, since it is in each case just the integral over the short time autocorrelation and corresponding energy envelope introduced earlier. By construction this envelope is the same for every test state $|\phi_{\Gamma_0}(0)\rangle$. The equal energy dispersion of all the $\phi_{\Gamma_0}(0)$ assures a fair measure of population of phase space, and each Gaussian, for Γ_0, on the energy shell E_0 gives the same result. The ϕ_{Γ_0} are everywhere in phase space centered on the energy shell E_0, and their population summing over the stack of states within ΔE is uniform. Therefore, a sum over a sufficiently large range of eigenstates is uniform in phase space, as claimed.

Since there may be strongly scarred states among these eigenstates, the necessity that the average over a sufficiently large number of states is uniform requires the existence of antiscarred states, with low instead of high probability in the region of strong scars. It was realized that the existence of such anomalously small amplitudes means that some decay processes (for example, in billiards with small holes at the loci of strong periodic orbits) will have antiscarred states with anomalously long escape times [175].

Strongly scarred states can be detected by the projection of all eigenfunctions in an appropriate energy range onto a coherent state wavepacket centered on the scarring periodic orbit. The wavepacket becomes an even better scar detector by adjustment of the complex gaussian spreading parameter, aligning the position-momentum correlation with stable and unstable manifolds belonging to the periodic orbit. This is just what was done in figure 22.4, worth studying as an exemplar of what we have been discussing in this section.

Self-focal regions (figure 22.5) are common and correspond to places where the unstable mainfold lies vertically, projecting onto one region of coordinate space.

Chapter 23

Random Waves and Nodal Structure of Eigenstates

23.1 Random Waves

Random superposition of waves is a concept with a long history, longest and deepest in oceanography for fairly obvious reasons. Michael S. Longuet-Higgins, mathematician and oceanographer, brother of H. Christopher Longuet-Higgins, the well-known chemist and cognitive scientist, introduced Gaussian random waves into oceanography and worked out dozens of statistical implications.

A parallel idea is random superpositions of plane waves [154, 176] as a model for quantum chaology. What should be the quantum analog to fixed energy classical trajectories in a chaotic system flying by every spot in coordinate space in every direction uniformly? It is compelling that representative eigenfunctions are constructed of random superpositions of plane waves of the same wave vector magnitude traveling in every direction. To represent the random histories of the trajectories passing by, the plane waves should also have random amplitudes because of their disparate histories and stabilities, and certainly random phase shifts. Using random amplitudes is not critical to the final result, due to the central limit theorem. This is called the Berry conjecture.

Such random superpositions of plane waves are not *speckle patterns*. Laser speckle is evident for example by reflecting an expanded fixed frequency laser beam onto a rough diffusive surface, such as a white painted wall. An observer looking at the illuminated zone will see very bright pinpoints or "speckles" of light. These pinpoints will be unique to the observer, and someone standing nearby will see a different set of them. They are due to spots of constructive interference of the random wave arriving at the observer's retina. However, they are two-dimensional slices of the three-dimensional wave field. The three-dimensional wave would display tubes that, intersected by the retina, look like speckles. The analog for random superpositions of plane waves in two dimensions would be a one-dimensional slice through the random wave, that would indeed reveal hotspots or speckles. Otherwise, looking at the full wave in 2D, fascinating structures are seen that are not properly regarded as speckles. There are two features that catch the eye at least: quasi-linear higher-amplitude structures that persist for some distance, and "holes" in the wavefunction with nodal lines leading toward the center of the hole.

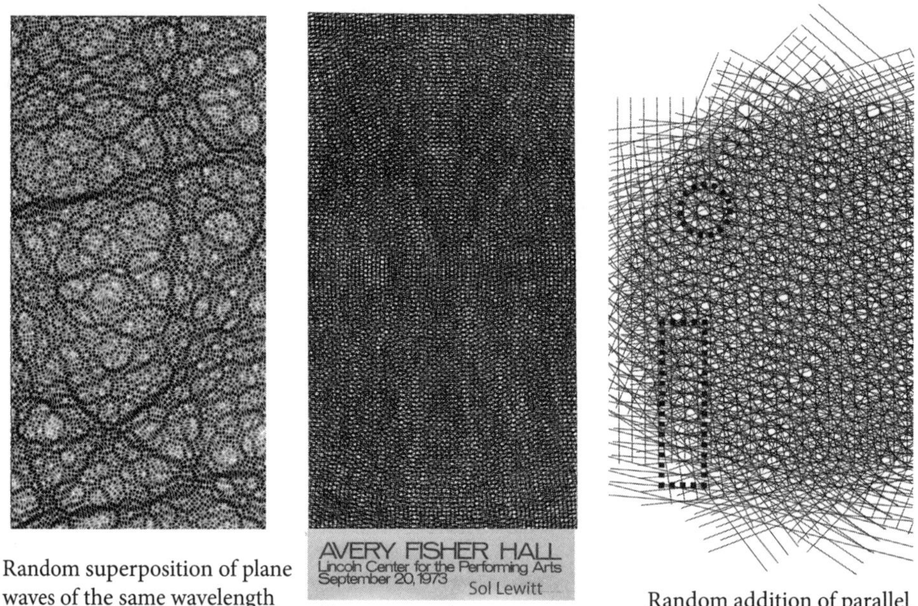

Random superposition of plane waves of the same wavelength

Commemorative poster

Random addition of parallel sets of lines

Figure 23.1. Three depictions of superpositions of plane waves, random in direction but all of the same wave vector magnitude (corresponding to a fixed spacing between all the adjacent parallel lines). On the left, we see a sum of smooth phase-shifted cosine plane waves of random amplitude and direction but fixed wavevector. A contour map of the result is plotted. This is locally the ideal random quantum eigenstate, according to Berry's hypothesis. In the center, a 1973 lithographic poster by the artist Sol LeWitt is shown. The poster was commissioned for the opening of Avery Fisher Hall at Lincoln Center, after reconstruction precipitated by its initial acoustical failures. The author found it very intriguing that the construction by Mr. LeWitt precisely simulated randomized acoustical waves in a room. On the right, a computer has been used to superimpose random sets of parallel lines, representing crests of plane waves, in a transparent way. It does not matter to the overall pattern or impression, as Mr. LeWitt himself recognized, whether concentric circles from various points are used, or whether plane waves are used. This precisely connects with the completeness and equivalence of a Bessel function basis and a plane wave basis.

By the central limit theorem, the statistics of wavefunction amplitude is Gaussian. The projection of the random wave onto Gaussian coherent states also yields Gaussian statistics in the real and imaginary parts as a function of the position and momentum of the coherent state. Other test states can be constructed that are sensitive to the linear structures just mentioned—for example, elongated Gaussian states with momentum along the long axis. Other test states could have well-defined angular momentum at a point. These also will have Gaussian statistics of their overlaps with the random wave superpositions.

Figure 23.1, left, shows a random superposition of many plane waves. This has become an important model for quantum chaology, first proposed by Sir Michael Berry [176]. In the middle and on the right, such random superpositions of many plane waves are represented by sets of parallel lines; each line can be regarded as a surrogate for a wave crest.

Sol LeWitt, a founder of both minimalist and conceptual schools of art, was one of the most prominent artists of the twentieth century. Among thousands of works, he made many consisting of sets of carefully ruled parallel lines and concentric circles with near perfect equal spacing between them. They never yield to uniformity no matter how many random sets are overlain. Incredibly, one set he made was for a poster created on commission for the dedication of Avery Fisher Hall at Lincoln Center (Avery Fisher donated \$10.5 million to the orchestra in 1973), renamed from Philharmonic Hall after its reconstruction due to dismal acoustics. Remarkably (private correspondence with Mr. LeWitt), Mr. LeWitt precisely but unknowingly simulated randomized acoustical waves in a room (see figure 23.1) on the Avery Fisher dedication poster.

Avery Fisher Hall was still not satisfactory, so in 2015, Lincoln Center removed Mr. Fisher's name and renamed the hall after the highest donor, David Geffen, who generously gave \$100 million toward another reconstruction, which will cost more than \$500 million.

RANDOM WAVES IN A HARMONIC OSCILLATOR

The title of this section may seem to be a non sequitur, since there can be no more regular, nonchaotic dynamics than in a harmonic oscillator. Yet in two and more dimensions, and well up in the spectrum of eigenvalues, a degenerate two-dimensional harmonic oscillator possesses manifolds of degenerate states—for example, $|N, 0\rangle, |N-1, 1\rangle, |N-2, 3\rangle, \ldots, |0, N\rangle$. Just as in Berry's original idea of random addition of degenerate plane waves differing only in direction, of which there are an infinite supply at any energy, we can pick a large N and add all within the degenerate set together, with random coefficients. The results are intriguing [177]. Deep in the classically allowed region, the superposition should look locally a lot like random addition of plane waves at a certain local wave vector magnitude, but interesting questions emerge. How far does one have to be from the classical turning surfaces where $V = E$ to look fairly random in the Berry sense? What happens near the classical turning surfaces? And what happens deep in the classically forbidden tunneling region, where naïvely one might not expect any nodal surfaces to penetrate? See the following and figure 23.2, which reveal two typical random superpositions of degenerate eigenstates.

Nodal surfaces of eigenfunctions can be extremely informative, often allowing a kind of deconstruction of the wavefunction in terms of its semiclassical components—that is, in terms of the classical ray fields that underlie the state.

23.2 Intersection of Nodal Surfaces and Role of Evanescent Waves

With symmetries or boundaries, or other imposed conditions, nodal lines of eigenstates can intersect. With the kinetic energy in 2D of the form $\hat{p}_x^2 + \hat{p}_y^2$, the intersection of two nodal lines is always at right angles, as we discuss shortly.

The question we address now is, can nodal lines ever intersect in a random wave eigenstate? We inspect a wave of the type in figure 23.2 more closely, in figure 23.3. In order for there to be an intersection of nodal lines, the value of the wave and two gradients have to vanish at the same point—three conditions—but there are only two dimensions in which to search for such a point. The problem is overdetermined and

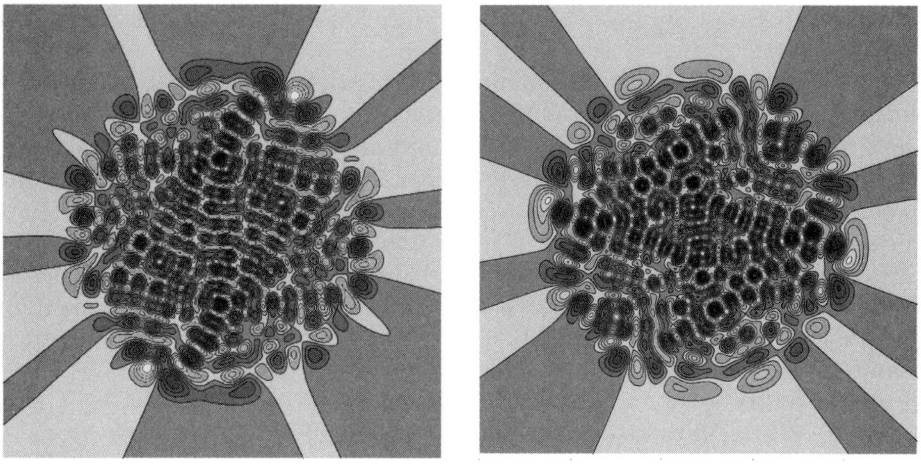

Figure 23.2. Contour map of $|\psi|^2$ and penetrating nodal lines of a high-lying eigenstate of the degenerate harmonic oscillator, constructed by adding all of the degenerate states $(N, 0), (N-1, 1), \cdots, (0, N)$ with random coefficients. Two different random superpositions of all 50 degenerate solutions of the 2D harmonic oscillator for this energy, with two different sets of randomly chosen amplitudes are shown. Typically, the resulting eigenfunction has $\sim \sqrt{N}$ nodal lines making their way to infinity [177]. On the left a nodal line is seen pinching off after a failed attempt to reach a large radius. The $\sqrt{N}$ result is peculiar to the harmonic oscillator, but it is interesting to think about nodal lines for more general potentials penetrating large distances into the tunneling regime.

there is no solution, or rather its probability is vanishingly small. Put another way, the observation of nodal lines crossing each other is an indication of a symmetry, or hidden symmetry if it was not known before, or a boundary condition.

RESTRICTIONS ON NODAL LINE INTERSECTIONS

Inspection of figure 22.1 reveals that the nodal lines approaching the wall in this hard-walled billiard appear to do so at right angles; the wall is effectively an imposed nodal line in the wavefunction, with Dirichlet boundary conditions. The billiard has vertical and horizontal reflection symmetry; this state is odd about both symmetry lines, and there too the nodes approach perpendicularly. Is it a requirement that nodal lines must intersect at right angles? The ideas can be generalized to 2D nodal surfaces in three dimensions, and so on.

We are considering cases where symmetries, walls, or other boundary conditions have supplied two of the three necessary conditions for a nodal intersection ($\Psi = 0$ and $\partial\Psi/\partial\hat{u} = 0$), leaving a one-dimensional search along the wall for instance with one remaining condition open to success. Figure 23.4 is a semi-random plane wave, forced to be odd about the line $y = 0$.

It is no accident that if two nodal lines cross, they must do so at 90-degrees as seen in figure 23.4. Without loss of generality, we can take one of the nodal lines to lie along $y = 0$ at the crossing, taken to be at (0, 0). Then, the only local solution of the wave equation with a crossing at (0, 0) and a nodal line locally along $y = 0$ reads, near (0, 0) and overall disregarding normalization,

$$\psi(x, y) = \sin(k_x x)\sin(k_y y), \tag{23.1}$$

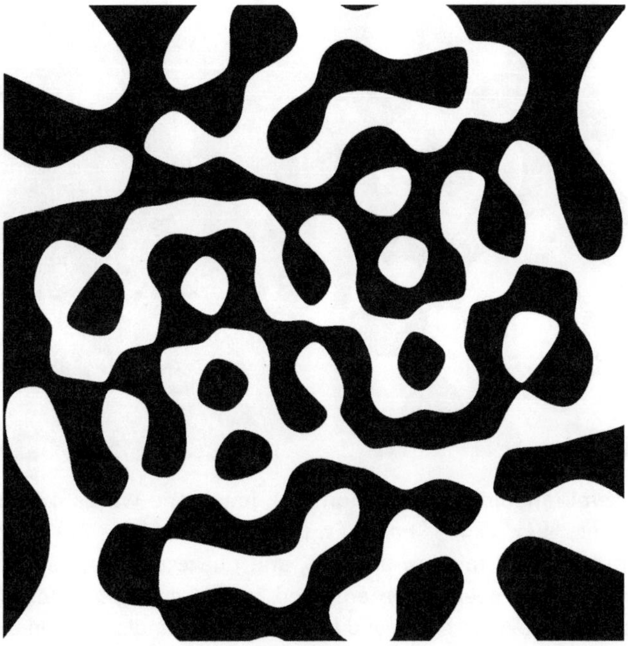

Figure 23.3. Map of the sign of a random superposition of degenerate harmonic oscillator states, with nodal lines appearing as boundaries between white and black regions of opposite signs. There are no crossing nodal lines, but a few narrowly avoided crossings. Note the near-perpendicular approach of the nodal lines, until they veer off at the avoided zone.

where $k^2 = k_x^2 + k_y^2$. The nodal lines cross perpendicularly. N nodal lines can also cross, but must do so with all interior angles of π/N.

One may wonder about two Dirichlet boundary lines crossing with an arbitrary interior angle. Is this not an enforcement of a violation of the theorem just proved? No, because no solution can then be continuous across the boundaries, even if they have the same k^2 energy. Inside one wedge, with one side given by the line $y = 0$ and the other a line passing through $x = 0$, $y = 0$ at an angle $\theta = \pi/\eta$, a class of solutions is

$$\psi(r, \theta) = J_\eta(kr) \sin(\eta\theta), \tag{23.2}$$

satisfying

$$r\frac{d}{dr}\left(r\frac{dJ_\eta(kr)}{dr}\right) + (k^2r^2 - \eta^2)J_\eta(kr) = 0. \tag{23.3}$$

This solution is the same for the similar wedge across the origin. However, the complementary wedges of interior angle $\pi(1 - \eta)$ bordering the other two will not have any solutions that are continuous with continuous gradient across the border. The infinitely many solutions (depending on boundary conditions) with the hard walls present are fine, since the discontinuity does not matter. But intersecting nodal lines cannot appear spontaneously at angles other than π/N, since the wave and its derivative would have to be continuous everywhere, including the nodal lines.

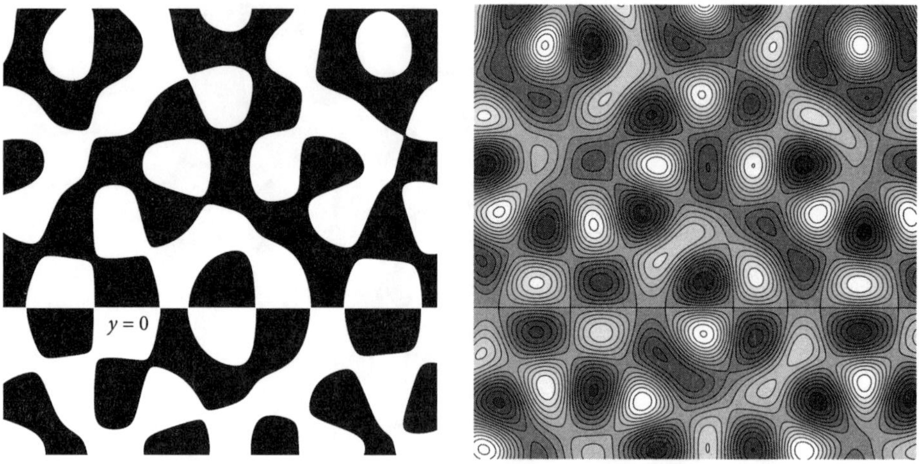

Figure 23.4. A semi-random superposition of a few plane waves of the same energy in two dimensions, $\Psi(x, y) = \sum_n a_n \psi_n(x, y) = \sin(k \cos\theta_n \; x + \delta_n) \sin(k \sin\theta_n \; y)$ with random amplitude a_n and random angle θ_n and phase shift δ_n, except the boundary condition $\Psi(x, 0) = 0$, is seen to be enforced in this superposition. On the left, the negative regions are shown in black, and the $y = 0$ line is indicated. On that line, several nodal line intersections occur, locally perpendicular to $y = 0$, as expected. In order for there to be a true intersection of nodal lines away from the imposed $y = 0$ nodal line, the value of the wave and two gradients have to vanish at the same point—three conditions—but there are only two dimensions to search for such a point. The result is that no intersections are expected anywhere away from $y = 0$. On the line $y = 0$, two of the three conditions are enforced ($\Psi(x, y) = 0$, $\partial\Psi(x, y)/\partial x = 0$), leaving a one-dimension search along the line successful. Avoided crossings are plainly seen in the interior, however. For reference, a contour map of $\Psi(x, y)$ is shown on the right .

WALL ROUGHNESS ON SMALL SCALES

We note that all plane waves of the form

$$\psi(x, y) = e^{ik_x x + ik_y y}, \tag{23.4}$$

with $k_x^2 + k_y^2 = k^2$, solve the wave equation in free space with (real) wave vector k. If both k_x and k_y are also real, the wave is ordinary, not evanescent. But what if k_y is pure imaginary, $k_y = i|k_y|$? We then have $k_x^2 - |k_y|^2 = k^2$, with $k_x > k$, with k_x real. This is an example of an evanescent wave, one that is decaying in the positive y direction. Since it increases without bound in the opposite direction, this solution cannot be used in unbounded domains. Far from walls and boundaries, the "free space" conditions prevail.

Figure 23.5 shows a wave of wavelength λ rebounding from an impenetrable corrugated wall. No linear combination of ordinary plane waves at this energy can possibly match the zero boundary condition at the corrugated wall, because their wavelength is too large to describe boundary oscillations smaller than themselves. An exact solution is attainable by using an evanescent wave of the type just mentioned, allowing faster oscillation than the ordinary plane waves in one direction at the expense of exponential decay in a perpendicular direction, leading to "healing" of the wave far

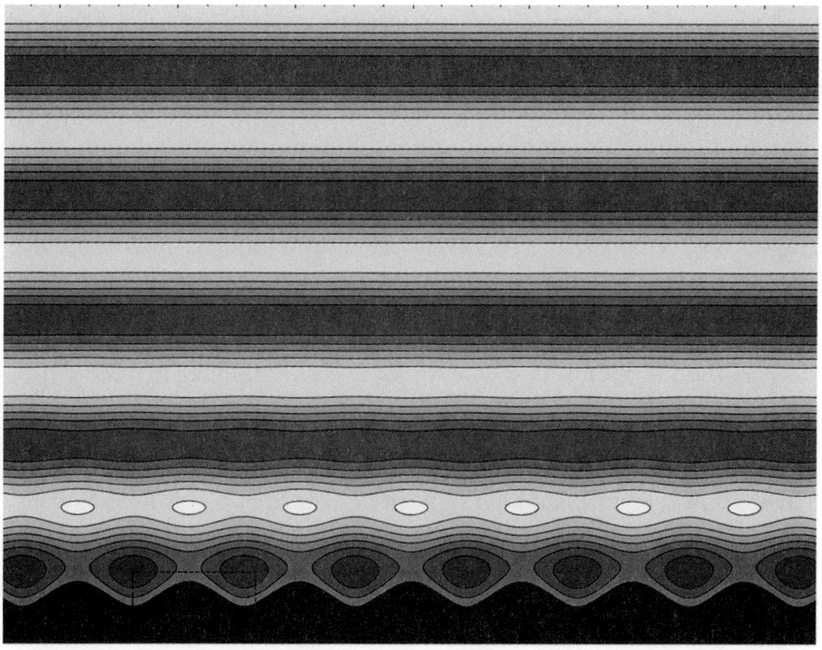

Figure 23.5. A wave is rebounding from a corrugated wall; however, its wavelength is slightly larger than the wavelength of the corrugation. The dotted line at the bottom shows the relative size of the two. The wave is seen to be recovering from the effect of the corrugations as distance from the wall increases. There are a number of ways to see that the wall must appear to be perfectly flat some distance away. Bragg scattering theory or Kirchhoff diffraction applied to this case would show there is no nonspecular diffraction in the scattering. No linear combination of periodic plane waves $\exp[ik_x x + ik_y y]$ with $k^2 = k_x^2 + k_y^2$ with real k_x and k_y could satisfy the zero boundary condition at the corrugated wall, since they are too broad a brush, so to speak, to follow the shorter corrugation. An exact solution is attainable by instead using evanescent waves, with pure imaginary k_y if y is the vertical direction. Then $k_x^2 > k^2$, allowing fast oscillation along the wall and satisfaction of the boundary condition.

from the wall. The same principles allow mirrors to be optically perfect in spite of having myriads of imperfections on a subwavelength scale.

What earns this topic a place in this section is that the corrugated wall might be considered as a nodal line, but it is not one that could have arisen in free space, because of the requirement of the evanescent waves on both sides, decaying away from the (impossible) nodal line, without any discontinuity at the node. A free-standing interior piece of corrugated "Dirichlet" wall can be present in free space, but the wave will be found to have discontinuities across the wall.

23.3 Fermionic Wavefunctions

The ultimate authority on this subject is D. M. Ceperley [178]. Here, we give a few essentials. The first and most important point is that *nodal surfaces in wavefunctions are by definition (and for good reason) surfaces of co-dimension 1*—that is, the surface occupies one dimension less than the full dimension of the coordinate space. This point is general, not specific to Fermionic wavefunctions. A nodal surface for the hydrogen atom is two dimensional, and for Helium, it is five dimensional, taking the nucleus to be the origin of the electronic coordinates.

The wavefunction for two or more electrons we know must vanish as any two electrons with the same spin approach each other. Assuming all spins up for simplicity, $\psi(\vec{r}_1, \vec{r}_2, \cdots, \vec{r}_N) \to 0$ as $\vec{r}_i \to \vec{r}_j, i \neq j$. In discussions about electron correlation and quantum energy eigenvalues, the "free ride" in lowering electron repulsion by having a Fermi hole is often emphasized. There is an old maxim though: "nodes cost energy"—kinetic energy, to be precise. This goes back to deBroglie, $p\lambda = h$, and the $p^2/2m$ term in the Hamiltonian, except if the momentum is imaginary, as in tunneling domains. So perhaps the kinetic energy cost of a "Fermi hole," of co-dimension 3, balances the reduction in Coulomb repulsion. However, for N electrons living in three-dimensional space, $\vec{r}_i = \vec{r}_j$ constitutes three constraints, whereas a co-dimension 1 nodal surface is equivalent to one constraint. The Fermi hole therefore does not constitute a nodal surface, save in a problem with one dimension per electron. Aligning spins might seem to win hands down, arguing for high spin species. Since $\vec{r}_i = \vec{r}_j$ is not a nodal surface, maybe the kinetic energy cost of the Fermi hole is minimal, while Coulomb repulsion is reduced.

The "follow the repulsion" school of electron energies in atoms has nonetheless a mixed record of success. Sometimes arranging the electrons to minimize their repulsion delivers a nasty surprise by raising the kinetic energy by a large amount. Indeed, Fermi holes save a surprisingly modest amount of repulsive energy. If the spins are opposite, two electrons may approach, but though the Coulomb repulsion increases as $1/|r_i - r_j|$, the spatial integral has a volume element proportional to $|r_i - r_j|^2$ in spherical polar coordinates. The energy cost of contact is quite finite.

It is important though that Fermi holes must reside *on* genuine nodal surfaces. This is easily shown. To explore the co-dimension 3 surface $\psi(\vec{r}_i = \vec{r}_j) = 0$, both electrons move around together. Moving any other electrons (or combination of others) while keeping $\vec{r}_i = \vec{r}_j$ leaves the wavefunction unchanged—that is, $\psi = 0$. Such movement is merely exploration of $3N - 3$-dimensional interior volume of the Fermi hole. This is not co-dimension 1 surface, thus not a nodal surface. To discover how to get off the $3N - 3$-dimensional interior volume $\psi = 0$ surface, and perhaps explore an associated $3N - 1$-nodal surface on which the $3N - 3$-dimensional Fermi hole sits, we take a gradient of ψ in the relative position of $\vec{r}_i$ and $\vec{r}_j$—namely,

$$\vec{\nabla}_{\vec{r}_{ij}} \psi(\vec{r}_1, \vec{r}_2, \cdots, \vec{r}_N)|_{\vec{r}_{ij}=0},$$

where $\vec{r}_{ij} = \vec{r}_i - \vec{r}_j$. The gradient, by its nature, not only gives the steepest direction to leave the surface on which $\psi = 0$, *but also, being a 3D gradient, it carries two perpendicular directions, along which the wavefunction does not change—that is, ψ remains 0.* This adds two coordinates along which ψ remains 0 to the $3N - 3$ implied by $\vec{r}_i = \vec{r}_j$. Thus, we have found that the Fermi hole resides on a patch of $3N - 1$-dimensional nodal surface. This does not imply, however, that all nodal surfaces have Fermi holes living on them.

23.4 Classically Forbidden Regions in Few-Body Eigenstates

The penetration of nodal surfaces into classically forbidden regions is an intriguing subject. As nodal surfaces extend deeper into classically forbidden zones, their presence becomes ever more surprising and unlikely. The reason is that they represent energy and momentum traveling perpendicular to the tunneling direction. Since tunneling exponentially diminishes wavefunctions, with slower exponential decay for higher energy in the penetrating direction, it would be expected that the surviving amplitude for deep tunneling would be found expending all of its energy in the direction of the tunneling. Eigenfunctions (or parts of them) that failed to do so would not penetrate as far, suggesting nodal surfaces would be disadvantageous and disappear or "pinch off" as the tunneling region is penetrated. That is, one would expect tunneling to filter out wave amplitude that "wastes" some of its energy in transverse motion. In the case of random linear combinations of 2D degenerate states in high-lying manifolds of the symmetric harmonic oscillator, it was shown that the number of known lines that penetrate to infinity goes (on average) as $\sqrt{N}$, as shown in figure 23.2 [177]. Although the number of penetrating lines is finite, they do grow more rare in that the distance between them goes to infinity as the radius from the center increases.

The situation can be cast clearly for a set of N short-range interacting particles, all confined to the same harmonic potential. Suppose an exact *excited* eigenstate $\Psi_K(\boldsymbol{q}_1, \boldsymbol{q}_2, \ldots \boldsymbol{q}_N)$ of this N-body system is examined deep in the tunneling region of a single particle $|\boldsymbol{q}_N|$. Barrier tunneling for a single particle in the many-body system is defined such that the penetrating particle's potential energy alone, due only to the confining potential (we assume the tunneling particle is remote from and not interacting with the other $N-1$ particles) requires more energy than the exact quantum energy E_K^N of the N-body eigenfunction Ψ_K minus the ground energy E_0^{N-1} of the $N-1$-body system left behind. (This is the maximum energy that the Nth particle can be given, out of the range of interaction with its $N-1$ compatriots.) As the particle penetrates deeper into the classically forbidden region the remaining $N-1$-body system is increasingly likely to be found in its ground state $\psi_0(\boldsymbol{q}_1, \boldsymbol{q}_2, \ldots \boldsymbol{q}_{N-1})$. This was first noted by Katriel and Davidson [179]. The N-body wavefunction is correlated, and the part correlating with the $N-1$-body ground state tunnels deepest.

23.5 Where Does "Tunneling" Start in Many Dimensions?

It is not widely advertised that beyond the usual one-dimensional oscillator with well-defined eigenstate energies defining the limits of classical motion for a given potential, things get a little murky about exactly where the classically forbidden regions begin in a multidimensional system. The problem is already evident in two dimensions, in a way that a shrewd undergraduate could have used to trip up her professor. After the usual discussion of where the $V = E$ tunneling region begins in a one-dimensional harmonic oscillator, the professor calculates how much of the quantum probability lies in the classically forbidden zone. For a Hamiltonian $H = p^2/2m + 1/2m\omega^2x^2$ with a normalized ground eigenstate of $\psi_0(x) = (m\omega/\pi\hbar)^{1/4}\exp[-m\omega/2\hbar\ x^2]$ with energy $E_0 = 1/2\hbar\omega$, the $V(x) = E_0$ point is given by $1/2m\omega^2x^2 = 1/2\hbar\omega$; i.e., $x^* = \sqrt{\hbar/m\omega}$. The integral $2\int_{x^*}^{\infty} |\psi_0(x)|^2\, dx = 0.1573$—that is, about 16% of the time the system will be found in the forbidden region.

Now consider two such oscillators, independent of each other. For what fraction of the time is at least one of them busy tunneling? There is a $p = (1 - 0.1573)^2$ chance that neither are, or about 29% that at least one is tunneling. For three oscillators, it is $p = (1 - 0.1573)^3$ that no oscillator is tunneling, or about 40% that at least one is. It is clear that this is headed for a 100% chance that at least one oscillator is tunneling at any given moment for a large number of them.

When the calculation is done in terms of independent oscillators, it is clear that the zero point energy determining the classically allowed region for an oscillator is just that of each oscillator—that is, we don't use the total zero point energy of both oscillators to say where tunneling starts for one of them.

The two independent harmonic oscillators problem is the same as one particle in the potential $V(x, y) = 1/2m(\omega_x^2 x^2 + \omega_y^2 y^2)$. This suddenly seems to cause us to regard the problem differently, for if we draw the elliptical $V = E$ contour line in coordinate space, do we now regard the probability outside this ellipse as tunneling? If the energy E is the total zero point energy, then only about 13.5% is tunneling in the case $\omega_x = \omega_y = \omega$, a far cry from 29%.

Which view is correct? We have to admit that an arbitrary construct called *barrier tunneling* has been imposed on the quantum world, and unless we are more precise in our definitions, problems like this will arise. For let's say 100 bonds in a molecule, setting E equal to the zero point energy could easily mean a few atoms could be put at infinity without crossing a "classically forbidden" boundary.

Some consideration leads to the following definition, which is consistent with what has been done for years for molecular reaction barriers and quantum point contacts, where particles squeeze through small zones with the zero point energy of motion perpendicular to the progress coordinate counted as part of the potential energy.

> ***Definition of barrier tunneling***: For a smooth potential of N degrees of freedom $V(\boldsymbol{q})$, and total quantum energy (including zero point energy) E, a point $\boldsymbol{q}_0$ resides in the classically forbidden zone if the quantity
>
> $$E - V(\boldsymbol{q}) - \sum_{1}^{N-1} \epsilon_k < 0,$$
>
> where the ϵ_k are the $N-1$ zero point energies of harmonic approximations to the potential along the $N-1$ locally normal coordinates perpendicular to the gradient of the potential at the point $\boldsymbol{q}_0$. If any $\epsilon_k < 0$, it is set to 0.

Clearly, this definition is still arbitrary, but it seems to avoid a lot of ridiculous situations, like the 100 chemical bond example just mentioned.

The role of what to do with the zero point energy in classical simulations of two or more degrees of freedom has been a longstanding thorn in the paw. Consider figure 23.6, showing the radically different quantum versus classical behavior when the classical motion is allowed to use all the zero point energy to expel a particle that in fact cannot be expelled quantum mechanically (for example, the fast, classical autoionization of helium seen at the right, at the perfectly stable quantum ground state energy of helium).

23.6 Many-Particle Random Waves and Statistical Mechanics

The standard tools of quantum chaos investigations include random matrix theory and periodic orbit theory (Gutzwiller trace formula), the Van Vleck-Morette-Gutzwiller (VVMG) propagator, and many techniques and phenomena derived from these approaches. Standing somewhat to the side as an inspired insight is Berry's

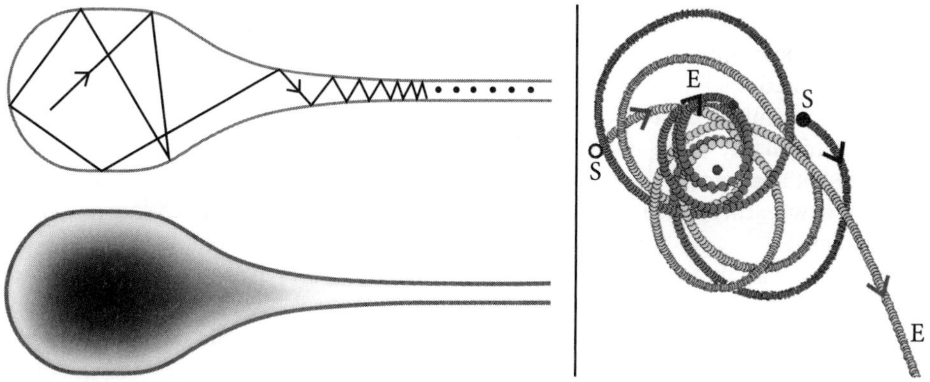

Figure 23.6. The classical autoionization of helium seen at the right, revealed in the track left behind of a typical simulation. "S" marks the start of each electron path, and "E" the end of this short time sequence. The electron remaining bound has dropped below the quantum zero point energy. The classical mechanics is of course oblivious to this, but it supplied the needed escape velocity to the second electron, causing a quantum mechanically illegal autoionization. In another example on the left, the confining wall allows classical escape at any energy, but the lowest energy quantum eigenstate is provably bound, with only an evanescent wave penetrating into the asymptotic corridor. The proof is simple: the large space available at the left of the wall boundary allows a low zero point ground energy, lower than the lowest energy required in the corridor. Only an evanescent wave can exist there at the ground state energy, making a nonescaping, exponentially decaying square integrable ground state wavefunction. What is the best measure to take against the embarrassing classical escape?

conjecture, which loosely stated is the idea that as $\hbar \to 0$ eigenstates of classically chaotic systems will be locally indistinguishable from superpositions of infinitely many plane waves with random amplitude, direction, and phase, but with fixed wavelength appropriate to the local kinetic energy. This mimics the uniformly random direction, fixed energy trajectories passing through any point in a chaotic system of fixed energy. In two dimensions these assumptions result in strictly Gaussian statistics of the eigenfunctions and the autocorrelation function $\langle \psi^*(\vec{x})\psi(\vec{x}+\vec{R})\rangle = J_0(ka)$, where k is the local wavenumber and $|\vec{R}| = a$.

The Berry random plane wave (RPW) [180] hypothesis is free of any specific dynamical information, except fixed total energy, which defines the "ensemble" (that is, microcanonical). By extending the RPW hypothesis we can accommodate constraints, incorporating specific information about real systems. Hard walls have been included [181], as well as soft boundaries [177]. Related work has been done by Urbina and Richter [182].

The idea of random waves is not confined to one particle in two dimensions. Indeed, Berry gave the N-dimensional formula for free particles in his 1977 paper [176]. The underlying idea in the RPW hypothesis is uniform randomness within a quantum context. Thus, we expect to encounter some familiar territory in quantum statistical mechanics as the RPW hypothesis is extended to the large N limit. In 1994 Srednicki had suggested that the Berry random wave hypothesis was indeed a foundation for quantum statistical mechanics [183], and showed that the appropriate canonical ensemble was reached for large N, depending on particle statistics. Here, we see what happens as the number of particles increases, through a nonstandard asymptotic form for Bessel functions [184], which encodes the equivalence of the canonical and microcanonical ensembles of statistical mechanics. In making the connections to quantum statistical mechanics, one also needs procedures for incorporating constraints, which are an essential aspect of the theory. Generalizing the RPW to include constraints is an essential new feature, since the constrained eigenstates are no longer random in Berry's original sense.

It must be said, however, that an RPW approach cannot be a *complete* replacement of quantum statistical mechanics. The reason is an all too glossed over fact about quantum as opposed to classical statistical mechanics: whereas classical mechanics is always ready to start counting classical phase space to get partition functions, quantum statistical mechanics assumes that Schrödinger equation energy eigenvalues have been determined at the outset—quite an assumption! While having infinitely many particles at high temperatures might allow one to skip this step, we must acknowledge that an RPW hypothesis by itself will not solve many-body problems or solve for the energies E_m going into the partition functions, and so on. Quantum statistical mechanics involves a statistical population of eigenstates, which include some very specific violations of the idea of "random local plane waves," especially for the lowest energies in the problem.

Given a continuum at energy E, such as an enclosure with walls very far away, we can perform the average over all random waves as a trace (using 2D as an example):

$$\begin{aligned}\langle\psi^*(\vec{q})\psi(\vec{q}\,')\rangle &= \mathrm{Tr}[\delta(E-H)|\vec{q}\rangle\langle\vec{q}\,'|] = \int d^2\vec{k}\,\langle\vec{k}|\delta(E-H)|\vec{q}\rangle\langle\vec{q}\,'|\vec{k}\rangle,\\ &= \int k dk \int_0^{2\pi} d\theta\;\delta(E-k^2/2)e^{i|\vec{q}-\vec{q}\,'|k\cos\theta},\\ &= \int_0^{2\pi} d\theta\; e^{ika\cos\theta} = \pi J_0(ka).\end{aligned} \tag{23.5}$$

This is Berry's result, apart from normalization that we choose differently here. A trace over a complete basis is independent of any unitary transformation on that basis, so it does not matter whether we use a trace over a complete set of random waves or of plane waves; both give $J_0(ka)$, where $a=|\vec{q}-\vec{q}\,'|$ for the case of one free particle in two dimensions. In this way the imaginary part of the retarded Green function $-\frac{1}{\pi}\mathrm{Im}[G^+(E)]=\delta(E-H)$ becomes central, formally convenient, and equivalent to Berry's RPW hypothesis.

The Green function completely characterizes a quantum system, whether it is interacting or not, or has few or many degrees of freedom. The retarded Green function G^+ is

$$G^+ = \mathcal{P}\frac{1}{E-H} - i\pi\delta(E-H). \tag{23.6}$$

Here, $\mathcal{P}$ stands for the principal value of the integral, and is the basis for wavefunction statistics and density matrix information, through the following relations, with a convenient choice of normalization:

$$\langle \psi(\boldsymbol{q})\psi^*(\boldsymbol{q}')\rangle = -\frac{1}{\pi}\mathrm{Im}\langle \boldsymbol{q}|G^+|\boldsymbol{q}'\rangle/\rho(E), \tag{23.7}$$

$$= \langle \boldsymbol{q}|\delta(E-H)|\boldsymbol{q}'\rangle/\rho(E), \tag{23.8}$$

where

$$\rho(E) = \mathrm{Tr}[\delta(E-H)], \tag{23.9}$$

and where $\langle\cdots\rangle$ stands for the average over the degenerate states. We take these degeneracies to be of dimension up to $ND-1$, where N is the number of particles and D the spatial dimension each particle lives in. (We use boldface notation—for example, $\boldsymbol{q}$—for the $N*D$ degrees of freedom.) If true degeneracies do not exist in a particular system, we can artificially open up the system to a continuum. For example, a two-dimensional closed billiard may not have a degeneracy, but it acquires one if we open a hole in it and let it communicate with the outside unbounded 2D space. Of course, this changes the billiard properties, and the size of the hole might be problematic, but in fact we shall never really have to open up a system in this way. The quantity $\delta(E-H)$ then implies the average over all scattering wavefunctions at fixed energy E.

The wavefunction correlation is equal to the coordinate space matrix element of the constant energy density matrix:

$$\langle \psi(\boldsymbol{q})\psi^*(\boldsymbol{q}')\rangle = \langle \boldsymbol{q}|\delta(E-H)|\boldsymbol{q}'\rangle/\rho(E) = \rho(\boldsymbol{q},\boldsymbol{q}',E). \tag{23.10}$$

Reduced density matrices can also be derived from wavefunction correlations—for example,

$$\tilde{\rho}(\vec{q}_1,\vec{q}_1{}',E) = \int d\vec{q}_2 d\vec{q}_3\cdots, d\vec{q}_N\,\rho(\vec{q}_1,\vec{q}_2,\cdots;\vec{q}_1{}',\vec{q}_2,\cdots;E), \tag{23.11}$$

the one-particle reduced density matrix.

We can approach the correlations via Fourier transform from the time domain, since

$$\delta(E-H) = \frac{1}{2\pi\hbar}\int_{-\infty}^{\infty} e^{iEt/\hbar}e^{-iHt/\hbar}\,dt. \tag{23.12}$$

Thus, the statistics, density matrices, and correlations are derivable without further averaging by knowing the time propagator. In the following, we define the Green function propagator $G(\boldsymbol{q},\boldsymbol{q}',t)$ and the retarded Green function propagator $G^+(\boldsymbol{q},\boldsymbol{q}',t)$ as

$$\begin{aligned} G(\boldsymbol{q},\boldsymbol{q}',t) &= \langle \boldsymbol{q}|e^{-iHt/\hbar}|\boldsymbol{q}'\rangle,\\ G^+(\boldsymbol{q},\boldsymbol{q}',t) &= \frac{-i}{\hbar}\Theta(t)\langle \boldsymbol{q}|e^{-iHt/\hbar}|\boldsymbol{q}'\rangle, \end{aligned} \tag{23.13}$$

where $\Theta(t)$ is the Heavyside step function $\Theta(t)=0,\ t<0,\ \Theta(t)=1,\ t>0$. It is very rewarding to expand the propagator in semiclassical terms, involving short time (zero length) and longer trajectories. We take $G_{direct}(\boldsymbol{q},\boldsymbol{q}+\boldsymbol{r},t) = \langle \boldsymbol{q}|\exp[-iHt/\hbar]|\boldsymbol{q}+\boldsymbol{r}\rangle$, the very short time semiclassical propagator, which for N

particles each in D dimensions reads

$$G_{direct}(\boldsymbol{q}, \boldsymbol{q}+\boldsymbol{r}, t) \approx \left(\frac{m}{2\pi i\hbar t}\right)^{ND/2} e^{imr^2/2\hbar t - iV(\boldsymbol{q}+\frac{\boldsymbol{r}}{2})t/\hbar}, \tag{23.14}$$

where $r^2 = |\boldsymbol{r}|^2$. It is not difficult to cast the Fourier transform of this short time version to fit the definition of a Hankel function—that is,

$$\begin{aligned} G^+_{cl}(\boldsymbol{q}, \boldsymbol{q}+\boldsymbol{r}, E) &= \frac{-i}{\hbar}\int_0^\infty \left(\frac{m}{2\pi i\hbar t}\right)^{ND/2} e^{imr^2/2\hbar t - iV(\boldsymbol{q}+\frac{\boldsymbol{r}}{2})t/\hbar} e^{iEt/\hbar}\, dt, \\ &= -\frac{im}{2\hbar^2}\left(\frac{k^2}{2\pi kr}\right)^d H_d^{(1)}(kr), \end{aligned} \tag{23.15}$$

where $d = ND/2 - 1$, $k = k(\boldsymbol{q}+\boldsymbol{r}/2, E)$, $H_d^{(1)}(kr) = J_d(kr) + iY_d(kr)$ is the Hankel function of order d, J_d is the regular Bessel function of the first kind of order d, and Y_d is a Bessel function of the second kind. The wavevector k varies with the local potential—that is, $\hbar^2 k(\boldsymbol{q}, E)^2/2m = E - V(\boldsymbol{q})$. Here, using only the extreme short time version of the propagator, we must suppose $\boldsymbol{r}$ is not large compared to significant changes in the potential, but this restriction can be removed by using the full VVMG semiclassical propagator rather than the short time version. For the case of one particle in two dimensions, $d = 0$, and we recover Berry's original result for one particle in 2D, $\langle \psi^*(\vec{q})\psi(\vec{q}+\vec{r})\rangle \propto J_0(kr)$.

According to the short time approximation, for any N,

$$\langle \psi(\boldsymbol{q})\psi^*(\boldsymbol{q}+\boldsymbol{r})\rangle \approx -\frac{1}{\pi}\frac{\mathrm{Im}[G^+_{cl}(\boldsymbol{q}, \boldsymbol{q}+\boldsymbol{r}, E)]}{\rho(E)} = \frac{1}{\rho(E)}\frac{m}{2\pi\hbar^2}\left(\frac{k^2}{2\pi kr}\right)^d J_d(kr), \tag{23.16}$$

where $k = k(\boldsymbol{q}, E)$. This result includes interparticle correlations through the potential $V(\boldsymbol{q})$ and the spatial dependence of $k = k(\boldsymbol{q}, E)$; the diagonal $r = 0$ limit (see the following section) is equivalent to classical statistical mechanics. It is interesting that although the short time Green function is manifestly semiclassical, the energy form—for example, equation 23.16—is uniformized, since we obtained it by supposing exact Fourier transform of the semiclassical propagator, rather than by stationary phase.

23.7 Diagonal Limit

The diagonal ($r \to 0$) N body Green function is obtained using the asymptotic form

$$\lim_{r\to 0} J_d(kr) = \frac{1}{\Gamma(d+1)}\left(\frac{kr}{2}\right)^d \approx \frac{1}{\sqrt{2\pi d}}\left(\frac{ekr}{2d}\right)^d, \tag{23.17}$$

from which we obtain

$$-\frac{1}{\pi}\mathrm{Im}[G^+_{cl}(\boldsymbol{q}, \boldsymbol{q}, E)] \approx \frac{m}{2\pi\hbar^2}\frac{1}{\Gamma(d+1)}\left(\frac{k^2}{4\pi}\right)^d \approx \frac{m}{2\pi\hbar^2}\frac{1}{\sqrt{2\pi d}}\left(\frac{ek^2}{4\pi d}\right)^d, \tag{23.18}$$

where the second form uses Stirling's approximation, $n! \sim n^n e^{-n}\sqrt{2\pi n}$, appropriate for large N. We note that this behaves as $k^{2d} \sim (E - V(\vec{q}))^d$. This factor is familiar from the computation of the classical density of states. Tracing over all $\vec{q}$ results in

$$\int d\boldsymbol{q} \frac{m}{2\pi\hbar^2} \frac{1}{\Gamma(d+1)} \left(\frac{k^2}{4\pi}\right)^d = \int \frac{d\boldsymbol{q}\, d\boldsymbol{p}}{h^{ND}} \delta(E - H_{cl}(\boldsymbol{p}, \boldsymbol{q})) = \rho_{cl}(E), \tag{23.19}$$

that is, the classical density of states. The Berry RPW hypothesis, the short time propagator, and the classical or Weyl (sometimes called Thomas-Fermi) term in the quantum density of states are all closely related.

The quantum integral, equation 23.19, is over all coordinates including classically forbidden values, so how does the classical partition function emerge, since the classical integral is clearly only taken over classically allowed coordinates? For forbidden positions, k is imaginary and can be written as $i\kappa$. An identity for Hankel functions can then be used ($i^{n+1} H_n^{(1)}(ix) = \frac{2}{\pi} K_n(x)$) to show that the imaginary part is zero, explaining why even equation 23.19 is really only over classically allowed values.

As long as $\boldsymbol{r} = 0$ (that is, diagonal Green function), the results obtained within the short time propagator approximation for any quantity in the presence of a potential (*including* interparticle potentials such as atom-atom interactions) will be purely classical.

23.8 Link to the Canonical Ensemble

BESSEL FUNCTIONS BECOME GAUSSIANS

We now consider the large N limit for the nondiagonal semiclassical short time Green function, $\boldsymbol{r} \neq 0$. Taking the large N limit of equation 23.16, we are confronted with an unusual question about Bessel functions [184]. The large d limit of $J_d(x)$ is well known, but this is not sufficient for our purposes. It reads

$$\lim_{d\to\infty} \frac{J_d(kr)}{(kr)^d} = \frac{1}{2^d\, \Gamma(d+1)} \approx \frac{1}{\sqrt{2\pi d}} \left(\frac{e}{2d}\right)^d. \tag{23.20}$$

Equation 23.20 should be the first term in a power seres for $J_d(kr)$ in kr. Another standard result is the power series expansion, valid for all d and kr:

$$J_d(kr) = \sum_{m=0}^{\infty} \frac{(-1)^m}{m!\Gamma(m+d+1)} \left(\frac{kr}{2}\right)^{2m+d}. \tag{23.21}$$

Our requirements are unusual in that we want the energy to increase in proportion to the number of particles, lest the temperature decline toward 0 as $N \to \infty$. Then $k \sim \sqrt{E} \sim \sqrt{N} \sim \sqrt{d}$, meaning that for fixed r the combination (kr) is increasing as $\sqrt{d}$ as $d \to \infty$. If the argument of the Bessel function increases without bound along with its order, we find the desired form using equation 23.21, after summing a series recognized as a Taylor expansion of a Gaussian,

$$\lim_{d\to\infty} \frac{1}{(kr)^d} J_d(kr) = \frac{1}{2^d\, d!} \sum_{m=0}^{\infty} \frac{1}{m!} \left(\frac{-k^2 r^2}{4(d+1)}\right)^m = \frac{1}{2^d\, d!} e^{-k^2 r^2/(4(d+1))}, \tag{23.22}$$

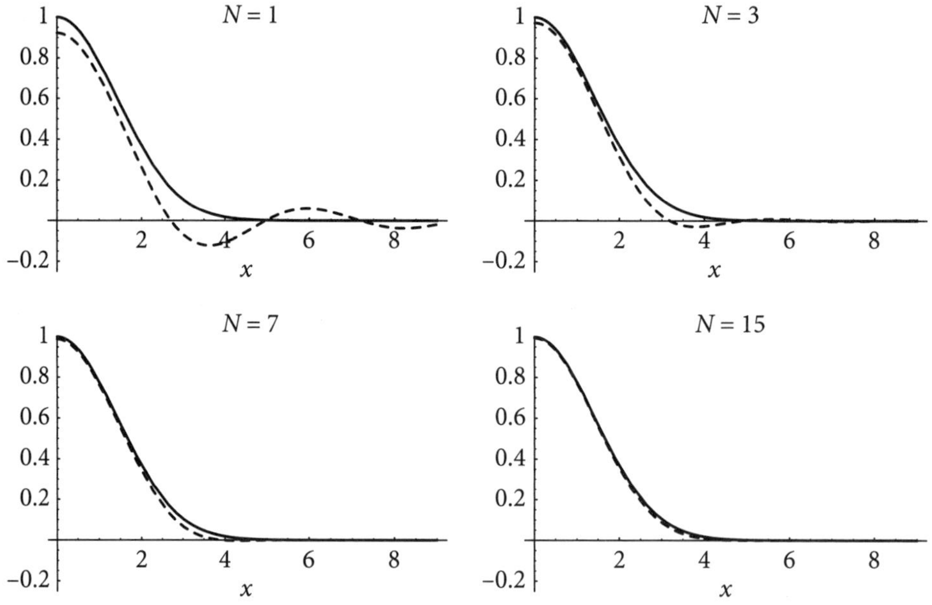

Figure 23.7. As N increases, the combination $\frac{1}{x^d}J_d(x)$, where $d = ND/2 - 1$, approaches a Gaussian. This is the key link between the quantum microcanonical and canonical ensembles.

where again $\hbar^2 k^2/2m = E - V(\boldsymbol{q})$. Note that as $d \to \infty$, the argument of the Gaussian holds fixed because of the factor of $d+1$ in the denominator of that argument. Figure 23.7 illustrates the convergence to the Gaussian as N increases. The asymptotic limit in equation 23.22 is not in the usual references, although related results have been given for N-bead polymer random chain end-to-end distributions [185]. The connection between the path integral for the propagator and polymer chains is well known [186].

It is interesting that a Gaussian emerges from Bessel functions in the large N limit. We can put equation 23.22 together with equation 23.16 and equation 23.8, and express the result, as $N \to \infty$, as

$$\langle \psi(\boldsymbol{q})\psi^*(\boldsymbol{q}+\boldsymbol{r})\rangle = \rho(\boldsymbol{q}, \boldsymbol{q}', E) \to \frac{1}{\rho(E)}\frac{m}{2\pi\hbar^2 d!}\left(\frac{k^2}{4\pi}\right)^d e^{-k^2r^2/4(d+1)}. \quad (23.23)$$

The canonical ensemble result for the propagator has "dropped out" of the asymptotic large N limit of a microcanonical Green function, at least for noninteracting particles, and an unusual asymptotic form for the Bessel function has emerged as the link. With some caveats, the statement

$$\delta(E - H) \sim e^{-\beta H} \quad (23.24)$$

has meaning in the large N limit, where it is understood that E grows as N, and a temperature is extracted. At a qualitative level, equation 23.24 merely expresses the known equivalence of the ensembles. In the case of an interaction potential, the relation between E and temperature is of course problematical.

Chapter 24

Branched Flow

Branched flow results when exactly or approximately unidirectional waves or rays impinge on a weakly randomly refracting medium, the waves or rays eventually propagating over many correlation lengths of the randomness. Backscattering is taken to be absent or play a minor role. Beautiful, unexpected, and important things happen.

It is only recently that branched flow has been realized to be a common phenomenon in many disparate systems found in nature, in man-made structures, and in the laboratory. We now know it is important on the micron scale in semiconductors (electron waves), on the scale of kilometers in the atmosphere (acoustic waves), on the scale of thousands of kilometers in the oceans (internal acoustic waves, surface ocean waves leading to freak waves [187] and tsunami waves generated by earthquakes), and we suspect on the galactic scale involving microwave emission from pulsars traveling through the interstellar medium, just to name a few.

Branched flow is a mainline topic for this book, because an interesting new realm of nonlinear classical dynamics leads to branching structure and chaotic flow, in turn raising deep quantum-classical correspondence issues, with implications for many physical phenomena. For example, experimental branched flow is usually wave based and can be difficult to simulate, but can a vastly easier ray tracing analysis be trusted? Can semiclassical methods be implemented readily?

24.1 Classical Branched Flow

The agent of refraction is ultimately group velocity speed variation from place to place in the medium; this can be caused by variations in density, composition, temperature, potential energy, or velocity of the medium or a combination of these. For example, for deep water waves or air, with wave vector $\vec{k}$, and velocity $\vec{U}(\vec{x})$ at position $\vec{x}$ of the medium, we have

$$\begin{aligned} \omega_w(\vec{k}, \vec{x}) &= \sqrt{g|\vec{k}|} + \vec{k} \cdot \vec{U}(\vec{x}) \text{ for deep water waves and} \\ \omega_s(\vec{k}, \vec{x}) &= c|\vec{k}| + \vec{k} \cdot \vec{U}(\vec{x}) \text{ for sound waves,} \end{aligned} \tag{24.1}$$

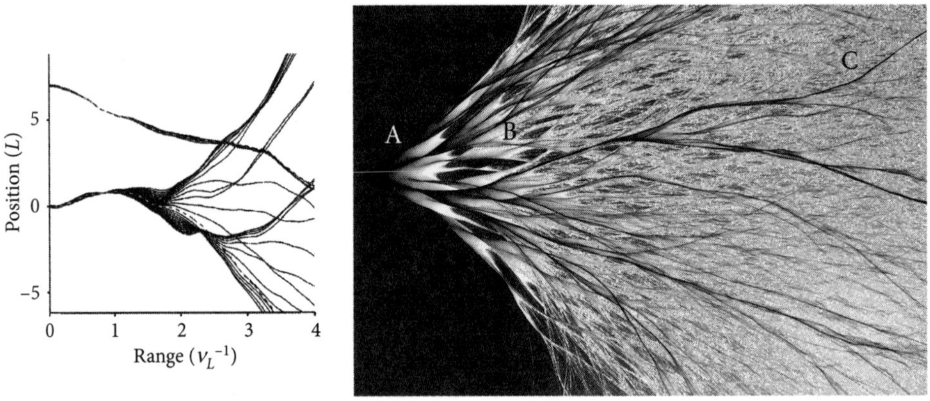

Figure 24.1. (Left) Ray tracing of acoustic waves propagating in the ocean, by Wolfson and Tomsovic [188]. A stable tube is shown at the top, and unstable motion at the bottom. (Right) A similar situation is shown, with more rays and for longer time and distance. The zones of lower flux are white; zones of no flux are colored gray; and zones of high flux, black. Still, the same structures are seen—namely, stable tubes (C), the early first caustic regime (A), and the regime of multiple caustics and the beginning of branched flow (B).

with the Hamilton-like equations

$$\frac{d\vec{k}}{dt} = -\frac{\partial \omega_{w,s}(\vec{k}, \vec{x})}{\partial \vec{x}}; \quad \frac{d\vec{x}}{dt} = \frac{\partial \omega_{w,s}(\vec{k}, \vec{x})}{\partial \vec{k}}. \tag{24.2}$$

If $\vec{U}(\vec{x})$ is turbulent or possesses eddies on various scales, and so on, the situation for random small angle refraction of the rays, and thus branched flow, may be set up.

Wolfson and Tomsovic [188] ventured well beyond the realm of the first focal caustic and examined the consequences. Their context was ocean acoustics. Independently, Topinka et al. discovered branched flow in electron waves traveling over the smooth random weak potential barriers present in semiconductor heterostructures like GaAs-AlGaAs [189]. With donor atoms present, these structures have a 2D electron rich layer with quasi-ballistic, branched flow of electrons if injected through a narrow "quantum point contact" at energy just above the Fermi level, or at holes propagating just below it. Topinka et al. also simulated the phenomenon numerically and explained its morphology. See figure 24.1. The paper by Wolfson and Tomsovic clearly delineated two key features of what we now call branched flow: the proliferation of caustics through the formation of subsequent generations of cusp caustics beyond the first, and the persistence of stable "tubes" of flow, the number of which decreases exponentially from the source.[1]

The regime just before the onset of the branched flow phenomenon is very common (figure 24.2, and region A in figure 24.1, right): an example is starlight twinkling in

[1] We want to clarify that some aspects of the literature of what we now call *branched flow* (including structure beyond the first focal caustic) is not without antecedent in the literature—in particular, the work of Yu A. Kravtsov is extensive and noteworthy. The Wolfson and Tomsovic article [188] lay on the *JASA* editor's desk for several years, and finally did not receive a proper submission date. The discovery of the stable branches and the understanding of their survival belongs to Wolfson and Tomsovic.

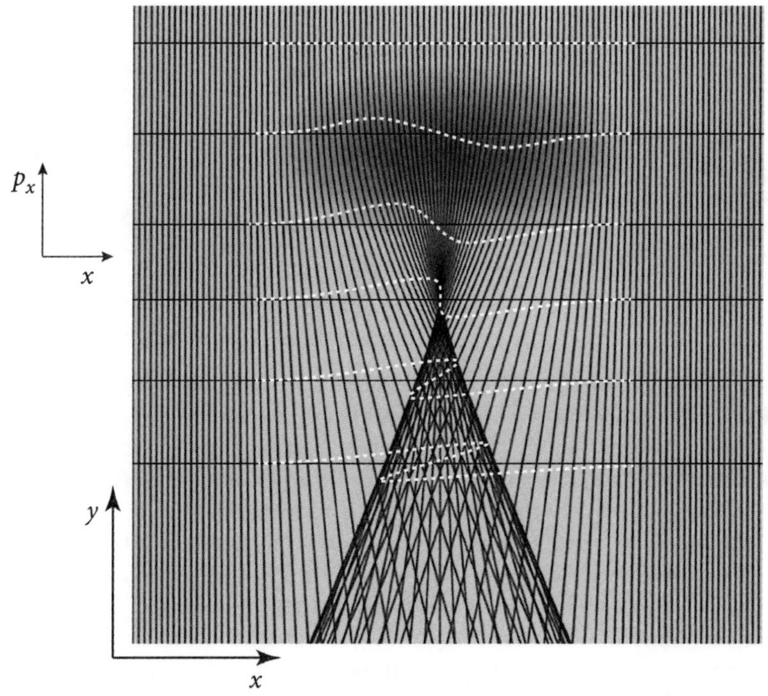

Figure 24.2. A cusp formed by parallel rays coming from the top impinging on a poorly focusing lens (darker gray) with variable refractive index shown. This is the generic morphology of rays near a degraded focal point; notice that only the rays passing near the center of the lens converge near a single point. Along six surface-of-section slices starting from the top, the momentum of the rays along the slice (x) coordinate is plotted against the position of rays crossing the slice, as white dashed lines. At first there is no x component of the momentum; then it develops as refraction occurs. Later, free drift to the right for positive momentum components, and to the left for negative, folds the dashed manifolds over on themselves. When a section of the manifold lies vertically over the x axis, as in the fourth slice starting at the top, a singularity develops in the density of rays. Two more singularities head out from there, as the manifold develops twofold caustics following from the momentum kicks and free drift motion forming a "V" shape.

a fairly stable atmosphere. This is the regime of the first focal caustic, a so-called cusp, loosely known as an imperfect focal point or focal point of a bad lens. As such, it is a structurally stable object—its basic morphology (coming up) will not change with small changes of the lens. A perfect lens, with all rays coming to a single point, is degraded (to a cusp) under the slightest change, and is structurally unstable. There are several classes of stable caustics in various co-dimensions, with characteristic diffraction patterns, a literature aptly discussed and enlarged by Berry and Upstill [190]. Regarding focusing of waves, discussed later in figure 24.4, their refraction and formation of first caustics by bottom depth variations was first discussed by Berry [191].

The beginning of branched flow is seen in figure 24.1, left, or in region B of figure 24.1, right. Here, the second and higher generation of caustics are beginning to form. Stable branches, first noted by Wolfson and Tomsovic [188], are seen in

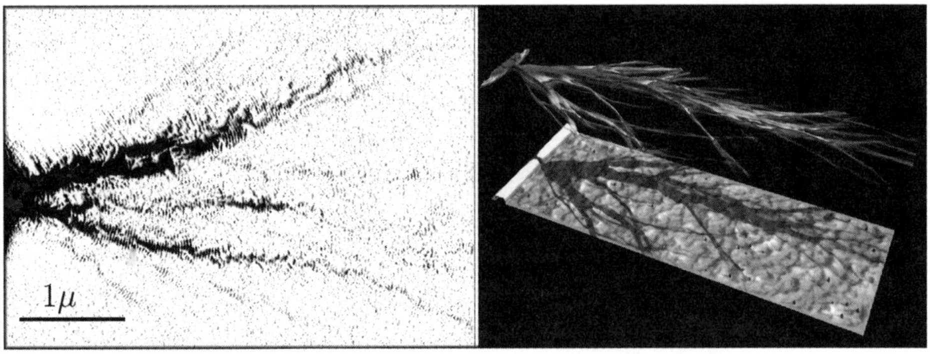

Figure 24.3. (Left) Experimental image of electrons emerging from a quantum point contact (QPC), as revealed by a charged scanning tip modulating the current through a GaAs-AlGaAs heterostructure. The data is current versus tip position; the current is reduced (colored darker for larger reduction) if the potential barrier caused by the tip backscatters electrons through the QPC; this can't happen unless the electrons are present and specularly backscatter at a given tip position; thus the tip images the presence of electrons. Focal cusps are seen along the branches, and quantum oscillations are present, owing to interference between two sources of returning electron amplitude: (1) a small amount of backscattering owing to sharp defects in the sample, and (2) backscattering from the rather large repulsive potential set up in scanned locations by the charged tip. (Right) Classical ray dynamics with the ray density lifted off the random potential, which was tuned to resemble the potential felt by the electron gas, for clarity; the shadow on the random potential shows the actual location of the branches relative to the potential undulations [4, 189]. Note that there is no simple relation between them.

figure 24.1, top left, or in region C of figure 24.1, right. The cusp in figure 24.2 is the result of two effects: (1) a change of direction due to refraction of rays, and (2) propagation some distance in the new direction. Outside the lens, there is only free propagation, but inside both refraction and drift are simultaneously taking place. Some hint about the ubiquity of branched flow is evident when inspecting figures 24.3 and 24.4.

We now begin to delve into branched flow in more detail, first in a classical context. We introduce a map in the (x, p_x) surface of section possessing all the essential properties of branched flow, called the *kick and drift map* [19]. As the name suggests, the momentum p_x is randomly raised and lowered (with a given correlation length of the randomness) in one step (the kick step), and then the displacements due to the new momenta are allowed to develop (free drift step). The process is then repeated, with independent random momentum changes. Successive kick-drift cycles are seen at the right in figure 24.5 applied to a horizontal strip with finite thickness, and an array of filled circles, the latter shown so that the local stability can be followed readily.

The random nature of the effective potential for electron flow in two-dimensional semiconductor electron gases was extremely well known for years, but it was treated statistically and no one made images of its effects; if they had, there would overnight have been much interesting new theory required! We start, however, with the older and very solid statistical arguments and the nature of small-angle random scattering. Under a succession of uncorrelated encounters with barriers of a height much less than

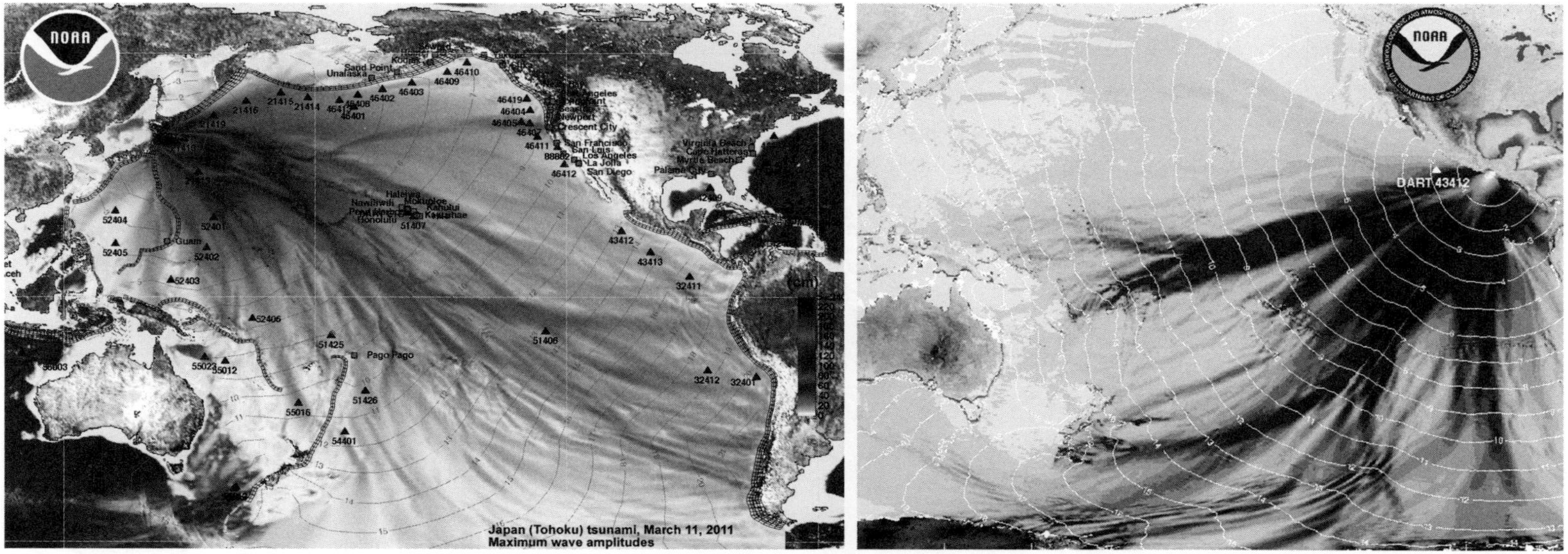

Figure 24.4. (Left) The wave height over the whole event arising in the Pacific from the Sendai earthquake, computed by the NOAA Center for Tsunami Research using the 9.0 Richter quake off Sendai and the contours of the ocean bottom. Branched flow is evident. The ocean bottom contours refract the wave, since the average 4 km deep Pacific Ocean is shallow water for the tsunami wave, which had a wavelength 100–500 km. The speed at which shallow water waves travel is proportional to the square root of the water depth. (See https://svs.gsfc.nasa.gov/30013 for an animation of the wave propagating, at over 400 miles per hour.) It was later noted [192] that the bathymetry data on the ocean depth is not yet sufficient to make accurate long range predictions of the disposition of high energy branches. See also [193]. A 25-year-old man was swept into the Pacific Ocean and drowned near the Klamath River in Del Norte County, near the odd branch seen striking the Northern California coast. (Right) The NOAA model for the November 7, 2012, 7.4 magnitude Guatemalan tsunami.

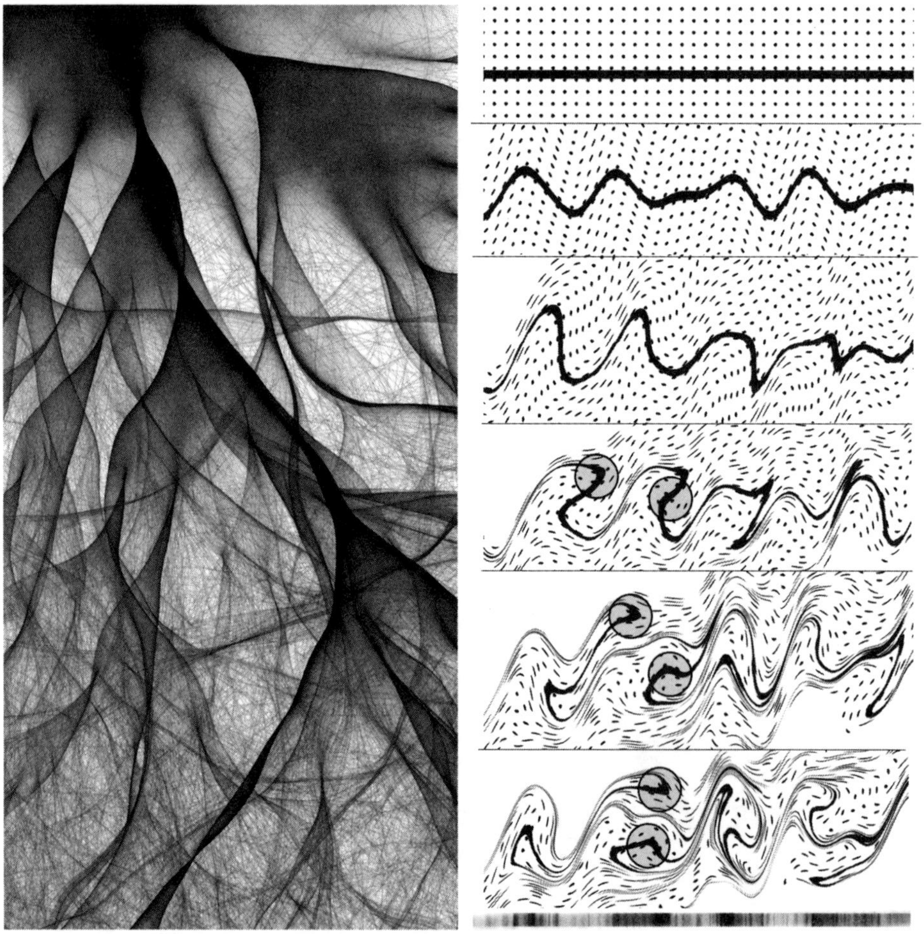

Figure 24.5. (Left) Ray trajectories launched from a point above the image, riding over a two-dimensional potential with random smooth barriers. First through fourth (at least) generation cusps are seen, along with a healthy stable branch. (Right) Kick and drift model applied to a horizontal rectangle with some thickness, and an array of filled circles. Five kick and drift cycles were followed. This image was easily produced using area-preserving distortion tools in Adobe Photoshop, called Wave (vertical random shear with horizontal component suppressed) and Shear (smooth ballistic drift proportional to momentum vertical height, with 0 at the center). Stable regions that have formed and survived until the fifth cycle are shown. Note the extreme nonuniformity of what has become of the initial rectangle. Local stability can be read off the shape and stretching of the mapped circles. The projection of these images onto the *x* axis corresponds to the coordinate space density such as on the left. These two cases have received different random kicks however. The bottom right shows the vertical projection of phase space—that is, the coordinate space density corresponding to the frame above it.

the kinetic energy of the trajectory, small left and right deflections start to add up to a diffusion and loss of memory of the initial direction of the trajectory. Clearly, this is a Gaussian random process like coin flipping. The momentum distribution of an ensemble of trajectories with initial direction $\theta = 0$ evolves as

$$P(\theta) = \frac{1}{\sqrt{4\pi t/\tau}}\, e^{-\tau\theta^2/4t}. \tag{24.3}$$

Since

$$\langle \cos(\theta(t)) \rangle = \int P(\theta)\cos(\theta)\, d\theta = e^{-t/\tau}, \tag{24.4}$$

and

$$\langle \vec{p}(0)\cdot\vec{p}(t)\rangle = |\vec{p}(0)|^2 \langle \cos(\theta(t))\rangle = |\vec{p}(0)|^2 e^{-t/\tau}, \tag{24.5}$$

we see that exponential decay of momentum autocorrelation results. This is of much importance, but still does not tell us what the flow looks like.

KICK AND DRIFT MODEL

A very simple way to gain some intuition for branched flow is though a kick and drift map, which simulates the flow over a smooth random potential. The kick corresponds to the manifold passing over random smooth weak barriers at a certain range, lowering and raising the momentum transverse to the flow along the manifold, according to which side of a barrier or dip in the medium that part of the manifold encounters. Displacement due to the kicks follows in the next drift step, just as motion follows acceleration in classical mechanics.

The map can be implemented in Adobe Photoshop. First, define a manifold with some thickness greater than a few pixels. Use a canvas with many pixels. Initial colored strips and circles can be informative. The thick manifolds can then be made fuzzy if you want, corresponding to a weighted distribution of initial conditions. Select the image and apply the Wave distortion—vertical as near as possible, with horizontal set to a minimum. Experiment with the number of generators, their amplitude, and wavelength range. Apply the transformation and then apply a Shear, setting the shear window. Then repeat—a new random kick will be generated—and you have an area-preserving kick and drift map, which was used to produce figure 24.5, right, for example, on a dot matrix and a horizontal band.

To obtain coordinate space density, the vertical projection (integrating over momentum at each position) of the kick and drift phase space plots is needed. This is called a radon transform of the image. A free tool operating under Java called ImageJ is available, and Matlab has radon transforms if you are a user. Otherwise, you can do a "poor person's radon transform" by using the motion blur tool with a large vertical blur, then increase the contrast again with Image → Levels.

Kick and drift can easily be implemented quantum mechanically as well, by alternately applying the potential operator $\exp[-iV(x)\tau]$ for a smooth random $V(x)$ followed by $\exp[-ip^2/2]$, using the split operator method on a grid of points (see section 7.3), and repeating the process for independent choices of $V(x)$ each iteration.

CHARACTERISTIC PARAMETERS

Figure 24.6 shows branched flow labeled by two important parameters that we need to discuss: the typical distance d normal to the average flow between the first focal caustics, and the typical distance L downstream from the onset of refraction to the first caustic. The distance d between caustics, transverse to the mean flow, is easiest to pin down: by moving one correlation length of the potential across the flow, we have

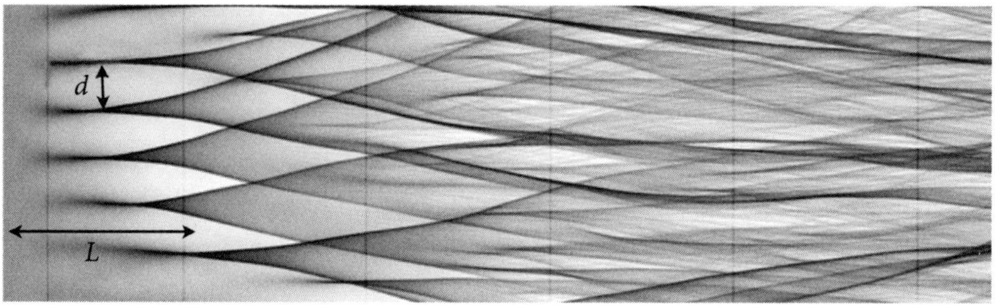

Figure 24.6. A typical branched flow labeled by two important parameters—namely, the typical distance d between the first focal caustics, normal to the average flow, and the typical distance L downstream from the onset of refraction to the first caustic.

arrived at an independent force field gauntlet for trajectories initiated there. Eventually, a totally independent focal caustic will be formed. Therefore, the transverse distance between first caustics is just the correlation length of the potential. It might be objected that some closer distance away would be sufficient, given the sensitivity of trajectories to initial conditions. This is certainly true for higher generations of caustics, but if the potential is fairly strong, the first caustic will form quickly.

We now discuss the distance L to the first caustic starting at the onset of parallel rays. We cast this question in the form of the ratio L/d, which is dimensionless. The barrier height ϵ and the trajectory energy E must both be key, so we are looking for some positive power of the dimensionless ratio E/ϵ, (positive since weaker barriers and faster trajectories must both lead to greater distance to the first caustic). But what power? We must be mindful that this question applies many correlation lengths downstream. Writing

$$\frac{E}{\epsilon} \propto \left(\frac{L}{d}\right)^{\eta},$$

with η to be determined, suppose we increase the correlation length d, leaving E snd ϵ alone. The number of encounters with barriers per unit time will decrease as $1/d$, and since the momentum diffusion goes as the square root of the number of encounters, the transverse momentum growth suffers a $1/\sqrt{d}$ slowdown. The time τ taken for folds from adjacent cusps to overlap as seen in figure 24.6 is approximately $v \cdot \tau = d$ where $v \propto 1/\sqrt{d}$ so $\tau \propto$ distance traveled to first caustic $\propto d^{3/2}$. Thus, the power $\eta = 3/2$, and

$$\frac{L}{d} \propto \left(\frac{E}{\epsilon}\right)^{2/3}. \tag{24.6}$$

FUZZY MANIFOLDS

The epicenter of a typical tsunami is usually nearly a point source relative to the waves it will produce. A point source of rays creates a thin manifold when plotted in phase space, just like a set of parallel rays does. The context of ocean storm waves, however, raises new questions. The storm is not a small source, and waves born of a storm have typically a 10- or 20-degree range of propagation directions even before any refraction. This corresponds to a "fuzzy" manifold with some thickness in phase space. It might be thought that the sensitivity of ray paths to small changes in initial

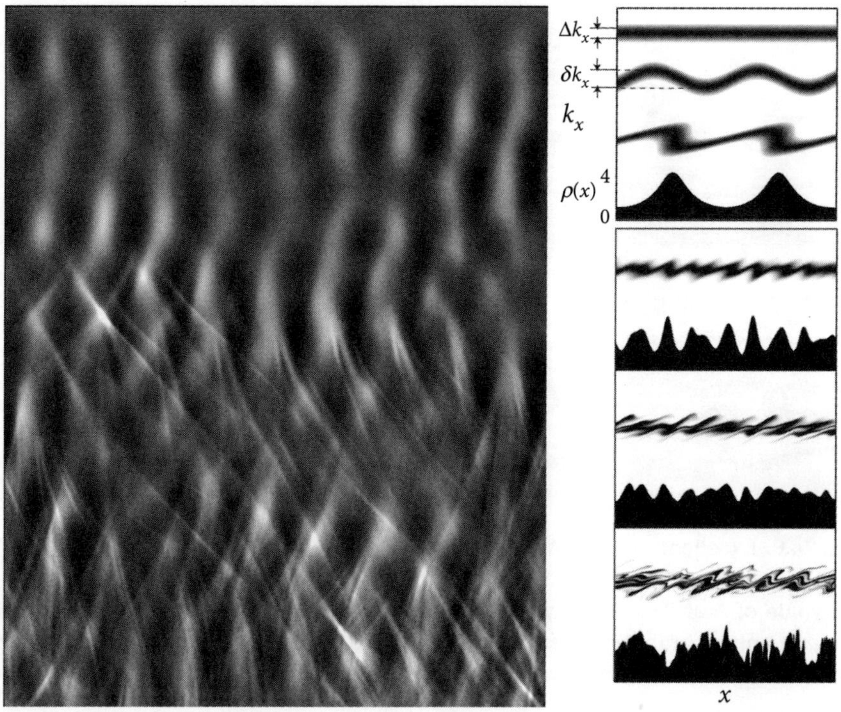

Figure 24.7. An almost self-explanatory image showing the evolution of a fuzzy manifold under weak random refraction, propagating from the top down. The fuzzy manifold is shown at the top right, with a wave vector spread of Δk_x. The density spread at first smooths over the sharpest caustic structures, but later, density evolution develops ever more dramatic or sinister structures, depending on your perspective. Not only are strong contrasts surviving the averaging of a fuzzy manifold, but the detail and contrast *increases*, at least for some time and distance, beyond the regime of the first caustic focal points. Note the transverse motion of the sharper features that are suspected agents of the "freak wave from the side quarter" phenomenon repeatedly reported.

condition in a chaotic system would neutralize the caustics. However, we have been careful not to designate the branched flow phenomenon as chaos—think of the stable branches, for example. True, over long times and distances the stable branches vanish and the usual criteria for chaos—namely, exponential separation of nearby trajectories applies—but the interesting phenomena seen in the preceding figures are not in that long range limit, even over the range of the whole Pacific Ocean in the case of tsunami waves.

So how does a fuzzy manifold behave under the random refraction? Figure 24.7 reveals volumes about the issue.

The reason for the sharpening of coordinate space features in spite of the initially smoothed phase space distribution has to do with incursion and folding of empty phase space into the occupied fuzzy zone, together with that zone being promoted to higher momentum in some places. In retrospect, this is almost obvious, but its implications are profound for the energy density (or probability, depending on the medium) of the medium. If you are a mariner, this is of the highest importance.

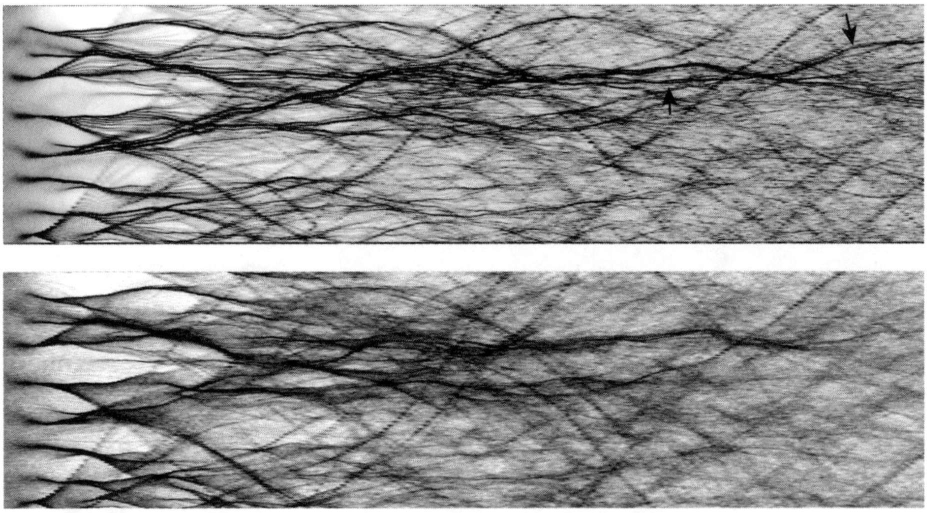

Figure 24.8. Excellent overall agreement between classical (bottom) and quantum (top) probability distributions for a kick and drift map. The quantum wave solution is for a small value of $\hbar$ at fixed energy for an initial plane wave. The arrows in the top panel show a quantum branch that seems to have appeared that is not present classically. Other regions of disagreement are seen farther downstream on the right.

SURVIVAL OF STABLE BRANCHES

Stable branches form by accident as the classical manifolds travel over the random refracting landscape. Certain paths just happen to travel through the right combination of dispersive and focusing zones to end up with no net instability, or very weak instability, up to some time or distance from the source. It is easy to see that such stable regions should soon begin to vanish exponentially, since the "special" zone of stable motion in phase space has no special privilege going forward over territory that is completely independent of what has come before. Therefore, each stable branch suffers a finite probability of encountering very unstable behavior in the next time interval. Once significant instability has affected a zone in phase space, it is unlikely (though not impossible) that future random encounters will undo the instability exactly along the local unstable axis and render it into a stable branch. As Wolfson and Tomsovic show [188], the probability of finding a stable branch a distance r from the source goes as $P(r) = a\sqrt{r}\exp[-br]$, where the $\sqrt{r}$ factor is due to "accidental" formation of new stable regions that then face exponential decay of survival probability.

24.2 Wave Branched Flow

We would not have spent so much time on the ray dynamics of branched flow, smoothed or unsmoothed, unless there were close connections with the wave dynamics in the same systems. First, we compare exact quantum with exact classical distributions for the same initial conditions and Hamiltonian. Figure 24.8 shows the close correspondence in one case. However, this does not preclude strong deviations in other cases, especially for long time propagation. After all, the development of multiple overlapping classical manifolds presents the opportunity for constructive and destructive interference shifting the quantum mechanical probability density away from the classical.

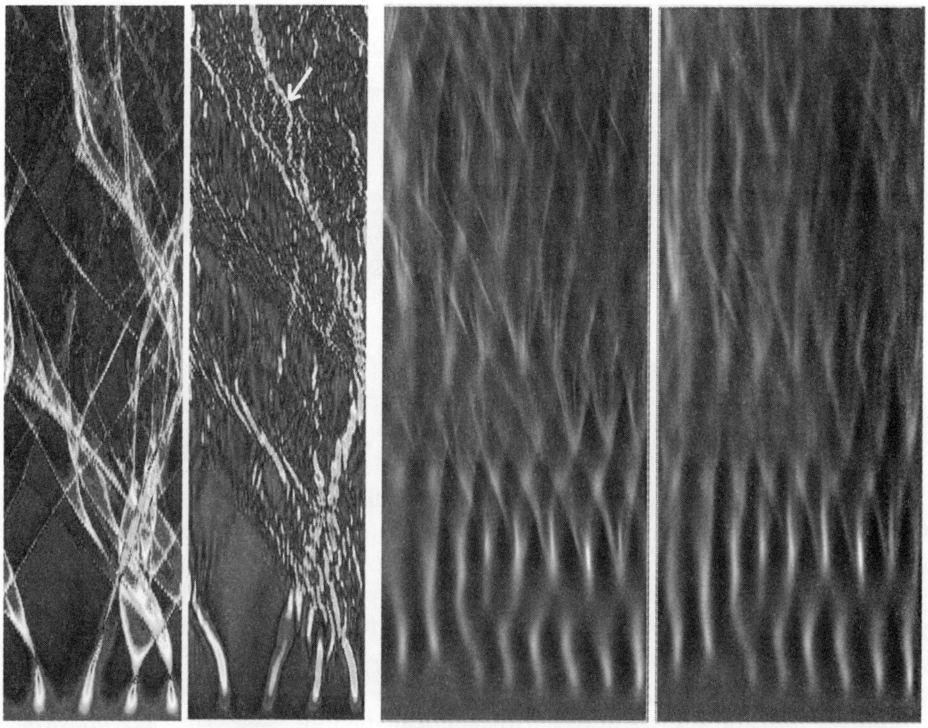

Figure 24.9. The classical (far left) and quantum solutions (adjacent) for a particular value of $\hbar$ in a kick and drift map, with the probability distribution plotted every step and stacked above the results of the previous step. A classical-quantum deviation is seen near the arrow in the second panel. Next is shown the classical distribution for a smooth random 2D potential, with an average over a 10-degree angle of incidence. On the far right, the quantum probability density over the same average is shown. Classical-quantum deviations have nearly disappeared under the averaging.

There is a revealing question about such deviations: are they temporary or permanent? If the semiclassical approximation is working well, there is no reason why the quantum probability distribution should always agree with the classical, since vertically overlapping multiple tendrils of the phase space flow must be added coherently, with resultant constructive and destructive interference shifting the quantum and semiclassical distribution from the classical. This was demonstrated dramatically in connection with a quantum revival in a Morse oscillator, captured correctly semiclassically and giving a very different probability distribution from the classical one (see figure 8.2). However, the very strong deviation is temporary, although the classical and quantum will never again agree in detail (and have not since the Ehrenfest time). If semiclassics is working well in a branched flow situation, a quantum branch might deviate (shift or break up) for a short time from the classical but would recover. This seems to have happened just below the arrow in the second panel of figure 24.9.

We have spent even less time wondering about the semiclassical evaluation of branched flow, and in fact this has been addressed with special techniques [19] involving uniformizations, which the semiclassical evaluation is fairly quickly in need of after a few generations of kicks and drifts or their equivalent.

Chapter 25

Decoherence

The notions of decoherence, entanglement, and so on have no bearing on the fundamental quantum measurement problem, often called the "reduction of the wavepacket." It is starkly framed in terms of the famous Schrödinger cat paradox. It is just as true today as when Feynman is supposed to have said, "If you think you understand quantum mechanics, you don't understand quantum mechanics." The mystery is clearly demonstrated by monoenergetic electrons emerging with fixed energy from a small port less than a wavelength across. The electrons diffract on leaving the port and set up a nearly hemispherical outgoing wave. The electron wave is coherent: if it is passed through two slits some distance away, a screen even farther away would show a two-slit interference pattern after many electrons have passed through. The electron wave, even for a single electron at a time, must have passed through both slits coherently. Removing the slits and allowing the electrons to fall directly on a large screen-detector, each electron will show up somewhere on the screen. No contradiction with quantum theory in the sense of predictions is implied, but we ask: what happened to the amplitudes for arriving at all the other places on the screen? Of course, we do not see more than one detection event for one electron. This is collapse or "reduction" of the wavefunction. It has been proven that there cannot be "hidden" variables that we are not yet privy to that do the deciding. In 1964 John Bell gave a theorem showing that if local hidden variables exist, they could be detected by quantum entanglement. Based on this, Alain Aspect and Paul Kwiat have performed experiments ruling out local hidden variable theories, but not nonlocal ones. So far, all "explanations" of wavefunction collapse have been pure speculation with no basis in experimental fact.

25.1 Calculating the Density Matrix

It is clear that for a system and bath, the system-only density matrix cannot evolve in closed form without approximation, since the response of the bath depends on its full description in general. Although no panacea, we may *always*, in principle at least, start out in different fully coherent wavefunctions for system and bath chosen to represent the experimental conditions, run the dynamics correctly, and arrive at the experimental measurables by appropriate averaging. It is important to gauge how many independent and totally coherent versions of these "runs" are needed to average over in order to reflect the experimental conditions. In some instances, the whole system-bath may effectively self-average in one run; in others, it may take many runs with well-chosen initial condition variations. The full dynamics may of course be daunting. If approximations to the coherent system-bath dynamics are made, they might be more realistic and accurate than many approximations commonly used for density matrix dynamics. The problem may become more tractable by treating the bath and system-bath interactions semiclassically, especially with wavepackets.

Suppose we begin in the wavefunction $|\psi_0^k\rangle = |s_1, s_2, \ldots\rangle|\eta_1, \eta_2, \ldots\rangle_k$, where s are system variables and η are bath variables for the kth trial run; this evolves into an entangled state $|\psi_t^k\rangle = |s_1, s_2, \ldots, \eta_1, \eta_2, \ldots\rangle_k$. We can write the outcome in density matrix form, extracting the measured system density matrix for the kth run as

$$\rho^k(s_1, s_2, \ldots; s_1', s_2', \ldots) = \int d\eta_1, d\eta_2, \ldots |\psi_t^k\rangle\langle\psi_t^{k'}|. \tag{25.1}$$

Now we obtain the full, in principle exact density matrix as the average over the trials k:

$$\rho(s_1, s_2, \ldots; s_1', s_2', \ldots) = \langle\rho^k(s_1, s_2, \ldots; s_1', s_2', \ldots)\rangle. \tag{25.2}$$

MEASURE OF COHERENCE

The density matrix ρ must maintain the normalization $\mathrm{Tr}[\rho] = 1$, since the system coordinates contained in ρ must be found somewhere. But $\gamma = \mathrm{Tr}[\rho]^2$ is another matter. If $\rho = |\psi\rangle\langle\psi|$—that is, a pure state density matrix, then $\gamma = 1$ and we say the system is pure or the purity is 1. But suppose $\rho = \frac{1}{2}|\psi_1\rangle\langle\psi_1| + \frac{1}{2}|\psi_2\rangle\langle\psi_2|$. This has $\mathrm{Tr}[\rho] = 1$, $\mathrm{Tr}[\rho^2] = 0.5$; the system is impure with a purity of 0.5. The density matrix records the coherence between different states, if any, in its off-diagonal elements. The one just considered is

$$\rho = \begin{pmatrix} \rho_{11} & \rho_{12} \\ \rho_{21} & \rho_{22} \end{pmatrix} = \begin{pmatrix} 0.5 & 0 \\ 0 & 0.5 \end{pmatrix}.$$

That is, there is no coherence between states 1 and 2.

25.2 Coherence in Photoemission

In section 5.3, we noted that spectroscopy treats spectral broadening due to a bath by including the bath variables in the autocorrelation expression for the spectrum, as light-matter interaction theory requires. In fact, there need be no clear distinction between "system" and "bath" (see section 5.3). An example is a molecule with a chromophore with a localized electronic transition with the associated local bonds well coupled to the rest of the molecule. One might want to think of the chromophore as the system and approximate the rest of the molecule as a bath, but to get the spectrum correctly, nothing is "traced over," and we need all the degrees of freedom. Even if there is a "bath," if it is displaced by the system, that affects the autocorrelation function too. We may discover that some degrees of freedom are not involved at all and can then be traced over, but since they are not correlated with the system, this has no effect.

In section 13.1, we discussed how decoherence can quench resonant tunneling, but does little harm to nonresonant tunneling.

Sometimes we are directly interested in measuring only a part of a larger system and, assuming we can isolate it and measure it, the remainder, the bath, is ignored—that is, not measured. Suppose we start with an excited hydrogen atom in a large cavity; the whole system may be described by the pure state wavefunction $|\psi_{2p_z}\rangle|0\rangle$, where $|0\rangle$ represents the cavity field state. This has a density matrix Ω,

$$\Omega = |\psi_{2p_z}\rangle|0\rangle\langle 0|\langle\psi_{2p_z}|, \tag{25.3}$$

and a "system" or atomic density matrix,

$$\rho = \mathrm{Tr}_{field}\left[|\psi_{2p_z}\rangle|0\rangle\langle 0|\langle\psi_{2p_z}|\right] = |\psi_{2p_z}\rangle\,\langle\psi_{2p_z}|. \tag{25.4}$$

This has purity 1. Suppose after some time the atom has a probability η of decaying to ψ_{1s}, exciting the field,

$$|\Psi\rangle = \sqrt{\eta}|\psi_{2p_z}\rangle|0\rangle + \sqrt{(1-\eta)}|\psi_{1s}\rangle|\mathrm{k}\rangle. \tag{25.5}$$

It is no longer possible to say that the atom is in some definite quantum state, without measuring the field too. In the absence of knowledge of the field, we can form a density matrix for the atom as

$$\rho = \mathrm{Tr}_{field}|\Psi\rangle\langle\Psi| = \eta|\psi_{2p_z}\rangle\langle\psi_{2p_z}| + (1-\eta)|\psi_{1s}\rangle\langle\psi_{1s}|. \tag{25.6}$$

Now, $\mathrm{Tr}[\rho] = 1$ and $\mathrm{Tr}[\rho^2] = \eta^2 + (1-\eta)^2 = 1 - 2\eta + 2\eta^2 < 1$ if $\eta > 0$. The density matrix for the atom alone is now impure. The atomic density matrix is

$$\rho = \begin{pmatrix} \eta\,|\psi_{2p}\rangle\,\langle\psi_{2p}| & 0 \\ 0 & (1-\eta)\,|\psi_{1s}\rangle\,\langle\psi_{1s}| \end{pmatrix},$$

and we see that whatever the purity, there is never any coherence between ψ_{2p_z} and ψ_{1s} in this case, which would be carried by nonvanishing off-diagonal elements of the density matrix. There is no wavefunction for the atom alone that properly describes its decay.

25.3 Decoherence in a Collision with a Bath Particle

It is a fundamental paradigm to examine the interaction of a system with a bath. We take here a bath of free atoms, too sparse to interact with each other, colliding with a 2D oscillator in its ground, lowest state initially, centered around $x = y = 0$. The initial system (oscillator) state is fully coherent, but it will collide with bath particles. What happens to the system coherence after the bath is traced over?

We treat the collisions as independent, since the bath is not interacting with itself. This reduces to a single bath particle with coordinates (u, v) interacting (colliding) with the system, concatenated over many collisions. We take the bath as a Gaussian wavepacket heading for the system, without loss of generality. The system begins in its ground state (a pure state of the system). The initial state

$$\begin{aligned}\Psi(x, y, u, v, 0) &= \phi_{sys}(x, y)\phi_{bath}(u, v), \\ &= \eta_{xy}\eta_{uv}\exp[-\alpha x^2 - \alpha y^2] \\ &\quad\times \exp[-\beta(u-u_0)^2 + ip_u(u-u_0) - \beta(v-v_0)^2 + ip_b(v-v_0)],\end{aligned} \tag{25.7}$$

is therefore a direct product of the system and bath wavefunctions, a Gaussian in four coordinates with no correlation or entanglement between them. After a single collision between the system and the bath, this is a different matter.

We suppose the interaction potential is smooth on the scale of the wavepacket widths for the whole collision, which can be quite true if $\hbar$ is small enough or masses large enough, and so on. In any case, we suppose that the thawed Gaussian semiclassical propagation is sufficiently accurate. After the collision, all the Gaussian parameters will have changed, including the appearance of nonvanishing elements everywhere in the 4×4 matrix $\boldsymbol{A}_t$ (equation 10.34), which began as a block diagonal with no correlation connecting the (x, y) and (u, v) coordinates.

If the initial, precollision density matrix $\rho = |\Psi(x', y', u', v', 0)\rangle\langle\Psi(x, y, u, v, 0)|$ is traced over the bath coordinates, the remaining system reduced density matrix has purity 1 and is fully coherent, since the initial state is a direct product of two pure state densities.

After the collision, because of the existence of system-bath correlation terms like $A_{xu}(x - x_0)(u - u_0)$ in the matrix $\boldsymbol{A}_t$, the reduced system purity will fall below 1. This is most easily shown for two rather than four degrees of freedom, revealing the essential point without complicating the formulas.

So we first consider a normalized two-dimensional Gaussian (y is now the bath)

$$\psi_0(x, y) = \left(\frac{2a}{\pi}\right)^{1/4} e^{-ax^2} \left(\frac{2}{\pi}\right)^{1/4} e^{-y^2}$$

and a rotated version

$$\psi_\theta(x, y) = \left(\frac{2a}{\pi}\right)^{1/4} \left(\frac{2}{\pi}\right)^{1/4} \exp[-a(x\cos\theta + y\sin\theta)^2] \exp[-(y\cos\theta - x\sin\theta)^2];$$

now the system x and bath y are correlated and entangled. Forming the density matrix for $\psi_0(x, y)$ and tracing over y, we get the reduced system density matrix

$$\rho^{red}(x', x) = \sqrt{\frac{2a}{\pi}}\, e^{-a\left(x^2+x'^2\right)}, \tag{25.8}$$

having trace 1 and purity 1 as we should for no rotation, where there is no entanglement.

Doing the same for the rotated version $\psi_\theta(x, y)$, we find the reduced system density matrix to be

$$\rho^{red}(x', x) = \frac{2\sqrt{a}\exp\left[\frac{\left(a^2+14a+1\right)x^2+\left(a^2+14a+1\right)x'^2-(a-1)^2\cos(4\theta)(x-x')^2-2(a-1)^2xx'}{8(a-1)\cos(2\theta)-8(a+1)}\right]}{\sqrt{\pi}\sqrt{-(a-1)\cos(2\theta)+a+1}}, \tag{25.9}$$

with trace 1 and purity

$$\frac{2\sqrt{2}a}{\sqrt{-a\left((a-1)^2\cos(4\theta)+a(a+6)+1\right)}}. \tag{25.10}$$

For example, the purity for $a = 6$ (the system wavepacket narrower than the bath) and $\theta = 0$ is 1, as it should be, and if $\theta = \pi/4$, the purity is 0.7. Thus, correlation or entanglement in even a single Gaussian (which could have many coordinates) causes decoherence when the "bath" is traced over. Rotations in higher dimensions would cause even faster purity decay. Adding more collisions taken from a thermal bath distribution would soon lead to a thermal system, with no coherence between its eigenstates—that is, a diagonal reduced system density matrix in the eigenstate basis.

25.4 Diffraction and Interference of Buckyballs

In 1999 Anton Zeilinger and his group succeeded in getting de Broglie wave interference of C_{60} Buckyball molecules emerging from an oven by using velocity selection and a 100 nm diffraction grating [194]. This impressive accomplishment is a veritable playground for questions about decoherence.

- The beam was velocity selected, but the Buckyballs came out hot, between 900 and 1000 K. It is easy to show that the density of vibration-rotation states is so high at these temperatures that no two Buckyballs ever emerged from the oven in the same internal state during the experiment. Should this have affected the diffraction pattern?
- It has been shown that the Buckyballs at that temperature would typically radiate several or many infrared photons, both before and after reaching the grating. Since photoemission can be decohering (see again the $2p \to 1s$ decay discussion at the start of this chapter), should the fringes survive?
- It would seem to be extremely easy to change the internal state of a hot Buckyball upon colliding with the diffraction grating. Even though each Buckyball arrives at the grating in a coherent spatially extended state (except for the radiation problem), Buckyballs emerging from the grating at different spots would not interfere if they came off the grating in internal states that varied from one grating groove to the next.

The first point is easy to address. The Buckyballs are distinguishable particles because of their different internal states. They can easily interfere with themselves, but there is no interference with other Buckyballs. Their mass is almost exactly the same no matter what the internal state, and the diffraction pattern is consistent with point particles of the mass of C_{60}. The fact that they all come out of the oven in different states makes no difference to the existence of an interference pattern.

The second point can be addressed by the "which-way principle." That is, if the event in question (photoemission) can be used to distinguish where the particle is—that is, where it bounced off the diffraction grating within a grating spacing, then there should be no interference. But the infrared photons are far too long a wavelength to distinguish the diffraction grating grooves one from the other. Imagine optics aimed at the apparatus designed to image the position of the infrared photon emission. The image would be so fuzzy that it would be unable to distinguish grating slits, even many grooves away. The photons thus did not provide a which-way avenue, and the coherent interference survives.

This brings up an interesting point. Suppose the experiment had been the classic two slit variety (this would be difficult). Then, if a mirror were placed between two separated coherent Buckyball beams—that is, one beam heading for each slit—a right-side infrared photon must have come from the right, and left from left, revealing which slit the radiating Buckyball went through in spite of the long wavelength of the

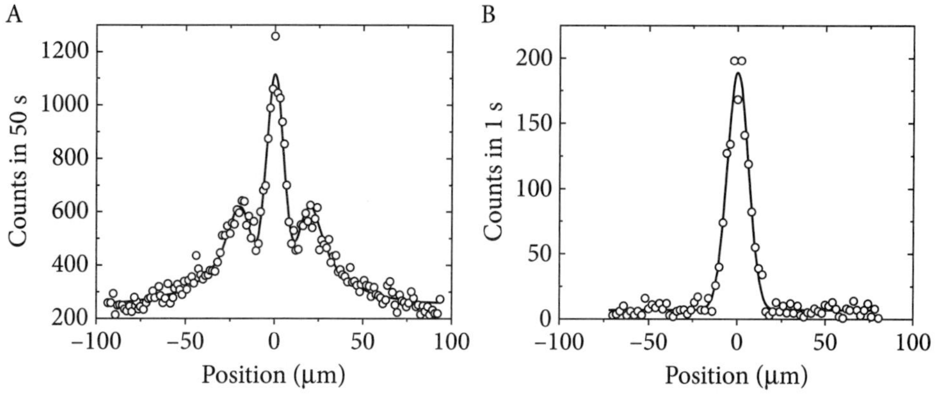

Figure 25.1. Interference pattern produced by C_{60} molecules. (Left) The experiment is shown by open circles, and a fit using Kirchhoff diffraction theory (see section 6.4) is given by the solid line. (Right) The molecular beam profile without the grating, to give a sense of the diffraction peak resolution. From reference [194].

radiation. This little modification destroys the interference pattern, even though the Buckyball never need touch the mirror.

The final point, about changing the internal Buckyball state at the grating, is more subtle. If the diffraction grating had been solid with periodic undulations, it seems likely that the hard collision with highly excited Buckyballs would be a localized detection of their presence and would destroy the diffraction interference. However, the grating slits were in fact 100 nm apart and open, making elastic passage through them possible (see figure 25.1)

25.5 Split Beam Coherence

Some years ago Prof. D. Pritchard and his group at MIT made a beautiful (and surprising, as least to the author) measurement of the refractive index of a gas of sodium atom matter waves passing through monatomic He, Ne, Ar, Kr, and Xe and molecular gases (N_2, CO_2, NH_3, and H_2O) [195]. This was done with a matter interferometer arrangement, with the sodium atoms split into two parallel but coherent beams. If the beams had been immediately recombined, there would be healthy interference fringes. Instead, one beam of the sodium atoms was passed through a dense enough gas of argon or other atoms or molecules (the other sodium beam traveling freely) to shift interference fringes of the recombined beams compared to when the gas arm was evacuated. It seemed surprising for the interference fringes to remain healthy, the sodium having interacted so strongly with the Ar atoms that the fringes shifted. There must have been a good chance of passing through the gas, suffering a healthy phase shift, most often without the slightest disturbance of any gas atom, which would have meant a which-way detection and ruination of the fringes. The fringes shifted due to the change in sodium atom wave speed and phase in the gas arm relative to the free arm.

This is a remarkable thing: a Na atom passes through an Ar gas, at least weakly interacting with many of the atoms (else there would be no measurable refractive index), *and yet standing a good chance of not disturbing, in the slightest, any of the extremely (one would think) vulnerable and essentially free Ar atoms.* The slightest change of momentum of any of them would leave behind a calling card that Na had been present—sure death to any fringes when the beams are recombined, due to a which-way measurement! Yet somehow the Na atoms often get through without the slightest disturbance of even a single Ar atom.

Professor Pritchard knew the cross section for Na atoms impinging on, for example, Ar atoms and used an Ar pressure that gave a good chance of Na atoms getting through without an inelastic collision. The reason Prof. Pritchard was confident the fringes would persist can be summed up by saying he believed in the standard concept of finite total (elastic + inelastic) quantum cross section for Na atoms impinging on Ar. Even an elastic scattering event causes both Na and Ar to recoil and change path, revealing the presence of the Na atom. An inelastic event would be much more drastic and means one or both atoms change internal state.

There must be a phase shift of the Na matter wave, due to the optical theorem (equation 26.32 in 2D): the forward moving flux must be reduced to account for the elastic collisions deflecting the Na beam. This is accomplished by sending scattered amplitude forward *but out of phase with the background plane wave amplitude passing by.* In three dimensions as in the Pritchard experiments, the optical theorem reads:

$$\sigma_{\text{tot}} = \frac{4\pi}{k}\,\text{Im}\, f(0), \tag{25.11}$$

where $f(0)$ is the scattered amplitude in the forward direction. The density of gas was low enough that, according to the cross section, a Na atom had a good chance of getting through without revealing its presence, but each forward scattering event phase shifted the Na matter wave without recoil (because it was purely forward!) and finally led to a shift of fringes in an interference pattern with the reference beam. A detailed study of all the physics confirmed what Prof. Pritchard, Dr. Schmiedmayer, and co-workers knew all along, that the experiment should work [196].

All this is in stark contrast to purely classical mechanics. The atoms interact with weak, long-range forces that slightly perturb their trajectories, in principle quite detectable classically. Put another way, the classical cross section for Na scattering from Ar or other atoms and molecules is infinite. Given the extremely unstable classical mechanics of such a many body system, the state of the Ar gas would be vastly different in detail classically in a very short time as the Na passed through. The sodium atom would leave its calling card every time, classically.

Why is the cross section finite quantum mechanically? At large impact parameter (that is, the closest distance the atoms would approach if there were no interaction between them), the classical deflection is very small, but not 0. Quantum mechanically, collision and measurable deflection happen for small impact parameter, but must shut down for large impact parameter, lest the inelastic cross section get huge. This raises the question of how to detect that a collision had occurred at large impact parameter. Suppose a wavepacket of Na atoms is created that has only large impact parameters in it. This would require a coherent superposition of plane waves of slightly different momenta. If the required range of momenta in the superposition is larger than the classical deflection of momentum at large impact parameters, the classical deflection becomes too small to be detectable even in principle; the effect on the cross section would be nil.

This argument can be used to quickly estimate atomic and molecular quantum cross sections if a Born-Oppenheimer interaction potential and some classical trajectories are available. The decomposition of a plane wave into "partial waves" of different angular momentum is

$$e^{ikz} = \sum_{\ell=0}^{\infty}(2\ell+1)i^{\ell} j_{\ell}(kr)P_{\ell}(\cos\theta), \tag{25.12}$$

where θ is the angular deviation from the straight-ahead z axis passing through the origin, and $j_{\ell}(kr)$ is a spherical Bessel function that behaves asymptotically at large r as

$$j_{\ell}(kr) \to \frac{1}{r}\left(\exp(i(kr - l\pi/2)) - \exp(-i(kr - l\pi/2))\right). \tag{25.13}$$

The angular momentum is $\ell_z = b \times p_z$—that is, the relevant component of the angular momentum vector $\vec{\ell} = \vec{r} \times \vec{p}$. We can create an incoming wave possessing only large impact parameters b by leaving out the low-ℓ terms in equation 25.12, from zero up to $\ell_{min} = b_{max} \times p_z$. The full plane wave equation 25.12 has no momentum uncertainty, but the incoming wave with only large impact parameters is

$$\varphi_{\ell_{min}} = \sum_{\ell=\ell_{min}}^{\infty}(2\ell+1)i^{\ell} j_{\ell}(kr)P_{\ell}(\cos\theta). \tag{25.14}$$

What is the momentum uncertainty of the "hole" of missing low ℓ in $\varphi_{\ell_{min}}$? This hole is $2b_{max}$ across, so

$$\delta p = \frac{\hbar}{2b_{max}}. \tag{25.15}$$

Thus, starting at the large impact parameter b_{max}, defined to the impact parameter beyond which the classical momentum deflection becomes less than δp, we can discount any higher b or ℓ. Then we have the elastic cross section σ_{in}

$$\sigma_{el} \approx \pi b_{max}^2, \tag{25.16}$$

quite finite because the classical momentum deflection starts to decay rapidly, due to a $1/b^6$ or $1/b^7$ van der Waals potential fall-off, as opposed to the $1/b_{max}$ momentum uncertainty. The elastic phase shift for a single long range encounter of Na with Ar will depend on the impact parameter.

25.6 Sticking or No Sticking?

For quite a few years a controversy simmered: Do very low energy atoms approaching a surface aways stick to it, or always bounce off, or somewhere in between? It turned out that the answer lay in decoherence [197]. First, we must entreat the reader to review section 12.5 on quantum reflection.

Classically, the incoming atom, no matter how slowly it approached from a distance, would be traveling fast and hit the repulsive wall hard after accelerating in the van der Waals attractive region. Any phonon created and radiated away in the solid would leave it bereft of the energy needed to escape. It would stick. Quantum mechanically, the one-dimensional potential used in section 12.5 is representative of the potential an atom feels approaching a surface, if the semi-infinite solid forming the surface had its atoms frozen. So the answer would seem to be that atoms always reflect as they approach with energy $E \to 0$. There was an exception though: a bound state exactly at threshold would draw the atom in—but this is a rather rare occurrence in 1D. It leads to a zero-energy resonance and a long delay inside the potential well, plenty of time to relax to a lower level in the well by emitting energy to the surface and become trapped.

If the collision is with an atom on a spring, with an attractive potential well, another type of threshold resonance comes in to play: an inelastic "Feshbach" resonance, where the well depth is tuned so the atom drops into a bound state of the well if the incoming approach energy is 0. In a semi-infinite solid there will be Feshbach resonances galore at threshold, since phonons with a continuum of energies are available to drop it into a bound state within the well. This is sticking! Creating the right energy phonon would be a resonant process, causing it to dominate. So it would seem we have arrived at the opposite answer—very low energy atoms would always stick to the surface.

The question was partially answered experimentally [198] with a result of partial sticking of about 0.2 probability, and around the same time the explanation was given [197] that there should be no sticking. The experiment involved dropping helium atoms from a low altitude onto a superfluid ^{4}He surface. A superfluid surface, however, is a new ballgame. Later, excited neon and helium atoms were reflected from a flat silicon surface [199], and the reflection probability agreed with a nonresonant one-dimensional picture—that is, reflection probability going to 1 as $E \to 0$.

Returning to the picture of the formation of a threshold resonant buildup (sticking) in one dimension described in section 12.5, we recall that to get a buildup past the quantum reflection limit, small bits of amplitude that managed to penetrate must meet incoming amplitude in phase and entice it to pass inside by constructive interference. The precise constructive interference condition is just that of a bound state at threshold. For an inelastic atom-oscillator collision, this picture still holds as the oscillator alternately gives up its energy and presents a constructively interfering wave to the incoming wave. However, this is not going to work if there are many types of phonons produced of the same energy but different wave vector and different phonon band. The incoming particle wave becomes entangled with many slightly different phonons and coherence is lost with the incoming wave. The coherence of the atom wavefunction after tracing over the solid does not persist, threshold resonance buildup does not happen, and quantum reflection prevails.

25.7 Semiclassical Wavepacket Approaches to Decoherence

Decoherence is possible whenever the "system" that we choose to measure has interacted with degrees of freedom that we do not or cannot measure or control—that is, the "bath." If the bath is not measured, then after the system has interacted with it we must integrate over its coordinates, and perhaps average over different initial states of the bath.

The destruction of coherence in the system is measurable: interference fringes diminish or disappear; the system reveals less or no quantum-like behavior. Frozen Gaussian wavepackets (section 11.1) were successfully used by Neria and Nitzan [200] to investigate nonadiabatic rates in condensed phase systems. In a study closer to decoherence issues, Prezhdo and Rossky tied solvent relaxation times as proportional to decoherence times of the solute [201]. This followed earlier work by these authors on quantum-classical molecular dynamics simulations [202]. Other important work on electronic transitions and decoherence followed [203].

A more formal but general Gaussian wavepacket approach was taken later in reference [204]. The distinction is made between the limit of a classical bath that does not readily act as a which-way detector and a quantum bath that does. In a two-arm coherent matter interferometer, suppose one beam goes through a thermal radiation filled cavity and the other one does not. We assume those frequencies are not resonant with the atom, but the fields do modulate its phase. (Atomic energy levels are generally raised or lowered by the presence of an electromagnetic field, causing the phase evolution.) The classical cavity field is robust in the sense that its quantum mechanical state is nearly unaffected by the presence of the atom, and is certainly not thrown into an orthogonal state. This follows from the coherent state description of classical limit electromagnetic fields, described next. Highly excited coherent states are particularly resistant to decoherence. The classical electromagnetic field consists of highly excited cavity modes, each of which at frequency ω is its own quantum harmonic oscillator of that frequency. If the field is strong and random, the atomic phase change can become random within 2π from shot to shot, washing out the interference with a reference beam that did not pass through the cavity. This is not due to the changes in the bath, but rather the bath has become a source of random noise.

It is interesting that we routinely encounter perfectly safe statements like "a molecule is exposed to a time-dependent electromagnetic field causing a force $\vec{F}(t)$ to be exerted on its center of mass." We take that statement for granted and think about the molecule and the applied force. Yet the field is ultimately a quantum electrodynamic entity; why don't we have to worry about changes to quantum state of the field? Might photons be scattered by the molecule and lead to decoherence of the field "bath"? The field however is ultimately a high-energy coherent state or states, and very robust against decoherence. It is safe to take it to imply a purely classical external force, provided the field is highly populated with photons.

Chapter 26

Multiple Scattering Theory

We cannot provide a well-rounded discussion of scattering theory here. Even a start would take a book the size of this one. Instead, we choose to focus on a topic that is self-contained, rarely treated, very useful, and very physical, including recent experiments.

Multiple scattering theory can serve as a numerical tool for treating arbitrary potentials, by representing the potential as a dense set of small obstacles, closer together than a wavelength. Here, we primarily have in mind another limit of isolated scatterers that could be points or objects, from less than a wavelength up to many wavelengths across, separated in free space by any distance. The larger objects are still to be represented as many dense point scatterers. Travel between the objects leads to deflection when they are encountered (like bouncing off a wall, as in Kirchhoff scattering, section 6.4). Such deflection is often accompanied by diffraction or even is primarily diffractive scattering at the objects. Point scattering methods are ideal for representing this diffraction. We have mentioned that *dynamical tunneling* (see sections 6.4 and 13.1, for example) is a type of diffraction.

The energy E is a free parameter, there being no boundaries leading to quantization. We start with a Hamiltonian H_0 with known solutions in the continuum, like plane waves in a two-dimensional flat potential, for example. To this we add a localized perturbation, which does not have to be terribly small in physical extent, but is of small strength, at least for the moment anyway. The problem is

$$(H_0 + \lambda V)\psi = E\psi. \tag{26.1}$$

Solutions to $H_0\psi_0 = E\psi_0$ are presumed known. Expanding $\psi = \psi_0 + \lambda\psi_1 + \ldots$, we have, to first order,

$$(H_0 - E)\psi_1 = -V\psi_0;\ \psi_0 + \lambda\psi_1 = \psi_0 + \lambda\frac{1}{E - H_0 + i\epsilon}V\psi_0. \tag{26.2}$$

We write this as

$$\psi = \psi^0 + G_0^+ V\psi_0, \tag{26.3}$$

where

$$(H_0 - E)G_0^+ = \mathbf{1}. \tag{26.4}$$

Apparently, we are free to choose any ψ_0 as long as it is an eigenfunction of H_0: this freedom corresponds to the choice of boundary condition. For example, ψ_0 might be a plane wave coming from some direction, or it might be a Green function source emanating from a particular point. In 3D or 2D, $\psi(\vec{r})$ satisfies

$$-\frac{\hbar^2}{2m}\nabla^2\psi(\vec{r}) + (V(\vec{r}) - E)\psi(\vec{r}) = 0, \tag{26.5}$$

while the free Green functions satisfy

$$\nabla^2 G_0^+(\vec{r},\vec{r}') + k^2 G_0^+(\vec{r},\vec{r}') = \delta(\vec{r}-\vec{r}'), \tag{26.6}$$

with the solution

$$G_0^+(\vec{r},\vec{r}') = -\frac{i}{4}H_0^{(1)}(k|\vec{r}-\vec{r}'|), \tag{26.7}$$

in 2D (angular momentum $\ell = 0$, s-wave solution only). The asymptotic form of the Hankel function in equation 26.7 is

$$H_0^1(kr) \rightarrow \sqrt{\frac{2}{\pi kr}}e^{ikr-i\pi/4}, \quad r\rightarrow\infty. \tag{26.8}$$

In three dimensions,

$$G_0^+(\vec{r},\vec{r}') = -\frac{e^{ik|\vec{r}-\vec{r}'|}}{4\pi|\vec{r}-\vec{r}'|}. \tag{26.9}$$

It is instructive to iterate equation 26.3, obtaining

$$\psi = \psi^0 + G_0^+V\psi^0 + G_0^+VG_0^+V\psi^0 + \ldots. \tag{26.10}$$

This can be written as an integral equation for ψ,

$$\psi = \psi^0 + G_0^+V\psi, \tag{26.11}$$

with ψ given by the expansion, equation 26.10. We can also define the "T-matrix," an operator, as

$$T = V + VG_0^+V + VG_0^+VG_0^+V + \ldots. \tag{26.12}$$

Then from equation 26.10, we have

$$\psi = \psi^0 + G_0^+T\psi^0. \tag{26.13}$$

The "uniterated" form of T is

$$T = V + VG_0^+T, \tag{26.14}$$

which may be used to formally solve for T as

$$T = (1 - VG_0^+)^{-1}V. \tag{26.15}$$

It is seen from equation 26.13 that the T matrix is a real workhorse that, once determined, converts any incident wave ψ^0 to the complete solution ψ. There is an important aspect to take note of: from equation 26.13, we write

$$\psi(\boldsymbol{q}) = \psi^0(\boldsymbol{q}) + \int\int d\boldsymbol{q}''d\boldsymbol{q}' G_0^+(\boldsymbol{q}, \boldsymbol{q}'')T(\boldsymbol{q}'', \boldsymbol{q}')\psi^0(\boldsymbol{q}'), \tag{26.16}$$

which has a clear interpretation: $T(\boldsymbol{q}'', \boldsymbol{q}')$ takes as input the amplitudes given by $\psi^0(\boldsymbol{q}')$. T knows what happens to this "initial contact" amplitude at $\boldsymbol{q}'$: *It knows what the final amplitude is for the last interaction with V to emanate from the point* $\boldsymbol{q}''$. This amplitude is $T(\boldsymbol{q}'', \boldsymbol{q}')$. From there, the free Green function takes over and brings the amplitude to the point $\boldsymbol{q}$. $\psi(\boldsymbol{q})$ is the summation of all amplitude leaving from different spots $\boldsymbol{q}''$ and arriving at $\boldsymbol{q}$. Both $\boldsymbol{q}'$ and $\boldsymbol{q}''$ must reside within the range of V, or else T vanishes.

Even a smooth potential could be treated as a collection of small pieces, with appropriate modulation of density or strength to reproduce the smooth undulations of the potential. This works for the same reason that small structures less than a wavelength across on the surface of a mirror give rise to perfectly specular scattering some short distance away, rather than the diffusion one might have expected (see figure 23.5). So suppose

$$V(\vec{r}) = \sum_i V_i(\vec{r}), \tag{26.17}$$

where each piece $V_i(\vec{r})$ occupies its own region in coordinate space. It may correspond to one scatterer or one region of the potential.

Suppose we break up the full T matrix into pieces $T = \sum_i T_i$, where each T_i is responsible for one region, so to speak. Substitute $T = \sum_i T_i$ and $V = \sum_i V_i$ in equation 26.14. Assume now that we solve for each T_i separately, as in

$$T_i = V_i + V_i G_0^+ T = V_i + V_i G_0^+ \left(\sum_j T_j\right) \tag{26.18}$$

Note that each T_i must necessarily incorporate the effect of all the other scatterers. If we sum over all the T_i, we recover $T = V + VG_0^+T$—that is, the whole solution, so solving for all the individual equations 26.18 is justified.

Bringing the diagonal i^{th} term across to the left-hand side permits us to write

$$\left(1 - V_i G_0^+\right) T_i = V_i + V_i G_0^+ \left(\sum_{j\neq i} T_j\right), \tag{26.19}$$

and multiplying through by $\left(1 - V_i G_0^+\right)^{-1}$ gives

$$T_i = \left(1 - V_i G_0^+\right)^{-1} V_i + \left(1 - V_i G_0^+\right)^{-1} V_i G_0^+ \left(\sum_{j\neq i} T_j\right), \tag{26.20}$$

or

$$T_i = t_i + t_i G_0^+ \left(\sum_{j\neq i} T_j\right), \tag{26.21}$$

where $t_i = \left(1 - V_i G_0^+\right)^{-1} V_i$ (see equation 26.15) is the t-matrix that solves the scattering problem for the ith region or scatterer *without the other scatterers present.* In analogy to the full T operator,

$$t_i = V_i + V_i G_0^+ V_i + V_i G_0^+ V_i G_0^+ V_i + \cdots . \tag{26.22}$$

Equation 26.22 shows that the nonvanishing domain of $t_i(\boldsymbol{r}, \boldsymbol{r}') = \langle \boldsymbol{r}|t_i|\boldsymbol{r}'\rangle$ is entirely confined in both $\boldsymbol{r}$ and $\boldsymbol{r}'$ to the domain of $V_i(\boldsymbol{r})$.

The domain of T_i is revealed by iterating equation 26.21:

$$T_i = t_i + \sum_{j \neq i} t_i G_0^+ t_j + \sum_{j \neq i, k \neq j} t_i G_0^+ t_j G_0^+ t_k + \cdots . \tag{26.23}$$

Therefore, $T_i(\boldsymbol{r}, \boldsymbol{r}') = \langle \boldsymbol{r}|T_i|\boldsymbol{r}'\rangle$ spans the domain of $V_i(\boldsymbol{r})$ in $\boldsymbol{r}$, while $\boldsymbol{r}'$ roams over all the domains $V_j(\boldsymbol{r}')$. Clearly then, $T(\boldsymbol{r}, \boldsymbol{r}') = \sum_i T_i(\boldsymbol{r}, \boldsymbol{r}')$ spans the full domain occupied by scatterers, but it vanishes anywhere for $\boldsymbol{r}$ or $\boldsymbol{r}'$ in between scatterers, if they indeed have gaps separating them.

We can collect the equations for the T_i and t_i in equation 26.21 in matrix form as

$$\begin{pmatrix} T_1 \\ T_2 \\ \vdots \\ T_{N-1} \\ T_N \end{pmatrix} = \begin{pmatrix} t_1 \\ t_2 \\ \vdots \\ t_{N-1} \\ t_N \end{pmatrix} + \begin{pmatrix} 0 & t_1 & t_1 & \cdots & t_1 \\ t_2 & 0 & t_2 & \cdots & t_2 \\ t_3 & t_3 & 0 & \cdots & t_3 \\ \vdots & \vdots & \vdots & \vdots & \vdots \\ t_N & t_N & t_N & t_N & 0 \end{pmatrix} \begin{pmatrix} G_0^+ T_1 \\ G_0^+ T_2 \\ \vdots \\ G_0^+ T_{N-1} \\ G_0^+ T_N \end{pmatrix} . \tag{26.24}$$

All the symbols are operators; equations 26.24 involve matrices and vectors of operators. Note that $t_i G_0^+ T_j$, $i \neq j$, sandwiches G_0^+ between the domains of V_i on the left and V_j on the right, so we may as well write $t_i G_0^+ T_j = t_i G_{ij}^0 T_j \equiv a_{ij} T_j$. Then a more compact and useful matrix relation is

$$\begin{pmatrix} T_1 \\ T_2 \\ \vdots \\ T_{N-1} \\ T_N \end{pmatrix} = \begin{pmatrix} t_1 \\ t_2 \\ \vdots \\ t_{N-1} \\ t_N \end{pmatrix} + \begin{pmatrix} 0 & a_{12} & a_{13} & \cdots & a_{1N} \\ a_{21} & 0 & a_{23} & \cdots & a_{2N} \\ a_{31} & a_{32} & 0 & \cdots & a_3 N \\ \vdots & \vdots & \vdots & \vdots & \vdots \\ a_{N1} & a_{N2} & a_{N*} & a_{NN-1} & 0 \end{pmatrix} \begin{pmatrix} T_1 \\ T_2 \\ \vdots \\ T_{N-1} \\ T_N . \end{pmatrix} . \tag{26.25}$$

With $T = \sum_i T_i$, and $t = \sum_i t_i$, equation 26.25 implies

$$T = t + AT. \tag{26.26}$$

Solving for T,

$$(1 - A)T = t; \quad T = (1 - A)^{-1} t, \tag{26.27}$$

where the last form suggests a practical scheme to determine T by inverting $(1 - A)$ with the elements of A represented in coordinate space. This is made concrete next.

26.1 Point Scatterer Model

The point scattering model is extremely versatile, able to describe scattering by disjoint pieces of wall, objects of disparate shapes, smooth potentials (approximated by many points lying closer together than a wavelength), and much more. It is clearly connected with (but more general than) the Kirchhoff approximation; see section 6.4.

We define a point scatterer at $\vec{r} = 0 = F'$ as a potential with a t-operator given as

$$t(\vec{r}, \vec{r}') = \delta(\vec{r})\, \tau\, \delta(\vec{r}'). \tag{26.28}$$

Using $\psi = \psi^0 + G_0^+ t \psi$, or more explicitly, in coordinate space,

$$\begin{aligned}\psi(\vec{r}) &= \psi^0(\vec{r}) + \int d\vec{r}\, d\vec{r}'\, G_0^+(\vec{r}, \vec{r}')\left[\delta(\vec{r})\, \tau\, \delta(\vec{r}')\right] \psi(\vec{r}'),\\ &= \psi^0(\vec{r}) + G_0^+(\vec{r}, \vec{0})\tau\psi(\vec{0}).\end{aligned} \tag{26.29}$$

If the incident wavefunction is a plane wave $\psi^0 = e^{i\vec{k}\cdot\vec{r}}$, we get in 2D:

$$\psi(\vec{r}) = e^{i\vec{k}\cdot\vec{r}} + \tau G_0^+(\vec{r}, \vec{0}) \to e^{i\vec{k}\cdot\vec{r}} - \tau\sqrt{\frac{2}{\pi k r}}e^{ikr - i\pi/4}, \tag{26.30}$$

where $r = |\vec{r}|$, and we used the asymptotic form of the Green function $G_0^+(\vec{r}, \vec{0})$—that is, the Hankel function $H_0^{(1)}(kr)$.

A point scatterer is not simply a Dirac delta function potential with a weight or strength. Indeed, in two and higher dimensions, point Dirac delta potentials do not scatter at all! For example, in three dimensions, it is well known that a hard sphere scatterer of radius a has cross section (the effective size of the object, in the sense of the probability of hitting it if a uniform sheet of flux is sent at it)

$$\sigma = 4\pi a^2, \tag{26.31}$$

which vanishes as $a \to 0$. A delta function potential of strength η can be constructed as the limit of a circular well or barrier of radius a with depth or height going as $3\eta/4\pi a^3$ as $a \to 0$. Any incoming wave will be forced to vanish at such an abrupt and indefinitely strong potential change (a fact that seems to cause it to scatter, even if in the limit $a \to 0$ the wave has to vanish only at a point). But the nascent delta potential becomes a hard sphere of radius a, and since $\sigma = 4\pi a^2$, there is no scattering from it as $a \to 0$!

In 2D, as M. V. Berry noted, a Bessel function of the second kind Y_0 has a logarithmic singularity anywhere it is located and a vanishingly small Y_0 may be used to satisfy the boundary condition required of a delta function barrier, as follows: Starting with a finite radius barrier, an ever smaller amplitude of the Bessel function is needed to make use of its logarithmic singularity in setting the solution to 0 at the required radius as the radius diminishes. Ultimately, the amplitude of the extended, nonsingular part of the Bessel function, and the scattering it would imply, vanishes along with the whole Bessel function amplitude, as the radius of the object shrinks on the way to becoming a delta function.

Interestingly, there are zero range potentials that have do have finite cross sections in 2D and 3D. As pointed out by J. Hersch [205], a potential well may be constructed that deepens as it reduces radius, in such a way that the single bound state energy in the well remains a predetermined finite energy below $E = 0$. This potential retains a finite cross section as $a \to 0$.

Not all values of t are physically permissible. Normally, t should satisfy the optical theorem, which embodies the requirement that a scatterer should neither add nor subtract particles incident on it; they should only "bounce off" so to speak, just as in classical mechanics. A point scatterer emits only spherical (or circular in 2D) waves. If a plane wave $\psi^0 = e^{i\vec{k}\cdot\vec{r}}$ impinges on the scatterer, the spherical scattered wave heads out uniformly in every direction. All the directions are new, except for one: the direction the incident plane wave was heading. Somehow, the scattered wave has to destructively interfere with the incident wave, to account for the flux heading out in new directions. The incident wave is traveling only in the forward direction, along $\vec{k}$. The "optical theorem" in 2D for a point scatterer reads

$$-2\text{Im}[\tau] = |\tau|^2; \tag{26.32}$$

this guarantees the requirement of no net flux added or subtracted. Exceptions to the optical theorem arise when the scatterer is absorptive, or when the scatterers are artificial if for example they arise from a mathematical discretization intended to enforce certain boundary conditions at a continuous wall or object.

If

$$t_i(\vec{r}, \vec{r}\,') = \delta(\vec{r} - \vec{r}_i)\; \tau_i\; \delta(\vec{r}\,' - \vec{r}_i), \tag{26.33}$$

then by our "domain of T" arguments earlier, we must have

$$\begin{aligned} T(\vec{r}, \vec{r}\,') &\equiv \sum_{i,j} \delta(\vec{r} - \vec{r}_i)\; \mathrm{T}_{i,j}\; \delta(\vec{r}\,' - \vec{r}_j), \\ T_i(\vec{r}, \vec{r}\,') &\equiv \sum_{j} \delta(\vec{r} - \vec{r}_i)\; \mathrm{T}_{i,j}\; \delta(\vec{r}\,' - \vec{r}_j), \end{aligned} \tag{26.34}$$

which can be checked explicitly with the preceding formulas. Note that the $\mathrm{T}_{i,j}$ are parameters, not operators, as are the t_i. This leads to the matrix equation involving scalars,

$$\mathrm{T}_i = \tau_i \delta_{ij} + \tau_i \sum_k G_0^+(\vec{r}_i, \vec{r}_k)\mathrm{T}_{k,j}. \tag{26.35}$$

Now set

$$A_{ij} \equiv \tau_i\; G_0^+(r_i, r_k), \tag{26.36}$$

and then

$$\mathbf{T} = \boldsymbol{\tau} + \mathbf{A} \cdot \mathbf{T}, \tag{26.37}$$

$\mathbf{T}$ (bold symbol notation for the matrix of $\mathrm{T}_{i,j}$'s) and $\boldsymbol{\tau}$ are now ordinary matrices ($\boldsymbol{\tau}$ is diagonal), and solving for $\mathbf{T}$ we have

$$\mathbf{T} = (\mathbf{1} - \mathbf{A})^{-1}\, \boldsymbol{\tau}. \tag{26.38}$$

Thus, both the scalar matrix $\mathbf{T}$ and the operator matrix $\boldsymbol{T}$ are solved in terms of a matrix inverse.

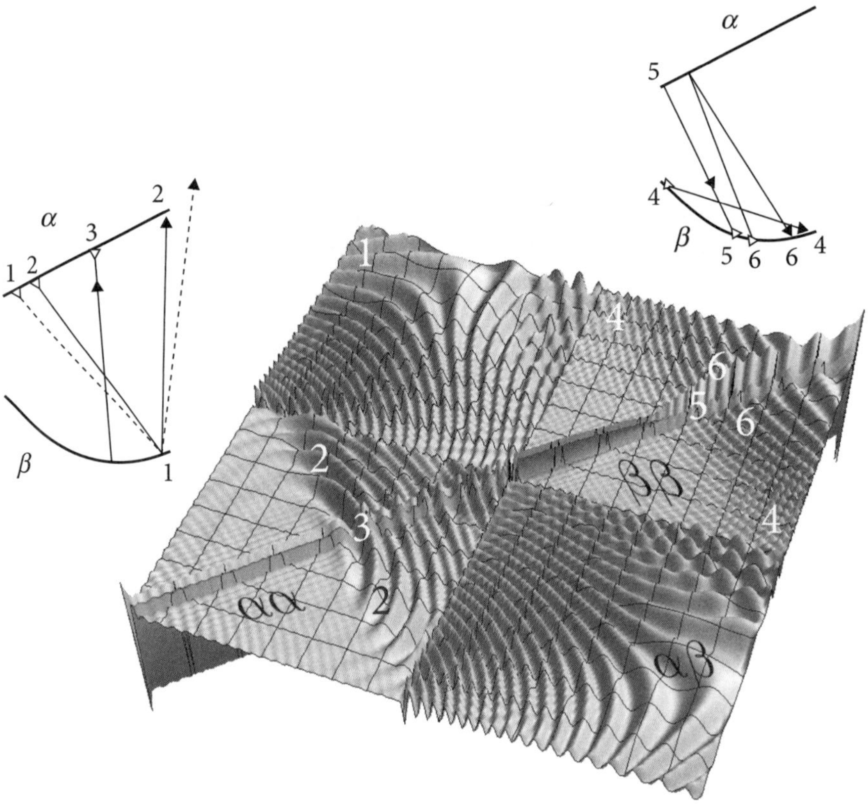

Figure 26.1. T-matrix for two walls, each made of 100 point scatterers, depicted in the satellite diagrams. These show various relevant classical paths relating to featured regions of T. The α wall is at the top, the β wall is curved, below the α. See text for details.

The wavefunction $\psi = \psi^0 + G_0^+ \boldsymbol{T} \psi^0$ in the presence of all the point scatterers becomes

$$\begin{aligned} \psi(\vec{r}) &= \psi_0(\vec{r}) + \sum_{i,j} G_0^+(\vec{r}, \vec{r}_i) \mathrm{T}_{ij} \psi^0(\vec{r}_j), \\ &\equiv \psi_0(\vec{r}) + \sum_i f_i G_0^+(\vec{r}, \vec{r}_i). \end{aligned} \tag{26.39}$$

From the second form we can see that the scattered wave can be regarded as a sum of point sources with amplitude f_i.

26.2 Walls as Point Arrays

We consider the scattering T-matrix for two proximate walls in two dimensions, each with 100 point scatterers $\delta((\vec{r} - \vec{r}_i)\tau_i \delta((\vec{r}\,' - \vec{r}_i)$ defining the wall. Figure 26.1 displays the real part of the T-matrix and shows the two walls. Figure 26.2 shows the real space wave for a certain incident plane wave. In figure 26.1, the "$\alpha\alpha$" block

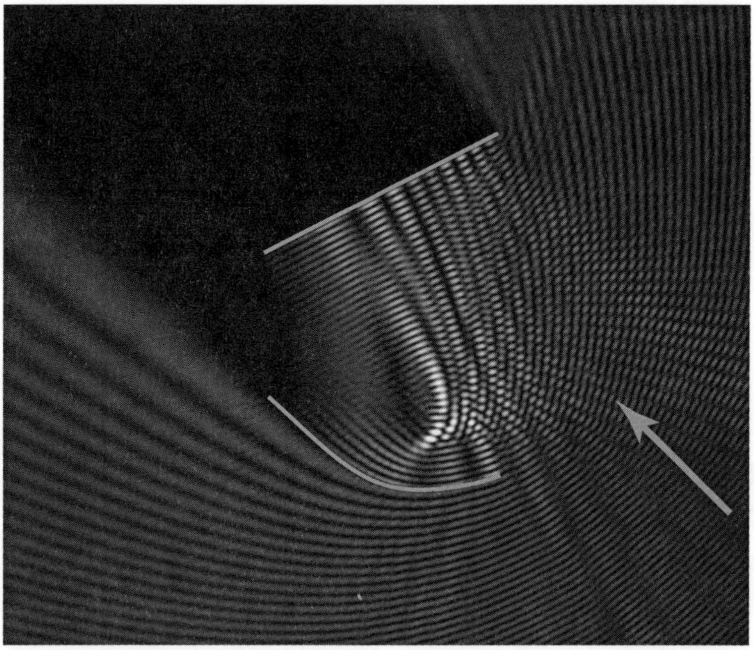

Figure 26.2. Absolute value of the wavefunction for a plane wave incident on the α and β walls shown in figure 26.1. Equations 26.38 and 26.39 were used to compute the wavefunction over a large region. The arrow shows the direction of the plane wave incident on the walls.

describes the T-matrix elements corresponding to arriving at a point on the α line as the first interaction with the walls, and leaving from another point on the α line as the last interaction, after any number of interactions along both walls. There are no classical specular bounce paths from the left end of the α line back to the α line, or vice versa, so the amplitude is weak in this part of the block. Only diffractive paths from the right end of α to the left end of β and back to α are evident as weak undulations. Paths emerge connecting the left part of the α line to the far right part of the α line about 1/8th of the way along it—for example path 2 in the $\alpha\alpha$ block, and the pair of white arrows. Similar stories are told for the $\beta\beta$ block. It is always possible here to reach anywhere on the α line to anywhere on the β line and vice versa, thus the filled $\alpha\beta$ block. The phase (essentially $\exp[ik \cdot \text{length}]$) of paths from spots on α to spots on β, or vice versa, is seen in the undulations.

The diagonal line in the T-matrix corresponds to scattering from a given point to near neighboring points along the line. They are complete with end point diffractive structures. Region 3 in the $\alpha\alpha$ block near the diagonal corresponds to ray paths of the same label that are able to start along α and return to the same spot after a bounce off β. Region 6 in the $\beta\beta$ block near the diagonal corresponds to ray paths of the same label that are able to start along β and return to the same spot after a bounce off α. Corresponding rays are shown in the satellite diagrams.

26.3 Quantum Corrals

In the 1990s the Don Eigler group at IBM Almaden pioneered the positioning of atoms, one-by-one, into predetermined patterns on surfaces using a scanning

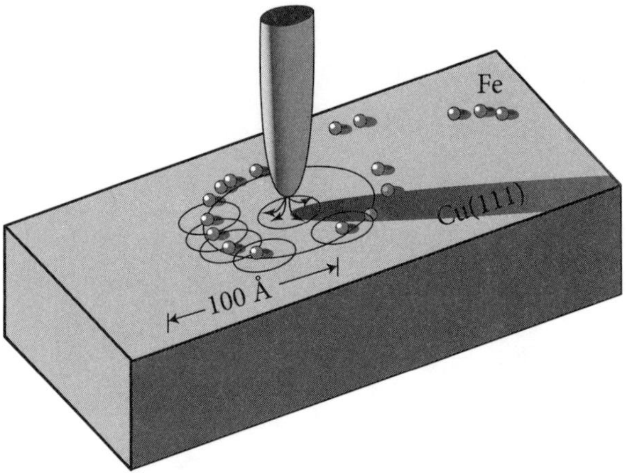

Figure 26.3. Schematic diagram showing the STM tip and electron flow to the surface, where the electron amplitude propagates and encounters Fe adatoms scattering electron waves in all directions, including toward other atoms, where they scatter again. Some amplitude sooner or later returns to the tip by many paths. A single electron has amplitude coherently inside the tip, emerging from it into the vacuum, propagating and scattering on the surface, and returning to the tip to interfere *with itself.* If constructive interference occurs, the rate of tunneling is enhanced. The tip is almost literally a movable Green function source, and maps such as figure 26.6 (later) reveal $\langle \vec{x}_T | \mathrm{Im}[G^+(E)] | \vec{x}_T \rangle$.

tunneling microscope (STM) tip (figure 26.3). Then they imaged the atoms, and the electron sea around them, using the same tip. They placed Fe adatoms on the Cu (111) substrate surface, nestling them in the lattice of triangular lattice pockets. The (111) face of noble metals have an electron conduction band living only near this surface, within a band gap of the bulk material (see figure 26.4). The electrons move freely on the two-dimensional surface, having quadratic dispersion in crystal momentum k—that is, $T \sim |\vec{k}|^2$ with $\vec{k}$ lying any direction along the surface. The surface electron wavefunction decays exponentially (evanescent) in the z direction normal to the surface, heading into the bulk as befits the bulk band gap. It decays with another exponent into the vacuum, starting at the surface.

The data is typically taken by modulating the STM tip height as it is scanned across the surface, keeping the current constant by changing the tip height (the current is coming from electrons tunneling through free space from the tip onto the surface, or possibly in the reverse direction, depending on bias voltage). Plotting the tip height versus position on the surface, the atoms show up prominently since they bring the electron sea up with them to higher elevations. But in a surprise discovery, beautiful undulations were found well away from the adatoms or defects at high tip elevations (too high to sense individual Cu atoms). The atoms or defects were nonetheless responsible for the undulations. Long-range surface electron scattering from tip to atoms or defects and back to the tip is responsible for the undulations. The phase of the returning wave depends on wavelength and the geometry of the scatterers relative to the tip position.

The energy and wavelength of the electrons arriving at the surface is selected by varying the tip potential voltage. The change in current with small changes in voltage—that is, the derivative current dI/dV, is sensitive to only a narrow band of energies

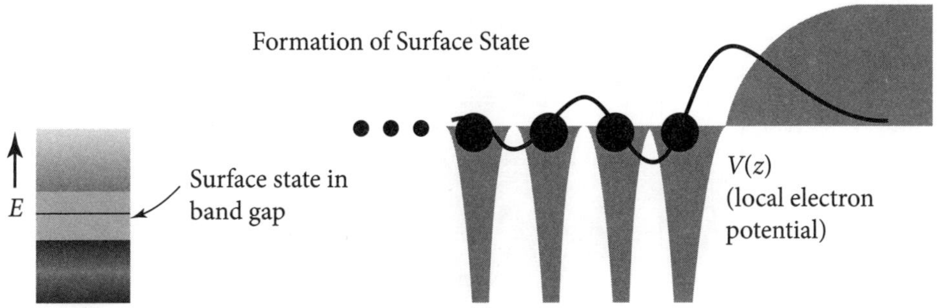

Figure 26.4. Formation of an electron surface state in a bulk band gap, showing the surface wave (black undulating line) in the direction perpendicular to the surface. The electron potential energy in the vicinity of atoms and beyond the surface is shown as dark shading. Band gaps in solids are not devoid of solutions to the Schrödinger equation; rather, they have solutions that blow up in one direction or another and thus are unphysical in unbounded perfectly periodic crystals. Given the opportunity they will take advantage of defects or surfaces and become evanescent waves with energies in the band gap. Here, we see a band gap wave diverging from left to right, only to meet the surface and match smoothly onto another evanescent wave penetrating into the vacuum.

near the top of the Fermi sea with the voltage properly adjusted. The Fe adatoms are small compared to the wavelength of the electrons, which is about 50 atomic units. The iron atoms therefore cause only point-like, s-wave scattering. The atoms are also nonoverlapping in the sense that the smallest distance between them is 9.5Å, much bigger than their atomic diameter. The quantum corrals are thus an ideal case of multiple scattering from separate pointlike scatterers.

The boundary conditions are interesting. The tip is only one atom wide and is small compared to a Fermi wavelength. There are new boundary conditions each time the tip is moved—a source of outgoing waves at the tip position (see figure 26.5). It acts like a (movable) Green function source. The images (see figure 26.6) are obtained by plotting height of the tip at constant current versus position on the surface. The images are not a standing electron wave caught in the corral, but rather the tunnel current as a function of tip position.

If the cavity efficiently confined the electrons (high "Q"), the resonances belonging to states of the cavity would be sharp and isolated from each other, with a resonance width less than the typical spacing between resonances. Then, if the energy of the electrons coming from the tip is selected to correspond to such a resonance, the STM raster scan image would reveal a single standing wave in the corral, a nearly bound eigenstate of the cavity. This is usually not the case in the quantum corral studies, because Q is low.

A simple picture for the appearance of current oscillations with tip position among the Fe adatom structures goes as follows. For positive bias voltages (similar arguments can be made for negative bias using holes), electrons tunnel to the Cu(111) surface from the STM tip. This produces electron amplitude traveling away from the tip along the surface. If the electron hits something (step edge, adatom, and so on), then amplitude may be scattered and return to the region of the tip, *where it will interfere constructively or destructively with the amplitude for the same electron still*

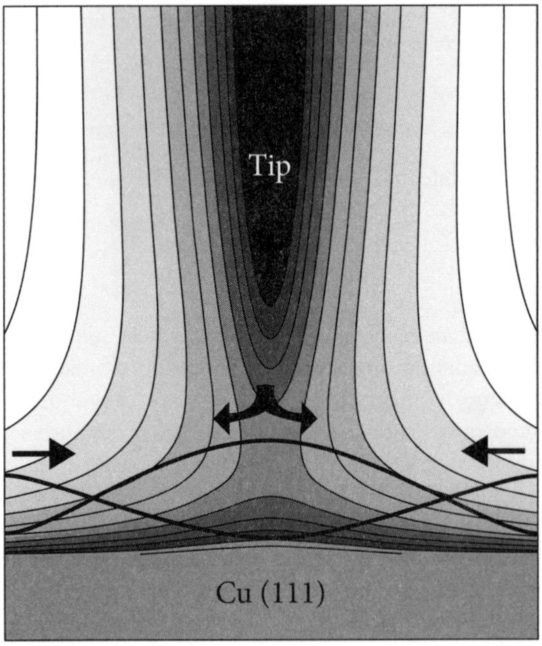

Figure 26.5. Schematic illustration of the causes of tunneling current undulations in STM tip scans near a metal with a surface state such as the one belonging to Cu (111). Electron waves tunnel from the tip to the surface and are then free to travel parallel to the surface. The STM tip and Cu surface are shown surrounded by potential energy contours. Darker shades are lower potential energy. The most direct path from the tip to the Cu surface involves a saddle point barrier; indirect paths through free space have even higher barriers. The electrons must tunnel onto the Cu, entering the surface state when they reach the surface; they are shown spreading right and left. The electron waves reflect off barriers or atoms not visible in the image, returning in phase or out of phase with the waves emerging from the tip. Here, the reflected waves are shown returning out of phase, suppressing the flux from tip to copper surface by destructive interference. If they come back in phase, they enhance the tunneling by constructive interference. The returning amplitude and phase is determined by the distance to the reflecting objects, the electron wavelength, and possible multiple scattering effects. The distances may be changed by moving the tip horizontally; the wavelength may be changed by the applied voltage difference that affects the surface electron energy.

leaving the tip. The flux of electrons leaving the tip is proportional to the square of the net amplitude. The interference therefore will cause oscillations of the current either as a function of (1) distance of the tip to the defects, or (2) electron energy and therefore wavelength, determined by the bias voltage, both of which affect the phase of the returning wave (see figure 26.5). The experiment is sensitive to electrons within only a narrow range of energies because dI/dV is recorded at each tip position and mean voltage. The amplitude may scatter from multiple defects before returning, complicating the calculation of the interference but not the basic principle. The tip, being a point source at $\vec{x} = \vec{x}_T$, generates a Green function wave, $\psi^0 = G_0^+(\vec{q}, \vec{q}_T)$ (on a perfect surface with no adatoms, defects—that is, no scattering).

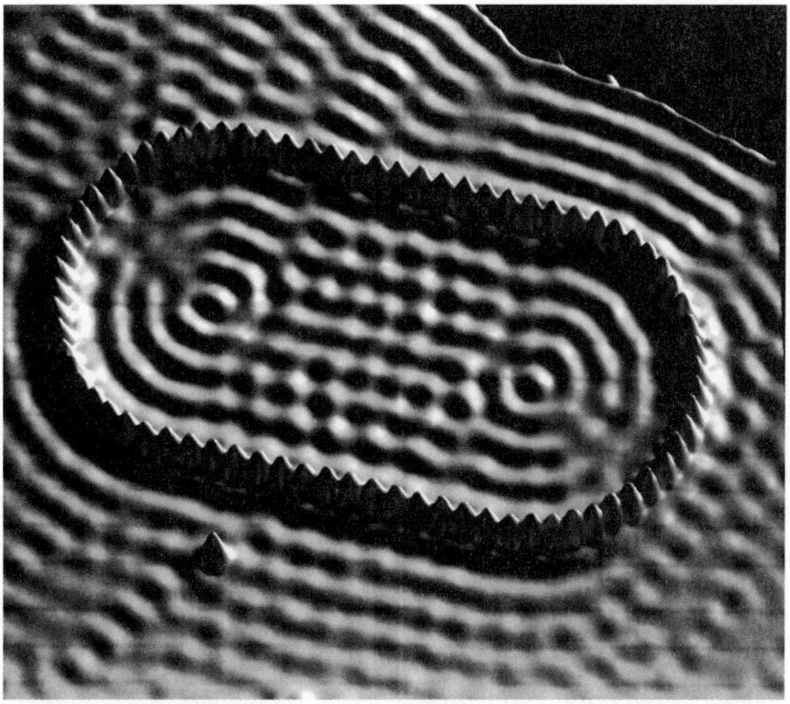

Figure 26.6. A 76 Fe atom quantum corral STM tunnel current image, consisting of iron atoms on the surface of copper (111), arranged by the tip into the shape of a stadium, and imaged by the tip in constant current mode as a raster scan across the sample. The data were rendered in a 3D perspective plot. The Bunimovich stadium is a well-studied billiard shape in the field of quantum chaos; it is classically chaotic under specular bouncing from smooth stadium-shaped walls.

The full wavefunction with this boundary condition in the presence of scatterers is (equation 26.13)

$$\begin{aligned}\psi(\vec{x}) &= \psi^0 + G_0^+ T\psi^0,\\ &= \psi^0(\vec{x}) + \int d\vec{x}' d\vec{x}''\ G_0^+(\vec{x},\vec{x}')\ T(\vec{x}',\vec{x}')\ \psi^0(\vec{x}''),\\ G^+(\vec{x},\vec{x}_T) &= G_0^+(\vec{x},\vec{x}_T) + \sum_{i,j} G_0^+(\vec{x},\vec{x}_m)\ \mathrm{T}_{i,j}\ G_0^+(\vec{x}_{m'},\vec{x}_T),\\ G^+ &= G_0^+ + G_0^+ T G_0^+, \end{aligned} \tag{26.40}$$

that is, the wavefunction created by the tip is in fact the full Green function G^* with all the scatterers present! That's nice to know, but how do we calculate the current through the tip?

The Green function indeed holds the key. We encircle the tip position and measure the flux heading away from the tip. Adding up all the flux passing though an enclosing ring gives us the current. As discussed in section 9.2, flux at a point is calculated as

$$\vec{j}(\vec{q}) = \frac{\hbar}{m}\mathrm{Im}\left[\psi^*(q)\nabla\psi(q)\right]. \tag{26.41}$$

We have just seen that here, the wavefunction $\psi(\vec{q})$ is the Green function, $G^+(\vec{q},\vec{q}_T) = G_0^+(\vec{q},\vec{q}_T) + \sum_{i,j} G_0^+(\vec{q},\vec{q}_j)\mathrm{T}_{i,j}G_0^+(\vec{q},\vec{q}_i)$. When we take the gradient of G^+, a great simplification happens: only the imaginary part of $G_0^+(\vec{q},\vec{q}_T)$ is singular ($H_0^1 = J_0 + iY_0$; $Y_0(kr) \sim 2/\pi \log[kr]$ for small kr) near $\vec{q} = \vec{q}_T$; the term involving $G_0^+ T G_0^+$ represents amplitude returning to the tip from distant sites and is not singular at $\vec{q} = \vec{q}_T$. When we "tighten the noose" around the site at $\vec{q} = \vec{q}_T$, adding up the flux on a small circle surrounding $\vec{q}_T$, only the singular part survives, since the circle circumference around $\vec{q}_T$ is proportional to r, the radius of the circle. The flux heads out radially from the singularity, and adding up around the circle, the gradient of the singular part gives

$$\frac{-i}{4}\frac{2}{\pi}(i)\int_0^{2\pi} r/r\,d\theta = 1,$$

and thus

$$\text{flux from tip} \propto -\mathrm{Im}[G^+]. \tag{26.42}$$

The imaginary part of the Green function is connected to the local density of states, or LDOS, at $\vec{x}_T$:

$$\begin{aligned}
-\frac{1}{\pi}\langle\vec{x}_T|\mathrm{Im}[G^+(E)]|\vec{x}_T\rangle &= \sum_\nu |\psi_\nu(\vec{x}_T)|^2\delta(E_\nu - \epsilon),\\
&= \langle\vec{x}_T|\delta(E-H)|\vec{x}_T\rangle,\\
&= \mathrm{LDOS}(\vec{x}_T,\epsilon).
\end{aligned} \tag{26.43}$$

(The total density of states is $\rho(E) = \mathrm{Tr}[\delta(E-H)]$.) We have

$$\begin{aligned}
-\frac{1}{\pi}\langle\vec{x}_T|\mathrm{Im}[G^+(E)]|\vec{x}_T\rangle &= -\frac{1}{\pi}\mathrm{Im}[G_0^+(\vec{x}_T,\vec{x}_T)] - \frac{1}{\pi}\mathrm{Im}[G_0^+TG_0^+],\\
&= \frac{1}{4\pi}J_0(0) - \frac{1}{\pi}\mathrm{Im}[G_0^+TG_0^+],\\
&= \frac{1}{4\pi}\mathrm{Re}\left[\left(1 + \mathrm{a}_T[\,1 - \mathrm{A}]^{-1}\,\mathrm{a}\right)\right],
\end{aligned} \tag{26.44}$$

where we have used equation 26.15, and from equation 26.25, using the asymptotic forms of the Hankel function, $A_{ij} = t\,G_{ij}$, $a_T(r_j) = \sqrt{\frac{2}{\pi k r_j}}e^{ikr_j - i\pi/4}$,

$$a(r) = \sqrt{\frac{2}{\pi k}}e^{i\pi/4}\frac{(\alpha_0 e^{2i\delta_0} - 1)}{2i}\frac{e^{ikr}}{\sqrt{r}}, \tag{26.45}$$

where k is the energy determined by the bias voltage and the Fermi level.

A fraction of the incident electron wave may be absorbed at the adatom due to inelastic scattering and elastic scattering into bulk states. We shall call this the absorptive channel. To account for this absorption we use the standard procedure of taking η_0 to be complex—that is, $\exp[2i\eta_0] \equiv \alpha_0 \exp[2i\delta_0]$, where α_0 and δ_0 are real. It

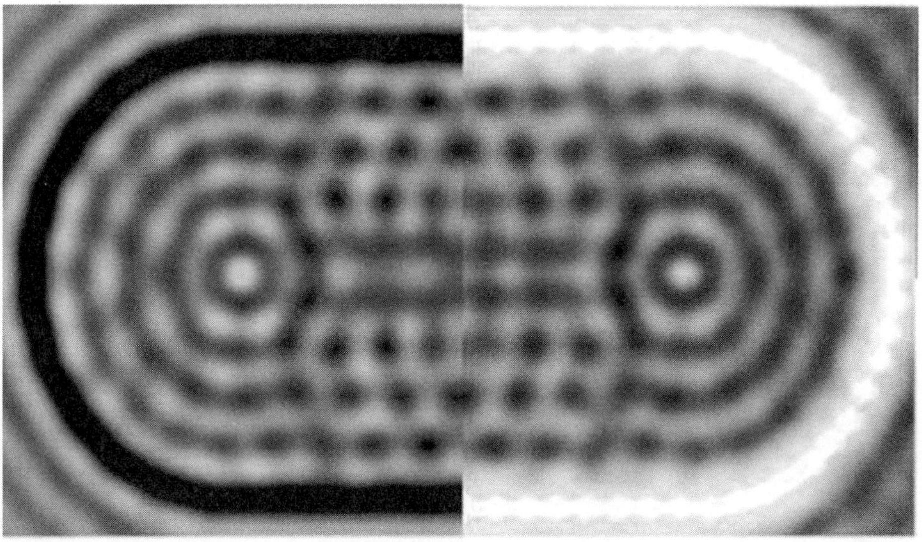

Figure 26.7. Comparison of experimental and theoretical STM signal in the interior of a quantum corral made of 76 iron atoms in the shape of a stadium. The theory, shown in the left half, does not treat the Fe atoms as finite-size objects, and this is incorrect very close to them.

is interesting that *maximal* "black dot" attenuation, $\mathrm{Im}[\eta_0] = i\infty$, leads to $\alpha_0 = 0$ and the scattered wave

$$a(r) = \sqrt{\frac{2}{\pi k}} e^{i\pi/4} \frac{(\alpha_0 e^{2i\delta_0} - 1)}{2i} \frac{e^{ikr}}{\sqrt{r}} \rightarrow \sqrt{\frac{2}{\pi k}} e^{i\pi/4} \frac{e^{i\pi/2}}{2} \frac{e^{ikr}}{\sqrt{r}}. \tag{26.46}$$

Thus, there is still a scattered wave even for black dot attenuation (a well-known fact in scattering theory), and interference with the incident amplitude still happens, but the optical theorem, equation 26.32, is of course violated.

In figure 26.7, experimental and theoretical results for a 76 Fe adatom "stadium" of dimension 141 × 285 Å is shown. It is important, for this level of agreement, to use the precise positions of the Fe atoms—they must register to the triangular grid of allowed binding sites on the Cu(111) face [206], and do not make perfect semicircular end caps.

26.4 Fermi Pseudopotential

Even though we know that a delta function cannot scatter in 2D and 3D, if we nonetheless use a potential of the form

$$V(\vec{r}) = V_0 \delta(\vec{r} - \vec{r}_0)$$

in the first order (Born) expansion of the scattered wavefunction, namely,

$$\psi \approx \psi^0 + G_0^+ V \psi^0, \tag{26.47}$$

we get a finite scattered wave. This is a lucky double error (first, supposing a δ function should scatter, and second, using perturbation theory for an infinitely sharp potential) that turns out alright. Fermi recognized that with V_0 replaced by a proper t-matrix scattering amplitude (for a single scatterer present), one can represent not only the scattered amplitude from a point scatterer but also the scattered amplitude from any smaller (much less than a wavelength across) scatterer, outside the range of any evanescent waves present.

The idea is that the Fermi pseudopotential

$$V_f = t_i \delta(\boldsymbol{r}' - \boldsymbol{r}_i) \tag{26.48}$$

can be used in a faux first-order perturbation theory to yield

$$\psi(\boldsymbol{r}) = \psi^0(\boldsymbol{r}) + \int d\boldsymbol{r}'\, G_0^+(\boldsymbol{r}, \boldsymbol{r}') t_i \delta(\boldsymbol{r}' - \boldsymbol{r}_i) \psi^0(\boldsymbol{r}') = \psi^0(\boldsymbol{r}) + G_0^+(\boldsymbol{r}, \boldsymbol{r}_i) t_i \psi^0(\boldsymbol{r}_j). \tag{26.49}$$

This is the exact one (small) scatterer result, if the right t_i is employed. This is especially useful in the presence of many scatterers, provided the scattering cross section for each is small enough, as in thermal neutron scattering from a solid or crystal.

Appendix

A.1 Partial Derivative Identities

The following method is useful for deriving partial derivative identities: Consider the Jacobian matrix of a transformation (that has unit determinant for canonical transformations, but this is more general). Considering p and q to be a function of P and Q—that is, $p(P, Q), q(P, Q)$—we form the matrix $\boldsymbol{A}$:

$$\boldsymbol{A} = \begin{pmatrix} \left(\dfrac{\partial p}{\partial P}\right)_Q & \left(\dfrac{\partial p}{\partial Q}\right)_P \\ \left(\dfrac{\partial q}{\partial P}\right)_Q & \left(\dfrac{\partial q}{\partial Q}\right)_P \end{pmatrix}. \tag{A.1}$$

The right inverse matrix—that is, $\boldsymbol{A}\boldsymbol{A}^{-1} = \mathbf{1}$—is (considering now $P(q, p)$ and so on)

$$\boldsymbol{A}^{-1} = \begin{pmatrix} \left(\dfrac{\partial P}{\partial p}\right)_q & \left(\dfrac{\partial P}{\partial q}\right)_p \\ \left(\dfrac{\partial Q}{\partial p}\right)_q & \left(\dfrac{\partial Q}{\partial q}\right)_p \end{pmatrix}. \tag{A.2}$$

This can be shown, since, for example,

$$dp = \left(\frac{\partial p}{\partial P}\right)_Q dP + \left(\frac{\partial p}{\partial Q}\right)_P dQ. \tag{A.3}$$

Thus, we have, dividing by dp and holding q constant,

$$1 = \left(\frac{\partial p}{\partial P}\right)_Q \left(\frac{\partial P}{\partial p}\right)_q + \left(\frac{\partial p}{\partial Q}\right)_P \left(\frac{\partial Q}{\partial p}\right)_q, \tag{A.4}$$

and dividing by dq and holding p constant,

$$0 = \left(\frac{\partial p}{\partial P}\right)_Q \left(\frac{\partial P}{\partial q}\right)_p + \left(\frac{\partial p}{\partial Q}\right)_P \left(\frac{\partial Q}{\partial q}\right)_p. \tag{A.5}$$

Equations (A.4) and (A.5) are just the (1, 1) and (1, 2) elements of $\mathbf{1} = \boldsymbol{A} \cdot \boldsymbol{A}^{-1}$, respectively. This already gives eight identities of this type, considering $\mathbf{1} = \boldsymbol{A} \cdot \boldsymbol{A}^{-1}$ and $\mathbf{1} = \boldsymbol{A}^{-1} \cdot \boldsymbol{A}$ for the left inverse.

We have other identities from the matrix inverse formula,

$$\begin{pmatrix} a & b \\ c & d \end{pmatrix}^{-1} = \begin{pmatrix} \dfrac{d}{J} & -\dfrac{b}{J} \\ -\dfrac{c}{J} & \dfrac{a}{J} \end{pmatrix}, \tag{A.6}$$

where $J = ad - bc$. Thus,

$$\begin{pmatrix} \left(\dfrac{\partial P}{\partial p}\right)_q & \left(\dfrac{\partial P}{\partial q}\right)_p \\ \left(\dfrac{\partial Q}{\partial p}\right)_q & \left(\dfrac{\partial Q}{\partial q}\right)_p \end{pmatrix} = \frac{1}{J} \begin{pmatrix} \left(\dfrac{\partial q}{\partial Q}\right)_P & -\left(\dfrac{\partial p}{\partial Q}\right)_P \\ -\left(\dfrac{\partial q}{\partial P}\right)_Q & \left(\dfrac{\partial p}{\partial P}\right)_Q \end{pmatrix}. \tag{A.7}$$

For example,

$$\left(\frac{\partial P}{\partial p}\right)_q = \frac{1}{J}\left(\frac{\partial q}{\partial Q}\right)_P, \tag{A.8}$$

and

$$\left(\frac{\partial Q}{\partial q}\right)_p = \frac{1}{J}\left(\frac{\partial p}{\partial P}\right)_Q, \tag{A.9}$$

where $J = \det \boldsymbol{A}$, the Jacobian of the transformation.

A.2 Gaussian Integrals

We first derive the integral $I = \int_{-\infty}^{\infty} e^{-x^2} dx$ in a familiar way. We square it and go to polar coordinates

$$\begin{aligned} I^2 &= \int_{-\infty}^{\infty}\int_{-\infty}^{\infty} e^{-x^2-y^2}\, dx\, dy = 2\pi \int_0^{\infty} e^{-r^2} r\, dr, \\ &= \pi \int_0^{\infty} e^{-u}\, du = \pi, \end{aligned} \tag{A.10}$$

so $I = \sqrt{\pi}$. Then by simple substitution or scaling,

$$\int_{-\infty}^{\infty} e^{-ax^2}\, dx = \sqrt{\frac{\pi}{a}}. \tag{A.11}$$

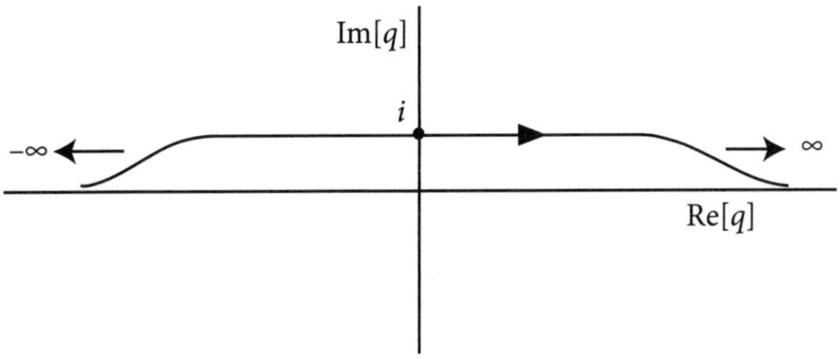

Figure A.1. Contour path for the integrand $\exp[-(x-i)^2]$.

Also clearly

$$\int_{-\infty}^{\infty} e^{-a(x-\beta)^2}\, dx = \sqrt{\frac{\pi}{a}}, \tag{A.12}$$

which works even with β complex; this can be seen by distorting an integration path into the complex plane (figure A.1).

By either expanding the preceding quadratic or completing the following square, we have

$$\int_{-\infty}^{\infty} e^{-ax^2+2bx}\, dx = \sqrt{\frac{\pi}{a}}\, e^{\frac{b^2}{a}}. \tag{A.13}$$

Differentiation under the integral sign is simple:

$$\int_{-\infty}^{\infty} x^n\, e^{-ax^2+bx}\, dx = \frac{\partial^n}{\partial b^n}\int_{-\infty}^{\infty} e^{-ax^2+bx}\, dx,$$

$$= \sqrt{\frac{\pi}{a}}\frac{\partial^n}{\partial b^n}\, e^{\frac{b^2}{4a}}. \tag{A.14}$$

MULTIDIMENSIONAL GAUSSIAN INTEGRALS

Consider

$$I_2 = \int\!\!\int e^{-a_1x_1^2-a_2x_2^2+b_1x_1+b_2x_2}\, dx_1\, dx_2 = \sqrt{\frac{\pi^2}{a_1a_2}}\, e^{\sum_{i=1}^{2} b_i^2/4a_i},$$

$$= \sqrt{\frac{\pi^2}{\det a}}\, e^{\frac{1}{4}b^t\cdot a^{-1}\cdot b}, \tag{A.15}$$

where

$$b = \begin{pmatrix} b_1 \\ b_2 \end{pmatrix};\ b^t = (b_1\ \ b_2) \text{ and so on} \tag{A.16}$$

and

$$a = \begin{pmatrix} a_1 & 0 \\ 0 & a_2 \end{pmatrix}. \tag{A.17}$$

Suppose the integrand is of the form

$$\exp[-u^T \cdot \Lambda \cdot u + B^T \cdot u], \tag{A.18}$$

with Λ nondiagonal. By finding the transformation that diagonalizes Λ, we reduce the problem to a product of one-dimensional ones. The result, given next, shows that we don't have to diagonalize Λ, but rather take its determinant and if there is a linear term—that is, $B \neq 0$—Λ^{-1} is required.

$$I_2 = \int \cdots \int e^{-u^T \cdot \Lambda \cdot u + B^T \cdot u}\, du_1 \cdots du_n = \sqrt{\frac{\pi^n}{\det \Lambda}}\, e^{\frac{1}{4} B^T \cdot \Lambda^{-1} \cdot B}. \tag{A.19}$$

This is the general form for N-dimensional Gaussian integrals.

There is a related useful formula that can be derived by diagonalizing Λ and applying a unitary transformation:

$$\det[\Lambda] = e^{\mathrm{Tr} \log \Lambda}. \tag{A.20}$$

Further Reading

Below, I list major references that the reader may find helpful. Many have already been mentioned in context but it is good to collect them here.

The best classical mechanics text I have encountered is certainly not the longest, and should perhaps not be one's first mechanics text: L. D. Landau and E.M. Lifshitz, *Mechanics* (Pergamon Press, 1960) [2].

The book closest in spirit to this one is D. J. Tannor, *Introduction to Quantum Mechanics: A Time-dependent Perspective* (University Science Books, 2007) [1]. David is a long-time collaborator with the author, and we share many interests and perspectives. The author wrote the present book to complement Tannor's book, and ideally both would be on the shelf!

A very valuable book with much related material is M. S. Child, *Semiclassical Methods in Molecular Scattering and Spectroscopy* (Springer Science, 1980) [97].

A gem of a book that originally sparked my curiosity in advanced topics in mechanics, and taught me a great deal, is I. C. Percival and D. Richards, *Introduction to Dynamics* (Cambridge University Press, 1982) [8].

As background level of physics for this book, it is impossible to too highly recommend R. P. Feynman, R. Leighton, and M. Sands, *The Feynman Lectures on Physics* (Addison-Wesley, 1965) [207].

Feynman and Hibb's book on path integrals is indispensible. The first edition was a little sloppy; Prof. Dan Stryer of Oberln produced a list of 879 errors that extended to 39 pages. So there is now an emended edition, R. P. Feynman, A. R. Hibbs, and D. F. Styer, *Quantum Mechanics and Path Integrals* (Courier Corporation, 2010) [208].

A fine work extending path integrals to new domains and explaining the fundamentals too is L. S. Schulman, *Techniques and Applications of Path Integration* (Wiley, 1981) [209].

For a higher level book on quantum mechanics, one cannot go wrong (except for a lack of some modern topics) with L. D. Landau and E. M. Lifshitz, *Quantum Mechanics* (Pergamon Press, 1958) [210].

A very good book on dynamical tunneling is S. Keshavamurthy and P. Schlagheck, *Dynamical Tunneling: Theory and Experiment* (CRC Press, 2011) [172].

A classic reference on time-correlation functions and spectroscopy is R. G. Gordon, Correlation functions for molecular motion (*Advances in Magnetic Resonance*, 3: 1–42, 1968) [211].

John Klauder's book with B. Skagerstam on coherent states is an original classic, out of print but available as an eBook: *Coherent States: Applications in Physics and Mathematical Physics* (World Scientific, 1985) [212].

The whole field owes a great deal to M. Gutzwiller. His book on the trace formula, quantum chaology, and much more is a classic: M. Gutzwiller, *Chaos in Classical and Quantum Mechanics* (Springer Science, 2013) [213].

The semiclassical field also owes a great deal to Prof. William H. Miller. While he has not written a text per se, there are key iconic articles and review articles, some listed now, but hardly exhaustively. In particular, the field of semiclassical reactive molecular collisions "belongs" to Miller and his many collaborators, but since that subject is not emphasized here we do not list most of that literature. W. H. Miller, "Semiclassical Methods in Chemical Physics" [214]; classic for molecular collisions and the revival of semiclassical methods for dynamical systems: W. H. Miller, "Semiclassical Theory of Atom-Diatom Collisions: Path Integrals and the Classical *S* Matrix" [215] and "Classical-limit Quantum Mechanics and the Theory of Molecular Collisions" [61] (containing a different perspective on canonical transformations and their relation to quantum amplitudes); W. H. Miller and T. F. George, "Semiclassical Theory of Electronic Transitions in Low Energy Atomic and Molecular Collisions Involving Several Nuclear Degrees of Freedom" [216]; and H. D. Meyer and W. H. Miller, "A Classical Analog for Electronic Degrees of Freedom in Nonadiabatic Collision Processes" [15]. In 1970 Miller introduced the important concept of *initial value representation* (IVR) [62, 63]; classically forbidden (dynamical tunneling) processes were discussed in connection with collisions in W. H. Miller and T. F. George, "Analytic Continuation of Classical Mechanics for Classically Forbidden Collision Processes" [92] and other papers.

No one has been more inspirational to the subject matter of this book than Sir Michael Berry, who has written a much needed compendium of his contributions, *A Unique Collection of Michael Berry's Works* (World Scientific, 2017). In particular I want to draw attention to the important work, "Semiclassical Theory of Spectral Rigidity" [217], which I tried in vain to summarize compactly in this book, and ended up omitting it.

There are diverse entries into the world of classical chaos. Recommended are M. Tabor, *Chaos and Integrability in Nonlinear Dynamics* (Wiley, 1988); H.-J. Stöckmann, *Quantum Chaos, an Introduction* (Cambridge University Press, 1999); L. Reichl, *The Transition to Chaos* (Springer, 1987) (this book contains much material about quantum manifestations of classical chaos); A. J. Lichtenberg and M. A. Lieberman, *Regular and Stochastic Motion* (Springer, 2013).

A unique work found online at ChaosBook.org, by P. Cvitanović, R. Artuso, R. Mainieri, G. Tanner, and G. Vattay, *Chaos: Classical and Quantum*, is a comprehensive work covering a wide range of topics relating to chaos theory, including quantum chaology.

It was mentioned in the introduction that proof of the mastery of classical mechanics by the old quantum theorists is found in M. Born, *The Mechanics of the Atom*, a profound book for its time, and today.

Bibliography

[1] D. J. Tannor. *Introduction to quantum mechanics: a time-dependent perspective*. University Science Books, 2007.

[2] L. D. Landau and E. M. Lifshitz. *Mechanics*. Pergamon Press, 1960.

[3] L. Kaplan, N. T. Maitra, and E. J. Heller. Quantizing constrained systems. *Physical Review A (Atomic, Molecular, and Optical Physics)*, 56: 2592, 1997.

[4] Scot Elmer James Shaw. Propagation in smooth random potentials. PhD thesis, Physics Department, Harvard University, 2002.

[5] Eric J. Heller and Scot Shaw. Branching and fringing in microstructure electron flow. *International Journal of Modern Physics B*, 17(22–24): 3977, 2003.

[6] Zhixin Lu, Christopher Jarzynski, and Edward Ott. Apparent topologically forbidden interchange of energy surfaces under slow variation of a Hamiltonian. *Physical Review E*, 91(5): 052193, 2015.

[7] V. I. Arnold. *Mathematical methods of classical mechanics*, Springer, 1978.

[8] Ian C. Percival and Derek Richards. *Introduction to dynamics*. Cambridge University Press, 1982.

[9] Albert Einstein. Zum quantensatz von sommerfeld und epstein (English translation Jila 116 [1980]). *Verhandl. Dtsc. Phys. Ges.*, 19: 82–92, 1917.

[10] Robert G. Littlejohn. The Van Vleck formula, Maslov theory, and phase space geometry. *Journal of Statistical Physics*, 68(1–2): 7, 1992.

[11] Boris V. Chirikov. A universal instability of many-dimensional oscillator systems. *Physics Reports*, 52(5): 263, 1979.

[12] A. J. Lichtenberg and M. A. Lieberman. *Regular and chaotic dynamics*. Springer-Verlag, 2nd ed., 1992.

[13] Vladimir I. Arnold. Instability of dynamical systems with several degrees of freedom (instability of motions of dynamic system with five-dimensional phase space). *Soviet Mathematics*, 5: 581, 1964.

[14] Claude Froeschlé, Massimiliano Guzzo, and Elena Lega. Graphical evolution of the Arnold web: from order to chaos. *Science*, 289(5487): 2108, 2000.

[15] Hans-Dieter Meyer and William H. Miller. A classical analog for electronic degrees of freedom in nonadiabatic collision processes. *Journal of Chemical Physics*, 70(7): 3214, 1979.

[16] E. J. Heller. Time-dependent approach to semiclassical dynamics. *Journal of Chemical Physics*, 62: 1544, 1975.

[17] E. J. Heller. Classical s-matrix limit of wave packet dynamics. *Journal of Chemical Physics*, 65: 4979, 1976.

[18] E. J. Heller. Time dependent variational approach to semiclassical dynamics. *Journal of Chemical Physics*, 64: 63, 1976.

[19] Jiri Vanicek and Eric J. Heller. Uniform semiclassical wave function for coherent two-dimensional electron flow. *Physical Review E (Statistical, Nonlinear, and Soft Matter Physics)*, 67(1): 016211, 2003.

[20] P. Pechukas and M. S. Child. Is semi-classical scattering theory accurate for transitions from low-lying vibrational states? *Molecular Physics*, 31(4): 973, 1976.

[21] Harold Levine and Julian Schwinger. On the theory of electromagnetic wave diffraction by an aperture in an infinite plane conducting screen. *Communications on Pure and Applied Mathematics*, 3(4): 355, 1950.

[22] Steven Tomsovic and Eric J. Heller. Semiclassical dynamics of chaotic motion: unexpected long-time accuracy. *Physical Review Letters*, 67(6): 664, 1991.

[23] Michael V. Berry and Michael Tabor. Closed orbits and the regular bound spectrum. In *Proceedings of the Royal Society of London A: Mathematical, Physical and Engineering Sciences*, vol. 349, 101. Royal Society, 1976.

[24] Steven Tomsovic, Maurice Grinberg, and Denis Ullmo. Semiclassical trace formulas of near-integrable systems: resonances. *Physical Review Letters*, 75(24): 4346, 1995.

[25] M. L. Du and J. B. Delos. Effect of closed classical orbits on quantum spectra: ionization of atoms in a magnetic field. I. Physical picture and calculations. *Physical Review A*, 38(4): 1896, 1988.

[26] Richard Phillips Feynman. Space-time approach to non-relativistic quantum mechanics. *Reviews of Modern Physics*, 20(2): 367, 1948.

[27] Cécile Morette. On the definition and approximation of Feynman's path integrals. *Physical Review*, 81(5): 848, 1951.

[28] Martin C. Gutzwiller. Phase-integral approximation in momentum space and the bound states of an atom. *Journal of Mathematical Physics*, 8(10): 1979, 1967.

[29] Martin C. Gutzwiller. Phase-integral approximation in momentum space and the bound states of an atom. II. *Journal of Mathematical Physics*, 10(6): 1004, 1969.

[30] E. J. Heller. Cellular dynamics: a new semiclassical approach to time-dependent quantum mechanics. *Journal of Chemical Physics*, 94: 2723, 1991.

[31] M. A. Sepulveda and E. J. Heller. Semiclassical calculation and analysis of dynamical systems with mixed phase space. *Journal of Chemical Physics*, 101: 8004, 1994.

[32] Zhe Xian Wang and Eric J. Heller. Semiclassical investigation of the revival phenomena in a one-dimensional system. *Journal of Physics A: Mathematical and Theoretical*, 42(28): 285, 2009.

[33] L. Kaplan. Semiclassical dynamical localization and the multiplicative semiclassical propagator. *Physical Review Letters*, 81(16): 3371, 1998.

[34] Hermann Weyl. *The theory of groups and quantum mechanics*. Courier Corporation, 1950.

[35] Eugene Wigner. On the quantum correction for thermodynamic equilibrium. *Physical Review*, 40(5): 749, 1932.

[36] Douglas J. Mason, Mario F. Borunda, and Eric J. Heller. Quantum flux and reverse engineering of quantum wave functions. *EPL (Europhysics Letters)*, 102(6): 600, 2013.

[37] E. J. Heller. Bound-state eigenfunctions of classically chaotic Hamiltonian systems: scars of periodic orbits. *Physical Review Letters*, 53: 1515, 1984.

[38] S.-Y. Lee and E. J. Heller. Time-dependent theory of Raman scattering. *Journal of Chemical Physics*, 71: 4777, 1979.

[39] David J. Tannor and Eric J. Heller. Polyatomic Raman scattering for general harmonic potentials. *Journal of Chemical Physics*, 77(1): 202, 1982.

[40] A. B. Myers, R. A. Mathies, D. J. Tannor, and E. J. Heller. Excited state geometry changes from preresonance Raman intensities: isoprene and hexatriene. *Journal of Chemical Physics*, 77: 3857, 1982.

[41] E. J. Heller. The semiclassical way to molecular spectroscopy. *Accounts of Chemical Research*, 14: 368, 1981.

[42] J. von. Neumann. Die eindeutigkeit der schrödingerschen operatoren. *Mathematische Annalen*, 104(1): 570, 1931.

[43] Michael J. Davis and Eric J. Heller. Semiclassical Gaussian basis set method for molecular vibrational wave functions. *Journal of Chemical Physics*, 71(8): 3383, 1979.

[44] Asaf Shimshovitz and David J. Tannor. Phase-space approach to solving the time-independent Schrödinger equation. *Physical Review Letters*, 109(7): 070402, 2012.

[45] Thomas Halverson and Bill Poirier. Accurate quantum dynamics calculations using symmetrized Gaussians on a doubly dense Von Neumann lattice. *Journal of Chemical Physics*, 137(22): 224101, 2012.

[46] Robert G. Littlejohn. The semiclassical evolution of wave packets. *Physics Reports*, 138(4–5): 193, 1986.

[47] S.-Y. Lee and E. J. Heller. Time-dependent theory of Raman scattering. *Journal of Chemical Physics*, 71: 4777, 1979.

[48] E. J. Heller. Time-dependent approach to semiclassical dynamics. *Journal of Chemical Physics*, 62: 1544, 1975.

[49] Eric J. Heller. Generalized theory of semiclassical amplitudes. *Journal of Chemical Physics*, 66(12): 5777, 1977.

[50] Yitzhak Weissman. Semiclassical approximation in the coherent states representation. *Journal of Chemical Physics*, 76(8): 4067, 1982.

[51] Daniel Huber and Eric J. Heller. Generalized Gaussian wave packet dynamics. *Journal of Chemical Physics*, 87(9): 5302, 1987.

[52] Daniel Huber, Eric J. Heller, and Robert G. Littlejohn. Generalized Gaussian wave packet dynamics, Schrödinger equation, and stationary phase approximation. *Journal of Chemical Physics*, 89(4): 2003, 1988.

[53] Troy Van Voorhis and Eric J. Heller. Nearly real trajectories in complex semiclassical dynamics. *Physical Review A*, 66(5): 050501, 2002.

[54] H. Pal, M. Vyas, and S. Tomsovic. Generalized Gaussian wavepacket dynamics: integrable and chaotic systems. *Physical Review E*, 93(1): 012213, 2014.

[55] Noa Zamstein and David J. Tannor. Communication: overcoming the root search problem in complex quantum trajectory calculations. *Journal of Chemical Physics*, 140(4): 041105, 2014.

[56] S. Tomsovic and E. J. Heller. Semiclassical dynamics of chaotic motion: unexpected long-time accuracy. *Physical Review Letters*, 67: 664, 1991.

[57] P. W. O'Connor, S. Tomsovic, and E. J. Heller. Accuracy of semiclassical dynamics in the presence of chaos. *Journal of Statistical Physics*, 68: 131, 1992.

[58] M. A. Sepulveda, S. Tomsovic, and E. J. Heller. Semiclassical propagation: how long can it last? *Physical Review Letters*, 69: 402, 1992.

[59] S. Tomsovic and E. J. Heller. Long-time semiclassical dynamics of chaos: the stadium billiard. *Physical Review E (Statistical Physics, Plasmas, Fluids, and Related Interdisciplinary Topics)*, 47: 282, 1993.

[60] Harinder Pal, Manan Vyas, and Steven Tomsovic. Generalized Gaussian wave packet dynamics: integrable and chaotic systems. *Physical Review E*, 93(1): 012213, 2016.

[61] William H. Miller. Classical-limit quantum mechanics and the theory of molecular collisions. *Advances in Chemical Physics*, 25(1): 69, 1974.

[62] William H. Miller. Classical S matrix: numerical application to inelastic collisions. *Journal of Chemical Physics*, 53(9): 3578, 1970.

[63] William H. Miller. The semiclassical initial value representation: a potentially practical way for adding quantum effects to classical molecular dynamics simulations. *Journal of Physical Chemistry A*, 105(13): 2942, 2001.

[64] E. J. Heller. Frozen Gaussians: a very simple semiclassical approximation. *Journal of Chemical Physics*, 75: 2923, 1981.

[65] Michael F. Herman and Edward Kluk. A semiclassical justification for the use of non-spreading wavepackets in dynamics calculations. *Chemical Physics*, 91(1): 27, 1984.

[66] Michael F. Herman. Time reversal and unitarity in the frozen Gaussian approximation for semiclassical scattering. *Journal of Chemical Physics*, 85(4): 2069, 1986.

[67] Kenneth G. Kay. The Herman-Kluk approximation: derivation and semiclassical corrections. *Chemical Physics*, 322(1): 3, 2006.

[68] Frank Grossmann and Tobias Kramer. Spectra of harmonium in a magnetic field using an initial value representation of the semiclassical propagator. *Journal of Physics A: Mathematical and Theoretical*, 44(44): 445309, 2011.

[69] Lucas Kocia and Eric J. Heller. Directed HK propagator. *Journal of Chemical Physics*, 143(12): 124102, 2015.

[70] Andrew R. Walton and David E. Manolopoulos. A new semiclassical initial value method for Franck-Condon spectra. *Molecular Physics*, 87(4): 961, 1996.

[71] Mark L. Brewer, Jeremy S. Hulme, and David E. Manolopoulos. Semiclassical dynamics in up to 15 coupled vibrational degrees of freedom. *Journal of Chemical Physics*, 106(12): 4832, 1997.

[72] Dafna Zor and Kenneth G. Kay. Globally uniform semiclassical expressions for time-independent wave functions. *Physics Review Letters*, 76: 1990, 1996.

[73] Eric J. Heller. Time dependent variational approach to semiclassical dynamics. *Journal of Chemical Physics*, 64(1): 63, 1976.

[74] A. D. McLachlan. A variational solution of the time-dependent Schrodinger equation. *Molecular Physics*, 8(1): 39, 1964.

[75] J. Frenkel. *Wave mechanics, advanced general theory*. Oxford University Press, 1934.

[76] Robert Heather and Horia Metiu. Time-dependent theory of Raman scattering for systems with several excited electronic states: application to a h[sup +][sub 3] model system. *Journal of Chemical Physics*, 90(12): 6903, 1989.

[77] Robert Heather and Horia Metiu. A numerical study of the multiple Gaussian representation of time dependent wave functions of a Morse oscillator. *Journal of Chemical Physics*, 84(6): 3250, 1986.

[78] Shin-Ichi Sawada and Horia Metiu. A Gaussian wave packet method for studying time dependent quantum mechanics in a curve crossing system: low energy motion, tunneling, and thermal dissipation. *Journal of Chemical Physics*, 84(11): 6293, 1986.

[79] S. Sawada and H. Metiu. A multiple trajectory theory for curve crossing problems obtained by using a Gaussian wave packet representation of the nuclear motion. *Journal of Chemical Physics*, 84(1): 227, 1986.

[80] Shin-Ichi Sawada, Robert Heather, Bret Jackson, and Horia Metiu. A strategy for time dependent quantum mechanical calculations using a Gaussian wave packet representation of the wave function. *Journal of Chemical Physics*, 83(6): 3009, 1985.

[81] E. J. Heller. Time-dependent variational approach to semiclassical dynamics. *Journal of Chemical Physics*, 64: 63, 1976.

[82] H.-D. Meyer, Uwe Manthe, and Lorenz S. Cederbaum. The multi-configurational time-dependent Hartree approach. *Chemical Physics Letters*, 165(1): 73, 1990.

[83] Cristian Predescu, Pavel A. Frantsuzov, and Vladimir A. Mandelshtam. Thermodynamics and equilibrium structure of Ne 38 cluster: quantum mechanics versus classical. *Journal of Chemical Physics*, 122(15): 154305, 2005.

[84] R. D. Coalson and M. Karplus. Extended wave packet dynamics: exact solution for collinear atom, diatomic molecule scattering. *Chemical Physics Letters*, 90(4): 301, 1982.

[85] Rob D. Coalson and Martin Karplus. Multidimensional variational Gaussian wave packet dynamics with application to photodissociation spectroscopy. *Journal of Chemical Physics*, 93(6): 3919, 1990.

[86] M. Ben-Nun, Jason Quenneville, and Todd J. Martínez. Ab initio multiple spawning: photochemistry from first principles quantum molecular dynamics. *Journal of Physical Chemistry A*, 104(22): 5161, 2000.

[87] Michal Ben-Nun and Todd J. Martinez. Ab initio quantum molecular dynamics. *Advances in Chemical Physics*, 121: 439, 2002.

[88] Robert G. Littlejohn and Jonathan M. Robbins. New way to compute Maslov indices. *Physics Review A*, 36: 2953, 1987.

[89] Y. C. Ge and M. S. Child. Nonadiabatic geometrical phase during cyclic evolution of a Gaussian wave packet. *Physics Review Letters*, 78: 2507, 1997.

[90] M. V. Berry. Classical adiabatic angles and quantal adiabatic phase. *Journal of Physics A (Mathematical and General)*, 18: 15, 1985.

[91] Olivier Brodier, Peter Schlagheck, and Denis Ullmo. Resonance-assisted tunneling. *Annals of Physics*, 300(1): 88, 2002.

[92] William H. Miller and Thomas F. George. Analytic continuation of classical mechanics for classically forbidden collision processes. *Journal of Chemical Physics*, 56(11): 5668, 1972.

[93] Michael J. Davis and Eric J. Heller. Quantum dynamical tunneling in bound states. *Journal of Chemical Physics*, 75(1): 246, 1981.

[94] Lev Davidovich Landau and Evgenii Mikhailovich Lifshitz. *Quantum mechanics: non-relativistic theory*, vol. 3. Elsevier, 2013.

[95] N. T. Maitra and E. J. Heller. Semiclassical perturbation approach to quantum reflection. *Physical Review A*, 54(6): 4763, 1996.

[96] R. T. Lawton and M. S. Child. Local and normal stretching vibrational states of H_2O: classical and semiclassical considerations. *Molecular Physics*, 44(3): 709, 1981.

[97] Mark Child. *Semiclassical methods in molecular scattering and spectroscopy*, vol. 53. Springer Science, 1980.

[98] William G. Harter and Chris W. Patterson. Rotational energy surfaces and high-j eigenvalue structure of polyatomic molecules. *Journal of Chemical Physics*, 80(9): 4241, 1984.

[99] S. M. Colwell, N. C. Handy, and W. H. Miller. A semiclassical determination of the energy levels of a rigid asymmetric rotor. *Journal of Chemical Physics*, 68(2): 745, 1978.

[100] David M. Harland and Ralph Lorenz. *Space systems failures: disasters and rescues of satellites, rocket and space probes*. Springer Science & Business Media, 2007.

[101] T. Uzer, D. W. Noid, and R. A. Marcus. Uniform semiclassical theory of avoided crossings. *Journal of Chemical Physics*, 79(9): 4412, 1983.

[102] Alfredo M. Ozorio de Almeida. Tunneling and the semiclassical spectrum for an isolated classical resonance. *Journal of Physical Chemistry*, 88(25): 6139, 1984.

[103] Floyd L. Roberts and Charles Jaffe. The correspondence between classical nonlinear resonances and quantum mechanical Fermi resonances. *Journal of Chemical Physics*, 99(4): 2495, 1993.

[104] B. Ramachandran and Kenneth G. Kay. The influence of classical resonances on quantum energy levels. *Journal of Chemical Physics*, 99(5): 3659, 1993.

[105] Grayson H. Walker and Joseph Ford. Amplitude instability and ergodic behavior for conservative nonlinear oscillator systems. *Physical Review*, 188(1): 416, 1969.

[106] David Farrelly and T. Uzer. Semiclassical quantization of slightly nonresonant systems: avoided crossings, dynamical tunneling, and molecular spectra. *Journal of Chemical Physics*, 85(1): 308, 1986.

[107] Steven Tomsovic and Denis Ullmo. Chaos-assisted tunneling. *Physical Review E*, 50(1): 145, 1994.

[108] N. T. Maitra and E. J. Heller. Barrier tunneling and reflection in the time and energy domains: the battle of the exponentials. *Physical Review Letters*, 78: 3035, 1997.

[109] Ugo Fano. Effects of configuration interaction on intensities and phase shifts. *Physical Review*, 124(6): 1866, 1961.

[110] Mordechai Bixon and Joshua Jortner. Intramolecular radiationless transitions. *Journal of Chemical Physics*, 48(2): 715, 1968.

[111] Stefano Longhi. Bound states in the continuum in a single-level Fano-Anderson model. *European Physical Journal B*, 57(1): 45, 2007.

[112] Perttu Luukko, Byron Drury, Anna Klales, Kaplan Lev, Eric J. Heller, and Esa Räsänen. Strong quantum scarring by local impurities. *Scientific Reports*, 6, 2016.

[113] Robert B. Laughlin. Anomalous quantum hall effect: an incompressible quantum fluid with fractionally charged excitations. *Physical Review Letters*, 50(18): 1395, 1983.

[114] Shaohong L. Li, Donald G. Truhlar, Michael W. Schmidt, and Mark S. Gordon. Model space diabatization for quantum photochemistry. *Journal of Chemical Physics*, 142(6): 064106, 2015.

[115] Christian Evenhuis and Todd J. Martínez. A scheme to interpolate potential energy surfaces and derivative coupling vectors without performing a global diabatization. *Journal of Chemical Physics*, 135(22): 224110, 2011.

[116] J. Z. Zhang, E. J. Heller, D. Huber, and D. G. Imre. Spectroscopy and photodissociation dynamics of a two-chromophore system: Raman scattering as a probe of nonadiabatic electronic coupling. *Journal of Physical Chemistry*, 95(16): 6129, 1991.

[117] Jinzhong Zhang, Eric J. Heller, Daniel Huber, Dan G. Imre, and David Tannor. Ch_2I_2 photodissociation: dynamical modeling. *Journal of Chemical Physics*, 89(6): 3602, 1988.

[118] Robert Heather, Xue-Pei Jiang, and Horia Metiu. The use of Raman spectroscopy to study photodissociation dynamics for systems where curve crossing is important. *Chemical Physics Letters*, 142(5): 303, 1987.

[119] Song Ling, Dan G. Imre, and Eric J. Heller. Effects of the transition dipole in Raman scattering. *Journal of Physical Chemistry*, 93(20): 7107, 1989.

[120] M. Ben-Nun, Jason Quenneville, and Todd J. Martínez. Ab initio multiple spawning: photochemistry from first principles quantum molecular dynamics. *Journal of Physical Chemistry A*, 104(22): 5161, 2000.

[121] John C. Tully and Richard K. Preston. Trajectory surface hopping approach to nonadiabatic molecular collisions: the reaction of H+ with D_2. *Journal of Chemical Physics*, 55(2): 562, 1971.

[122] John von Neumann and Eugene Wigner. On the behavior of eigenvalues in adiabatic processes. *Phys. Z*, 30: 467, 1929.

[123] Michael V. Berry. Quantal phase factors accompanying adiabatic changes. *Proceedings of the Royal Society of London A: Mathematical, Physical and Engineering Sciences*, 392: 45, 1984.

[124] C. Alden Mead and Donald G. Truhlar. On the determination of Born-Oppenheimer nuclear motion wave functions including complications due to conical intersections and identical nuclei. *Journal of Chemical Physics*, 70(5): 2284, 1979.

[125] G. Herzberg and H. C. Longuet-Higgins. Intersection of potential energy surfaces in polyatomic molecules. *Discussions of the Faraday Society*, 35: 77, 1963.

[126] H. C. Longuet-Higgins. The intersection of potential energy surfaces in polyatomic molecules. *Proceedings of the Royal Society of London A: Mathematical, Physical and Engineering Sciences*, 344: 147, 1975.

[127] Jeffrey A. Cina. Wave-packet interferometry and molecular state reconstruction: spectroscopic adventures on the left-hand side of the Schrödinger equation. *Annual Review of Physical Chemistry*, 59: 319, 2008.

[128] William Louisell. *Quantum statistical properties of radiation*. Wiley, 1973.

[129] Joel S. Bader and B. J. Berne. Quantum and classical relaxation rates from classical simulations. *Journal of Chemical Physics*, 100(11): 8359, 1994.

[130] J. L. Skinner and Kisam Park. Calculating vibrational energy relaxation rates from classical molecular dynamics simulations: quantum correction factors for processes involving vibration-vibration energy transfer. *Journal of Physical Chemistry B*, 105(28): 6716, 2001.

[131] Benjamin J. Schwartz, Eric R. Bittner, Oleg V. Prezhdo, and Peter J. Rossky. Quantum decoherence and the isotope effect in condensed phase nonadiabatic molecular dynamics simulations. *Journal of Chemical Physics*, 104(15): 5942, 1996.

[132] E. J. Heller. Photofragmentation of symmetric triatomic molecules: time dependent picture. *Journal of Chemical Physics*, 68: 3891, 1978.

[133] K. Hirai and E. J. Heller. Topological angles near a periodic orbit. *Physical Review Letters*, 79(7): 1249, 1997.

[134] Jeffrey R. Reimers, Kent R. Wilson, and Eric J. Heller. Complex time dependent wave packet technique for thermal equilibrium systems: electronic spectra. *Journal of Chemical Physics*, 79(10): 4749, 1983.

[135] T. Darrah Thomas, Leif J. Saethre, Stacey L. Sorensen, and Svante Svensson. Vibrational structure in the carbon 1s ionization of hydrocarbons: calculation using electronic structure theory and the equivalent-cores approximation. *Journal of Chemical Physics*, 109(3): 1041, 1998.

[136] Soo-Y. Lee. Placzek type polarizability tensors for Raman and resonance Raman scattering. *Journal of Chemical Physics*, 78(2): 723, 1982.

[137] E. J. Heller, R. L. Sundberg, and D. J. Tannor. Simple aspects of Raman scattering. *Journal of Physical Chemistry*, 86: 1822, 1982.

[138] Jinzhong Zhang and Dan G. Imre. Spectroscopy and photodissociation dynamics of H_2O: time-dependent view. *Journal of Chemical Physics*, 90(3): 1666, 1989.

[139] Clarence Zener. Non-adiabatic crossing of energy levels. *Proceedings of the Royal Society of London A: Mathematical, Physical and Engineering Sciences*, 137: 696, 1932.

[140] S. Kallush, Bilha Segev, A. V. Sergeev, and E. J. Heller. Surface jumping: Franck-Condon factors and Condon points in phase space. *Journal of Physical Chemistry A*, 106(25): 6006, 2002.

[141] S. Kallush, B. Segev, A. V. Sergeev, and E. J. Heller. Surface jumping: Franck-Condon factors and Condon points in phase space. *Journal of Physical Chemistry A*, 106: 6006, 2002.

[142] E. J. Heller, B. Segev, and A. V. Sergeev. Hopping and jumping between potential energy surfaces. *Journal of Physical Chemistry B*, 106: 8471, 2002.

[143] E. J. Heller and Doug Beck. Nonclassical Franck-Condon processes. *Chemical Physics Letters*, 202: 350, 1993.

[144] E. J. Heller, Bilha Segev, and A. V. Sergeev. Hopping and jumping between potential energy surfaces. *Journal of Physical Chemistry B*, 106(33): 8471, 2002.

[145] Lee Tutt, David Tannor, Eric J. Heller, and Jeffrey I. Zink. The MIME effect: absence of normal modes corresponding to vibronic spacings. *Inorganic Chemistry*, 21(10): 3858–3859, 1982.

[146] Lee W. Tutt, Jeffrey I. Zink, and Eric J. Heller. Simplifying the MIME: a formula relating normal mode distortions and frequencies to the MIME frequency. *Inorganic Chemistry*, 26(13): 2158, 1987.

[147] Eric J. Heller and William M. Gelbart. Normal mode spectra in pure local mode molecules. *Journal of Chemical Physics*, 73(2): 626, 1980.

[148] Eric J. Heller and Michael J. Davis. Molecular overtone bandwidths from classical trajectories. *Journal of Physical Chemistry*, 84(16): 1999, 1980.

[149] J. H. Frederick, E. J. Heller, J. L. Ozment, and D. W. Pratt. Ring torsional dynamics and spectroscopy of benzophenone: a new twist. *Journal of Chemical Physics*, 88: 2169, 1988.

[150] Karl W. Holtzclaw and David W. Pratt. Prominent, and restricted, vibrational state mixing in the fluorescence excitation spectrum of benzophenone. *Journal of Chemical Physics*, 84(8): 4713, 1986.

[151] Eric J. Heller. Wigner phase space method: analysis for semiclassical applications. *Journal of Chemical Physics*, 65(4): 1289, 1976.

[152] E. J. Heller and M. J. Davis. Molecular overtone bandwidths from classical trajectories. *Journal of Physical Chemistry*, 84: 1999, 1980.

[153] Robert C. Brown and Eric J. Heller. Classical trajectory approach to photodissociation: the Wigner method. *Journal of Chemical Physics*, 75(1): 186, 1981.

[154] Michael V. Berry. The Bakerian Lecture, 1987: quantum chaology. *Proceedings of the Royal Society of London A: Mathematical, Physical and Engineering Sciences*, 413(1844): 183, 1987.

[155] Michael Berry. Quantum chaology, not quantum chaos. *Physica Scripta*, 40(3): 335, 1989.

[156] Eric J. Heller and Steven Tomsovic. Postmodern quantum mechanics. *Physics Today*, 46: 38, 1993.

[157] A. I. Schnirelman. Ergodic properties of eigenfunctions. *Uspekhi Matematicheskikh Nauk*, 29: 181, 1974.

[158] Y. Colin de Verdiere. Ergodicité et fonctions propres du LaPlacien. *Communications of Mathematical Physics*, 102: 497, 1985.

[159] S. Zelditch. Uniform distribution of eigenfunctions on compact hyperbolic surfaces. *Duke Mathematics Journal*, 55: 919, 1987.

[160] L. Kaplan and E. J. Heller. Weak quantum ergodicity. *Physica D*, 121: 1, 1998.

[161] Patrick W. O'Connor and Eric J. Heller. Quantum localization for a strongly classically chaotic system. *Physical Review Letters*, 61(20): 2288, 1988.

[162] Eugene P. Wigner. Characteristic vectors of bordered matrices with infinite dimensions. In *The Collected Works of Eugene Paul Wigner*, 524–540. Springer, 1993.

[163] Madan Lal Mehta. *Random matrices*, vol. 142. Academic Press, 2004.
[164] O. Bohigas and M. J. Giannoni. *Mathematical and Computational Methods in Nuclear Physics, Proceedings of the Sixth Granada Workshop, Granada, Spain, 1983*. Springer, 1984.
[165] Oriol Bohigas, Marie-Joya Giannoni, and Charles Schmit. Characterization of chaotic quantum spectra and universality of level fluctuation laws. *Physical Review Letters*, 52(1): 1, 1984.
[166] O. Bohigas, R. U. Haq, and A. Pandey. Fluctuation properties of nuclear energy levels and widths: comparison of theory with experiment. In *Nuclear data for science and technology*, 809–813. Springer, 1983.
[167] R. Bigwood, M. Gruebele, D. M. Leitner, and P. G. Wolynes. The vibrational energy flow transition in organic molecules: theory meets experiment. *Proceedings of the National Academy of Sciences*, 95(11): 5960, 1998.
[168] G. M. Stewart and J. D. McDonald. Intramolecular vibrational relaxation from C-H stretch fundamentals. *Journal of Chemical Physics*, 78(6): 3907, 1983.
[169] Evan Hudspeth, David A. McWhorter, and Brooks H. Pate. Intramolecular vibrational energy redistribution in the acetylenic C-H and hydroxyl stretches of propynol. *Journal of Chemical Physics*, 109(11): 4316, 1998.
[170] M. Gruebele and P. G. Wolynes. Vibrational energy flow and chemical reactions. *Accounts of Chemical Research*, 37(4): 261, 2004.
[171] E. J. Heller and M. J. Davis. Quantum dynamical tunneling in large molecules. A plausible conjecture. *Journal of Physical Chemistry*, 85: 307, 1981.
[172] Srihari Keshavamurthy and Peter Schlagheck. *Dynamical tunneling: theory and experiment*. CRC Press, 2011.
[173] E. J. Heller. In *1989 NATO Les Houches summer school on chaos and quantum physics*, ed. M.-J. Giannoni, A. Voros, and J. Zinn-Justin, 547. Elsevier, 1991.
[174] L. Kaplan and E. J. Heller. Linear and nonlinear theory of eigenfunction scars. *Annals of Physics*, 264: 171, 1998.
[175] L. Kaplan. Scar and antiscar quantum effects in open chaotic systems. *Physics Review E*, 59: 5325, 1999.
[176] Michael V. Berry. Regular and irregular semiclassical wavefunctions. *Journal of Physics A: Mathematical and General*, 10(12): 2083, 1977.
[177] W. E. Bies and E. J. Heller. Nodal structure of chaotic eigenfunctions. *Journal of Physics A: Mathematical and General*, 35: 5673, 2002.
[178] D. M. Ceperley. Fermion nodes. *Journal of Statistical Physics*, 63(5–6): 1237, 1991.
[179] Jacob Katriel and Ernest R. Davidson. Asymptotic behavior of atomic and molecular wave functions. *Proceedings of the National Academy of Sciences*, 77(8): 4403, 1980.
[180] Gerard Iooss, Robert H. G. Helleman, and Raymond Stora. *Chaotic behaviour of deterministic systems*. North-Holland, 1983.
[181] M. V. Berry and H. Ishio. Nodal densities of Gaussian random waves satisfying mixed boundary conditions. *Journal of Physics A: Mathematical and General*, 35(29): 5961, 2002.
[182] Juan Diego Urbina and Klaus Richter. Semiclassical construction of random wave functions for confined systems. *Physical Review E*, 70(1): 015201, 2004.
[183] Mark Srednicki. Chaos and quantum thermalization. *Physical Review E*, 50(2): 888, 1994.
[184] Eric J. Heller and Brian R. Landry. Statistical properties of many particle eigenfunctions. *Journal of Physics A: Mathematical and Theoretical*, 40(31): 9259, 2007.
[185] H. M. Kleinert. *Path integrals in quantum mechanics, statistics, and polymer physics*, 2nd ed., World Scientific, 1995.
[186] D. Chandler and P. G. Wolynes. Exploiting the isomorphism between quantum theory and classical statistical mechanics of polyatomic fluids. *Journal of Chemical Physics*, 74: 4078, 1981.
[187] E. J. Heller, L. Kaplan, and A. Dahlen. Refraction of a Gaussian seaway. *Journal of Geophysical Research: Oceans*, 113(C9), 2008.
[188] M. A. Wolfson and S. Tomsovic. On the stability of long-range sound propagation through a structured ocean. *Journal of the Acoustical Society of America*, 109: 2693, 2001.

[189] M. A. Topinka, B. J. LeRoy, R. M. Westervelt, S. E. J. Shaw, R. Fleischmann, E. J. Heller, K. D. Maranowski, and A. C. Gossard. Coherent branched flow in a two-dimensional electron gas. *Nature*, 410(6825): 183, 2001.

[190] M. V. Berry and C. Upstill. Catastrophe optics: morphologies of caustics and their diffraction patterns. *Progress in Optics*, XVIII: 257, 1980.

[191] M. V. Berry. Focused tsunami waves. *Proceedings of the Royal Society of London A: Mathematical, Physical and Engineering Sciences*, 463: 3055, 2007.

[192] Henri Degueldre, Jakob J. Metzger, Theo Geisel, and Ragnar Fleischmann. Random focusing of tsunami waves. *Nature Physics*, 12(3): 259, 2016.

[193] Eric Heller. Physics of waves: warning from the deep. *Nature Physics*, 12(9): 824, 2016.

[194] Markus Arndt, Olaf Nairz, Julian Vos-Andreae, Claudia Keller, Gerbrand Van der Zouw, and Anton Zeilinger. Wave-particle duality of C_{60} molecules. *Nature*, 401(6754): 680, 1999.

[195] Jörg Schmiedmayer, Michael S. Chapman, Christopher R. Ekstrom, Troy D. Hammond, Stefan Wehinger, and David E Pritchard. Index of refraction of various gases for sodium matter waves. *Physical Review Letters*, 74(7): 1043, 1995.

[196] Scott N. Sanders, Florian Mintert, and Eric J. Heller. Coherent scattering from a free gas. *Physical Review A*, 79(2): 023610, 2009.

[197] Areez Mody, Michael Haggerty, John M. Doyle, and Eric J. Heller. No-sticking effect and quantum reflection in ultracold collisions. *Physical Review B*, 64(8): 085418, 2001.

[198] John M. Doyle, Jon C. Sandberg, A. Yu Ite, Claudio L. Cesar, Daniel Kleppner, and Thomas J. Greytak. Hydrogen in the submillikelvin regime: sticking probability on superfluid ^{4}He. *Physical Review Letters*, 67(5): 603, 1991.

[199] Hilmar Oberst, Yoshihisa Tashiro, Kazuko Shimizu, and Fujio Shimizu. Quantum reflection of He* on silicon. *Physical Review A*, 71(5): 052901, 2005.

[200] Eyal Neria and Abraham Nitzan. Semiclassical evaluation of nonadiabatic rates in condensed phases. *Journal of Chemical Physics*, 99(2): 1109, 1993.

[201] Oleg V. Prezhdo and Peter J. Rossky. Relationship between quantum decoherence times and solvation dynamics in condensed phase chemical systems. *Physical Review Letters*, 81(24): 5294, 1998.

[202] Oleg V. Prezhdo and Peter J. Rossky. Evaluation of quantum transition rates from quantum-classical molecular dynamics simulations. *Journal of Chemical Physics*, 107(15): 5863, 1997.

[203] László Turi and Peter J. Rossky. Critical evaluation of approximate quantum decoherence rates for an electronic transition in methanol solution. *Journal of Chemical Physics*, 120(8): 3688, 2004.

[204] Gregory A. Fiete and Eric J. Heller. Semiclassical theory of coherence and decoherence. *Physical Review A*, 68(2): 022112, 2003.

[205] Jesse Hersch. Scattering resonances in the extreme quantum limit. PhD thesis, Department of Physics, Harvard University, 1999.

[206] M. A. Topinka, B. J. LeRoy, S. E.J. Shaw, E. J. Heller, R. M. Westervelt, K. D. Maranowski, and A. C. Gossard. Imaging coherent electron flow from a quantum point contact. *Science*, 289: 2323, 2000.

[207] R. Feynman, R. Leighton, and M. Sands. *The Feynman Lectures on Physics*, vol. 3. Addison-Wesley, 1965.

[208] Richard P. Feynman, Albert R. Hibbs, and Daniel F. Styer. *Quantum mechanics and path integrals*. Courier Corporation, 2010.

[209] L. S. Schulman. *Techniques and applications of path integration*, vol. 198. Wiley, 1981.

[210] L. D. Landau and E. M. Lifshitz. *Quantum mechanics*. Pergamon Press, 1958.

[211] R. G. Gordon. Correlation functions for molecular motion. *Advances in Magnetic Resonance*, 3: 1–42, 1968.

[212] John Klauder and B. Skagerstam. *Coherent states: applications in physics and mathematical physics*. World Scientific, 1985.

[213] Martin C. Gutzwiller. *Chaos in classical and quantum mechanics*, vol. 1. Springer Science & Business Media, 2013.

[214] W. H. Miller. Semiclassical methods in chemical physics. *Science*, 233: 171, 1986.

[215] W. H. Miller. Semiclassical theory of atom-diatom collisions: path integrals and the classical S matrix. *Journal of Chemical Physics*, 53: 1949, 1970.

[216] W. H. Miller and T. F. George. Semiclassical theory of electronic transitions in low energy atomic and molecular collisions involving several nuclear degrees of freedom. *Journal of Chemical Physics*, 56: 5637, 1972.

[217] M. V. Berry. Semiclassical theory of spectral rigidity. *Proceedings of the Royal Society of London, Series A (Mathematical and Physical Sciences)*, 400: 229, 1985.

Index